D1709189

# SHORTWAVE DIRECTORY

*Seventh Edition*

*by Robert B. Grove*

## Copyright Notice

Printed in the United States of America

**ISBN 0-944543-00-6**
Library of Congress Catalog Card No. 87-051078

# Table of Contents

*NOTE: All time references are in Universal Coordinated Time (UTC).*

# Introduction

## DXing the Utilities

"Utilities" monitoring? To the uninitiated, the terminology probably conjures up visions of some demented radio buff tuning in water and sewer repair crews! Nothing could be further from the truth. While many shortwave hobbyists are quite content listening to foreign broadcasts, the rest of us are busily tapping in on the two-way action.

It could be a pilot radioing his position report or describing a mechanical problem he is experiencing over the North Atlantic; a sinking ship's radioman frantically calling for help as the Coast Guard gears up for a rescue operation; NASA ground and air support crews anxiously coordinating their observations as the Space Shuttle thunders into a blue Florida sky; the strained voices of Navy personnel cautiously navigating the treacherous Persian Gulf.

Many listeners enjoy tuning in ham radio operators as they describe their equipment and discuss a wide variety of technical issues. Specialty nets (networks) can be heard assisting craft at sea, providing weather alert information, passing personal messages between military personnel and their families back home or reassuring relatives of disaster victims in distant lands.

The HF (high frequency or "shortwave") portion of the spectrum, a swath of frequencies beginning just above the top end of the AM medium wave broadcast band (1.7 MHz) and extending to just above the CB and amateur ten meter band (30 MHz), is the most signal-saturated chunk of the radio spectrum.

The variety of signals which can be heard here is almost endless: radioteletype, facsimile, single sideband, amplitude modulation, frequency modulation, slow-scan television, propagational beacons, telemetry, pulse signals, independent sideband, multiplex, continuous wave (Morse code), ionosonde bursts, packet and ASCII data, and more.

And the users? You name them -- they'll probably be found here. Drug smugglers have taken to the airwaves to coordinate shipments of illicit cargo from Central and South America to the United States; on nearby frequencies the U.S. Coast Guard and Drug Enforcement Agency (DEA) will be in hot pursuit.

One listener reported hearing shouts from a drug runner as machine gun fire punctuated his last transmission. When the vessel Achille Lauro was hijacked, utilities listeners heard the negotiations -- and preparations for rescue -- over the air.

As President Reagan was rushed to the hospital suffering from the gunshot wound from a would-be assassin's bullet, an HF communications link was immediately established between the White House and Air Force One for Vice President Bush to assume a leadership role. Listeners monitoring that link knew the President's condition nearly an hour before the news was broadcast over the networks!

During the ill-fated Reykjavik summit conference, as Air Force One was again in flight back to the United States, it was the "ute DXer" who had the news first -- by listening to the press corps relaying their information via HF radio back to the United States.

But the shortwave spectrum is an enormous range to tune blindly. Where does a prospective utility listener find the right information? Certainly, experience helps, but you have to start somewhere; fortunately, there are some excellent sources of this vital information.

The *Shortwave Directory* is a leading source of accurate information. Concentrating primarily on U.S. (but with worldwide entries as well), the *Shortwave Directory* includes listings for military, government, aircraft, space programs, maritime, public safety, scientific, private radio systems, common carriers, and broadcasting.

Many listings are very unusual such as terrorist networks, freebanders, spy numbers stations, hurricane networks, petroleum platform helicopters and emergency nets.

But to keep up with daily changes in the dynamic radio spectrum, the serious listener needs a reliable monthly magazine like *Monitoring Times*, the first and largest magazine to cover all aspects of radio communications and broadcasting from the lowest frequencies, through shortwave and scanners, and on to satellites.

A subscription to this leading magazine is only $18 a year (U.S.) and you can request a sample by sending $2.00 to *Monitoring Times*, PO Box 98, Brasstown, NC 28902.

While every attempt has been made to ensure accuracy, some errors are inevitable and schedules do change. Corrections and additions from our readers are encouraged and appreciated.

In general, frequencies are listed in kilohertz (kHz), center carrier. In some cases upper or lower sideband offsets may be shown; these usually amount to 1.5 kHz displacement (1.4 kHz in maritime), and 1.2-2 kHz for facsimile and radioteletype.

Except for radio amateurs, shortwave broadcast feeders, and some military flights, virtually all voice transmissions in the HF spectrum will be in upper sideband (USB). Frequencies which are underlined are most commonly reported in use.

Yes, the HF spectrum radiates an incredible variety of entertainment and intrigue for every listening taste, and it is all at your fingertips when you know how and where to listen. Next time a newsbreaking story of international importance is developing, meet me on shortwave -- I'll be listening!

# Communications Act of 1934

In order to preserve the privacy of communications, this regulatory section directs U.S. listeners who intercept communications not intended for their listening.

Section 705 has been generally interpreted to mean that a person who listens in to a radio message not intended for him may not reveal the contents of the message to a third party, nor may he use those contents for his own personal gain.

Section 501 of the Penal Code is invoked for serious infractions, enforced by the Federal Bureau of Investigation.

Prosecutions have been effected against individuals who have violated the provisions and sold the information or services to unauthorized persons.

So far as we know, no arrest has ever been made of an American hobbyist listening to shortwave for his own entertainment.

## Federal Communications Act of 1934, as Amended
### (Enforced by the Federal Bureau of Investigation)

SECTION 705A. No person receiving or assisting in receiving, or transmitting, or assisting in transmitting, any interstate or foreign communicatin by wire or radio shall divulge or publish the existence, contents, substance, purport, effect, or meaning thereof, except through authorized channels of transmission or reception, to any person other than the addressee, his agent, or attorney, or to a person employed or authorized to forward such communication to its destination, or to proper accounting or distributing officers of the various communicating centers over which the communication may be passed, or to the master of a ship under whom he is serving, or in response to a subpoena issued by a court of competent jurisdiction, or on demand of other lawful authority; and no person not being authorized by the sender shall intercept any communication and divulge or publish the existence, contents, substance, purport, effect, or meaning of such intercepted communication to any person; and no person not being entitled thereto shall receive or assist in receiving any interstate or foreign communication by wire or radio and use the same or any information therein contained for his own benefit or for the benefit of another not entitled thereto; and no person having received such intercepted communication or having become acquainted with the contents, substance, purport, effect, or meaning of the same or any part thereof, knowing that such information was so obtained, shall divulge or publish the existence, contents, substance, purport, effect, or meaning of the same or any part thereof, or use the same or any information therein contained for his own benefit or for the benefit of another not entitled thereto: PROVIDED, that this section shall not apply to the receiving, divulging, publishing, or utilizing the contents of any radio communication broadcast, or transmission by amateurs or others for the use of the general public, or relating to ships in distress.

# The Electronic Communications Privacy Act of 1986 (ECPA)

In an effort to legislate privacy for mobile telephone users, as well as safeguard against government surveillance abuse, the U.S. Senate ratified and the President signed into law this controversial bill, effective January 19, 1987.

Providing strict penalties including stiff fines and imprisonment for violators, the Act is primarily of interest to scanner owners rather than shortwave listeners.

## ECPA Protected Transmissions

Encryped, scrambled or other privacy techniques
Paid-subscription Subsidiary Carrier Authorization (SCA)
Voice/data Radio Common Carrier (mobile telephone, paging)
Remote broadcast pickup; studio to transmitter links
Private microwave systems

## Communications Legal to Monitor

Federal, state or local government
Military
Police, fire, medical, civil defense
Private land mobile services
Marine and aeronautical
Amateur and CB
Cordless telephones
Tone-only paging
Broadcasting
Any transmission readily accessible to the general public

Many of the provisions may appear ambiguous; court test cases will be required to resolve them.

# The Top 100 Shortwave Frequencies

Nowhere in the radio spectrum is there as much intrigue and fascination to be found as in the 2-30 MHz spectrum. The following list of recent loggings includes some of the most active frequencies (in kHz). The voice utilities (two-way communicators) are virtually all upper sideband (USB), while the broadcasters run full carrier amplitude modulation (AM). Since most listening is done in the evening hours, this representative sampling reflects that time.

## UTILITIES

| | | | |
|---|---|---|---|
| 2670 | U.S. Coast Guard | 8291.1 | Ship to shore, Pacific |
| 4063 | Mississippi River barges | 8418 | "Spy numbers" broadcast |
| 4069.2 | Ship working channel | 8740.6 | Ship to shore |
| 4087.8 | Mississippi River barges | 8778 | U.S. Navy |
| 4112.6 | Ship calling channel | 8784 | Ship to shore, Pacific |
| 4125 | Ship calling channel | 8805.7 | Ship to shore |
| 4143.6 | Mississippi River barges | 8808.8 | Ship to shore |
| 4413.4 | Ship working channel | 8825 | International airlines |
| 4419.4 | Ship working channel | 8846 | International airlines |
| 4467.5 | Civil Air Patrol | 8879 | International airlines |
| 4517 | Air Force MARS | 8891 | International airlines |
| 4593.5 | Air Force MARS | 8921 | International airlines |
| 4637.5 | Offshore petroleum | 8972 | U.S. Navy, Atlantic |
| 4670 | "Spy numbers" broadcasts | 8984 | U.S. Coast Guard, air/ground |
| 4742 | "Spy numbers" broadcasts | 8989 | U.S. Air Force, air/ground |
| 5015 | Army Corp of Engineers | 8993 | U.S. Air Force, air/ground |
| 5320 | U.S. Coast Guard | 9014 | U.S. Air Force, air/ground |
| 5598 | International airlines | 9027 | U.S. Air Force, air/ground |
| 5616 | International airlines | 10000 | WWV time |
| 5680 | U.S. Coast Guard | 10493 | FEMA, emergency net |
| 5692 | U.S. Coast Guard air/ground | 10780 | NASA air/ground |
| 5696 | U.S. Coast Guard air/ground | 11176 | U.S. Air Force, air/ground |
| 5703 | Tactical Air Command | 11182 | Scott Air Force Base, air/gnd |
| 5812 | "Spy numbers" broadcasts | 11200 | RAF, flight weather |
| 6506.4 | U.S. Coast Guard ships | 11233 | Canadian Air Force air/ground |
| 6518.8 | Inland waterways | 11234 | RAF, air ground |
| 6521.9 | Mississippi River barges | 11243 | Strategic Air Command |
| 6577 | International airlines | 11246 | U.S. Air Force, air/ground |
| 6586 | International airlines | 11282 | International airlines |
| 6604 | Flight weather | 12429.2 | Ship to shore |
| 6673 | NOAA hurricane hunters | 13113.2 | U.S. Coast Guard marine weather |
| 6683 | Andrews AFB, VIP | 13181 | U.S. Navy |
| 6697 | U.S. Navy | 13201 | U.S. Air Force air/ground |
| 6705 | U.S. Air Force air/ground | 13215 | U.S. Air Force air/ground |
| 6723 | U.S. Navy | 13241 | U.S. Air Force air/ground |
| 6738 | U.S. Air Force air/ground | 13270 | Flying weather |
| 6753 | Canadian Air Force | 13282 | Flight weather |
| 6761 | Strategic Air Command | 13306 | International airlines |
| 6802 | "Spy numbers" broadcast | 13354 | NOAA, hurricane hunters |
| 6927 | Andrews AFB, VIP | | |

## The Hottest Shortwave Frequencies for Two-Way Voice Intrigue

| | |
|---|---|
| 2180-2900 | 8700-9050 |
| 4300-4750 | 11175-11400 |
| 5300-5900 | 13200-13340 |
| 6600-6900 | 15010-15100 |
| 7500-8000 | |

# INTERNATIONAL BROADCASTERS
*(Schedules subject to change)*

| | | | |
|---|---|---|---|
| 5960 | Canada | 9710 | Cuba |
| 6085 | Germany | 9745 | Ecuador |
| 6175 | England | 9755 | Canada |
| 7150 | USSR | 11775 | Ecuador |
| 7325 | England | 11820 | Cuba |
| 7355 | USA | 11880 | Spain |
| 9435 | Israel | 15155 | Ecuador |
| 9630 | Spain | 17795 | Australia |

## Time Signals

| WWV | CHU |
|---|---|
| 5000 | 3330 |
| 10000 | 7335 |
| 15000 | 14670 |

## United States

### Air Force Bases in the United States

| | |
|---|---|
| Gunter[9] | Alabama |
| Maxwell[9] | Alabama |
| Davis-Monthan[5] | Arizona |
| Luke[5] | Arizona |
| Williams[7] | Arizona |
| Eaker[1] | Arkansas |
| Little Rock[1] | Arkansas |
| Beale[1] | California |
| Castle[1] | California |
| Edwards[6] | California |
| George[5] | California |
| March[1] | California |
| Mather[1] | California |
| McClellan[10] | California |
| Norton[4] | California |
| Onizuka[8] | California |
| Travis[4] | California |
| Vandenberg[1] | California |
| Cheyenne Mt.[8] | Colorado |
| Colorado Springs[2] | Colorado |
| Falcon[8] | Colorado |
| Lowry[7] | Colorado |
| Peterson[8] | Colorado |
| Dover[4] | Delaware |
| Bolling[4] | D.C. |
| Eglin[6] | Florida |
| Homestead[5] | Florida |
| Hurburt Fld[4] | Florida |
| MacDill[5] | Florida |
| Patrick[6] | Florida |
| Tyndall[5] | Florida |
| Dobbins[3] | Georgia |
| Moody[5] | Georgia |
| Robins[10] | Georgia |
| Mt. Home[5] | Idaho |
| Chanute[7] | Illinois |
| Chicago O'Hare IAP[3] | Illinois |
| Scott[4] | Illinois |
| Grissom[1] | Indiana |
| McConnell[1] | Kansas |
| Barksdale[1] | Louisiana |
| England[5] | Louisiana |
| New Orleans NAS[3] | Louisiana |
| Loring[1] | Maine |
| Andrews[4] | Maryland |
| Martin State Airport | Maryland |
| Hanscom[6] | Massachusetts |
| Westover[3] | Massachusetts |
| KI Sawyer[1] | Michigan |
| Selfridge[11] | Michigan |
| Wurtsmith[1] | Michigan |
| Minneapolis-St Paul[3] | Minnesota |
| Columbus[7] | Mississippi |
| Keesler[7] | Mississippi |
| Richard-Gebaur[3] | Missouri |
| Whiteman[1] | Missouri |
| Malmstrom[1] | Montana |
| Offutt[1] | Nebraska |
| Nellis[5] | Nevada |
| Pease[1] | New Hampshire |
| Mcguire[4] | New Jersey |
| Cannon[5] | New Mexico |
| Holloman[5] | New Mexico |
| Kirtland[4] | New Mexico |
| Griffiss[1] | New York |
| Niagara Falls IAP[3] | New York |
| Plattsburgh[1] | New York |
| Stewart Int'l Airport | New York |
| Pope[4] | North Carolina |
| Seymour Johnson[5] | North Carolina |
| Grand Forks[1] | North Dakota |
| Minot[1] | North Dakota |
| Newark[10] | Ohio |
| Rickenbacker[11] | Ohio |
| Wright Patterson[10] | Ohio |
| Youngstown Municipal[3] | Ohio |
| Altus[4] | Oklahoma |
| Tinker[10] | Oklahoma |
| Vance[7] | Oklahoma |
| Portland IAP | Oregon |
| Gtr Pittsburgh[3] | Pennsylvania |
| State College | Pennsylvania |
| Willow Grove NAS[3] | Pennsylvania |
| Charleston[4] | South Carolina |
| Myrtle Beach[5] | South Carolina |
| Shaw[5] | South Carolina |
| Ellsworth[1] | South Dakota |
| Arnold[6] | Tennessee |
| Bergstrom[5] | Texas |
| Brooks[6] | Texas |
| Carswell[1] | Texas |
| Dyess[1] | Texas |
| Goodfellow[7] | Texas |
| Kelly[10] | Texas |
| Lackland[7] | Texas |
| Laughlin[7] | Texas |
| Randolph[7] | Texas |
| Reese[7] | Texas |
| Sheppard[7] | Texas |
| Hill[10] | Utah |

# AIR FORCE

*SAC headquarters with Underground Command Post at Offutt AFB (SAC photo)*

| | |
|---|---|
| Langley[5] | Virginia |
| Fairchild[1] | Washington |
| McChord[4] | Washington |
| Gen.Mitchell Field[3] | Wisconsin |
| F.E.Warren[1] | Wyoming |

| | |
|---|---|
| (1) SAC | (7) ATC |
| (2) US AF Academy | (8) AFSPACECOM |
| (3) AFRES Base | (9) AU |
| (4) MAC | (10) AFLC |
| (5) TAC | (11) ANG |
| (6) AFSC | |

## United States Air Force Installations Outside CONUS

AB = Air Base (major installation)
AS = Air Station (support base)
AFS = Air Force Station (radar)

| | |
|---|---|
| Albrook AFS | Panama |
| Andersen AFB | Guam |
| Ankara AS | Turkey |
| Ascension AB | Ascension Is. |
| Aviano AB | Italy |
| Bellows AFS | Hawaii |
| Bitburg AB | FRG |
| Camp New Amsterdam | Netherlands |
| Clark AFB | Philippines |
| Clear AFS | Alaska |
| Comiso AS | Italy |
| Eielson AFB | Alaska |
| Elmendorf AFB | Alaska |
| Florennes AB | Belgium |
| Fylingdales Moor AFS | UK |
| Galena Airport | Alaska |
| Hahn AB | FRG |
| Hellenikon AB | Greece |
| Hessisch-Oldendorf AB | FRG |
| Hickam AFB | Hawaii |
| Howard AFB | Panama |
| Incirlik AB | Turkey |
| Iraklion AS | Crete |
| Izmir AS | Turkey |
| John Hay AS | Philippines |
| Kadena AB | Okinawa, Japan |
| Keflavik NAS | Iceland |
| King Salmon Airport | Alaska |
| Kunsan AB | KOR |
| Kwangju AB | KOR |
| Lajes Field | Azores |
| Lindsey AS | FRG |
| Misawa AB | Japan |
| Osan AB | KOR |
| Pirinclik AFS | Turkey |
| RAF Alconbury | UK |
| RAF Brentwaters | UK |
| RAF Chicksands | UK |
| (RAF Croughton | UK) |
| RAF Fairford | UK |
| RAF Greenham Common | UK |
| RAF Lakenheath | UK |
| RAF Mildenhall | UK |
| RAF Upper Heyford | UK |
| RAF Woodbridge | UK |
| Ramstein AB | FRG |
| Rhein-Main AB | FRG |
| San Miguel AFS | Philippines |
| San Vito AS | Italy |
| Sembach AB | FRG |
| Shemya AFB | Alaska |
| Sondrestrom AB | Greenland |
| Spangdahlem AB | FRG |
| Suwon AB | KOR |
| Taegu AB | KOR |
| Tempelhof Central Airport | W Berlin FRG |
| Thule AB | Greenland |
| Torrejon AB | Spain |
| Wallace AS | Philippines |
| Wheeler AFB | Hawaii |
| Yokota AB | Japan |
| Zaragoza AB | Spain |
| Zweibrucken AB | FRG |

*The Underground Command Post showing part of the control consoles and the 16x16 foot projection screens. The command post is manned by a battle staff 24 hours a day (SAC photo).*

## USAF Strategic Air Command

Fully 30% of SAC aircraft are on alert status at all times. This effort requires considerable coordination between aircraft, ground stations and nuclear missile silos. As part of the H.F. failsafe system ("Giant Talk"), routine Foxtrot ("Skyking") mission status broadcasts are made throughout each hour ("alpha monitor" periods). Should the United States be involved in armed conflict, the go-code would be transmitted via this network (as well as others).

Primary "Skyking" channels are underlined. Nine transmitter sites maintain the net: Andrews, Clark, Croughton, Elmendorf, Incirlik, McClellan, Offutt, Thule and Yokota. Andrews, Offutt and McClellan are control stations. Andrews transmits from Davidsonville, MD, and receives at Brandywine, MD. Emission is USB.

Headquartered at Offutt AFB, Nebraska, and comprising the forces of Barksdale AFB, Louisiana, and March AFB, California, SAC can also take command of 20 reserve and Air National Guard units with their 15,800 members.

On a worldwide basis, SAC involves more than 120,000 military and civilian personnel operating on 26 air bases and with tenant privileges on 47 more. Western Pacific operations are conducted from Anderson AFB, Guam, and in Europe from Ramstein AB, Germany.

The tones which accompany voice transmissions key up remote terrestrial and satellite circuits. Occasionally, two tones will be heard keying up two locations; an "echo" effect will often be heard as a result.

## Emergency Action Messages (EAM)
### (*"Skyking"* Broadcasts)

Skyking broadcasts are made every 60 minutes via encoded high-priority SSB voice. Additionally, "CAPSULE" transmissions periodically update operational instructions to MAC aircraft.

Sample EAM: "Quebec Tango four five zero, standby; Quebec Tango four five zero, standby,

# AIR FORCE

message follows (series of letters and numbers); I say again (series repeated). This is Sandwich; out". QT450 is the message identifier; it is repeated to give the aircrew time to prepare to copy the message. If they had copied that message previously, they ignore it.

Another format is "Skyking, Skyking, do not answer; Oscar, Charlie, Golf, time 53, authentication Sierra Juliett; I say again (etc.)" These are the go-code or cancellation messages for nuclear strike. If they are repeated three times, monitor 6761 or 11243.

Skyking broadcasts are frequently heard on U.S. Navy frequencies as well (6697, 11267, and others) as they are recorded by the Navy for playback on their frequencies.

### SAC Pacific
10344  12129  15734

### SAC Europe (SSB)
| | | | |
|---|---|---|---|
| 3096 | 3923 | 4465 | 4497 |
| 4595 | 4702 | 4725 | 4744 |
| 4853 | 5132 | 5477 | 9025 |

## SAC Frequencies and Channel Identifiers

| Freq | Channel | Usage |
|---|---|---|
| 3113* | BRAVO UNIFORM, SIERRA 302 | Airborne Command Post Intercommunication |
| 3292* | (varies) | |
| 3295* | ALPHA MIKE, SIERRA 303 | |
| 3369* | ALPHA SIERRA | |
| 4373* | ZULU ZULU | |
| 4492* | (varies) | |
| 4495* | ECHO, SIERRA 304 | Airborne Command Post Intercommunication |
| 4725* | VICTOR, SIERRA 390 | Primary Air-to-ground/refueling/Airborne Command Post |
| 4896* | (varies) | |
| 5020* | FOXTROT | Airborne Command Post Intercommunication |
| 5026* | FOXTROT (alt.) | Airborne Command Post Intercommunication |
| 5110* | (varies) | |
| 5171* | (2 letters; varies) | |
| 5215 | (varies) | |
| 5243 | | Airborne Command Post Intercommunication |
| 5328 | (varies) | |
| 5684 | FOXTROT QUEBEC | |
| 5700 | BRAVO QUEBEC, PAPA 381 | Airborne Command Post Intercommunication |
| 5800* | WHISKEY 101 | |
| 5826 | BRAVO UNIFORM, PAPA 382 | Airborne Command Post Intercommunication |
| 6680 | FOXTROT XRAY | |
| 6712 | (varies) | Also Alpha Two in PACAF |
| 6730 | XRAY 903 | |
| 6757* | WHISKEY 103 | |
| 6761 | QUEBEC; SIERRA 391 | Primary Night Air-to-ground Channel |
| 6826 | GOLF | |
| 6840 | | |
| 6863 | OSCAR | |
| 6870 | KILO | Airborne Command Post Intercommunication |
| 6886 | | |
| 6969 | | |
| 7330 | YANKEE and XRAY | |
| 7475* | WHISKEY 104 | Airborne Command Post Intercommunication |
| 7831* | WHISKEY 105 | |
| 7983 | FOXTROT CHARLIE | |
| 8101 | ALPHA PAPA | Airborne Command Post Intercommunication |
| 9017* | XRAY 904 | |
| 9023 | | SAC/NORAD Intercommunications/AWACS |
| 9027 | ROMEO; SIERRA 392 | Primary Air-to-ground Channel |
| 9057 | PAPA | Airborne Command Post Intercommunication |
| 9220 | (varies) | |
| 9234 | (varies) | SAC/NORAD Intercommunications |
| 10153* | FOXTROT 311 | Airborne Command Post Intercommunication |
| 10452 | OSCAR | PACAF Designator |
| 10498* | FOX 28 | Also used by FEMA |
| 10510 | | Possible PACAF Channel |

| STATION | EAM | CAPSULE |
|---|---|---|
| Albrook | unlisted | H+25,+55 |
| Andersen | H+23 | H+20,+50 |
| Ascension | unlisted | H+15,+45 |
| Clark | H+53 | H+25,+55 |
| Croughton | H+08 | H+00,+30 |
| Elmendorf | H+43 | H+15,+45 |
| Hickam | H+33 | H+00,+30 |
| Incirlik | H+18 | H+10,+40 |
| Lajes | H+38 | H+05,+35 |
| Loring | H+48 | |
| MacDill | H+58 | H+20,+50 |
| McClellan | H+13 | H+10,+40 |
| Scott | unlisted | upon receipt |
| Thule | H+28 | H+15,+45 |
| Yokota | H+03 | H+05,+35 |

| | | |
|---|---|---|
| 11100 | ALPHA 21 | |
| 11118 | FOXTROT 315 | Airborne Command Post Intercommunication |
| 11179* | WHISKEY 108 | |
| 11220 | BRAVO | |
| 11226* | XRAY 905 | |
| 11243 | SIERRA 393 | Primary Day Air-to-ground Channel |
| 11408 | YANKEE QUEBEC | Airborne Command Post Intercommunication |
| 11494 | LIMA, SIERRA 311 | Airborne Command Post Training |
| 11607 | ALPHA ZULU | |
| 12070* | WHISKEY 108 | |
| 12075 | FOXTROT FOXTROT | |
| 12107 | FOXTROT 915 | |
| 13205 | ALPHA XRAY | SAC Special Operations Channel |
| 13211 | SIERRA 312 | Airborne Command Post Intercommunication |
| 13217 | XRAY 906 | |
| 13241 | SIERRA 394 | Primary Air-to-ground Channel |
| 13247* | WHISKEY 109 | |
| 13547 | (varies) | |
| 13907 | ALPHA CHARLIE | |
| 14448 | ALPHA JULIET | Airborne Command Post Intercommunication |
| 14716 | SIERRA ECHO | |
| 14744 | ALPHA TANGO | |
| 14775 | (varies) | "Mike" in PACAF |
| 14955 | CHARLIE | |
| 15035 | CHARLIE QUEBEC | Shared with Canadian forces |
| 15041 | MIKE | Primary Air-to-ground Channel |
| 15044* | FOXTROT 383 | |
| 15091 | BRAVO XRAY | TAC-to-SAC Intercommunication |
| 15449 | GOLF GOLF | |
| 15544 | | Possible SAC Point-to-point Channel |
| 15962 | INDIA | |
| 17617 | BRAVO HOTEL | |
| 17972* | WHISKEY 111 | |
| 17975 | SIERRA 395 | Primary Air-to-ground Channel |
| 18005 | TANGO | PACAF Designator |
| 18046 | JULIETTE | |
| 18594 | ZULU ONE | |
| 20124 | WHISKEY 115 | |
| 20167 | WHISKEY 116 | |
| 20631 | WHISKEY | Primary Air-to-ground Channel |
| 20737 | | Possible PACAF Channel |
| 20740 | LIMA | PACAF Designator |
| 20846 | CHARLIE ALPHA | SAC-to-CAP Intercommunication |
| 20890 | DELTA | |
| 21815 | FOXTROT SIERRA | |
| 23337 | UNIFORM | |
| 23419 | | Possible SAC-NORAD Intercommunication |
| 27870 | DELTA QUEBEC | |

(Underlined frequencies carry simulcast Foxtrot broadcasts)
*USAF (non-SAC) command frequencies

# AIR FORCE

## Regency Net ("Scope Force")

Consisting of several subordinate networks like the Cemetery, Gang Buster and Inform nets, the Regency net is intended to pass Emergency Action Messages to NATO and US nuclear forces at European bases in England, Germany, Spain, Netherlands, Italy, Greece and Crete. Two-frequency simulcasting provides anti-jam redundancy.

Messages are first transmitted from Mildenhall, England, to Permasens, Germany, on 3060 kHz for subsequent rebroadcast on "Gang Buster" and "Cemetery" net frequencies.

### Cemetery and Gang Buster Net Frequencies

2452 2760 3060 3200 3688 3760 3931
3973.5 3996 4020.5 4035 4458 4480 4496
4545 4560 4821 4933 4980 4987 6506
6796 6937 7368 8118 9126 9244 10139
10478 11446 14705 11515 13341

### Inform Network Frequencies (USB)

| Ch | Freq | Ch | Freq |
|---|---|---|---|
| A1 | 2403 | A21 | 10139 |
| A2 | 2452 | A22 | 11641 |
| A4 | 3215 | A23 | 13479 |
| A6 | 3946 | A24 | 13545 |
| A7 | 3958 | A25 | 14374 |
| A8 | 4477 | A26 | 14423 |
| A9 | 4612 | A27 | 14505 |
| A10 | 5092 | A28 | 14682 |
| A11 | 5105 | A29 | 15476 |
| A12 | 5146 | A30 | 15560 |
| A13 | 6819 | A31 | 18179 |
| A14 | 7305 | A32 | 18744 |
| A15 | 7384 | A33 | 20539 |
| A16 | 7424 | A34 | 20604 |
| A17 | 7895/8085 | A35 | 20609 |
| A18 | 7919 | A36 | 20849 |
| A20 | 9414/9477 | A38 | 2031 |

*Air Force One*

## U.S. Air Force
## Tactical Air Command (TAC)
*(All frequencies USB)*

| | |
|---|---|
| 3032 | 8964 |
| 3060 | 9014 |
| 4449 | 11214 |
| 4742 | 13204 |
| 5703 | 15091 |
| 5800 | 18019 |
| 6730 | 23206 |
| 6753 | |

## TAC Ground Stations

| Callsign | Location |
|---|---|
| Fireside 1 | TAC HQ, Langley AFB, VA |
| Fireside 2 | (Not in use) |
| Fireside 3 | Shaw AFB, SC, 9th AF HQ, 3630 TFW, |
| Fireside 4 | Mountain Home AFB, ID |
| Fireside 5 | 12th AF HQ, Bergstrom AFB, TX |
| Raymond | TAC Command Posts |
| Raymond 1 | HQ TAC, ALCC Langley AFB, VA |
| Raymond 6 | George AFB, CA |
| Raymond 7 | 27th & 35th TFW, Cannon AFB, NM |
| Raymond 8 | Davis-Monthan AFB, Tucson, AZ |
| Raymond 9 | Dyess AFB, TX |
| Raymond 10 | 1st SOW, Hurlburt Field, FL |
| Raymond 11 | Eglin AFB, FL |
| Raymond 12 | England AFB, LA |
| Raymond 13 | Homestead AFB, FL |
| Raymond 14 | 49th TFW, Holloman AFB, NM |
| Raymond 15 | 31st TFW, Homestead AFB, FL |
| Raymond 16 | 1st TFW, 316th Unit, Langley, VA |
| Raymond 17 | 347th TFW, Moody AFB, GA |
| Raymond 18 | Luke AFB, Phoenix, AZ |
| Raymond 19 | 56th TFW, MacDill AFB, FL |
| Raymond 20 | Patrick AFB, Cape Canaveral, FL |
| Raymond 21 | 345th TFW, Myrtle Beach AFB, SC |
| Raymond 22 | 57th & 474th TFW, Nellis AFB, NV |
| Raymond 23 | Pope AFB, NC |
| Raymond 24 | 552nd AWACS, Tinker AFB, OK |
| Raymond 25 | 4th TFW, Seymour Johnson AFB, NC |

Raymond 26 363rd TFW, Shaw AFB, SC
Raymond 27 366th TFW & 662 TACW, Mountain Home AFB, ID
Raymond 28 Bergstrom AFB, TX

## U.S. Air Force
## Global Command & Control System
## (GCCS) Stations-AFCC/ATOS

Formerly known as the Airways Stations and In-Flight Stations, these stations are now used for command and control of U.S. military operations. Regular ATC and position reporting now take place on commercial ATC frequencies. Most transmissions will be USB; LSB usually denotes VIP flight. Hours of operation are shown where applicable.

| | |
|---|---|
| 3067 | Croughton,UK (2300-0500) |
| 3081 | Lajes,Azores(2100-1000) |
| 3144 | Hickam,HI(0600-1700) |
| 4721 | Andrews,MD(2400-1200) |
| 4746 | Lajes,Azores(2100-1000) |
| | MacDill,FL(0001-0900) |
| | McClellan,CA(0400-1600) |
| 4747 | Yokota,J (1000-2100) |
| 5302 | Croughton,UK |
| 5688 | Loring,ME(0900-2400) |
| | MacDill,FL(0001-0900) |
| 5703 | Croughton,Eng(2100-0800) |
| 5710 | Albrook,PNZ(0200-1200) |
| | Elmendorf,AK |
| | Thule,GRL |
| | Yokota,Japan |
| 6683 | Albrook(0001-1400) |
| 6730 | Albrook, PNZ |
| 6738 | Andersen,Guam(0700-2200) |
| | Clark,Phil(1200-2200) |
| | Elmendorf,AK |
| | Hickam,HI(0400-1900) |
| | McClellan,CA(0400-1600) |
| | Incirlik,Turkey |
| | Thule,GRL |
| | Yokota(0900-2400) |
| 6750 | Croughton,UK |

| | |
|---|---|
| | Lajes,Azores |
| | MacDill,FL(0001-0900) |
| 6753 | Ascension(2000-0800) |
| 6756 | Andrews,MD |
| 6757 | Croughton,UK |
| 8964 | Hickam,HI |
| 8967 | Andersen,Guam |
| | Hickam,HI |
| | Lajes,Azores |
| | Thule,GRL |
| | Yokota,Japan |
| 8989 | Elmendorf,AK |
| | Loring,ME |
| | MacDill,FL |
| | McClellan,CA |
| 8993 | Albrook,PNZ |
| | Ascension Is |
| | Clark,Phillipines |
| | MacDill,FL |
| 9011 | Croughton(0500-2300) |
| 9014 | Loring,ME(0900-2400) |
| 9018 | Andrews,MD |
| 11176 | Andersen,Guam |
| | Clark,Phil(1500-0200) |
| | Ascension(1800-1000) |
| | Croughton,Eng |
| | Diego Garcia(USN)(1500-0200) |
| | Elmendorf |
| | Incirlik,Turkey |
| | Seymour Radio Station - Maintenance (radio check) |
| 11179 | Croughton,UK |
| | Hickam,HI |
| | Loring,ME(0100-0900) |
| | MacDill,FL(0900-2400) |
| 11207 | Croughton,UK |
| 11226 | Elmendorf,AK |
| | Lajes,Azores(1000-2100) |
| 11228 | Thule,GRL |
| 11236 | Yokota,Japan |
| 11239 | McClellan,CA |
| 11243 | Croughton,UK |
| 11246 | MacDill,FL |
| 13201 | Andersen,Guam |
| | Clark,Phil |
| | Elmendorf,AK |
| | Hickam,HI(1700-0600) |
| | McClellan, CA |
| | Thule,GRL |
| | Yokota,J(2100-1000) |
| 13214 | Incirlik, Turkey |
| | Loring,ME(1839, 1849, 1858 UTC) |
| | Croughton,UK(0800-2100) |
| 13215 | Hickam,HI |
| | Andersen, Guam |
| | Lajes, Azores |
| | Yokota, J |
| 13244 | Ascension(1000-1800) |

# AIR FORCE

|  | | |  | |
|---|---|---|---|
|  | Croughton,UK |  | MEDEVAC | Hospital aircraft |
|  | Lajes,Azores(1000-2100) |  | COPTER/tail # | Helicopter |
|  | MacDill,FL(0900-2400) |  | AIR FORCE RESCUE/ | |
| 13247 | Andrews,MD(1200-2400) |  | tail # | Search & rescue aircraft |
| 15014 | Incirlik, Turkey |  | PEDRO or | |
| 15015 | Albrook,PNZ(1200-0200) |  | SAVE/01-99 | USN/USMC rescue craft |
|  | Ascension(0800-2000) |  |  | not on rescue mission |
| 15031 | McClellan,CA(1600-0400) |  | 3-letter word/3 d. | Undergraduate pilot |
| 15036 | Croughton,UK |  |  | training |
| 16454 | Croughton,UK |  | 6-letter word/1 d. | TAC combat crew training |
| 18002 | Andersen,Guam(2200-0700) |  | 3-letter word/4 d. | MAC/TAC civil distur- |
|  | Clark,Phil(2200-1200) |  |  | bance airlift |
|  | Hickam,HI(1900-0400) |  | 6-15 letter word(s) | SAC ground station |
|  | Loring,ME(0100-0900) |  | 3-5 letter word/2 d. | SAC bomber or tanker |
|  | McClellan,CA(1600-0400) |  | US AIR FORCE | Foreign over-flights |
|  | Yokota,J(0001-0900) |  | 6-15 letter word(s) | SAC ground station |
| 18019 | Albrook,PNZ(1400-2400) |  |  | |
|  | MacDill,FL(0900-2400) |  |  | |
| 23227 | Clark,Phil(0200-1300) |  |  | |
|  | Diego Garcia, Indian Ocean (USN) |  |  | |

## U.S. Air Force In-Flight Callsigns (DV Code)

| AIR FORCE 1 | US President (USAF #27000) |
|---|---|
| AIR FORCE 2 | US Vice President (USAF #31682) |
| EXECUTIVE 1 | US President aboard non-military aircraft |
| EXECUTIVE 2 | US Vice President aboard non-military aircraft |
| EXECUTIVE 1 or 2 FOXTROT | US Presidential or Vice Presidential family member aboard any aircraft |
| AFKAI 01-99 | Tactical ops |
| AFKAI (Word) 01-99 | Non-tactical courier or tactical airlift SAM, 89 MAW |
| SAM/5 digits | Special Air Mission (VIP/Diplomatic flight) (SAM86971 Secretary of State) (SAM86972 National Security Advisor) |
| SAM 01 | Heads of state of foreign countries |
| VENUS | SAM aircraft without VIP aboard |
| MAC/tail # | Military Airlift Command |
| SPAR/01-99 | Other Special Air Mission (diplomatic, attache, MAAG, 89MAW, etc) |
| AIR EVAC/tail # | Air evacuation C-9 aero-medical aircraft |
| AIR RESCUE/tail # | SAR aircraft on mission |

## USAF Tactical Callsigns and Code Words

| ACROBAT | Andrews Air Force Base |
|---|---|
| AFKAI | 97th MAW flight |
| AGAR | 4050th Test Wing, Wright-Patterson AFB, OH |
| ALLIGATOR | AWACS SECOM (See abbreviations) |
| ARMY 1 | US Army Helicopter, President on board |
| ARMY 2 | US Army Helicopter, Vice President on board |
| BENT SPEAR | Significant nuclear accident |
| BIG FOOT | McChord AFB, WA |
| BROKEN ARROW | Damaged nuclear weapon; emergency 2000 foot perimeter |
| CAPSULE | General call sign for all MAC aircraft |
| COVERED WAGON | Confirmed hostile act |
| CROWN | White House communications agency (WHCA) |
| DENALI | Alaskan ALCC (MAC) center at Elmendorf AFB (See SENTRY) |
| DETONE | (See SENTRY) |
| DISCARD | 22nd Air Force Operations Center (MAC Travis AFB, CA) |
| DRAGNET | (See SENTRY) |
| DROPKICK | SAC HQ, Offutt AFB, NE |
| DULL SWORD | Nuclear incident |
| EDGY | AWACS Aircraft, Training flight from Tinker AFB, OK |
| ELECTRIC | NEACP (National Emergency Airborne Command Post) |
| FADED GIANT | Radiological incident or accident |

# AIR FORCE

| | |
|---|---|
| FAST CHARGER | Air Force Joint Command, Europe |
| FERTILE | SAGE/NORAD net center (Ft. Lee, VA) |
| FLYNEST | Nuclear-CBW emergency disaster team |
| FORMAT | 21st AF (MAC McGuire AFB, NJ) |
| FURIOUS | S. American ALCC (MAC, Albrook AFB) |
| GOLD COIN | MacDill AFB, FL, Strike Command Control |
| GOLDEN | General call from TAC aircraft to ground station |
| GOLIATH | NORAD, Duluth (MN) International Airport |
| GULL | 53rd WRS Keesler AFB, MS |
| HELPING HAND | Threat of hostile act |
| HILDA | MAC Operations Center, Scott AFB, IL |
| JOLLY 01-99 | SAR helicopter (HH3/HH53) not on mission |
| KING 01-99 | SAR aircraft (HC-130) not on mission |
| LARK | 55th WRS McClellan AFB, CA (RC-135/WC-135 aircraft) |
| LETTERMAN | Hickam AFB, HI, weather forecasts |
| MAINSAIL | General call to any command control station |
| MARINE 1 | USMC Helicopter, President on board |
| MARINE 2 | USMC Helicopter, Vice President on board |
| MINUTEMAN | Andrews AFB ANG |
| NIGHT HAWK | Presidential helicopters (Marine Corps) |
| NO-LONE ZONE | Lone individuals prohibited (nuclear weapons area) |
| PHANTOM | European theater airlift control |
| QUEEN | Rescue Operations Ctr, Eglin AFB ARRS |
| SAFE WIND | Emergency security order from higher command |
| SENTRY | AWACS aircraft (E3A) from Tinker AFB (OK), NCS Raymond 24 |
| SKYBIRD | General call to any SAC ground station |
| SKYKING | General call to any SAC SIOP forces aircraft |
| SPAR | 89th MAW VIP flight |
| SWAN | 54th WRS Andersen AFB, Guam (WC-130 aircraft) |
| TEAL | 815th WRS Keesler AFB, MS (WC-130 aircraft) |
| TONIGHT | MAC - Pacific Airlift Control Center |
| TRICE | 911th TAC Airlift Group, Pittsburgh, PA |

## U.S. Air Force Miscellaneous USB

| | |
|---|---|
| 6780 | Andrews AFB |
| 6812 | Andrews AFB |
| 7840 | Andrews AFB |
| 9270 | Andrews AFB |
| 9320 | Andrews AFB |
| 13207 | Palmerola AFB, Honduras |
| 20153 | Andrews AFB |

## U.S. Air Force Commando Escort Net (Hawaii)

**Callsigns:** American Eagle, Fivespot, Fletcher
**Frequencies:** 10452 10510 20737 26515

## U.S. Air Force Pacific Point to Point

| 6712 | USB | "Alpha 2" |
|---|---|---|
| 10452 | USB | "Oscar" |
| 14775 | LSB | "Mike" |
| 18005 | USB | "Tango" |
| 20740 | USB | "Lima" |

## U.S. Air Force Atlantic Practice Range

Communications may be heard from fighter aircraft as they duel with a string of eight towers 50 miles off the U.S. east coast (approximate coordinates: 21°N/80°W) on a frequency of 4295 kHz (USB). Tactical call - "Sea Tac."

## U.S. Air Force Military Airlift Command (MAC)

5197 18027 USB

## U.S. Military Advisory Group (MAG) U.S. Office of Defense Cooperation Mission Radio Network

The U.S. Air Force operations a single sideband communications network to provide logistic and morale phone patches among the military attaches located at U.S. Embassies in South & Central America. The network control station is at Howard AFB, Republic of Panama.

Stations identify with the last letter and numerals of call sign; for instance, AHF4 identifies as "Fox 4". AHF1B has been heard airborne.

| | |
|---|---|
| 3503 | Ch 1 |
| 7430 | Ch 2  Night Pri |
| 10935 | Ch 3 |
| 13937 | Command |
| 13950 | Ch 4 Day Pri |
| 20885 | Ch 1 Day Sec |

| | |
|---|---|
| AHF4 | Howard AFB, Panama |
| LQU21 | Buenos Aires, Argentina |
| VPL1D5 | Belize City, Belize |
| CPP67 | La Paz, Bolivia |
| CEF5U1 | Santiago, Chile |
| 5K0225 | Bogota, Colombia |
| ACB | Rio de Janeiro, Brazil |
| TI2USA | San Jose, Costa Rica |
| HIP491 | Santo Domingo, Dominican Republic |
| HCUS1 | Quito, Ecuador |
| YS1HUKE | San Salvador, El Salvador |
| TDMG3 | Guatemala City, Guatemala |
| ACH54 | Port Au Prince, Haiti |
| HR1MM | Tegucigalpa, Honduras |
| YN1AFM | Managua, Nicaragua |
| ZPM261 | Asuncion, Paraguay |
| OAE21 | Lima, Peru |
| CXC20 | Montevideo, Uruguay |
| YWA6 | Caracas, Venezuela |
| AHF1B | Unidentified |
| AHF1A | Unidentified, possibly El Salvador |
| AHF5 | Unidentified |
| TGHM1 | Unidentified |

## North American Aerospace Defense Command (NORAD)

Secured nearly a mile beneath Cheyenne Mountain near Colorado Springs is the "war room," home of NORAD. Intelligence from satellites, radar, surveillance aircraft and other sources is gathered continuously.

Encharged with the responsibility to detect the intrusion of hostile aircraft, NORAD is staffed by Air Force and Army personnel. Other transmitters are located at North Bend, OR and Hamilton, CA.

USB radio checks are made approximately on the hour; callsigns change often, with "Ringmaster" reportedly used at NORAD HQ.

| | | |
|---|---|---|
| 5297 | 10452 | 14364 |
| 9023 (Delta) | 11214 (Charlie) | 14894 |
| 9793 | 11284 | 18027 |
| 10194 | 11441 | 20855 |

## Aerospace Rescue and Recovery Service (ARRS)

Operating from Air Force bases worldwide, the ARRS provides emergency assistance to downed aircraft. During rescue exercises, ARRS aircraft will identify as "Air Force Rescue" followed by their tail numbers.

During non-rescue missions, C-130 aircraft will identify as "King," HH53 helicopters as "Jolly." The Rescue Operations Center at Eglin AFB, Florida, is "Queen."

| | |
|---|---|
| 6715 | Homestead AFB, FL |
| 11440 | Eglin AFB, FL |

## Air National Guard Nationwide

| | | |
|---|---|---|
| 4760 | 6714 | 7460 |
| 5703 | 6769 | 7853 |
| 5745 | 6800 | 8731 |
| 5805 | 6890 | 14650 |
| 5827 | 6895 | 17975 |
| 6100 | 6940 | |

# AIR FORCE

Louisiana Air National Guard F-15 (photo by Johnny Autery)

## U.S. Air Force Reserves

4457    4812

## U.S. Air Force Security Service

2010      Medina, TX
26710     Kelly AFB, TX

## U.S. Air Force Crash Investigation

4277

"Redstone": Communications teams
"Stardust": Commanders

## U.S. Air Force
(DATA-AUTODIN)

Frequencies:   10185   12201   13473
                15620.5   19955
Callsign: AFA, Davidsonville, MD

## U.S. Air Force Mulcast
### (Multichannel broadcast)

A single wideband channel may be selected for several simultaneous transmissions of information. A typical example (10205 kHz) shows a channel breakdown as follows:

| | |
|---|---|
| 10118.8 | USB voice |
| 10205 | LSB voice |
| 10205.4-10208 | 16 RTTY channels (85 Hz shift, 85 Hz separation) |
| 10211.3 | LSB voice |

**AFA--Andrews Air Force Base, MD, Mulcast**
7870  7922.5  10205  10315  14417.5  14646

**CUW--Lajes AFB, Azores, Mulcast**
9198  10305  13728  15757.5  19285  21767

**AJE--(Croughton), England, Mulcast**

5224  5246  5370.5  9236  10531.5  10537.8
16035  16448

**AJE--(Croughton) to CUW (Lajes) Mulcast**

5224      5242.5   7567   9172.5   11682.5
14879.5   15575.5  16140  18226    23360

Mulcast includes 50 baud AP news, 45 baud UPI news, 75 baud weather (all USB) and LSB AFRTS voice (6.3 kHz above center frequency).

## Ground Wave Emergency Network (GWEN)

As part of the Survivable Low Frequency Communications Systems (SLFCS), the Air Force is installing hundreds of transmitter and receiver sites in the 150-175 kHz range across the nation. Some frequencies support more than one transmitter. Channel spacing is 1.25 kHz and the mode is encrypted FSK, similar to packet. Transmitting individually every 5-10 seconds, the transmitters all come on in unison approximately every 20 minutes.

## GWEN Transmitter Sites

| Freq | State | Location |
|---|---|---|
| 154.375 | AL | Hackleburg |
| 158.125 | AL | Slapout (Holtville) |
| 158.125 | AR | Fayetteville (Spring Valley) |
| 170 | AZ | Flagstaff (Navajo Army Depot) |
| 166.875 | CA | Bakersfield |
| 165.625 | CA | Biggs (Gridley) |
| 171 | CA | Essex |
| 161.875 | CA | Roseville |
| 163.125 | CO | Aurora (Lowry AFB Annex) |
| 169.375 | CO | Pueblo |
| 165.625 | CO | Pueblo Army Depot |
| 171.875 | CO | Denver (Rocky Flats) |
| 150.625 | GA | Huber (Macon) |
| 169.375 | GA | Pembroke (Savannah Beach) |
| 154.375 | IA | Mechanicsville |
|  | IA | Glenwood/Pacific Jct. (St. Marys) |
| 155.625 | KS | Oberlin |
| 173 | KS | Maple Hill (Topeka) |
|  | MA | Acushnet |
|  | MA | Barre Falls (Amherst) |
|  | ME | Herseytown (Sherman) |
|  | ME | Penobscot |
| 163.125 | MD | Crownsville (Annapolis) |
| 170.625 | MD | Lappans (Hagerstown) |
| 159.375 | MD | Beantown (Waldorf) |
| 156.875 | MI | Alaeidon Twp. (Lansing) |
| 165.625 | MS | Alligator Twp. (Clarksdale) |
|  | MT | Billings |
|  | MT | Great Falls (Stockett) |
|  | MT | Ronan |
|  | NE | Ainsworth |
| 154.375 | NE | Omaha (SAC) (Bellevue) |
| 174 | NM | Albuquerque (Kirtland AFB) |
|  | NJ | Little Egg Harbor |
| 160.625 | NY | Pine Valley (Elmira) |
| 150.625 | NY | Hudson Falls (Glen Falls) |
| 165.625 | NY | Remsen (Utica) |
| 173.125 | NC | Antioch (Beaufort) |
|  | ND | Devils Lake |
|  | ND | Edinburg (Langdon) |
| 155 | ND | Medora |
| 158.125 | OK | Canton |
| 163.125 | OR | Klamath Falls (Kingsley Field) |
| 156.875 | OR | Seneca |
| 163.125 | PA | Harborcreek (Erie) |
| 159.375 | PA | Gettysburg |
| 171.875 | PA | Phillipsburg (Hawk Run) |
|  | RI | Little Compton |
| 150.625 | SC | Dunbar (Kensington) |
| 169 | SD | Clark (Raymond) |
| 171.875 | TX | Summerfield (Westway/Hereford) |
| 153.125 | VA | Carrollton (Driver) |
| 160.625 | WA | Appleton |
| 150.625 | WA | Spokane (Lyons) |
|  | WA | Wenatchee |
| 160.625 | WI | Mequon (Milwaukee) |

# AIR FORCE

## Civil Air Patrol (CAP)

A civilian search and rescue effort sponsored by the Air Force, CAP conducts training nets and rescue missions nationwide via shortwave radio. Aircraft, ships, mobile units and base stations communicate using tactical callsigns followed by "mobile", "air mobile" or "air".

### Tactical Net

| Region | Headquarters Location | Callsign | Tactical Call |
|---|---|---|---|
| 1)North East | Mineola, NY | KGC632 | Northeast 10 |
| 2)Middle East | Shaw AFB, SC | KIL769 | Middle East 4 |
| 3)Great Lakes | Wright Patterson AFB,OH | KSF248 | Great Lakes 55 |
| 4)South East | Nashville, TN | KIJ960 | Southeast 4 |
| 5)North Central | Minneapolis, MN | KAJ506 | North Central 8 |
| 6)South West | Dallas, TX | KKQ226 | Southwest 11 |
| 7)Rocky Mountain | Lowrey AFB, CO | KAI562 | Rocky Mountains 28 |
| 8)Pacific | Hamilton AFB, CA | KMG664 | Northwind 8 |
| Net Control | Maxwell AFB, AL | KJ9885 | Headcap22 |

### Frequencies and Regions
(USB unless indicated otherwise)

| Freq | Regions | Freq | Regions |
|---|---|---|---|
| 2371 | USA | 4627 | 6 |
| 2374 | USA | 4630 | 6 |
| 4273 | USA (packet/RTTY) | 7375 | USA (packet/RTTY) |
| 4464.5 | (1) (to become 4466) | 7635 | USA pri. |
| 4467.5 | 1 4 (to become 4469) | | (w/USAF & FEMA) |
| 4504.5 | (8) (to become 4506) | 7918.5 | USA (to become 7917) |
| 4507.5 | 3 5 (to become 4509) | | (w/USAF) |
| 4582 | Emergency/ | 11975 | USA |
| | calling (to become 2,8) | 14905 LSB | USA (to become 14902 USB) |
| 4585 | 2 | | (w/USAF & FEMA) |
| 4599.5 | (7) | 20873 | USA |
| 4602.5 | 3 (to become 4604) | 26617 | USA |
| | | 26620 | USA (AM/USB) |

*Military maneuvers are popular monitoring targets.*

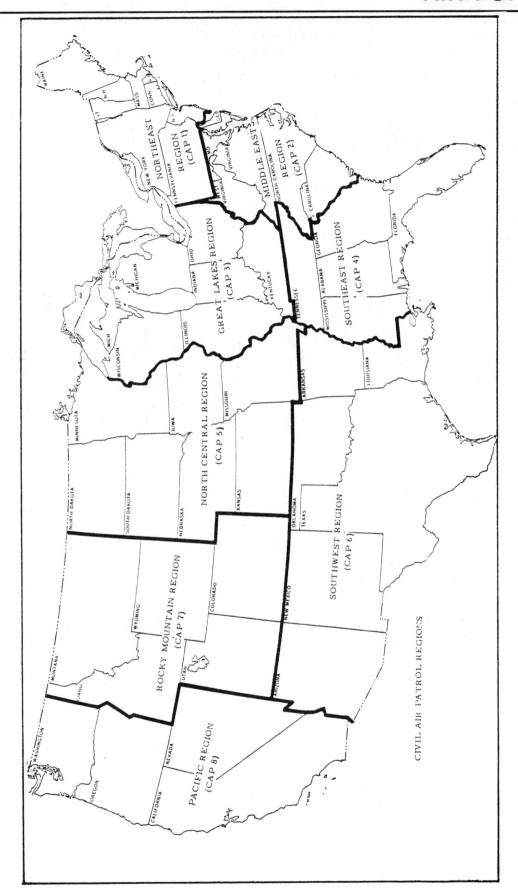

# AIR FORCE

## Training Net Schedule
(daily, USB)

| Region | Frequencies | | Net Times | | |
|--------|-------------|--------|-----------|-----------|-----------|
| 3 | 4602.5 | 4507.5 | 0000-0230 | 1330-1600 | 2130-2400 |
| 2 | 4585 | 4467.5 | 0000-0200 | 1230-1600 | 2100-2400 |
| 5 | 4705.5 | 4585 | 0000-0400 | 1200-1600 | 1800-1830 |
| 1 | 4464.5 | 4585 | 0000-0230 | 1300-1700 | 2100-2400 |
| 8 | 4585 | 4504.5 | 0130-0630 | 1450- | 2130-2400 |
| 7 | 4599.5 | 4602.5 | 0100-0400 | 1345-1530 | 1600-1800 |
| 4 | 4467.5 | 4585 | 0000-0230 2230-2400 | 1230-1400 | 1430-1500 |
| 6 | 4627 | 4630 | 0000-0400 | 1300-1430 | 2300-2400 |

*National Command Net 1115 Eastern time daily 7635 USB/1130 Eastern time 14905 LSB*

## Callsign Identifications

| State | Callsign | Identifier |
|-------|----------|------------|
| AL | KIG442 | Goldenrod |
| AK | KWA677 | Sourdough |
| AR | KKI719 | Dogwood |
| AZ | KOF424 | Thunderbird |
| CA | KME284 | Eagle |
| CO | KAF357 | Pikes Peak |
| CT | KCC590 | Nutmeg |
| DE | KGC462 | Gabby |
| GA | KIG443 | Red Star |
| FL | KIG444 | Sparrow |
| HI | KUA341 | Firebrand |
| ID | KOP334 | Magpie |
| IL | KSC952 | Red Fox |
| IN | KSC953 | Red Fire |
| IA | KAF358 | Corn State |
| KS | KAF359 | Jayhawk |
| KY | KIG445 | Middleground |
| LA | KKI720 | Magnolia |
| ME | KCC591 | Pine Tree |
| MD | KGC464 | Plant |
| MA | KCC592 | Freedom |
| MI | KQD405 | Red Robin |
| MN | KAF360 | Starfish |
| MO | KAF361 | Blue Bird |
| MS | KKI721 | Mockingbird |
| MT | KOF426 | Father |
| NB | KAF362 | Wigwam |
| NC | KIG446 | Red Dog |
| ND | KAF363 | Blackfoot |
| NH | KCC593 | Profile |
| NJ | KEC994 | Zig Zag |
| NM | KKI722 | Pueblo |
| NV | KOP335 | North Wind |
| NY | KEC995 | Empire |
| OH | KQD406 | Black Hawk |
| OK | KKI723 | Sooner |
| OR | KOF428 | Beaver Fox |
| PA | KCC465 | Keystone |
| RI | KCC594 | Rhody |
| SC | KIG447 | Crescent |
| SD | KAF364 | Dacotah |
| TN | KIG448 | Blue Chip |
| TX | KKI724 | Eagle Nest |
| UT | KOF429 | Uncle Mike |
| VA | KIG449 | Blue Flight |
| VT | KCC595 | Pico |
| WA | KOF430 | Fir |
| WI | KSC954 | Badger |
| WV | KQD407 | Lowland |
| WY | KOF431 | King |
| PR* | WWA353 | Pineapple |
| HQ** | KGC463 | Aero |

*Puerto Rico
**National CAP, Washington, DC

## Military Affiliate Radio System (MARS)

The Department of Defense sponsors a worldwide network of radio stations utilizing amateur radio operators to carry personal messages between military personnel and their families.

Voice phone patches, radioteletype and CW are all used to sustain this morale effort among Army, Air Force and Navy men and women.

Frequency assignments come from the Department of Defense and some frequencies are used by more than one service branch on a time sharing basis.

USAF MARS REGIONS

REGION 8 - PACIFIC
REGION 9 - ALASKA

REGION 1
REGION 2
REGION 3
REGION 4
REGION 5
REGION 6

ME
NH
VT
MA
RI
CT
NY
NJ
PA
DE
MD
VA
WVA
OH
KY
NC
SC
MI
IN
TN
GA
FL
WI
IL
MS
AL
MI
MN
IA
MO
AR
LA
ND
SD
NE
KS
OK
TX
MT
WY
CO
NM
ID
UT
AZ
WA
NV
OR
CA

# AIR FORCE

## U.S. Air Force Mars

| Primary Use | Frequency | Freq Code | Emission |
|---|---|---|---|
| ALASKA GATEWAY | 7938.0 | AK-2 | USB/CW/RTTY/DATA |
| ALASKA GATEWAY | 9047.0 | AK-3 | USB/CW/RTTY/DATA |
| ALASKA GATEWAY | 9224.0 | AK-4 | USB/CW/RTTY/DATA |
| ALASKA GATEWAY | 13985.0 | AK-1  WSDN-3 | USB/CW/RTTY/DATA |
| ALASKA GATEWAY | 15712.0 | AK-5 | USB/CW/RTTY |
| ALASKA GATEWAY | 19200.0 | AK-6 | USB/CW/RTTY/DATA |
| ALASKA GATEWAY | 24573.5 | AK-7 | USB/CW/RTTY |
| ALASKA INTERTHEATER | 4061.0 | ALS IT | USB/CW/RTTY |
| ALASKA INTERTHEATER | 4758.0 | ALS IT | USB/CW/RTTY |
| ALASKA INTERTHEATER | 4815.0 | ALS IT | CW |
| ALASKA INTERTHEATER | 4882.0 | ALS IT | USB/CW/RTTY |
| ALASKA INTERTHEATER | 4885.0 | ALS IT | LSB/CW/RTTY |
| ALASKA INTERTHEATER | 6995.0 | ALS IT | USB/LSB/CW/RTTY |
| ALASKA INTERTHEATER | 7331.0 | ALS IT | USB/CW/RTTY |
| ALASKA INTERTHEATER | 10140.0 | ALS IT | USB/CW/RTTY |
| AZORES PHONE PATCH | 7633.5 | ACJ | USB/CW/RTTY |
| AZORES PHONE PATCH | 13927.0 | ACB | USB/CW/RTTY |
| AZORES PHONE PATCH | 14606.0 | ACF | USB/CW/RTTY/DATA |
| CENTRAL/SOUTH AMER GATEWAY | 7540.0 | CSA GW | LSB |
| CENTRAL/SOUTH AMER INTERTHEATER | 13927.0 | CSA IT | LSB |
| CENTRAL/SOUTH AMER INTERTHEATER | 13977.0 | CSA IT | LSB |
| EUROPE PHONE PATCH | 20188.5 | ACG | USB/CW/RTTY/DATA |
| EUROPEAN DIAMOND NT | 14372.5 | ER-3 | USB/CW/RTTY/AM |
| EUROPEAN DIAMOND NT | 14505.0 | ER-4 | USB |
| EUROPEAN RUBY NET | 3875.0 | ER-1 | USB |
| GERMANY GATEWAY | 11098.5 | D GW | USB/CW/RTTY |
| GERMANY INTERTHTR | 3802.5 | D IT | CW |
| GERMANY INTERTHTR | 3888.0 | D IT | CW/RTTY/AM |
| GERMANY INTERTHTR | 3888.0 | ER-3 | USB |
| GERMANY PHONE PATCH | 14390.5 | ACE | USB/CW/RTTY |
| GERMANY RTTY GATEWY | 7457.0 | D GW | USB/RTTY |
| GERMANY RTTY GATEWY | 18155.0 | D GW | CW/RTTY |
| GERMANY/PAC PHONE PATCH | 11407.0 | ACA PCK-1 | USB/CW/RTTY |
| GREECE INTERTHEATER | 3966.0 | GRC IT | USB |
| GREENLAND GATEWAY | 7799.0 | GRL GW | USB |
| GRL/GERMANY GATWAY | 6775.0 | GRL/D GW | USB |
| GUAM PHONE PATCH | 20991.0 | PCG-2 PS-6 | USB/CW/RTTY |
| GUAM PHONE PATCH | 27736.0 | PCG-3 | USB/CW/RTTY |
| GUAM PHONE PATCH | 27829.0 | 0CG-4 | USB/CW/RTTY |
| GAUM/HI INTERISLAND | 7360.0 | GUM II HWA II | USB/CW/RTTY |
| HAWAII INTERISLAND | 4054.0 | HWA II | USB/CW/RTTY |
| HAWAII INTERTHEATER | 7312.0 | HWA IT | USB/CW/RTTY |
| HAWAII INTERTHEATER | 7315.0 | HWA IT | USB/CW/RTTY |
| JAPAN INTERISLAND | 7407.0 | J II | USB/CW/RTTY |
| JAPAN INTERTHEATER | 3841.0 | J IT | USB |
| JAPAN INTERTHEATER | 4633.0 | J IT | USB |
| JAPAN INTERTHEATER | 4638.0 | J IT | USB |
| JAPAN INTRAISLAND | 8177.0 | J II | USB/CW/RTTY |
| JAPAN PHONE PATCH | 13614.0 | PCJ-2 | USB/CW/RTTY |
| JAPAN PHONE PATCH | 14829.0 | PCJ-3 PRR-1 | USB/CW/RTTY |

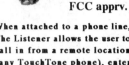
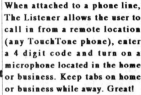

# AIR FORCE

| | | | |
|---|---|---|---|
| JAPAN PHONE PATCH | 20807.0 | PCJ-4 PS-4 PRR-3 | USB/CW/RTTY |
| JAPAN PHONE PATCH | 23862.0 | PCJ-5  PRR-6 | USB/CW/RTTY |
| JAPAN/KOREA INTER- THEATER | 8178.5 | J/KOR IT | USB/CW/RTTY |
| KOREA INTERTHEATER | 3496.0 | KOR IT | USB/RTTY |
| KOREA PHONE PATCH | 14389.0 | PCK-2 WSDN-4 | USB/CW/RTTY |
| KOREA PHONE PATCH | 15632.0 | PCK-3 PRR-4 | USB/CW/RTTY |
| PACAF INTERTHEATER | 7357.0 | PAC IT | USB/CW/RTTY |
| PACAF INTRATHEATER | 15805.5 | PS-2 | USB/CW/RTTY |
| PACIFIC INTRATHTR | 19937.0 | PS-3 PCK-4 | USB/CW/RTTY |
| PACIFIC RTTY GATEWY | 15807.0 | PAC RATT GW | USB/CW/RTTY |
| PACIFIC RTTY GATEWY | 24019.0 | PAC RATT CW PRR-7 | USB/CW/RTTY |
| PANAMA INTERTHEATER | 3454.0 | PNR IT | LSB |
| PANAMA INTERTHEATER | 3545.0 | PNR IT | USB |
| PANAMA INTERTHEATER | 7305.0 | PNR IT | LSB |
| PANAMA INTERTHEATER | 7305.0 | PNR IT | USB |
| PANAMA PHONE PATCH | 27978.5 | TINKER AFB/ HOWARD AB | USB |
| PHILIPPINES INTER- | 4912.0 | PHL IT | USB/CW/RTTY |
| PHIL PHONE PATCH | 10267.0 | PCP-1 | USB/CW/RTTY |
| PHIL PHONE PATCH | 10270.0 | PCP-2 | USB/CW/RTTY |
| PHIL PHONE PATCH | 14402.0 | PCP-3 | USB/CW/RTTY |
| PHIL PHONE PATCH | 14530.0 | PCP-4 | USB/RTTY |
| PHIL PHONE PATCH | 14877.0 | PCP-6 | USB/CW/RTTY |
| PHIL PHONE PATCH | 16452.0 | PCP-78 | LSB/USB/CW/RTTY |
| PHIL PHONE PATCH | 17670.0 | PCP-8 | USB/CW/RTTY |
| PHIL PHONE PATCH | 19226.0 | PCP-9 | USB/CW/RTTY |
| PHIL PHONE PATCH | 27991.0 | PCP-11 PS-7 | USB/CW/RTTY |
| REGION 1 | 3315.0 | RA(1) | USB/CW/RTTY |
| REGION 1 | 4593.5 | RB(1) | USB/CW/RTTY/DATA |
| REGION 1 | 7324.0 | RC(1) | USB/CW/RTTY/DATA |
| REGION 2 | 3299.0 | RD(2) | USB/CW/RTTY/DATA |
| REGION 2 | 4577.0 | RE(2) | USB/CW/RTTY/DATA |
| REGION 2 | 7313.5 | RF(2) | USB/CW/RTTY |
| REGION 2-RTTY NET | 4500.0 | 2R1(2) | LSB/CW/RTTY/AM |
| REGION 3 | 3308.0 | RG(3) | USB/CW/RTTY/DATA |
| REGION 3 | 4517.0 | RH(3) | USB/DATA |
| REGION 3 | 7305.0 | RI(3)  RT(5) | USB/RTTY |
| REGION 4 | 3370.5 | RJ(4) | USB/CW/RTTY/DATA |
| REGION 4 | 4557.0 | RK(4) | USB/CW/RTTY/DATA |
| REGION 4 | 7302.0 | RL(4) | USB/CW/RTTY/DATA |
| REGION 5 | 3292.0 | RM(5) | USB/CW/RTTY |
| REGION 5 | 4450.0 | RN(5) | USB/CW/RTTY/AM |
| REGION 5 | 7329.0 | RO(5) | USB/CW/RTTY/DATA |
| REGION 6 | 3296.0 | RP(6) | USB/CW/RTTY/DATA |
| REGION 6 | 4575.0 | RQ(6) | USB/CW/RTTY/DATA |
| REGION 6 | 7457.0 | RR(6) | USB/CW/RTTY/DATA |
| SPACE DIV WORLDWIDE | 14405.0 | WSDN-5 | USB/CW/RTTY |
| SPACE DIV WORLDWIDE | 20870.0 | WSDN-5 | USB/CW/RTTY |
| SPACE DIV WORLDWIDE | 27985.0 | WSDN-7 | USB/CW/RTTY |
| TCON AD TECH TVRS-1 | 3312.0 | TVRS-1 | USB/CW/RTTY/DATA |
| TCON AD TECH TVRS-2 | 4588.5 | TVRS-2 | USB/DATA |
| TCON AD TECH TVRS-3 | 6775.0 | TVRS-3 | USB/CW/RTTY/DATA |
| TCON AD TECH TVRS-4 | 7680.0 | TVRS-4 | USB/CW/RTTY/DATA |
| TCON AD TECH TVRS-6 | 14529.5 | TVRS-6 | USB/CW/RTTY/DATA |
| TCON AD TECH TVRS-7 | 17487.0 | TVRS-7 | USB/CW/RTTY/DATA |

| | | | |
|---|---|---|---|
| TCON AD TECH TVRS-8 | 26910.0 | TVRS-8 | DATA/AM |
| TCON TRR/TRVRS/<br>  TRC-5 | 11619.5 | TRR-5 TVRS-5<br>TRC-5 | USB/CW/RTTY/DATA |
| TRANSCON CALLING | 2624.0 | CA(T) | USB/CW/RTTY |
| TRANSCON CALLING | 3311.0 | CB(T) | USB/CW/RTTY/DATA |
| TRANSCON CALLING | 4590.0 | CC(T) | USB/CW/RTTY/DATA |
| TRANSCON CALLING | 7540.0 | CD(T) | USB/CW/RTTY/DATA |
| TRANSCON CALLING | 13993.0 | CE(T) | USB/CW/RTTY/DATA |
| TRANSCON CALLING | 20737.0 | CF(T) | USB/CW/RTTY |
| TRANSCON CW | 3311.0 | TRC-1 | USB/CW/RTTY/AM |
| TRANSCON CW | 3347.0 | TRC-2 | CW/RTTY |
| TRANSCON CW | 4878.0 | TRC-3 | LSB/CW/RTTY |
| TRANSCON CW | 6994.0 | TRC-4 | USB/CW/RTTY |
| TRANSCON CW | 14527.J0 | TRC-7 | USB/CW/RTTY |
| TRANSCON CW | 20960.0 | TRC-8 PS-5<br>PRR-2 | USB/CW/RTTY |
| TRANSCON CW | 27994.0 | TRC-9 | USB/CW/RTTY |
| TRANSCON RTTY | 3228.0 | TRR-1 | USB/RTTY |
| TRANSCON RTTY | 4872.0 | TRR-2 | USB/CW/RTTY/DATA |
| TRANSCON RTTY | 7831.5 | TRR-3 | USB/CW/RTTY/DATA |
| TRANSCON RTTY | 7913.5 | TRR-5 | USB/CW/RTTY/DATA |
| TRANSCON RTTY | 15513.5 | TRR-6 | USB/CW/RTTY/DATA |
| TRANSCON RTTY | 23863.5 | TRR-7 | USB/CW/RTTY |
| TRANSCON RTTY | 27877.0 | TRR-8 | USB/CW/RTTY |
| TRANSCON TRAFFIC | 13977.0 | T1(T) ACC<br>APCN-1 PPCN-1 | USB/CW/RTTY |
| TRANSCON TRAFFIC | 4560.0 | T4(T) | USB/DATA |
| TRANSCON TRAFFIC | 4765.0 | TM(T) | USB/CW/RTTY/DATA |
| TRANSCON TRAFFIC | 4832.0 | TN(T) | USB/CW/RTTY |
| TRANSCON TRAFFIC | 4842.0 | TO(T) | USB/CW/RTTY |
| TRANSCON TRAFFIC | 7527.0 | TW(T) WSDN-2 | USB |
| TRANSCON TRAFFIC | 7545.0 | TR(T) | USB/CW/RTTY |
| TRANSCON TRAFFIC | 7632.0 | TS(T) | USB/CW/RTTY |
| TRANSCON TRAFFIC | 13996.0 | TU(T) TRC-6 | USB/CW/RTTY |
| TRANSCON TRAFFIC | 20740.0 | TV(T) | USB/CW/RTTY |
| TRANSCON TRAFFIC | 4602.6 | TX(T) | USB/DATA |
| UNASSIGNED | 2049.0 | | USB |
| UNASSIGNED | 3231.0 | | USB/CW/RTTY |
| UNASSIGNED | 4447.0 | | USB/CW/RTTY |
| UNASSIGNED | 4450.0 | | USB |
| UNASSIGNED | 4466.0 | | CW/RTTY |
| UNASSIGNED | 4593.5 | | USB/CW/RTTY |
| UNASSIGNED | 5740.0 | | USB |
| UNASSIGNED | 7862.5 | | USB/RTTY/AM |
| UNASSIGNED | 10273.5 | | USB/CW/RTTY |
| UNASSIGNED | 10576.0 | | USB/CW/RTTY |
| UNASSIGNED | 11121.0 | | USB/CW/RTTY/DATA |
| UNASSIGNED | 13498.0 | | USB/CW/RTTY |
| UNASSIGNED | 14408.0 | | LSB/CW/RTTY/DATA |
| UNASSIGNED | 14408.0 | | USB |
| UNASSIGNED | 14411.0 | | LSB/USB/CW/RTTY |
| UNASSIGNED | 15083.5 | | USB/CW/RTTY |
| UNASSIGNED | 19612.0 | | USB/RTTY |
| UNASSIGNED | 19615.0 | | USB/CW/RTTY |
| UNASSIGNED | 20763.0 | | USB/CW/RTTY |
| UNASSIGNED | 20873.0 | | USB |
| UNASSIGND | 20992.5 | | USB/CW/RTTY |
| UNASSIGNED | 20994.0 | | USB/CW/RTTY |

# AIR FORCE

| | | | |
|---|---|---|---|
| UNASSIGNED | 22947.0 | | USB/CW/RTTY |
| UNASSIGNED | 22950.0 | | USB/CW/RTTY |
| UNASSIGNED | 23865.0 | | USB/CW/RTTY |
| WORLDWIDE FREQ | 6996.0 | WSDN-1 PCJ-1 ER-2 | USB/CW/RTTY/DATA |

## U.S. Air Force MARS
## Callsign Identifications

### REGION 1

| | |
|---|---|
| AGA1AD | Andrews AFB DC |
| 1AN | Andrews AFB DC |
| 1AS | Andrews AFB DC |
| 1BA | Barnes MAP, Westfield MA |
| 1BC | Battle Creek ANGB MI |
| 1BG | Bangor IAP ME |
| 1BM | Baltimore MD |
| 1BR | Bradley ANGB CT |
| 1BU | Burlington IAP VT |
| 1BW | Rickenbacker AFB OH |
| 1CO | Coventry ANGS RI |
| 1DC | Pentagon DC |
| 1DO | Dover AFB DE |
| 1FA | Ft Wayne MAP IN |
| 1FW | Ft Wayne MAP IN |
| 1GR | Grissom AFB IN |
| 1HA | Hanscom AFB MA |
| 1HU | Hulman Field IN |
| 1KI | K I Sawyer AFB MI |
| 1MA | Mansfield Lahm Aprt OH |
| 1MC | McGuire AFB NJ |
| 1MG | McGuire AFB NJ |
| 1MI | Harrisburg IAP Middletown PA |
| 1MU | McGuire AFB NJ |
| 1NC | Griffiss AFB NY |
| 1NE | Newark AFS OH |
| 1NI | Niagara Falls IAP NY |
| 1NS | N Smithfield RI |
| 1OR | Orange ANGS CT |
| 1PE | Pease AFB OH |
| 1RB | Rickenbacker AFB OH |
| 1RI | Rickenbacker AFB OH |
| 1RO | Roslyn ANGS NY |
| 1SA | Springfield MAP OH |
| 1SC | Schenectady CO Aprt NY |
| 1SE | Selfridge ANGB MI |
| 1SM | Springfield MAP OH |
| 1SO | Springfield MAP OH |
| 1SP | S Portland ANGS ME |
| 1SU | Suffolk Co ANGB NY |
| 1SX | Deployed Element |
| 1SY | Hancock Field NY |
| 1TE | Toledo Express Arpt Swanton OH |
| 1TF | TF Green Arpt Warwick RI |
| 1WE | Wellesley ANGS MA |
| 1WG | Willow Grove NAS PA |

| | |
|---|---|
| 1WH | Westchester Co Aprt White Plains NY |
| 1WI | Gtr Wilmington Aprt DE |
| 1WO | Worcester ANGS MA |
| 1WP/B | Wright-Patterson AFB OH |
| 1WS | Westover AFB MA |
| 1WU | Wurtsmith AFB MI |
| 1WX | Deployed Element |
| 1YO | Youngstown MAP OH |
| AIR | Andrews AFB DC |

### REGION 2

| | |
|---|---|
| AGA2AT | Martin ANG Stn Gadsden AL |
| 2AX | Deployed Element |
| 2CH | Charleston AFB SC |
| 2DO | Dobbins AFB GA |
| 2DU | Charlotte NC |
| 2EA | Eastover SC |
| 2EG | Eglin AFB FL |
| 2HU | Hurlburt Fld FL |
| 2KA | Kanawha Co Aprt Charleston WV |
| 2LA | Langley AFB VA |
| 2MA | Maxwell AFB AL |
| 2MB | Myrtle Beach AFB SC |
| 2MC | McEntire ANGB Eastover SC |
| 2MD | MacDill AFB FL |
| 2ME | Memphis TN |
| 2MH | Myrtle Beach AFB SC |
| 2MO | Moody AFB GA |
| 2MR | E WV Reg Aprt(ANG) Martinsburg WV |
| 2MT | McGhee Tyson Aprt Knoxville TN |
| 2MY | McGhee Tyson Aprt Knoxville TN |
| 2NA | Nashville TN |
| 2NC | Badin NC |
| 2PA | Patrick AFB FL |
| 2PO | Pope AFB NC |
| 2PP | Pope AFB NC |
| 2PT | Patrick AFB FL |
| 2PX | Deployed Element |
| 2RB | Robins AFB GA |
| 2RH | Hurlburt Fld FL |
| 2RO | Robins AFB GA |
| 2RX | Quick Reaction Package (QRP) |
| 2SA | Savannah MAP GA |
| 2SH | Shaw AFB SC |
| 2SJ | Seymour-Johnson AFB NC |
| 2SN | Byrd IAP VA |
| 2SS | Birmingham Muni Aprt AL |
| 2ST | St Simons Is GA |
| 2SZ | Shaw AFB SC |
| 2TY | Tyndall AFB FL |

## REGION 3

| | |
|---|---|
| AGA3AF | HQ AFCC/XOPRM Scott AFB IL |
| 3BM | Billy Mitchell Fld Milwaukee WI |
| 3CA | Capital MAP Springfield IL |
| 3CH | Chanute AFB IL |
| 3CS | Scott AFB IL |
| 3DI | Duluth ANGB MN |
| 3FA | N Dakota St Univ Fargo ND |
| 3HQ | Scott AFB IL |
| 3LI | Lincoln ANGB NE |
| 3MA | Traux Fld Madison WI |
| 3MC | McConnell AFB KS |
| 3MI | Minn St Paul IAP MN |
| 3OH | O'Hare ARF Facility IL |
| 3OR | O'Hare IAP Chicago IL |
| 3PE | Gtr Peoria Aprt IL |
| 3RG | Richards-Gebaur AFB MO |
| 3SF | Joe Foss Field SD |
| 3SI | Sioux City MAP IA |
| 3SJ | Rosecrans Mem Aprt MO |
| 3SL | Bridgeton MO |
| 3SP | Minn St Paul IAP MN |
| 3VO | Volk Fld Camp Douglas W |
| 3WH | Whiteman AFB MO |

## REGION 4

| | |
|---|---|
| AGA4AL | Altus AFB OK |
| 4BE | Bergstrom AFB TX |
| 4CO | Columbus AFB MS |
| 4DY | Dyess AFB TX |
| 4DZ | Dyess AFB TX |
| 4EI | Bergstrom AFB TX |
| 4EL | Ellington AFB TX |
| 4EN | England AFB LA |
| 4FS | Ebbing ANGB Ft Smith MAP AR |
| 4GA | Garland ANGS TX |
| 4GO | Goodfellow AFB TX |
| 4GX | Deployed Element |
| 4HE | Hensley Fld Dallas TX |
| 4HS | Mem Aprt Hot Springs AR |
| 4JA | Jackson MS |
| 4KE | Kelly AFB TX |
| 4KL | Kelly AFB TX |
| 4KR | Keesler AFB |
| 4KS | Keesler AFB MS |
| 4LA | Laughlin AFB TX |
| 4LI | Little Rock AFB AR |
| 4LP | LaPort ANGB TX |
| 4LR | Little Rock AFB AR |
| 4LZ | Little Rock AFB AR |
| 4MA | Camp Mabry Austin TX |
| 4ME | Meridian MS |
| 4NO | USNAS New Orleans LA |
| 4RA | Randolph AFB TX |
| 4RE | Reese AFB TX |

| | |
|---|---|
| 4SH | Sheppard AFB TX |
| 4TI | Tinker AFB OK |
| 4TK | Tinker AFB OK |
| 4TN | Tinker AFB OK |
| 4TU | Tulsa OK |
| 4TX | Deployed Element |
| 4VA | Vance AFB OK |
| 4WR | Will Rogers ANGB OK |

## REGION 5

| | |
|---|---|
| AGA5BE | Bellingham WA |
| 5BO | Boise ID |
| 5BU | Buckley ANGB CO |
| 5CH | Cheyenne MAP WY |
| 5FA | Fairchild AFB WA |
| 5FI | Fairchild AFB WA |
| 5HI | Hill AFB UT |
| 5LO | Lowry AFB CO |
| 5MC | McChord AFB WA |
| 5MU | Camp Murray Tacoma WA |
| 5MX | Deployed Element |
| 5MZ | McChord AFB WA |
| 5PE | Peterson AFB CO |
| 5PO | Portland IAP OR |
| 5PT | Peterson AFB CO |
| 5SE | Seattle WA |
| 5SL | Salt Lake City UT |
| 5SP | Spokane IAP WA |
| 5TA | Camp Murray Tacoma WA |

## REGION 6

| | |
|---|---|
| AGA6CA | Cannon AFB NM |
| 6CM | Costa Mesa CA |
| 6CO | Compton CA |
| 6DM | Davis-Monthan AFB AZ |
| 6ED | Edwards AFB CA |
| 6GE | George AFB CA |
| 6HA | NAS Moffett Fld CA |
| 6HO | Holloman AFB NM |
| 6HY | Hayward ANGB CA |
| 6KI | Kirtland AFB NM |
| 6KR | Albuquerque NM |
| 6LA | Los Angeles CA |
| 6LU | Luke AFB AZ |
| 6MA | March AFB CA |
| 6MC | McClellan AFB CA |
| 6ML | McClellan AFB CA |
| 6MT | Mather AFB CA |
| 6NE | Nellis AFB NV |
| 6NH | N Highland CA |
| 6NO | Norton AFB CA |
| 6NX | Deployed Element |
| 6ON | Ontario IAP CA |
| 6PH | Phoenix AZ |
| 6PO | Phoenix AZ |
| 6TR | Travis AFB CA |

# AIR FORCE

| | |
|---|---|
| 6RE | Reno NV |
| 6TO | Torrence CA |
| 6TU | Tucson AZ |
| 6TV | Travis AFB CA |
| 6VA | Vandenberg AFB CA |
| 6VN | Van Nuys CA |
| 6VY | Sepulveda ANGS Van Nuys CA |
| 6WI | Williams AFB AZ |

REGION 7 (Europe/Middle East/Africa)

| | |
|---|---|
| AGA7AR | Araxos Greece |
| 7AT | Athens Greece |
| 7AU | Augsburg Germany |
| 7BI | Bitburg AB GErmany |
| 7CN | Camp New Amsterdam Netherlands |
| 7DR | Drama Greece |
| 7HA | Hahn AB Germany |
| 7HE | Hessisch-Oldendorf Germany |
| 7IR | Iraklion Crete |
| 7KA | Kalkar AS Germany |
| 7LE | Levkas Greece |
| 7LI | Lindsey AS Germany |
| 7MH | Mt Hortiatis Greece |
| 7MP | Mt Pateras Greece |
| 7RA | Ramstein AB Germany |
| 7RE | Reisenbach Germany |
| 7RM | Rhein Main AB Germany |
| 7SP | Spangdahlem AB Germany |
| 7ZW | Zweibrucken AB Germany |

REGION 8 (Pacific/Far East)

| | |
|---|---|
| AGA8CL | Clark AB Philippines |
| 8CJ | Choe Jong San Korea |
| 8EN | Hickam AFB HI |
| 8FR | Ft Ruger Hawaii |
| 8GO | Gozar AS Philippines |
| 8HI | Hickam AFB Hawaii |
| 8HY | Hampyeong, Korea |
| 8JI | Johnston Island |
| 8KJ | Kwang Ju Korea |
| 8KR | Kotar Korea |
| 8MI | Misawa AB Japan |
| 8MN | Mindanao Philippines |
| 8MS | Mangil San Korea |
| 8OS | Osan AB Korea |
| 8PH | Pohang Term Korea |
| 8PS | Palgong San Korea |
| 8RC | Camp Red Cloud Korea |
| 8TA | Taegu AB Korea |
| 8WA | Wallace AS Philippines |
| 8WI | Wake Island |
| 8YO | Yongmun San Korea |

REGION 9 (Alaska)

| | |
|---|---|
| AGA9CA | Campion AFS AK |

| | |
|---|---|
| 9CB | Cold Bay AFS AK |
| 9CE | Clear AFS AK |
| 9CL | Cape Lisburne AFS AK |
| 9CN | Cape Newenham AFS AK |
| 9CR | Cape Romanzof AFS AK |
| 9EI | Eielson AFB AK |
| 9EL | Elmendorf AFB AK |
| 9FA | Fairbanks AK |
| 9FY | Ft Yukon AFS AK |
| 9FR | Ft Richardson AK |
| 9GA | Galena Arpt AK |
| 9IM | Indian Mt AFS AK |
| 9JU | Juneau AK |
| 9KO | Kotzebue AFS AK |
| 9KS | King Salmon AFS AK |
| 9KU | Kulis ANGB Anchorage AK |
| 9MD | Murphy Dome AFS AK |
| 9MF | Miller Freeman (NOAA ship) |
| 9NG | Anchorage AK |
| 9SH | Shemya AFB AK |
| 9SP | Sparrevohn AFS AK |
| 9TA | Tatalina FS AK |
| 9TC | Tin City AFS AK |
| 9WA | Palmer AK |

REGION 10 (Caribbean, Central/ South America)

| | |
|---|---|
| AGA0AS | Paraguay |
| 0BR | Brazil |
| 0CO | Colombia |
| 0TG9 | Guatemala |
| 0HA | Port-au-Prince Haiti |
| 0HO | Howard AFB Balboa Panama |
| 0JA | Kingston Jamaica |
| 0MU | Muniz ANGB PR |
| 0NI | Nicaragua |
| 0PE | Peru |
| 0QU | Quito Ecuador |
| 0SS | El Salvador |
| 0TE | Honduras |
| 0UR | Uruguay |
| 0VE | Venezuela |

OTHER

| | |
|---|---|
| CUW20 | Lajes AB Azores |
| XP1AB | Sondrestrom AB Greenland |

## Special Call Signs

REGION 1

| | |
|---|---|
| AFA1L | Command MARS Director Air Force Logistics Command |
| AGA1Q | MARS Director Andrews AFB MD |

REGION 2

| | |
|---|---|
| AGA2R | Command MARS Director AFRES |

| AFA2T | Command MARS Director<br>Tactical Air Command |
|---|---|

REGION 3
| AGA3C | Chief MARS, USAF |
| AGA3D | USAF MARS Director |

REGION 4
| AGA4S | Command MARS Director,<br>USAF Security Service |
| AGA4T | Command MARS Director,<br>Air Training Command |

REGION 7
| AGA7E | Command, MARS Director,<br>USAFE |

REGION 8
| AGA8P | Command MARS Director,<br>Pacific Air Forces |

REGION 9
| AGA9A | Command MARS Director,<br>Alaska Air Command |

REGION 0
| AGA0S | MARS Director, Howard AFB<br>Panama |

| AIR | USAF HQ Pentagon (via Andrews<br>AFB) |

# Mystic Star Network

Controlled by Andrews AFB, the U.S. Air Force provides worldwide HF SSB support for presidential and vice presidential flights via remote transmitters on global USAF installations utilizing telephone voice channels, primarily for use when the aircraft are out of UHF range.

Since fox (channel) numbers are rotated periodically, they are not shown here, nor are the bases to which they are assigned.

**(A)**
4445 5152 5432 6993 7690 7873 7922 8029 8170 10202 11056 11466 11995 12106 12317 13752 14417 14863 15595 15733 16120 16320 17433 17480 18320 18390 19005 19050 19653 19668 20053 20400 22913 23687

**(B)**
4444 4487 5340 5398 6993 11413 11995 12029 12258 14670 14867 15687 16121 17496 18060 18320 18675 19513 19758 20124 20953 22723 23643 24274

**(C)**
4505 4865 4925 5078 5270 5395 5800 6812 7735 7907 7982 8045 8086 9320 9425 9885 10427 10590 10765 10780 11441 11615 12058 12105 12155 12270 13412 13485 14412 14715 14897 14957

**(D)**
3116 11229 13204 18023

**(E)**
6730 9017 11226 13217

**(F)**
6728 6731 9017 11220 11494 13211

**(G)**
7305 7469 8047 8162 9007 9991 11153 11466 11484 11995 12060 12106 12250 13565 14364 16011 16077 16320 16385 17460 17565 18100 18532 19008 20452 20650 22940 23243 24933 25433

**(H)**
4760 5340 5412 5437 5820 6993 8055 8162 8176 9270 9414 9958 10427 10586 11995 12090 12106 12240 12258 13440 15555 15672 15710 15742 16070 16100 16407 17480 17625 18095 18175 18478 18980 19180 19447 20165 20400 20972 21770 21843

**(I)**
6756 13440 13825 14412 14896 16246 17554 18331 20195 20340 20475 23035 25130

**(J)**
4455 4938 5026 5820 5937 6793 7497 7690 7827 8060 13247 13825 14448 15036 15041

**(K)**
4721 6756 6683 13247 13823 14576 14867 17457 17480 18320 18795 19160 20154 20400 23687

**(L)**
11056 16117

**(M)**
6927 6989 7466 7975 9475 9866 9921 10603 10881 11055 11498 11545 12170 12254 13457 13484 13585 14913 14957 15555 15680 15710 15760 16121 16320 16407 17386 17427 17480 18270 18490 19060 19373

**(N)**
8080 8162 9215 9320 10583 10720 10880 11055 11156 11448 11488 11627 12240 12324 13604 13710 13765 14572 14766 14829 14887 15664 15724 16080 16160 16186 17460 17480 17562 17679 17985 18146 18400 18629 18790 19047 19092 19270 19340 20150 20321 20425 20758 23397

**(O)**
19895 20016 20395 20665 23132 25227

# AIR FORCE

**(P)**
5305 5760 5910 6790 7316 7660 7813 7930 7955
9188 10153 10357 10530 10550 11118 11667
12242 13455 13582 13760 13825 14902 15500
15776 15821 15908 16041 17480 17547 18218
18590 19237 19800 20313 20415 20535 20640

**(Q)**
4760 4982 5434 5820 6817 7997 8039 8050 9320
10112 11058 11156 24483

**(R)**
4982 5775 5822 7605 7765 7910 9043 11407 11634
11988 12107 12277 13878 14497 14896 16246
17554 18331 20195 20340 20475 23035 25130

**(S)**
9006 9180 11667 16041

**(T)**
6716 15018

# Canada

## Canadian Forces Bases/Airfields

| | |
|---|---|
| Al Ismailiyah, Egypt | North Bay, ON |
| Alert | Ottawa, ON |
| Bagotville, PQ | Petawawa, ON |
| Chatham, NB | Portage, MN |
| Cold Lake, AL | Shearwater |
| Comox, BC | St Johns |
| Edmonton/Namao, AL | Summerside, PEI |
| Georgetown, NB | Trenton, ON |
| Greenwood,NS | Toronto, ON |
| Halifax, NS | Valcartier, PQ |
| Lahr, W Germany | Vancouver, BC |
| Montreal, PQ | Winnipeg, MN |
| Moose Jaw, Sask. | |

## Canadian Military Aeronautical Communications System (MACS)

### Frequencies USB

| kHz | User |
|---|---|
| 3046 | Halifax - VOLMET H+50 |
| 3092 | Edmonton |
| | Lahr |
| | St John's |
| | Trenton |
| 4704 | Al Ismailiyah |
| | Edmonton |
| | Lahr |
| | St John's |
| | Trenton |
| 4730 | Halifax |
| 4752 | Halifax |

| 5690 | Al Ismailiyah |
|---|---|
| | Lahr- wx 2000-0800, H+16 |
| 5703 | Shearwater |
| 5718 | Edmonton |
| | Halifax |
| | St John's |
| | Trenton |
| | Victoria |
| 6693 | Halifax |
| | Edmonton - wx H+20 |
| | Trenton - wx H+30 |
| | St John's - wx H+40 |
| 6705 | Alert |
| | Edmonton |
| | Lahr |
| | St John's |
| | Trenton |
| 6708 | Shearwater |
| 6716 | Halifax RCC |
| | St John's |
| | Vancouver |
| 6746 | *Halifax - wx |
| 6753 | Edmonton- wx 2300-1200, H+20 |
| | Lahr |
| | North Bay |
| | St John's - wx H+40 |
| | Trenton - wx H+30 |
| 8984 | North Bay |
| 9006 | Al Ismailiyah |
| | Edmonton |
| | Lahr |
| | St John's |
| | Trenton |
| 9010 | Halifax - wx H+20 |
| | Vancouver |
| 11233 | Al Ismailiyah |
| | Alert |
| | Edmonton |
| | Halifax |
| | Lahr |
| | St John's |
| | Trenton |
| 11249 | Halifax - wx |
| | Vancouver |
| 13231 | Al Ismailiyah |
| | Lahr- wx 0800-2000 H+16 |
| 13254 | *Halifax |
| 13257 | Al Ismailiyah |
| | Edmonton |
| | Lahr |
| | St John's |
| 13970 | Golan Heights |
| | (See CFAR) |
| | Trenton |
| 15031 | Edmonton |
| | Lahr |
| | St John's |
| | Trenton |

| | |
|---|---|
| 15035 | Edmonton- wx 1200-2300 H+20 |
| | Trenton- wx H+30 |
| | St John's - wx H+40 |
| 17995 | Edmonton |
| | St John's |
| | Trenton |
| 18012 | Edmonton |
| | Lahr |
| | St John's |
| | Trenton |
| 18027 | Halifax |
| | Edmonton |
| 23250 | Trenton |

\*   Tactical broadcasts H+51
\*\*   See "wx" station aviation weather schedule below.

## Weather Schedule/Call Signs

| | | |
|---|---|---|
| H+16 | Lahr | DHM95/VEG |
| H+20 | Edmonton | VXA |
| H+30 | Trenton | CHR |
| H+40 | St John's | CJX |
| H+50 | Halifax | CFX/CFH* |
| | ESQUIMALT | CKN |

## CFH Weather FAX (H+00/20)/RTTY

122.5   4271   6330   10536   13510

## CFH RTTY
(75 Bd/850 Hz Shift; cryptocovered)

Simulcasts:
(1) 4347 6452 8542 12814 17084
(2) 4225 6450 8662 12985 16880 22336.5

### CFH CW
4255 8697 12726 16926.5

## CKN Weather FAX (H+00/20)/RTTY

4268 6456 12753 16995

### CKN CW
4268 6384.5 8463

## Canadian Forces Coast Station
### Frequency Pairs:
### Ship - Shore

| Coast | Ship |
|---|---|
| 2054 | 2054 |
| 2134 | 2134 |
| 2237 | 2237 |
| 2458 | 2340 |
| 2530 | 2815 |
| 2538 | 2142 |
| 2514 | 2118 |
| 2558 | 2142 |
| 2582 | 2206 |

## Coast Guard Intership

| | |
|---|---|
| 4125 | 6221.6 |
| 8284.4 | 12424.5 |

## Northwest Territory Ship to Shore

| | |
|---|---|
| 4069.2 | 5803 |
| 4260 | 5875.5 |
| 4363.6 | 5940 |
| 5680 | |

**Submarine Distress Buoys:** 4340

## Canadian Forces Vancouver, BC (CKN)
USB/RTTY: 11236
CW 12702 12753 16960

# Cuban Air Force
9018 USB

# French Military

| Freq | Mode | Service |
|---|---|---|
| 2436 | | Toulon Navy Control Center |
| 6462 | CW | FUM |
| 11271 | | Air Force,RFFUID Biscarrose |
| 13239 | USB | Air Force;Air-Ground |
| 18001.5 | USB | French Air Force (Dakar) |
| 23254 | USB | Air Force "FDX" Paris (Ft de France) |

# Great Britain
## Royal Air Force/Navy Air to Ground Communications

(RAF Command Post; Alconbury, England)

| Freq. | Designator | User Location |
|---|---|---|
| 2428 | | RAF MARTELO*** |

# AIR FORCE

| | | | |
|---|---|---|---|
| 2591 | | RAF | Strike Command* |
| 3046 | | RAF | Alconbury (USAF only) |
| 3095 | | RAF | Salah, Oman |
| 3109 | | RAF | Alconbury |
| 3116 | | RAF | Gan Island, Maldives |
| 3120 | | RAF | Upavon |
| 3885 | | RN | Culdrose, Eng. |
| 3935 | | RAF | MARTELO*** |
| 3939 | | RAF | UKADGE** |
| 4140 | | RN | Culdrose, Eng. |
| 4464 | Kilo Romeo | RAF | UKADGE** |
| 4477 | | RAF | Alconbury (USAF only) |
| 4540* | Uniform Tango | | |
| 4707 | Delta | RAF | UKADGE** |
| 4710 | Tango Whiskey | RAF | UKADGE** |
| 4719 | | RAF | Architect |
| 4722 | | RAF | VOLMET |
| 4730 | | RAF | MARTELO*** |
| 4730 | | RAF | Akrotiri, Cyprus H24 |
| 4732 | | RAF | Upavon, Eng.,-H24 |
| | | RAF | Belize City |
| | | RAF | Luqa, Malta |
| | | RAF | Gibraltar 2000-0700 |
| 4739 | | RAF | UKADGE** |
| 4742* | Foxtrot Sierra | RAF | Strike Command* |
| | | RAF | Ascension |
| | | RAF | Gilbraltar |
| | Viper | RAF | Falklands |
| 4744 | | RAF | Strike Command* |
| 4749 | BT9P | RAF | Finningley |
| 5441 | | RAF | MARTELO*** |
| 5450 | | RN | Yeovilton |
| 5685 | BT9P | RAF | Finningley |
| 5695.5 | | RAF | Salah, Oman |
| 5729 | Romeo Delta | RAF | Strike Command* |
| 5747 | | RAF | UKADGE** |
| 6190 | | RAF | Strike Command* |
| 6562 | | RAF | Seychelles |
| 6686 | | RAF | Pitreave, Scotland UKADGE** |
| 6690 | | RAF | Strike Command* |
| 6693 | | RAF | Tynemouth UKADGE** |
| 6697 | | RAF | MARTELO*** |

| | | | |
|---|---|---|---|
| 6705 | | RN | Prestwick, Scotland |
| 6707 | | RAF | Combin, Cyprus |
| 6708 | | RN | Yeovilton |
| 6715 | | RAF | UKADGE** |
| 6730 | | RAF | Strike Command 0700-1800 |
| 6733 | | RAF | UKADGE** |
| 6738 | Bravo | RAF | Strike Command* |
| 6740 | | RAF | UKADGE** |
| 6741 | | RAF | Alconbury (USAF only) |
| 6748 | | RAF | UKADGE** |
| 6751 | | RAF | Akrotiri, Cyprus 0500-1600 |
| 6760 | | RAF | UKADGE** |
| 6765 | | RAF | UKADGE** |
| 6790 | | RAF | UKADGE** |
| 6825 | | RN | Culdrose, Eng. |
| 8190* | Romeo Alpha | | |
| 8967 | | RAF | Gan Island, Maldives |
| | | RAF | Seychelles |
| 8975 | | RAF | Tactical |
| 8977 | | RN | Yeovilton |
| 8984 | | RAF | Alconbury (USAF only) |
| 8993 | | RN | Portland |
| 8997 | | RAF | MARTELO*** |
| 8990 | | RAF | Strike Command 0700-1800 |
| 9011 | | RAF | UKADGE** |
| 9014 | | RN | Prestwick |
| 9024 | BT9P | RAF | Finningley |
| 9025 | | RAF | Alconbury (USAF only) |
| 9032* | Delta Whiskey | RAF | Upavon-H24 |
| | | RAF | Akrotiri, Cypress |
| | | RAF | Ascension |
| | | RAF | Bermuda |
| | Viper | RAF | Falklands |
| | | RAF | Luqa, Malta |
| | | RAF | Pitreave, Scot |
| | | RAF | Strike Command* |
| 9036 | | RAF | MARTELO*** |
| 9043 | | RAF | Pitreave, Scotland |
| 11185 | | | VIP flights |
| 11200 | | RAF | VOLMET |
| 11204 | Alpha | RAF | Strike Command* |

**THE NATION'S LARGEST SHORT WAVE RADIO DEALER**

# Chances are you're already on our mailing list for the new 1991 catalog

If you received our 1990 catalog, we'll automatically send you our 1991 catalog.

If you don't get your 1991catalog before November 1990, call, fax, or write us for your copy.

If you bought any SWL item from EEB, you're on the list.

- Forty pages packed with exciting items for the short wave listener
- We list more than 40 receivers, priced from $50 to $5000. Our catalog includes photos, specs, and prices.
- Over 100 accessories—antennas, filters, wire and cable, switches, rotors, towers, speakers, headsets, converters, clocks, recorders
- More than 50 different SWL books
- Computer radio control and demodulator

Our catalog is sent free in the USA. Send $1 for mailing to Canada. Send $3 (US) for AIRMAIL shipment everywhere else.

 **ELECTRONIC EQUIPMENT BANK**

Mail and Phone Orders
Phone: 800-368-3270
323 Mill Street, N.E.
Vienna, VA 22180

Local and Technical:
703-938-3350
Fax: 703-938-6911

GSA Contract
Agencies are invited
to inquire about our
pending GSA contract.

Prices subject to change • Shipping not included • Sorry, no CODs • Returns subject to 20% restocking fee

29

| | | | |
|---|---|---|---|
| 11212 | | RAF | MARTELO*** |
| 11212 | | RAF | Edinburgh |
| 11224 | | RAF | Upavon |
| 11234* | Hotel Whiskey | RAF | Upavon-H24 |
| | | RAF | Bermuda |
| | | RAF | Akrotiri, Cypress-H24 |
| | | RAF | Akrotiri, Cypress |
| | | RAF | Ascension |
| | Viper | RAF | Falklands |
| | | RAF | Luqa, Malta |
| | | RAF | Belize City |
| | | RAF | Gibraltar 0700-2000 |
| | | RAF | Bumpton |
| | | RAF | Strike Command* |
| 11244 | | RAF | Strike Command* |
| 11250 | BT9P | RAF | Finningley |
| 11257 | | RAF | Alconbury (USAF only) |
| 13205 | | RAF | Seychelles |
| 13216 | | RAF | Gan Island, Maldives |
| 13234 | | RAF | Tactical |
| 13237 | | RAF | MARTELO*** |
| 13250 | | RAF | Cyprus |
| 13257 | Foxtrot | RAF | Strike Command* 0700-1800 |
| 15013 | | RAF | Upavon |
| 15031 | | RAF | Strike Command* 0700-1800 |
| 15039 | | RAF | MARTELO*** |
| 17980 | | RAF | Tactical |
| 18018 | | RAF | Akrotiri, Cypress 0700-1800 |
| | | RAF | Strike Command* 0700-1800 |
| 20511 | | RAF | Akrotiri, Cypress |
| | | RAF | Gibraltar |
| 21886 | | RAF | Seychelles |
| 23220 | | RAF | Upavon-H24 |
| | | RAF | Strike Command* 0700-1800 |
| 23236 | | RAF | MARTELO*** |

*The H+30 broadcasts list the aerodrome followed by a color to indicate weather conditions (blue, white, green or yellow). Broadcast are on 4742, 6738, 9032, 11204 and 18018 kHz.*

*RAF Strike Command Integrated Communications System, (STCICS; callsign "Architect") transmits weather forecasts at H+00 and H+30 after the hour. The H+00 forecast lists the aerodrome with a three digit number denoting conditions.

**United Kingdom Air Defense Ground Environment (UKADGE) ground-based radar tracking of aircraft (like US NORAD).

***Maritime Telecommunications Organization (MARTELO; callsign MKL) conducts routine traffic from RAF Northwood.

## VIP Code Names

| | |
|---|---|
| ASCOT | RAF transport aircraft |
| RAINBOW | Prince Phillip |
| UNICORN | Prince Charles |
| KITTYHAWK | Queen Elizabeth |

# Australia and New Zealand

## Royal Australian Air Force & Royal New Zealand Air Force Air to Ground

| | | |
|---|---|---|
| 3032 | RAAF | Sydney 0900-2100 (Darwin, Perth, Townsville) |
| 3130 | RAAF | |
| 4703 | RNZAF | Auckland |
| | RNZAF | Ohakea |
| | RNZAF | Wigram |
| 5688 | RAAF | Darwin 0900-2100 |
| | RAAF | Pearce |
| | RAAF | Perth |
| | RAAF | Sydney |
| | RAAF | Townsville |
| 5695 | RAAF | Australia |
| | RAAF | Butterworth, Malaysia - Search & Rescue |
| 5707 | RNZAF | Auckland |
| 5835 | RAAF | Perth |
| 6676 | RAAF | Sydney (Volmet) |
| 6699.5 | RAAF | |
| 6720 | RAAF | |
| 6723 | RNZAF | Auckland |
| | RNZAF | Ohakea |
| | RNZAF | Wigram |
| 8971 | RAAF | |
| 8907 | RNZAF | Auckland |
| 8975 | RAAF | Sydney H24 Darwin, Perth, Townsville |
| 8980 | RAAF | Amberly |

|       |       |                          |
|-------|-------|--------------------------|
|       |       | (primary) 2200-0500      |
| 8984  | RNZAF | Auckland                 |
| 8992  | RNZAF | Auckland                 |
|       | RNZAF | Ohakea                   |
|       | RNZAF | Wigram                   |
| 9005  | RAAF  | Warranbool               |
| 11201 | RAAF  |                          |
| 11214 | RAAF  | Amberly                  |
|       |       | (secondary) 2200-0500    |
| 11235 | RAAF  | Darwin (day pri) 2100-0900 |
|       | RAAF  | Perth                    |
|       | RAAF  | Sydney                   |
|       | RAAF  | Townsville               |
| 11247 | RAAF  | Sydney                   |
|       | RNZAF | Auckland                 |
|       | RNZAF | Ohakea                   |
|       | RNZAF | Wigram                   |
| 13205 | RAAF  | Sydney 2100-0400         |
|       |       | Perth,Darwin,Townsville  |
| 13245 | RAAF  |                          |
| 13251 | RNZAF | Christchurch             |
| 17983 | RAAF  |                          |
| 18027 | RAAF  | Perth, Sydney            |

## Tactical Identifiers

| | |
|---|---|
| ADELPHI | RNZAF Aircraft |
| AUSY | RAAF on intl flights |
| AUSY 624,25, 27,29 | Boeing 707 No.34 Squadron |
| ALADDIN | Macchi |
| BUCKSHOT | Flight of F-111Cs |
| BUFFALO | Chinook helicopter |
| CONSORT | HS-748 |
| EAGLE | Iroquois UH-1 helicopter |
| ENFIELD | Caribou |
| ENVOY | BAC-111, HS-748 |
| FALCON | F-111C |
| KIWI | RNZAF Aircraft |
| MACH | F-18 Hornet |
| MARINER | Lockheed P-3 Orion |
| REGENT | Mystere Falcon, BAC-111 |
| ROLLER | CT-4 Airtrainer |
| SHEPHERD | Lockheed P-3C Orion |
| STALLION | Lockheed Hercules |
| STINGRAY | Lockheed P-3B Orion |
| STRIKER | Lockheed P-3A Orion |
| TESTER | Test flight |
| TORCH | Macchi |
| TROJAN | Lockheed Hercules |
| WALLABY | Caribou |
| WINDSOR | Boeing 707 within Australia |
| WISDOM | HS-748 |
| WOMBAT | Caribou |

## RAAF Facilities
(also Civil Airport)

| | |
|---|---|
| Alice Springs | Gingin |
| Amberley | Learmonth |
| Butterworth, Malaysia | Point Cook |
| Canberra | Sattler |
| Darwin | Townsville |
| Edinburgh | Williamtown |

## RNZAF Facilities

Ohakea
Whenuapai

# West German Air Force

Base Callsign: DHM91
Frequencies kHz: 5691 11187 13248 ("Oscar")
17992 ("Whiskey")
Munster Radio: 18006 13246 11266

# Philippines

| | |
|---|---|
| 6525 | Fernando Air Base, Luzon, Philippines Air Force |
| 6700.5 | Philippines Bureau of Air Transport |

# Saudi Arabian Air Force
## Ground to Air Communications

| kHz | GMT | Airport |
|-----|-----|---------|
| 3095 | 1900-0400 | Dhahran Int'l |
| | | Khamis Mushait |
| | | Bisha |
| | | Gizan |
| | | Taif |
| | | Nejran |
| | | Madinah |
| 5526 | 1900-0400 | Riyadh |
| | | Tabuk |
| | | Jouf |
| | | Hail |
| | | Wejh |
| 8967 | 0400-1900 | same as 3905 |
| 8990 | 0400-1900 | Dhahran Int'l |
| | | Ruydah |

# AIR FORCE

## South Africa

### Military In-Flight

| | |
|---|---|
| 3116 | Ground Use |
| 3932 | |
| 4703 | Search & Rescue |
| 5680 | Tower |
| 5695 | Internal, Secondary |
| 6687 | |
| 8975 | Internal, Primary |
| 8992 | |
| 11237 | Overseas, Primary |
| 15056 | |
| 17993 | Overseas, Secondary |
| 21860 | |
| 23287 | |

# United States

## U.S. Navy Radiocommunications Frequencies (USB/RTTY)

RTTY MULCAST consists of 16 channels 85Hz shift, 85Hz separation, 75 or 50 baud. NORMAL RTTY is 850 shift. Tactical ID's are typically <u>letter</u>-<u>number</u>-<u>letter</u>: ("Alpha 6 Uniform"). Frequency hopping is often heard when voice stations "go green".

| Freq | Use |
|------|-----|
| 2130 | U.S.Coastal |
| 2150 | Charleston Test Control |
| 2252 | Virginia Capes |
| 2368 | Harbor Control |
| 2434 | Harbor Control |
| 2550 | Disaster Net |
| 2586 | Harbor Control |
| 2630 | Harbor Control |
| 2714 | Harbor Control |
| 2716 | Harbor Control |
| 2732 | Lockheed/Nuclear Subs |
| 2745 | NAVFAC Grand Turk |
| 2836 | Harbor Control |
| 3023 | Search and rescue |
| 3050 | Air-Ground |
| 3053 | Ships/Tactical |
| 3088 | ASW |
| 3095 | Atlantic Fleet tactical |
| 3109 | Air-Ground/Gulf of Mexico, San Diego |
| 3130 | COMSTA Pacific |
| 3237 | PMFR Barking Sands/ Outrider Control |
| 3261 | NORATS Norfolk |
| 3265 | Atlantic Fleet tactical |
| 4014 | Air-Ground Atlantic |
| 4040 | West Pacific (HICOM) |
| 4045 | NORATS Norfolk |
| 4082 | PMFR Barking Sands/ Outrider Control |
| 4152 | Tactical |
| 4253 | Lockheed/Nuclear Subs |
| 4359 | NAS |
| 4373 | Virginia Capes |
| 4377 | Pacific Fleet tactical |
| 4416 | Pacific Fleet (HICOM) |
| 4491 | PMFR Barking Sands |
| 4515 | Charleston Test Control |
| 4622.5 | NAS |
| 4700 | ASW Pacific |
| 4702 | ASW Atlantic |
| 4704 | Atlantic/Pacific Fleets tactical |
| 4707 | ASW Pacific |
| 4710.5 | Air-Ground Atlantic |
| 4711 | Air-Ground Secondary |
| 4730 | ASW Atlantic |
| 4735 | PMFR Down range ships |
| 4813.5 | Guam (HICOM) |
| 4835 | Philadelphia Shipyard ship/shore |
| 5080 | PMFR Range Clearance |
| 5430 | COMSTA |
| 5446 | USMC Tactical |
| 5680 | Search and rescue |
| 5716.5 | Atlantic Fleet |
| 5718 | Atlantic/Pacific Fleets tactical |
| 5724 | Atlantic/Pacific Fleets |
| 6687 | US Marine Corps Special Convoys |
| 6693 | Pacific Fleet |
| 6697 | ASW Atlantic/Caribbean (HICOM) |
| 6705 | COMSTA |
| 6708 | Atlantic/Pacific Fleets tactical |
| 6720 | ASW Atlantic/Pacific (HICOM) |
| 6723 | Universal |
| 6742 | ASW Atlantic |

# NAVY

| | |
|---|---|
| 6746 | PMFR Pt Mugu NAS |
| 6799 | Atlantic Fleet |
| 6835 | Pensacola |
| 7485 | Charleston Test Control |
| 7507 | Atlantic Hurricane Warning (pri) |
| 7535 | Caribbean/Indian Ocean Fleet/Norfolk SESEF (HICOM) |
| 7645 | Disaster Net |
| 7839 | Tactical |
| 7885 | ASW Atlantic tactical |
| 7893 | ASW Atlantic |
| 8233.5 | NORATS Worldwide |
| 8757 | TACSFAC |
| 8771 | Pensacola |
| 8778 | Pacific Fleet (HICOM) |
| 8972 | ASW North Atlantic (guard) (HICOM) |
| 8976 | ASW Atlantic |
| 8981.5 | Air-Ground Pacific |
| 9002 | Pacific Fleet tactical |
| 9006 | Atlantic Fleet tactical |
| 9032 | Pacific Aircraft |
| 9037 | Atlantic Aircraft (sec) |
| 9124 | Tactical |
| 9257 | Tactical |
| 9260 | Tactical |
| 9380 | Atlantic/Pacific Fleets |
| 9950 | Charleston Test Control |
| 10730 | Ops |
| 11190 | Tactical |
| 11191 | ASW Atlantic |
| 11195 | Atlantic Fleet |
| 11198 | Atlantic Fleet |
| 11205 | ASW Atlantic |
| 11234 | COMSTA RTTY Coord. |
| 11252 | ASW Atlantic |
| 11255 | Atlantic/Pacific/Mediterranean Fleet (HICOM) |
| 11261 | CINCPAC |
| 11267 | Atlantic/Caribbean (HICOM) |
| 11410 | COMSTA |
| 11463 | Atlantic Fleet |
| 11570 | Tactical |
| 12215 | Indian Ocean/Guam (HICOM) |
| 12315 | Norfolk SESEF |
| 12761 | Philippines (HICOM) |
| 13147 | Tactical |
| 13169.5 | Pacific Fleet |
| 13181 | Pacific Fleet (HICOM) |
| 13224 | COMSTA |
| 13227 | Atlantic/Pacific Fleets tactical |
| 13237 | Atlantic/Pacific Fleets |
| 13251 | ASW Pacific |
| 13629.5 | Tactical |
| 15021 | Pacific Fleet |
| 15051 | Pacific Fleet tactical |
| 15067 | Tactical |
| 15077 | PMFR Down Range Ships |
| 15087 | Pacific Fleet |
| 15520 | Atlantic Fleet |
| 15522 | Caribbean/Guantanamo Bay (HICOM) |
| 17985 | Atlantic Fleet tactical |
| 18009 | Pacific Fleet (HICOM) |
| 18162.5 | Pacific tactical |
| 23177 | Pacific Fleet |
| 23224 | Pacific Fleet |
| 23227 | PMFR Down Range Ships |
| 23287 | Atlantic/Caribbean (HICOM) |
| 23315 | Indian Ocean (HICOM) |

## U.S. Navy Tactical Call Sign Identifiers

Although U.S. Navy callsigns are gradually being phased out in favor of NATO trigraphs ("Lima Five November"), many of the old callsigns are still in use. The more common of these are listed below.

| Callsign | Location |
|---|---|
| ANGEL | UH-1 Marines rescue helicopter |
| BARBARIC | US Navy Facility, Guantanamo Bay, Cuba |
| BATTER UP | Any NAVCOMSTA this Frequency |
| CLIMAX | USS Enterprise CVAN-65 |
| COMSERON | USCG Key West, Florida |
| COURAGE | USS America |
| EAGLE CLIFF | USS John F. Kennedy |
| FAIRFIELD | USS Saratoga |
| FISHER BODY | ATC Guantanamo Bay NAS, Cuba |
| GOLD EAGLE | USS Carl Vinson |
| GRAY EAGLE | USS Ranger |
| HAMPSHIRE | Commander Naval Forces Caribbean, Roosevelt Rds NAS, PR |
| HANDBOOK | USS Forrestal |
| HERSHEY | CINCLANT/Joint Air Reconnaissance Center, Key West |
| JITTERBUG | NAVCOMSTA Balboa, Canal Zone |
| MUSTANG | USS Coral Sea |
| OLD SALT | USS Nimitz |
| OTIS | KC-130 aircraft, Cherry Point MCAQS, NC |
| OVERWORK | Naval network general callsign |
| PANTHER | USS Kitty Hawk |
| PEDRO | Marines rescue helicopter |
| PLEAD CONTROL | Pacific Missile Firing Range |
| RAIDER | KC-130 aircraft, El Toro MCAS, GA |
| RASPBERRY | Naval air stations prefix |
| RAWHIDE BASE | Squadron VRC-40, Norfolk, Virginia |
| SCHOOL BOY | USS Midway |
| SPARTAN | USS Lexington CVT-16 |
| STONEWALL | CINCPAC |
| STRONGBOX | Naval network call designator |

34

| SWORDFISH | USCG Aircraft 01 |
| TOPHAND | Submarine command Atlantic, Norfolk |
| TORREADOR | NAVCOMSTA San Francisco, California |
| YELLOW BLOOD | HQ Fleet Marine Force, Pacific |

## U.S. Navy
## Convoy Voice Call Signs

Convoy internal call signs are used only between ships within a convoy, the escort force and supporting aircraft on short range communications channels. They are never used when communicating with shore stations or for contacting other forces.

| Ships or Authorities | Call Sign |
|---|---|
| Commodore | CHIEF |
| Vice Commodore | LUCK |
| Convoy Collective | TEAM |
| Commodore's section | VAMP |
| Vice Commodore's section | MERGE |
| Rear Commodore | CASEMATE |
| Rear Commodore's section | SWAP |
| Escort force commander | BOSS |
| Individual ships in convoy | DELTA + position no. |
| Escort ships, collective | GANG |
| Escort ships, individual | BIT + hull no. |
| Patrol/support group commander | FOREMAN |
| Patrol/support group collective | MOB |
| Escort carrier No. 1 | NEST |
| Escort carrier No. 2 | COTE |
| Escort carrier No. 3 | STARCAST |
| Escort carrier No. 4 | SACK |
| Rescue ships | STRETCHERS |
| HF DF ships | DUFFER |
| MF DF ships | METER |
| Ships operating aircraft | HAWKER |
| Ships in or near the van | VAN |
| Ships in or near the center | MID |
| Ships in or near the rear | REAR |
| AA cruiser(s) | CRACKER |
| Ship controlling aircraft | EAGLE |
| Even numbered ships | ROMP |

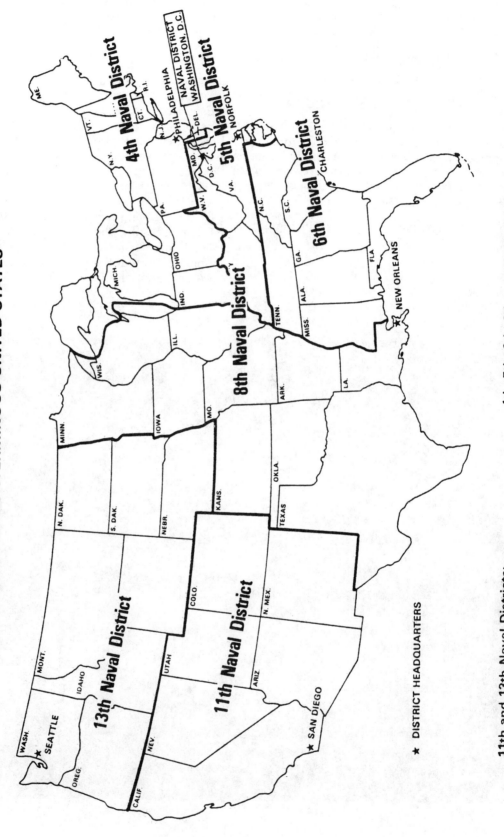

## U.S. NAVAL DISTRICTS
### WITHIN THE CONTERMINOUS UNITED STATES

4th, 5th, 6th, 8th and NDW Washington; Responsibility of CINCLANTFLT

11th and 13th Naval Districts; Responsibility of CINCPACFLT

★ DISTRICT HEADQUARTERS

| | |
|---|---|
| Odd numbered ships | TAX |
| Leading ships of columns | CANDY |
| Leading ships on the port side | BRIDEWELL |
| Leading ships on starboard side | PAGE |
| Rear ships on the port side | SODA |
| Rear ships of columns | SPRINT |
| Rear ships on the starboard side | ROLL |
| Even numbered columns | BLIGHT |
| Odd numbered columns | MIXED |
| Columns on the port side | SHIRT |
| Columns on the starboard side | PUPPY |
| Screen commander | COMMANDER |
| Escorting aircraft | ALFA |

## U.S. Navy Base Net - USB
Call Sign "Dude Ranch"
Newport, RI to Bath, ME

## International Voice Call Signs

| | |
|---|---|
| MAYDAY | Distress signal (SOS) |
| SECURITY | Safety signal |
| PAN | Urgent signal |

## U.S. Navy Encrypted RTTY
Cutler, Maine

(50 baud, 71.4 wpm/850 Hz shift)
Simulkey: 7455  10130  11689

## Radiotelephone Net (USB)
San Diego, CA
U.S. Navy Ship/Shore
Command Switching System (CSS)

| Ch | Duplex Freq |
|---|---|
| CSS1 | 4066.1/4360.5 |
| CSS2 | 8247.7/8771.6 |

## Norfolk, VA
## U.S. Navy Ship/Shore
## Intercommand Switchboard (ICSB)

**Duplex Freq**
4066.1/4360.0

## U.S. Navy
## Fleet Area Control and
## Surveillance Facility (FACSFAC)

FACSFAC provides USB communications support to military and civilian aircraft off U.S. coastlines and Hawaii.

| | |
|---|---|
| Virginia Capes, VA ("Giant Killer") | 2252 4373 |
| Jacksonville, FL ("Sea Lord") | 3130 6723 6742 11252 |
| Pensacola, FL ("Seabreeze") | 6835 |
| San Diego, CA ("Beaver") | 6723 |

## U.S. Naval Reserves
5158 USB

| Callsign | Location |
|---|---|
| NKE | Olathe, KS (Readiness Command) |
| NKE10 | Denver, CO |
| NKE11 | St. Louis, MO |
| NKE12 | Omaha, NE |
| NKE13 | Springfield, MO |
| NKE14 | Wichita, KS |
| NKE15 | Kansas City, KS |
| NKE16 | CP Girardeau, MO |
| NKE17 | Colorado Spring, CO |
| NKE18 | Pueblo, CO |
| NKE19 | Hutchinson, KS |

# NAVY

| | |
|---|---|
| NKE20 | Lincoln, NE |
| NKE21 | Joplin, MO |
| NKE22 | Cheyenne, WY |
| NKE23 | Topeka, KS |
| NKE24 | St.Joseph, MO |
| NKW | Diego Garcia |

## U.S. Navy Antarctic Stations

U.S. Navy Support Force ("Operation Deep Freeze")

| Station | Callsign | Mars Call | Ham Call |
|---|---|---|---|
| McMurdo | NGD | NNN0ICE | KC4USV |
| Amundson-Scott | NPX | NNN0 ? | KC4AAA |
| Palmer | NHG | NNN0NPA | KC4AAC |
| Siple | NQU | NNN0 ? | C4AAD |
| Williams Field | NZCM | NNN0 ? | KC4USX |
| Byrd Surface Camp | NBY | NNN0 ? | KC4USB |
| Dome Charlie | | | KC4AAK |
| Greenpeace | HOQ | | |

### Net No.
| | |
|---|---|
| | Harbor Common (McMurdo, Ships & Heli Control): 2716 |
| US-3 | Ship Tactical (Used by ships once under way): 2025 3237 |
| US-4 | Ships Conference (Long Range Ship Communications): 3319 8294.2 12429.2 |
| US-5 | Long Range Air/Ground Communications: 8997 Pri/ 13251 Sec |
| US-5 | Alternate Air/Ground Communications (SSB): 4718 5726 6708 11255 |
| US-9 | Air to Ship: 3102 5696 |
| | Forecaster to Pilot: 13251 |

US-14 Antarctic Broadcasts FAX & RTTY: 2650 4873 5810 6397 8090 11104 17361.5
Antarctic RTTY Working: 7340 7750 9073 9830 10235 12029 12280 13545 13551.5 15564

US-17 Antarctic Common (Field Party to Base): 2830 4768 7993 8974 11553
McMurdo to Christchurch, NZ (PTP):

| Call | Freq | Call | Freq |
|---|---|---|---|
| ZLK28 | 17450 | ZLK36 | 16152 |
| ZLK32 | 14768 | ZLK37 | 21804 |
| ZLK34 | 9785 | ZLK42 | 11580 |
| ZLK35 | 11460 | ZLK46 | 14700 |

Others include:
5080 7540 7929.5 9215 13826 13974 17403 21834 22842.5

## Antarctic MARS Frequencies

Modes: USB or RTTY (75 baud/170 Hz shift)

6970 7365 7370 7393 13826 13974 14470 14761.5

## Argentine Antarctic Stations

| Station | Callsign | Frequencies |
|---|---|---|
| Jubany | J25 | 3420 4490 4705 6040 8980 13280 13290 14440.5 14466.5 |
| Orcadass | LOK | 2422 4490 9985 12305 12660 |
| Esperanza | LTS | 13280 14402.5 14440 14697.5 15770 17290 |
| San Martin | LTS2 | 13280 13290 14402.5 14440 14697.2 |
| Belgrano | LTS4 | 13280 13290 14402.5 14440 14687.2 |
| Marambio | LUU | 4770 7530 9950 10870 10890 12230 12325 13280 14440 15619 15720 19395 22818 |
| Matienso | LUM | 3420 4490 4770 7530 12230 13235 13280 14440 |

## Australian Antarctic Stations

| Station | Official Call Sign |
|---|---|
| Mawson Base | VLV |
| Casey Base | VNJ |
| Davis Base | VLZ |
| Moore Pyramid Base(Summer) | VLV3 |
| Mt. Cresswell Base(Summer) | |
| MacQuarie Island | VJM |
| Wilkes Base (Closed) | VLL |

**Frequencies**

| | |
|---|---|
| Station to Station | 4040 5835 6850 |
| | 7922 9940 |
| | 12255 14415 |
| | 15845 17480 |
| Field Parties | 2720 4040 5400 |
| | 8110 |
| Aircraft | 6610 8938 |
| | 3023.5 |
| Additional | 7924 12215 |
| | 12255 |
| VNM Melbourne to Antarctic | 9941.5 12256.5 |
| | 14415 15846 |
| | 19256.6 17481.5 |

## Belgian Antarctic Station

Roi Baudouin Base    Callsign: ORU

| Callsign | Frequency |
|---|---|
| ORU27 | 7750 |
| ORU20 | 9272 |
| ORU30 | 10649 |
| ORU32 | 12152 |
| ORU50 | 20975 |
| ORU42 | 22960 |

6765 Belgian Air Force Kokside Rescue

## Brazilian Antarctic Station

Command. Ferraz

4143.6  6218.65  6997  8291.1  9265  12207
14365  15930

## Chilean Antarctic Stations

| Station | Callsign | Freq |
|---|---|---|
| Rodolpho Marsh | CAN6D | 5300 |
| | | 11660 |
| | | 15470 |
| Cap. Arturo Prat | CCZ CAD7P | 8828 |
| | | 10712 |
| | | 14630 |
| O'Higgins | CEF203, CAT2U | 9106 |

## Chinese Antarctic Station

Great Wall    Callsign: XSV

17870  15570

## French Antarctic Stations

### RTTY

| Station | Callsign |
|---|---|
| Pt.aux-Francaise | FJY2 |
| Dumont d'Orville | FJY3 |
| Martin de Vivies | FJY4 |
| Alfred Faure | FJY5 |

**Frequencies:**

5450 7420 9268 9788 10112 10113 11550
11553 11575 11985 14434 14440 14670
14940 16050 17480 17548 17695 19532.9
20112.5 20240 21845 23490 23986 24060
24460

## German Antarctic Station

### von Neumayer
Callsigns: DLA DAF5 DB

9020 9106

## Indian Antarctic Station

### Dakshin Gangotri
Callsign: VUA   Freq: 6275

## Italian Antarctic Station

### Terra Nova
2182 2773 4710.5 5400 8997

## Japanese Antarctic Stations

Syowa Base        Callsign: JGX

**Frequencies:**

2050 2570 3024 3025 3170 3195 4540
4575 5490 5947 7771 8161 8181 8186
8364 11532.5 11565 14450    14570 14895
18505 18675 20265 20475

Aircraft: 3035 4540 7771 5940 8186

Asuka Base        Callsign: JGY

**Frequencies:**
3024.5 4540

Aircraft: 3035 4540 7771 5940 8186

## New Zealand Antarctic Station

| Station | Callsign | Freq USB |
|---|---|---|
| Vanda Station | ZLEM | |
| Scott Base | ZLQ | 5400 |
| | | 13390 |
| | | 16065 |
| | ZLQ5 | 5785 |
| | ZLQ6 | 7490 |
| Channel #1 2773 | ZLQ7 | 10760 |
| #2 5400 | ZLQ8 | 14655 |
| #3 6708 | ZLQ9 | 19250 |
| #4 8997 | ZLQ24 | 11570 |
| | ZLQ25 | 10550 |
| | ZLQ26 | 10688 |
| | ZLQ27 | 14580 |
| | ZLQ29 | 9435 |

## Norwegian Antarctic Station

Frei Base        Callsign: LFB

5302 6600 8510 11602.5 FAX
11660 12850 14470 15470

## Polish Antarctic Station

### Arctowski

Callsign: 3ZL301   Freqs: 18124 20664

## South African Antarctic Stations

| Station | Callsign | Frequencies | | |
|---|---|---|---|---|
| Sanae III | ZRP | 3737 | 4059 | 5085 |
| | | 5300 | 5737 | 5812 |
| | | 6874 | 8182 | 9778 |
| | | 11000 | 11411 | 11987 |
| | | 13373 | 13403 | 14366 |
| | | 15458 | 16329 | 16425 |
| | | 18046 | 18515 | 18985 |

| Marion Island | ZRS | 2006 2020 4155 |
|---|---|---|
| | | 4764 4770 6402 |
| | | 6410 8185 8265 |
| | | 8300 12392 12460 |
| | | 13402 13425 14495 |
| | | 16300 16325 19010 |
| | | 19200 |
| Gough Island | ZOE33 | 2006 2020 4125 |
| | | 4155 4764 4800 |
| | | 6402 6455 6780 |
| | | 6820 8270 8257 |
| | | 9960 9985 13402 |
| | | 13425 16522 16552 |
| | | 17000 17100 |
| Tristan da Cunha | ZOE | 6780 |

## United Kingdom Antarctic Stations

| Station | Callsign |
|---|---|
| Signey, South Orkney IS. | ZHF33 |
| Faraday, Graham Land | ZHF44 |
| Rothera, Adelaide Is. | ZHF45 |
| Halley Bay (Base Z) | V5D |
| Bird Is., South Georgia Is. | ZBH22 |

**Frequencies:**

4067  5080  7775  9106  11055  14915

## U.S.S.R. Antarctic Stations

(CW/RTTY)

| Station | Callsign |
|---|---|
| Bellingshausen Base | UGE2 |
| Mirnyj Base | UUT |
| Molodezhnaya Base | RUZU |
| Novolazerevskaya Base | UDY |
| Vostok Station | RKIS |
| Russkaya Station | UDR3 |

**Frequencies:**

7665  8000  8105  12843  13385

## Navy/Marine Corps MARS

| Frequency | Region |
|---|---|
| 2025.0 | |
| 2026.5 | |
| '2049.5 | |
| 2150.0 | |
| 2470.0 | |
| 2576.0 | |
| 2656.0 | |
| 2820.0 | |
| 3190.5 | |
| 3192.5 | |
| 3319.0 | |
| 3322.0 | |
| 3350.5 | |
| 3391.5 | |
| 4001.0 | |
| 4001.5 | |
| 4005.0 | |
| 4007.0 | 1-2-3-5-7-HI-GUM |
| 4008.5 | |
| 4010.0 | |
| 4011.0 | |
| 4012.5 | |
| 4013.5 | 1-2-4-PR |
| 4015.0 | |
| 4016.5 | |
| 4040.0 | |
| 4041.0 | 1-4 |
| 4042.5 | |
| 4045.0 | |
| 4470.5 | 5-7-HI-PHL |
| 4472.0 | |
| 4515.0 | |
| 4625.0 | |

# NAVY

| Freq | Region |
|------|--------|
| 4803.5 | |
| 4805.5 | |
| 4820.0 | J-GUM |
| 4826.5 | |
| 5155.0 | |
| 5159.5 | ALL-HI |
| 5797.5 | |
| 5865.0 | |
| 6878.5 | |
| 6880.0 | |
| 6968.5 | |
| 6970.0 | |
| 7301.5 | 5-7-HI |
| 7346.5 | |
| 7350.0 | |
| 7360.0 | J |
| 7365.0 | 1-2-4-5-PR |
| 7367.5 | |
| 7370.0 | ALL-HI |
| 7372.5 | |
| 7375.0 | |
| 7380.0 | |
| 7382.5 | |
| 7385.0 | |
| 7387.5 | |
| 7393.0 | ALL-HI |
| 7495.0 | ALL-HI-J-PHL-GUM |
| 7500.0 | |
| 7353.0 | |
| 7684.0 | 1-2-4-5-7 |
| 7992.5 | |
| 8033.0 | 5-7-HI |
| 8150.0 | |
| 9052.0 | |
| 9210.0 | |
| 10255.0 | |
| 10259.5 | |
| 11401.0 | |
| 11402.5 | |
| 11538.0 | ALL-PR |
| 11539.0 | 4-7 |
| 11539.5 | HI-PAN |
| 11540.0 | 5 |
| 11655.0 | J |
| 12049.0 | 5-7-HI |
| 12124.0 | 1-2-3-4 |
| 12127.5 | |
| 12223.5 | |
| 13380.0 | 1-2-3-4 |
| 13485.0 | 5-7-HI-J-GUM |
| 13530.0 | 1-2-3-4 |
| 13540.0 | 1-2-3-4 |
| 13643.0 | ALL-HI-PHL-GUM |
| 13644.5 | |
| 13826.0 | ALL-HI-J-GUM |
| 13927.5 | |
| 13974.0 | ALL-PR-HI |
| 13975.5 | |
| 14383.5 | ALL-PR-HI-J-PHL-GUM |
| 14393.0 | |
| 14441.5 | ALL (Calling Frequency) |
| 14443.0 | |

| Freq | Region |
|------|--------|
| 14463.5 | |
| 14465.5 | |
| 14467.0 | 1-2-4 |
| 14468.5 | |
| 14470.0 | |
| 14471.5 | 1-2-4 |
| 14477.0 | 1-2-4 |
| 14478.5 | |
| 14480.0 | |
| 14483.5 | 1-2-3-4 |
| 14485.0 | |
| 14761.5 | |
| 14818.5 | 5-7-HI-PHL-GUM |
| 14840.0 | 1-2-3-4 |
| 14934.0 | ALL-HI |
| 15498.5 | |
| 16057.5 | |
| 16174.5 | |
| 16300.0 | ALL-PR-HI-J |
| 16387.5 | 1-2-3-4 |
| 19187.5 | 5-7-HI-J-PHL-GUM |
| 19430.0 | |
| 19671.5 | |
| 19882.0 | |
| 19918.5 | |
| 19956.5 | |
| 20375.0 | |
| 20525.0 | |
| 20580.0 | |
| 20625.0 | |
| 20680.0 | ALL-HI-GUM |
| 20936.0 | ALL-PR-HI |
| 20988.5 | 1-2-4 |
| 20998.5 | ALL-PR-HI |
| 23355.0 | |
| 24170.0 | |
| 24266.5 | |
| 24350.0 | |
| 26877.5 | |
| 27900.0 | |
| 27950.0 | |
| 27963.5 | 5-7-HI |
| 27975.5 | 5-7-HI |

## USN/USMC MARS Regions

| | |
|---|---|
| 1 | ME, NH, VT, CT, MA, RI, DE NJ, NY, PA, OH |
| 2 | DC, MD, VA, WV, KY, NC SC, GA, FL, TN, AL, MS, CARIBBEAN, EUROPE, MEDITERRANEAN, ATLANTIC, ICELAND, ANTARCTIC |
| 3 | LA, AR, OK, TX, NM, PANAMA |
| 4 | MI, IN, IL, WI, MN, IA, MO, ND, SD, NE, KS, CO, WY |
| 5 | AZ, NV, CA, UT |
| 6 | (NONE) |
| 7 | WA, OR, MT, ID, AK |
| 8 | HI, PACIFIC, INDIAN OCEAN |

**NAVY-MARINE CORPS MARS REGIONS (NMR)
WITHIN THE CONTERMINOUS UNITED STATES**

# NAVY

## Call Sign Procedure

NNN0C--:  Navy or Coast Guard ship
NNN0M--:  Marine Corps land station (NNN0MCL Camp Le Jeune, NC)
NNN0N--:  Naval base or ship

## MARS Afloat Specialty Net

(All call signs begin NNN0 and all ships begin USS unless otherwise noted)

### Key to Ship Type

| | | | | |
|---|---|---|---|---|
| AD | (as in USS Sierra and Puget Sound) | | CVN | Aircraft carrier (nuclear) |
| AE | Ammunition ship | | DD | Destroyer |
| AFS | Combat stores ship | | DDG | Guided missile destroyer |
| AGF | Miscellaneous command (flag) ship | | FF | Frigate |
| AGSS | Auxiliary research submarine | | FFG | Guided missile frigate |
| AO | Fleet oiler | | LCC | Amphibious command ship |
| AOE | Fast combat support ship | | LHA | (as in USS Belleau Wood) |
| AOR | Replenishment fleet tanker | | LKA | Amphibious cargo ship |
| AR | Repair ship | | LPD | Amphibious transport dock |
| ARL | Repair ship, landing craft | | LSD | Landing ship, dock |
| ARS | Salvage ship | | LST | Landing ship, tank |
| AS | Submarine tender | | MSO | (as in USS Fearless) |
| ASR | Submarine rescue ship | | SSN | Submarine (nuclear) |
| ATS | Salvage and rescue ship | | T-AF | Store ship (MSC operated) |
| AVT | (as in Lexington) | | T-AFS | Combat stores ship (MSC operated) |
| BB | Battleship | | T-AGS | Oceanographic surveying ship (MSC operated) |
| CG | Guided missile cruiser | | T-AO | Fleet oiler (MSC operated) |
| CGN | Guided misile cruiser (nuclear) | | T-ATF | Ocean tug fleet (MSC operated) |
| CV | Attack aircraft carrier | | | |

### Key to Homeports

| | | | | |
|---|---|---|---|---|
| A | Alameda, CA | | NHA | No homeport assigned |
| C | Charleston, SC | | NT | Newport, RI |
| CON | Concord, CA | | O | Oakland, CA |
| E | Weapons Station, Earle, NJ | | PA | Philadelphia, PA |
| G | Groton, CT | | PE | Pensacola, FL |
| GA | Gaeta, Italy | | PH | Pearl Harbor, HI |
| GM | Guam | | SB | Subic Bay, Philippines |
| LB | Long Beach, CA | | SD | San Diego, CA |
| LC | Little Creek, Norfolk, VA | | SO | Sajebo, Japan |
| M | Mayport, FL | | WSC | Weapons Station Charleston, SC |
| N | Norfolk, VA | | Y | Yokosuka, Japan |

### FPO Address

| | | | | |
|---|---|---|---|---|
| SE | Seattle, Washington | | MI | Miami, Florida |
| SF | San Francisco | | NY | New York, New York |

| Call | | Ship | Type/Hull # | FPO Address | Homeport |
|---|---|---|---|---|---|
| BLU | | BLUE RIDGE | LCC-19 | SF 96628-3300 | Y |
| CAA | | NEWPORT | LST-1179 | NY 09579-1800 | SF |
| CAB | | MC KEE | AS-41 | SF 96621-2620 | SD |
| CAC | | MOBILE BAY | CG-53 | MI 34092-1173 | M |
| CAE | | TAYLOR | FFG-50 | MI 34093-1504 | C |
| CAG | | MC ARTHUR | S-330 | | |
| CAI | | ANTIETAM | CG-54 | SF 96660-1174 | LB |
| CAK | | UNDERWOOD | FFG-36 | MI 34093-1491 | M |
| CAL | | ANCHORAGE | LSD-36 | SF 96660-1724 | LB |
| CAM | USNS | JOHN LENTHALL | T-AO-189 | NY 09587 | NHA |
| CBC | NOAAS | MILLER FREEMAN | R-223 | | |
| CBD | | THOMAS S. GATES | CG-51 | NY 09570-1171 | N |
| CBE | | LEYTE GULF | CG-55 | MI 34091-1175 | M |
| CBF | | HAWES | FFG-53 | MI 34091-1507 | C |
| CBG | | SAMUEL B. ROBERTS | FFG-58 | NY 09586-1512 | NT |
| CBH | | BUNKER HILL | CG-52 | SF 96661-1172 | SD |
| CBQ | | SIMPSON | FFG-56 | NY 09587-1510 | NT |
| CBR | | HALYBURTON | FFG-40 | MI 34091-1495 | C |
| CCE | NOAA | DAVID STARR JORDAN | R-444 | | |
| CCG | | CONQUEST | MSO-448 | SE 98799-1915 | SE |
| CFB | | FORD | FFG-54 | SF 96665-1508 | LB |
| CGA | | PIGEON | ASR-21 | SF 96675-3211 | LB |
| CGH | | BUCHANON | DDG-14 | SF 96661-1244 | SD |
| CGR | USNS | NARRAGANSETT | T-ATF-167 | SF 96673-4048 | NHA |
| CHB | | WICHITA | AOR-1 | SF 96683-3023 | O |
| CHG | | GARY | FFG-51 | SF 96666-1505 | LB |
| CHL | | SHASTA | AE-33 | SF 96678-3009 | CON |
| CHS | | VINCENNES | CG-49 | SF 96682-1169 | SD |
| CIL | | MARITIME PREPOSITION SQUADRON 2 | | | |
| CJB | | ROANOKE | AOR-7 | SF 96677-3029 | LB |
| CJH | | THACH | FFG-43 | SF 96679-1498 | SD |
| CJL | | JARRETT | FFG-33 | SF 96669-1489 | LB |
| CJN | | REID | FFG-30 | SF 96677-1486 | SD |
| CKK | | MISSOURI | BB-63 | SF 99689-1120 | LB |
| CKO | | COOK | FF-1083 | SF 96662-1443 | SD |
| CLB | | LONG BEACH | CGN-9 | SF 96671-1160 | SD |
| CLF | | VALLEY FORGE | CG-50 | SF 96682-1170 | SD |
| CLN | | ALAMO | LSD-33 | SF 96660-1721 | SD |
| CLT | | ELLIOT | DD-967 | SF 96664-1205 | SD |
| CMA | | JOHN L. HALL | FFG-32 | MI 34091-1488 | C |
| CMB | | TRUETT | FF-1095 | NY 09588-1455 | N |
| CMC | | AUBREY FITCH | FFG-34 | MI 34091-1490 | M |
| CMD | | IOWA | BB-61 | NY 09546-1100 | N |
| CME | USCG | NORTHLAND | WMEC-904 | | N |
| CMF | | JOHN KING | DDG-3 | NY 09576-1233 | N |
| CMG | | FLATLEY | FFG-21 | MI 34091-1477 | M |
| CMH | | MC CLOY | FF-1038 | NY 09578-1401 | N |
| CMI | | ROBERT G. BRADLEY | FFG-49 | MI 34090-1503 | C |
| CMJ | | CHARLES F. ADAMS | DDG-2 | MI 34090-1232 | M |
| CMK | | DE WERT | FFG-45 | MI 34090-1499 | C |
| CML | | MARITIME PREPOSITION SQUADRON 1 | | | |
| CMN | USNS | APACHE | T-ATF-172 | NY 09564 | NHA |
| CMO | | SNOOK | SSN-592 | | |
| CMP | NOAA | CHAPMAN | | | |
| CMQ | USCG | ALERT | WMEC-630 | | |
| CMS | USCG | COURAGEOUS | WMEC-622 | | |
| CMU | | SEATTLE | AOE-3 | NY 09587-3014 | N |
| CMV | USCG | TAMPA | WMEC-902 | | N |
| CMW | | GRAPPLE | ARS-53 | NY 09570-3223 | LC |
| CMX | USNS | HENRY J. KAISER | T-AO-187 | NY 09576-4086 | N |
| CMY | USNS | SATURN | T-AFS-10 | NY 09587-4052 | NHA |

| | | | | | |
|---|---|---|---|---|---|
| CMZ | USNS | PAWCATUCK | T-AO-108 | 09582-4045 | NHA |
| CNA | | TRIPOLI | LPH-10 | SF 96626-1645 | SD |
| CNB | | JUNEAU | LPD-10 | SF 96669-1713 | SD |
| CNC | | PAUL F. FOSTER | DD-964 | SF 96665-1202 | LB |
| CND | | JOSEPH STRAUSS | DDG-16 | SF 96678-1246 | PH |
| CNE | | ENGLAND | CG-22 | SF 96664-1146 | SD |
| CNF | | FORT FISHER | LSD-40 | SF 96665-1728 | SD |
| CNG | | BRISTOL COUNTY | LST-1198 | SF 96661-1819 | SD |
| CNH | | NEW JERSEY | BB-62 | SF 96688-1110 | LB |
| CNI | | DOLPHIN | AGSS-555 | SF 96663-3400 | SD |
| CNJ | | CARON | DD-970 | NY 0956-1208 | N |
| CNK | USNS | HASSAYAMPA | T-AO-145 | SF 96667 | NHA |
| CNL | | WORDEN | CG-18 | SF 96683-1142 | PH |
| CNM | | CLEVELAND | LPD-7 | SF 96662-1710 | SD |
| CNO | | REASONER | FF-1063 | SF 96677-1423 | SD |
| CNP | | BAINBRIDGE | CGN-25 | NY 09565-1161 | N |
| CNQ* | USNS | HITCHITI | | | |
| CNR | USCGC | MIDGETT | WHEC-726 | | |
| CNS | NOAAS | RAINIER S221 | MSS-21 | | |
| CNT* | USNS | ATAKAPA | | | |
| CNU | | PYRO | AE-22 | SF 96675-3003 | CON |
| CNV | | FREDERICK | LST-1184 | SF 96665-1805 | SD |
| CNW | | CHARLESTON | LKA-113 | NY 09566-1700 | N |
| CNX | | VIRGINIA | CGN-38 | NY 09590-1165 | N |
| CNY | | GLOVER | FF-1098 | NY 09570-1458 | N |
| CNZ | | WILLIAM V. PRATT | DDG-44 | MI 34092-1262 | C |
| COA | | RALEIGH | LPD-1 | NY 09586-1705 | N |
| COB* | | MULLINNIX | | | |
| COC | | SACRAMENTO | AOE-1 | SF 96678-3012 | A |
| COD | | DEWEY | DDG-45 | MI 34090-1263 | C |
| COE | | INDEPENDENCE | CV-62 | NY 09537-2760 | N |
| COF | | PATTERSON | FF-1061 | NY 09582-1421 | P |
| COG | | ENTERPRISE | CVN-65 | SF 96636-2810 | A |
| COH | | CAPE COD | AD-43 | SF 96649-2535 | SD |
| COI | | RECLAIMER | PH | SF 96677-3205 | PH |
| COJ | | BOWEN | FF-1079 | NY 09565-1439 | PH |
| COK | | MARS | AFS-1 | SF 96672-3030 | O |
| COL | USNS | TRUCKEE | T-AO-147 | NY 04586 | NHA |
| COM* | | POINT DEFIANCE | | | |
| CON* | | DUPONT | | | |
| COO | | TATTNALL | DDG-19 | MI 34093-1249 | M |
| COQ | | PENSACOLA | LSD-38 | NY 09582-1726 | LC |
| COR | | HUNLEY | AS-31 | NY 09559-2580 | N |
| COS | | W.S.SIMS | FF-1059 | MI 34093-1419 | M |
| COT | | FAIRFAX COUNTY | LST-1193 | NY 09569-1814 | LC |
| COU | | SARATOGA | CV-60 | MI 34093-1419 | M |
| COV | | HARRY E. YARNELL | CG-17 | NY 09594-1141 | N |
| COW | | TRENTON | LPD-14 | NY 09588-1716 | N |
| COY | | AUSTIN | LPD-4 | NY 09564-1707 | N |
| COZ | | FORRESTAL | CV-59 | MI 34080-2730 | M |
| CPB | USCGC | MUNRO | WHEC-724 | | |
| CPC | | BRONSTEIN | FF-1037 | SF 96661-1400 | SD |
| CPD | | OGDEN | LPD-5 | SF 96674-1708 | SB |
| CPE | USCGC | RUSH | WHEC-723 | | |
| CPF | | PEORIA | LST-1183 | SF 96675-1804 | SD |
| CPG* | | JOHN PAUL JONES | DDG-53 | | |
| CPH | USCGC | BASSWOOD | WAGL-388 | | |
| CPI | | LEAHY | CG-16 | SF 96671-1140 | LB |
| CPJ | | ROBISON | DDG 12 | SF 96677-1242 | SD |
| CPK | | HORNE | CG-30 | SF 96667-1153 | SD |
| CPL | | KISKA | AE-35 | SF 96670-3011 | CON |
| CPM | | TRUXTON | CGN-35 | SF 96679-1162 | SD |

| | | | | |
|---|---|---|---|---|
| CPN | KANSAS CITY | AOR-3 | SF 96670-3025 | O |
| CPO | MERRILL | DD-976 | SF 96672-1214 | SD |
| CPP | OKINAWA | LPH-3 | SF 96625-1630 | SD |
| CPQ | LOCKWOOD | FF-1064 | SF 96671-1424 | Y |
| CPR | HOEL | DDG-13 | SF 96667-1243 | SD |
| CPS | JOHN A. MOORE | FFG-19 | SF 96672-1475 | LB |
| CPU | USCGC YOCONA | | | |
| CPV | KINKAID | DD-965 | SF 96670-1203 | SD |
| CPW | JOHN YOUNG | DD-973 | SF 96686-1211 | SD |
| CPX | DURHAM | LKA-114 | SF 96663-1701 | SD |
| CPY | SCHENECTADY | LST-1185 | SF 96678-1806 | SD |
| CPZ | BENJAMIN STODDERT | DDG-22 | SF 96678-1252 | SD |
| CQA | ROARK | FF-1053 | SF 96677-1413 | SF |
| CQB | HEWITT | DD-966 | SF 96670-1203 | SD |
| CQC | BEAUFORT | ATS-2 | SF 96670-1203 | SD |
| CQD | USCGC JARVIS | WHEC-725 | | |
| CQE | WABASH | AOR-5 | SF 96683-3027 | LB |
| CQF* | SOMERS | | | |
| CQG | USNS CHAUVENET | T-AGS-29 | SF 96662-7104 | NHA |
| CQH | PROTEUS | AS-19 | SF 96646-2575 | GM |
| CQJ | DUBUQUE | LPD-8 | SF 96663-1711 | SO |
| CQK | COCHRANE | DDG-21 | SF 96662-1251 | Y |
| CQL | LYNDE MCCORMICK | DDG-8 | SF 96672-1238 | SD |
| CQM | FRESNO | LST-1182 | SF 96665-1803 | LB |
| CQN | STERETT | CG-31 | SF 96678-1154 | SB |
| CQO | WADDELL | DDG-24 | SF 96683-1254 | SD |
| CQP | MOUNT HOOD | AE-29 | SF 96672-3007 | CON |
| CQQ | MIDWAY | CV-41 | SF 96631-2710 | Y |
| CQR | DOWNES | FF-1070 | SF 96663-1430 | SD |
| CQS | BREWTON | FF-1086 | SF 96661-1446 | PH |
| CQT | USCGC IRONWOOD | WLB-297 | | |
| CQV | NOAA OCEANOGRAPHER | R-101 | | |
| CQW | WHITE PLAINS | AFS-4 | SF 96683-3033 | GM |
| CQX | RATHBURNE | FF-1057 | SF 96677-1417 | PH |
| CQY | ROBERT E. PEARY | FF-1073 | SF 96675-1433 | PH |
| CQZ | TUSCALOOSA | LST-1187 | SF 96679-1808 | SD |
| CRA* | USNS MOSOPELA | | | |
| CRB | DETROIT | AOE-4 | NY 09567-3015 | N |
| CRC | SIERRA | AD-18 | MI 34084-2505 | C |
| CRD | MOINESTER | FF-1097 | NY 09578-1457 | N |
| CRE | BOULDER | LST-1190 | NY 09565-1811 | LC |
| CRF* | NAUTILUS | | | |
| CRG | STUMP | DD-978 | NY 09587-1216 | N |
| CRH | MANITOWOC | LST-1180 | NY 09578-1801 | LC |
| CRI | SCULPIN | SSN-590 | NY 09587-2307 | G |
| CRK | PONCE | LPD-15 | NY 09582-1717 | N |
| CRL | CORONADO | AGF-11 | SF 96662-3330 | PH |
| CRM | JOSEPH HEWES | FF-1078 | MI 34092-1438 | C |
| CRN* | USCGC ACUSHNET | | | |
| CRQ | LEADER | MSO-490 | MI 34091-1917 | C |
| CRR* | BLANDY | | | |
| CRT | SKIPJACK | SSN-585 | NY 09587-2305 | G |
| CRU* | PLYMOUTH ROCK | | | |
| CRV | MACDONOUGH | DDG-39 | MI 34092-1257 | C |
| CRW | GUAM | LPH-9 | NY 09563-1640 | N |
| CRX | PAUL | FF-1080 | MI 34092-1440 | M |
| CRZ | AMERICA | CV-66 | NY 09531-2790 | N |
| CSA | LAWRENCE | DDG-4 | NY 09577-1234 | N |
| CSB | SPARTANBURG COUNTY | LST-1192 | NY 09587-1813 | LC |
| CSC | ORION | AS-18 | NY 09513-2570 | LM |
| CSE | ELMER MONTGOMERY | FF-1082 | MI 34092-1442 | M |
| CSF | SOUTH CAROLINA | CGN-37 | NY 09587-1164 | N |

| | | | | |
|---|---|---|---|---|
| CSG* | NEWMAN K.PERRY | | | |
| CSH* | MANLEY | | | |
| CSI | CONNOLE | FF-1056 | NY 09566-1416 | NT |
| CSK | BARNSTABLE COUNTY | LST 1197 | NY 09565-1818 | LC |
| CSL | DALE | CG-19 | MI 34090-1143 | PA |
| CSM* | DYESS | | | |
| CSN | MOOSBRUGGER | DD-980 | MI 34092-1218 | C |
| CSP | TRIPPE | FF-1075 | MI 34093-1435 | NT |
| CSQ | LA MOURE COUNTY | LST-1194 | NY 09577-1815 | LC |
| CSS | JESSE L.BROWN | FF 1089 | MI 34090-1449 | C |
| CST USCGC | DECISIVE | WMEC-629 | | |
| CSU | SAN DIEGO | AFS-6 | NY 09587-3035 | N |
| CSV* | BARRY | DDG-52 | | |
| CSW | SAIPAN | LHA-2 | NY 09549-1605 | N |
| CSX | KING | DDG-41 | NY 09576-1259 | N |
| CSY | SAMPSON | DDG-10 | MI 34093-1240 | M |
| CTA | STEIN | FF-1065 | SF 96678-1425 | SD |
| CTB USCGC | VENTUROUS | WMEC-625 | | |
| CTC USCGC | POLAR SEA | WAG-11 | | |
| CTE | FLORIKAN | ASR-9 | SF 96665-3207 | PH |
| CTF | DAVID R.RAY | DD-971 | SF 96677-1209 | SD |
| CTG | BERKELEY | DDG-15 | SF 96661-1245 | SD |
| CTH | RANGER | CV-61 | SF 96633-2750 | SD |
| CTI | HALSEY | CG-23 | SF 96667-1147 | SD |
| CTJ | TARAWA | LHA-1 | SF 96622-1600 | LB |
| CTK | BOLSTER | ARS-38 | | |
| CTL | MAUNA KEA | AE-22 | SF 96672-3001 | CON |
| CTM | REEVES | CG-24 | SF 96677-1148 | Y |
| CTN | BARBEY | FF-1088 | SF 96661-1448 | SD |
| CTO | VANCOUVER | LPD-2 | SF 96682-1706 | SD |
| CTP NOAAS | DISCOVER | R-102 | | |
| CTQ | O'BRIEN | DD-975 | SF 96674-1213 | SD |
| CTR NOAAS | SURVEYOR | S-132 | | |
| CTS | OLDENDORF | DD-972 | SF 96674-1210 | Y |
| CTT USNS | JUPITER | T-AKR-11 | SF 96601 | NHA |
| CTV | JASON | AR-8 | SF 96644-2560 | SD |
| CTW | BARBOUR COUNTY | LST-1195 | SF 96661-1816 | SD |
| CTX | DIXON | AS-37 | SF 96648-2605 | SD |
| CTY | FOX | CG-33 | SF 96665-1156 | SD |
| CTZ | BELLEAU WOOD | LHA-3 | SF 96623-1610 | SD |
| CUA | AYLWIN | FF-1081 | MI 34090-1441 | NT |
| CUB | ARTHUR W.RADFORD | DD-968 | NY 09586-1206 | N |
| CUC* | FORREST SHERMAN | | | |
| CUE | AINSWORTH | FF-1090 | NY 09564-1450 | N |
| CUF | MCCANDLESS | FF-1084 | NY 09578-1444 | N |
| CUG | TEXAS | CGN-39 | SF 96679-1166 | A |
| CUH | PETERSON | DD-969 | NY 09582-1207 | N |
| CUJ | VULCAN | AR-5 | NY 09548-2545 | N |
| CUK | VALDEZ | FF-1096 | NY 09590-1456 | NT |
| CUL | SEMMES | DDG-18 | MI 34093-1248 | C |
| CUN | MILLER | FF-1091 | NY 09578-1451 | NT |
| CUO | SPRUANCE | DD-963 | NY 09587-1201 | NT |
| CUP | NIMITZ | CVN-68 | SE 98780-2820 | |
| CUQ | FRANK CABLE | AS-40 | MI 34086-2615 | C |
| CUR | SAVANNAH | AOR-4 | NY 09587-3026 | N |
| CUS | INCHON | LPH-12 | NY 09529-1655 | N |
| CUT | NITRO | AE-23 | NY 09579-3001 | E |
| CUU USCGC | ACTIVE | WMEC-218 | | |
| CUV | COMTE DE GRASSE | DD-974 | NY 09566-1212 | N |
| CUW* | FISKE | | | |
| CUX | NASSAU | LAH-4 | NY 09557-1615 | N |
| CUZ | PUGET SOUND | AD-38 | NY 09544-2520 | N |

| | | | | |
|---|---|---|---|---|
| CVA | BELKNAP | CG-26 | NY 09565-1261 | GA |
| CVB | DAHLGREN | DDG-43 | NY 09567-1261 | N |
| CVC | KALAMAZOO | AOR-6 | NY 09576-3028 | N |
| CVD | SANTA BARBARA | AE-28 | MI 34093-3006 | WSC |
| CVE* | WELCH | | | |
| CVF | RICHMOND K.TURNER | CG-20 | MI 34093-1144 | C |
| CVG | DWIGHT D.EISENHOWER | CVN-69 | NY 09532-2830 | N |
| CVH | MOUNT BAKER | AE-34 | MI 34093-3010 | WSC |
| CVI | HARLAN COUNTY | LST-1196 | NY 09573-1817 | LC |
| CVJ | L.Y.SPEAR | AS-36 | NY 09547-2600 | N |
| CVK | NASHVILLE | LPD-13 | NY 09579-1715 | N |
| CVL | SHREVEPORT | LPD-12 | NY 09587-1714 | N |
| CVN | BRISCOE | DD-977 | NY 09565-1215 | N |
| CVO | GUADALCANAL | LPH-7 | NY 09562-1635 | N |
| CVP | CONOLLY | DD-979 | NY 09566-1217 | N |
| CVR USGCC | RELIANCE | WMEC-615 | | |
| CVS | MISSISSIPPI | CGN-40 | NY 09578-1167 | N |
| CVT | OLIVER HAZARD PERRY | FFG-7 | NY 09582-1465 | PA |
| CVU | SAGINAW | LST-1186 | NY 09587-1809 | LC |
| CVV | YELLOWSTONE | AD-41 | NY 09512-2525 | N |
| CVW | EMORY S.LAND | AS-39 | NY 09545-26120 | N |
| CVX | EL PASO | LKA-117 | NY 09568-1704 | N |
| CVY USNS | RIGEL | T-AF-58 | NY 09586- | NHA |
| CVZ | MOUNT WHITNEY | LCC-20 | NY 09517-3310 | N |
| CWA | PRAIRIE | AD-15 | SF 96639-2500 | LB |
| CWB | NIAGARA FALLS | AFS-3 | SF 96673-3032 | GM |
| CWC | SAMUEL GOMPERS | AD-37 | SF 96641-2515 | A |
| CWD | W.H. STANDLEY | CG-32 | SF 96678-1155 | SD |
| CWE | INGERSOLL | DD-990 | SF 96668-1228 | PH |
| CWF USCGC | CAPE SMALL | WPB-95300 | | |
| CWG | PELELIU | LHA-5 | SF 96624-1620 | LB |
| CWH | ST. LOUIS | LKA-116 | SF 96678-1703 | SO |
| CWI | WISCONSIN | BB-64 | | |
| CWJ | HAROLD E.HOLT | FF-1074 | SF 96667-1434 | PH |
| CWL | CALLAGHAN | DDG-994 | SF 96662-1266 | SD |
| CWM | WHIPPLE | FF-1062 | SF 96633-1422 | PH |
| CWN | DULUTH | LPD-6 | SF 96663-1709 | SD |
| CWO | FLINT | AE-32 | SF 96663-3008 | CON |
| CWP | FIFE | DD-991 | SF 96665-1229 | SD |
| CWQ* | PARSONS | | | |
| CWR | SAN BERNARDINO | LST-1189 | SF 96678-1810 | SO |
| CWS | GOLDSBOROUGH | DDG-20 | SF 9666-1250 | PH |
| CWT | CHANDLER | DDG-996 | SF 96662-1268 | SD |
| CWU USCGC | SASSAFRAS | WLB-401 | | |
| CWV | POINT LOMA | AGDS-2 | SF 96675-3406 | SD |
| CWW | FANNING | FF-1076 | SF 96675-3406 | SD |
| CWX | ACADIA | AD-42 | SF 96665-1436 | SD |
| CWY | CARL VINSON | CVN-70 | SF 96629-2840 | A |
| CWZ | GRIDLEY | CG-21 | SF 9666-1145 | SD |
| CXA | PREBLE | DDG-46 | NY 09582-1264 | N |
| CXB | JOHN F.KENNEDY | CV-67 | NY 09538-2800 | N |
| CXC | DOYLE | FFG-39 | MI 34090-1494 | M |
| CXD | JOHN RODGERS | DD-983 | MI 34092-1221 | C |
| CXE | NICHOLSON | DD-982 | MI 34092-1220 | C |
| CXF | LUCE | DDG-38 | MI 34092-1256 | M |
| CXG | O'BANNON | DD-987 | MI 34092-1225 | C |
| CXH | ESTOCIN | FFG-15 | NY 09569-1473 | PA |
| CXI | BUTTE | AE-27 | NY 09565-3005 | E |
| CXJ | SAMUEL E. MORISON | FFG-13 | MI 34092-1471 | C |
| CXK USCG | GALLATIN | WHEC-721 | | |
| CXL | BLAKELY | FF-1072 | MI 34090-1432 | C |
| CXM | MC INERNEY | FFG-8 | MI 34092-1466 | M |

# NAVY

| | | | | | |
|---|---|---|---|---|---|
| CXN | | PORTLAND | LSD-37 | NY 09582-1725 | LC |
| CXO | USNS | HARKNESS | T-AGS-32 | NY 09573 | NHA |
| CXP | USNS | WACCAMAW | T-AO-109 | MI 34093 | NHA |
| CXR | USNS | SIRIUS | T-AFS-8 | NY 09527 | NHA |
| CXS | USCG | DALLAS | WHEC-716 | | |
| CXT | | THORN | DD-988 | MI 34093-1226 | C |
| CXU | USNS | NEOSHO | T-AO-143 | NY 09579 | NHA |
| CXV | | NEWPORT | YACHT | NY 02840 | NT |
| CXX* | | THOMAS JEFFERSON | APA 30 | | |
| CXY | USCG | STEADFAST | WMEC-625 | | |
| CXZ | | WAINWRIGHT | CG-28 | MI 34093-1151 | C |
| CYB | | CLARK | FFG-11 | NY 09566-1469 | PA |
| CYC | | SHENANDOAH | AD-44 | NY 09551-2540 | N |
| CYD | | SUMTER | LST-1181 | NY 09587-1802 | LC |
| CYE | | LEXINGTON | AVT-16 | MI 34088-2700 | PE |
| CYF | | CALIFORNIA | CGN-36 | SF 96662-1163 | A |
| CYG | | TICONDEROGA | CG-47 | NY 09588-1158 | N |
| CYH | | FAHRION | FFG-22 | MI 34091-1478 | C |
| CYI | | FULTON | AS-11 | Y 09423-2565 | G |
| CYJ | | STARK | FFG-31 | MI 34093-1487 | M |
| CYK | | CONCORD | AFS-5 | NY 09566-3034 | N |
| CYL | | KIDD | DDG-993 | NY 09576-1265 | N |
| CYM | | JACK WILLIAMS | FFG-24 | MI 34093-1480 | M |
| CYO | | MARVIN SHIELDS | FF-1066 | SF 96678-1426 | SD |
| CYQ | | BRUNSWICK | ATS-3 | SF 96661-3219 | PH |
| CYR | | MOBILE | LKA-115 | SF 96672-1702 | LB |
| CYT | | YORKTOWN | CG-48 | NY 09594-1159 | N |
| CYU | | VIGILANT | WMEC-617 | CT 33701 | |
| CYV | USNS | DUTTON | T-AGS-22 | NY 09582 | NHA |
| CZA | | DEYO | DD-989 | MI 34090-1227 | C |
| CZB | | KITTIWAKE | ASR-13 | NY 09576-3208 | N |
| CZC | | SURIBACHI | AE-21 | NY 09587-3000 | E |
| CZD | | CONYNGHAM | DDG-17 | NY 09566-1247 | N |
| CZE | USCGC | CHILULA | WMEC-153 | | |
| CZG | | JOHN HANCOCK | DD-981 | MI 34091-1219 | M |
| CZI | | MERRIMACK | AO-179 | NY 09578-3020 | N |
| CZJ | | ARKANSAS | CGN-41 | SF 96660-1168 | SD |
| CZK | USCGC | VIGOROUS | WMEC-627 | | |
| CZL | | SYLVANIA | AFS-2 | NY 09587-3031 | N |
| CZM* | | FT SNELLING | | | |
| CZN | | BOONE | FFG-28 | MI 34093-1484 | M |
| CZO | | STEPHEN W.GROVES | FFG-29 | MI 34091-1482 | M |
| CZP | | GALLERY | FFG-26 | MI 34091-1482 | M |
| CZR | USNS | POWHATAN | T-ATF-166 | NY 09582 | NHA |
| CZS | | PLATTE | AO-186 | NY 09582-3022 | N |
| CZT | | ANTRIM | FFG-20 | MI 34090-1476 | M |
| CZU | | MONONGAHELA | AO-178 | NY 09578-3019 | N |
| CZV | | HAYLER | DD-997 | NY 09573-1231 | N |
| CZW | | ENGAGE | | | |
| CZX | | HOLLAND | AS-32 | MI 34079-2585 | C |
| CZY | | SCOTT | DDG-995 | NY 09587-1267 | N |
| CZZ | USNS | MISSISSINEWA | T-AO-144 | MI 34092 | NHA |

*Decommissioned*

## Other U.S. Navy MARS Stations

| Call | Station | Type/Hull # | Homeport |
|---|---|---|---|
| NAD | USS RECOVERY | | |
| NAE | WINTER HARBOR, ME | | |
| NAP | USS MILWAUKEE | AOR-2 | NY 09578-3024 |
| NAV | HQ WASHINGTON, DC | | |

| | | | | |
|---|---|---|---|---|
| NBG | MC MURDO STATION ANTARCTICA | | | |
| NCA | MARCUS ISLAND | | | |
| NCE | KURE ISLAND | | | |
| NCG | USCG ALEXANDRIA, VA | | | |
| NIG | PENSACOLA, FL | | | |
| NIK | MAYPORT, FL | | | |
| NIM | GULFPORT, MS | | | |
| NMC | WASHINGTON, DC | | | |
| NOF[1] | NORFOLK, VA | COMFEWSG, BLDG SP 71, NAS, NORFOLK, VA 23455 | | |
| NOK[1] | NORFOLK, VA | COMFEWSG, BLDG SP 71, NAS, NORFOLK, VA 23455 | | |
| NOT[1] | NORFOLK, VA | COMFEWSG, BLDG SP 71, NAS, NORFOLK, VA 23455 | | |
| NPA | PALMER STATION ANTARCTICA | | | |
| NPN | GUAM | | | |
| NQI | NORFOLK, VA | | | |
| NRD | ROOSEVELT ROADS, PUERTO RICO | | | |
| NRI | PORT HUANEME, CA | | | |
| NRJ | COMM. SCHOOL, SAN DIEGO, CA | | | |
| NRO | ROTA SPAIN | | | |
| NSD | SAN DIEGO, CA | | | |
| NSF | ALAMEDA, CA | | | |
| NTI | LONG BEACH, CA | | | |
| NTR | USS THEODORE ROOSEVELT CVN-71 | | NY 09599-2871 | N |
| NUZ | BYRD SURFACE CAMP, ANTARCTICA | | | |
| NVP | MIDWAY ISLAND | | | |
| NVT | JAPAN | | | |
| NWF | SUBIC BAY | | | |
| NXA | USCGC INGHAM | WHEC-35 | | |
| NXB | USS DONALD B.BEARY | FF-1085 | NY 09565-1445 | N |
| NXC | USS CAPODANNO | FF-1093 | NY 09566-1453 | NT |
| NXD | USCGC BRAMBLE | WAGL-392 | | |
| NXE* | USS ALBANY | SSN-753 | | |
| NXF | USCGC WESTWIND | WAGB-281 | | |
| NXG | USS PRESERVER | ARS-8 | NY 09582-3200 | LC |
| NXH* | USCGC BURTON ISLAND | | | |
| NXI* | USS MIRSCHER | | | |
| NXJ | USS BARNEY | DDG-6 | NY 09565-1236 | N |
| NXK | USS IWO JIMA | LPH-2 | NY 09561-1625 | N |
| NXL | USS PHARRIS | FF-1094 | NY 09582-1454 | N |
| NXM | USS CALOOSAHATCHEE | AO-98 | NY 09566-3016 | N |
| NXN | USS YOSEMITE | AD-19 | MI 34083-2510 | M |
| NXO | USCGC GLACIER | WAGB-4 | | |
| NXP* | USS JONAS INGRAM | | | |
| NXQ | USS OPPORTUNE | ARS-41 | NY 09581-3204 | LC |
| NXR | USS FEARLESS | MSO-442 | MI 34091-1907 | C |
| NXS | USCGC MALLOW | WLB-396 | | |
| NXT | USS RICHARD E.BYRD | DDG-23 | NY 09565-1253 | N |
| NXU | USS HOIST | ARS-40 | NY 09573-3203 | LC |
| NXV | USCGC MELLON | WHEC-717 | | |
| NXW | USS BIDDLE | CG-34 | NY 09565-1157 | N |
| NXX | USS EDENTON | ATS-1 | NY 09568-3217 | LC |
| NXY | USCGC CHASE | WHEC-718 | | |
| NXZ | USS SELLERS | DDG-11 | MI 34093-1241 | C |
| NYC | TRUK ISLAND | | | |
| NYN | CHESAPEAKE, VA | | | |
| NZA | USS THOMAS C.HART | FF-1092 | NY 09573-1452 | N |
| NZB | USS MAHAN | DDG-42 | 34092-1260 | C |
| NZC | USS JOSEPHUS DANIELS | CG-27 | NY 09567-1150 | N |
| NZD | USNS PONCHATOULA | AO-148 | SF 96675 | NHA |
| NZE* | USS SARGO | | | |
| NZF | USCGC BUTTONWOOD | WLB-306 | | |
| NZG* | USCGC BIBB | | | |

# NAVY

| NZH* | OCEANOGRAPHIC UNIT FIVE | | | |
|---|---|---|---|---|
| NZI | USS BADGER | FF-1071 | SF 96661-1431 | PH |
| NZJ | USS OUELLET | FF-1077 | SF 96674-1437 | PH |
| NZK | USS VREELAND | FF-1068 | MI 34093-1428 | M |
| NZL | USS CONSTELLATION | CV-64 | SF 96635-2780 | SD |
| NZM | USS NEW ORLEANS | LPH-11 | SF 96627-1650 | SD |
| NZN | USS MOUNT VERNON | LSD-39 | SF 96672-1727 | LB |
| NZO | USS JOUETT | CG-29 | SF 96669-1152 | LB |
| NZP | PT CANAVERAL, FL | | | |
| NZQ | USS CORAL SEA | CV-43 | NY J09550-2720 | N |
| NZR | USS LA SALLE | AGF-3 | NY 09577-3320 | PA |
| NZS | USS SHARK | SSN-591 | NY 09587-2308 | G |
| NZT | USS BLUEBACK | SS-581 | SF 96661-3403 | SD |
| NZU | USS MEYERKORD | FF-1058 | SF 96672-1418 | SD |

## Belgium

### Naval Communications (CW)

OSN Ostend Naval

## Canada

### Naval Communications

Halifax to ships (USB)
2716 Harbor Control "QHM"
(Queen's Harbor Master)
Day 13415   13515
Night 4560   5850

Submarine Distress Buoys: 4340

### Navy CW

5097  10945  15920  CFH Halifax
3287 CKN Vancouver

## Israel

### Naval Communications

12984.2   4XZ Haifa (Navy CW)

2800  4241.4  4289  8437.3
8518  10750.4  22330.5

## Japan

### Navy CW

12568   JJF (Tokyo)

## France

### Navy CW

| 6348 | HWN | Paris-Houilles |
|---|---|---|
| 6352 | FUG | LaRegine (Castelnaudary) |
| 8625 | FUM | |
| 11300 | FUM | Tahiti? |

```
12597  FUM   Papeete, Tahiti
12665  FUM
17380  FUM
22544  FUM   Tahiti
```

# Martinique

### French Navy (CW)

16961.5 13031.2 FUF Fort de France

# Mexico

### Mexican Navy (USB/RTTY)

11500

# Djibouti

### French Navy (CW)

16904.8    13042.5 FUV

# United Kingdom
### Navy CW

```
4221.2   6371.2   8625.2
         9065     10346
12824.2  GYU      Gibraltar
13473    15760    16987.3
```

# Australia
### Royal Australian Navy (CW)

| | |
|---|---|
| 4083 | Western Port Control |
| 4140 | Sydney Control |
| 4340 | Submarine Distress Beacons |
| 4600 | Jervis Bay Control |
| 6754.5 | NOWRA |
| 8530 | USN in Port |
| 16918.6 | VHP COMSTA Canberra |

# New Zealand
### Navy CW

8598.4      ZLO4 Irirangi, Wairou

*Royal Navy carrier "Invincible"*

8623      ZLP4  Irirangi, Wairou

# Rep. of South Africa
### Navy CW
COMCEN Cape (Silvermine)
4283    6386.5    ZSJ

# Portugal
### Navy (CW)
6389 CTP Oeiras (NATO)

# Spain
### Naval Communications (CW)
6388 EBA Madrid

# Italy
### Navy (CW)
13657 IDR5   Rome
4280 IDQ2 Rome
6390.3 IDQ3 Rome

# Denmark
### Danish Marine (CW)
8148 12329 OVG Frederikshavn Naval

# NAVY

*Soviet KIEV-class carrier "Novorossiysk"*

## USSR

### Navy (CW)

Frequencies: 7779  7965  9370.5
Callsign: RIW (Khiva)

## Cuba

### Soviet Navy (CW)

14968  CMU967  Santiago

## Inter-American Naval Telecommunications

### RTTY
#### (75 baud/850 Hz shift)

| | |
|---|---|
| 11000 | 16194 |
| 11570 | 18988  (Day) |
| 12182 | 19616 |
| 13371.5 | 20350 |
| 13550 | 21792 |
| 13700 | 22810 |
| 13704.7 | 25516 |
| 13904.5 (Night) | |

### USB

12183  15611.5  18990

### CW

| | | |
|---|---|---|
| 13555 | | Chile |
| 14430 | | |
| 17590 | CXR | Montevideo, Uruguay |
| | LOL | Buenos Aires, Argentina |
| | YWM1 | Bolivar, Venezuela |
| | 5KM | Colombia |

| Callsign | Location |
|---|---|
| NBA | US Navy - Balboa |
| LOL | Argentinian Navy - Buenos Aires |
| PWZ | Brazilian Navy - Rio de Janeiro |
| CCS | Santiago, Cuba |
| CCV | Valpariso, Chile |
| 5KM | Colombian Navy, Bogota |
| HDN | Ecuadorian Navy, Quito |
| HIWF-1 | Dominican Republic Navy |
| ZPK | Paraguay Navy |
| OBC | Peruvian Navy |
| YWM-1 | Venezuelan Navy |
| CXR | Uruguayan Armada, Montevideo |

## CW Beacons

Although not as intriguing as the spy numbers stations, single-letter high frequency beacons (SLHFB) have been reported for some 20 years, but are gradually disappearing into radio history.

Most appear to be land-based, 50 Hz A.C. operated (some use diesel generators) and use high accuracy frequency synthesis. Not all beacons are on 24 hours; they have scheduled operating periods.

Repeating their Morse letter identifications every few seconds, some send 5-digit weather code in Russian Cyrillic.

Determined by the FCC and military intelligence to originate in the USSR, these beacons are most probably hydrotelemetry from inland waters and reservoirs.

A representative list of beacons, frequencies and locations most recently reported follows.

| | | | |
|---|---|---|---|
| BEACON "E" | 14983 | | |
| BEACON "A" | 5758 | 7435 | |
| BEACON "V" | 3561 | 7394.5 | 10285.5 |
| BEACON "X" | 6771 | | |
| BEACON "S" | | | |
| Arkhangelsk | 3564.5 | 5305.5 | 6801.5 |
| | 8645.5 | 10643.5 | 13643.5 |
| | 17015.5 | 20991.5 | |
| BEACON "Z" | | | |
| Mukachevo | 3567.5 | 5308.5 | 6804.5 |
| | 8648.5 | 10646.5 | 13638.5 |
| | 17018.5 | 20994.5 | |
| BEACON I-O* | | | |
| Kzyl Orda | 3568 | 5309 | 6805 |
| | 8649 | 10647 | 13639 |
| | 17019 | 20995 | |

*I-O (or "YU") This Russian Cyrillic character is heard in Morse as ..--

## United States

### U.S. Army and National Guard
(Emission: USB)

| Freq | Use |
|------|-----|
| 2220 | NG |
| 2258 | NG |
| 2300 | NG |
| 2360 | NG |
| 2390 | NG |
| 2566 | NG |
| 2713 | Artillery range |
| 3175 | NG |
| 3205 | NG |
| 3261 | NG |
| 4000 | NG |
| 4001.5 | Statewide administrative |
| 4030 | NG Reg 1 |
| 4035 | NG |
| 4233 | NG Reg 1 |
| 4250 | NG |
| 4263.5 | Smart Field MO |
| 4280 | Jefferson MO Barracks |
| 4290 | NG |
| 4365 | NG |
| 4415 | NG Reg 1 |
| 4520 | NG Reg 1 |
| 4610 | NG Reg 1 |
| 4640 | NG Reg 1 |
| 4780 | NG Reg 1 |
| 4840 | NG |
| 4870 | NG |
| 4885 | NG |
| 4895 | Ft. Bragg NC |
| 4898 | NG |
| 4960 | Mississippi NG |
| 5090 | NG |
| 5235 | NG |
| 5397 | NG |
| 5850 | NG |
| 5997 | Jefferson MO Barracks |
| 6460 | Jefferson MO Barracks |
| 6766 | Training and Doctrine Command (TRADOC) |
| 6910 | NG |
| 6988 | NG |
| 6994 | NG |
| 7360 | NG Reg 1 |
| 7835 | NG Yoro, Honduras |
| 7861 | Tactical |
| 7932 | NG |
| 8040 | NG |
| 8060 | NG Reg 1 |
| 8170 | NG Reg 1 |
| 8180 | NG Reg 1 |
| 8500 | NG |
| 8565 | NG |
| 8603.5 | Ft. Leonard Wood MO |
| 8615.5 | Smart Field MO |
| 8624 | Jefferson MO Barracks |
| 9122.5 | NG Reg 1 |
| 9650 | Jefferson MO Barracks |
| 10586 | NG |
| 11510 | 320 FA Unit/1st Brigade |
| 11578 | Mississippi NG |
| 11764.5 | Smart Field MO |
| 12000 | NG |
| 12060 | NG Reg 1 |
| 12090 | NG Reg 1 |
| 12168 | Training and Doctrine Command (TRADOC) |
| 12240 | NG Reg 1 |
| 12255 | NG Reg 1 |
| 12270 | NG Reg 1 |
| 12355 | NG |
| 12908.5 | Jefferson Barracks MO |
| 13524.5 | NG Reg 1 |
| 13540 | NG |
| 13555 | NG Reg 1 |
| 15233 | Jefferson MO Barracks |
| 16320 | NG |
| 16340 | NG |
| 17460 | NG |

# ARMY

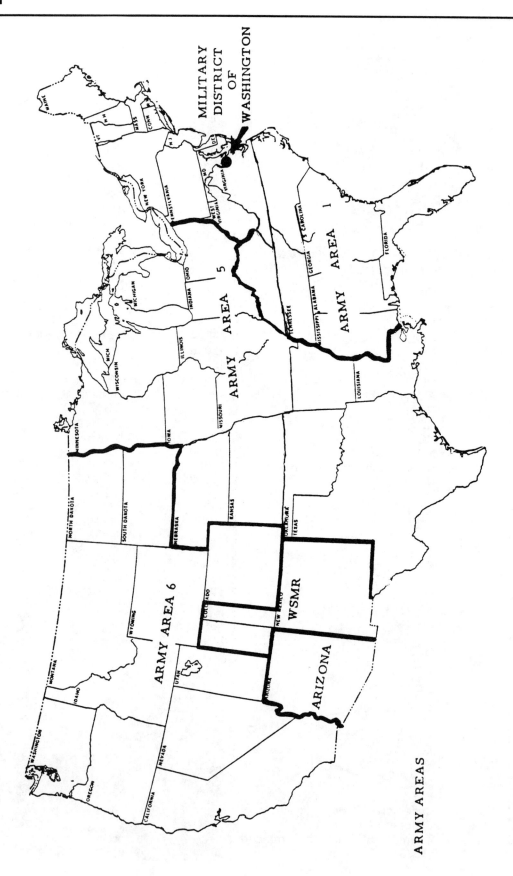

MILITARY
DISTRICT
OF
WASHINGTON

ARMY AREA 1

ARMY AREA 5

ARMY AREA 6

WSMR

ARIZONA

ARMY AREAS

56

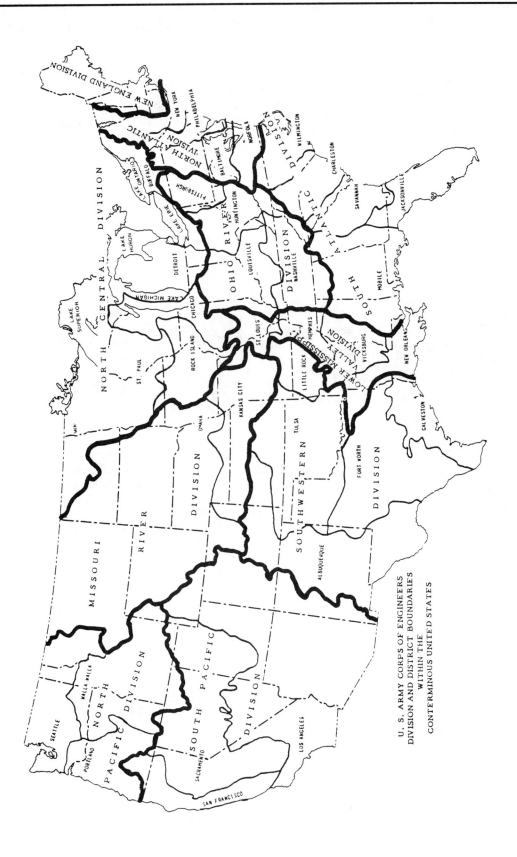

U. S. ARMY CORPS OF ENGINEERS
DIVISION AND DISTRICT BOUNDARIES
WITHIN THE
CONTERMINOUS UNITED STATES

# ARMY

| | | |
|---|---|---|
| 19090 | Tactical | |
| 22126 | NG | |

### Callsigns and Locations

| | |
|---|---|
| AAC20 | Ft. George Meade, MD |
| AAC25 | Ft. Detrick, MD |
| AAC35 | Ft. Belvoir, VA |
| AAC46 | Ft. Eustis, VA |
| AAD20 | Ft. Benning, GA |
| AAD21 | Ft. Gordon, GA |
| AAD32 | Ft. Rucker, AL |
| AAD60 | Ft. McClellan, AL |
| AAD62 | Ft. Jackson, SC |
| AAJ | Ft. Leavenworth, KS |

## Texas State Guard
### KKE397   Austin, TX

| | | | |
|---|---|---|---|
| 2710 | 4440 | 5820 | 8710 |
| 2726 | 4520 | 6385 | 13191.5 |
| 3201 | 5214 | 7360 | 15212.5 |

**United States Army Corps of Engineers**
*...Serving the Army*
*...Serving the Nation*

## U.S. Army Corps of Engineers
Nationwide Network (USB/LSB/RTTY/CW)

| | | |
|---|---|---|
| 2064 | 3302 | 6020 |
| 2300 | 3305 | 6785 (ch 5) |
| 2326 | 4850 | 6790 |
| 2348.5 | 5011 | 9122.5 (ch 8) |
| 2350 | 5015 | 11693.5 |
| 2602 | 5327 | 12070 |
| 2605 | 5346 | 12267 |
| 3287 | 5400 | 16077 |
| 3290 | 5437 | 16382 |

## COE Callsigns, Prefixes and Regions

| | |
|---|---|
| WAU | Waltham, MA |
| WUB2 | New York, NY |
| WUB3 | Philadelphia, PA |
| WUB4 | Baltimore, MD |
| WUB5 | Norfolk, VA |
| WUC | Atlanta, GA |
| WUC2 | Wilmington, NC |
| WUC3 | Charleston, SC |
| WUC4 | Savannah, GA |
| WUC5 | Jacksonville, FL |
| WUC6 | Mobile, AL |
| WUD2 | Buffalo, NY |
| WUD3 | Detroit, MI |
| WUD4 | Chicago, IL |
| WUD6 | Saint Paul, MN |
| WUD7 | Rock Island, IL |
| WUE | Cincinnati, OH |
| WUE3 | Pittsburgh, PA |
| WUG2 | Memphis, TN |
| WUG3 | Vicksburg, MS |
| WUG4 | New Orleans, LA |
| WUG5 | Saint Louis, MO |
| WUH4 | Omaha, NE |
| WUH5 | Omaha, NE |
| WUI | Dallas, TX |
| WUI2 | Little Rock, AR |
| WUI3 | Tulsa, OK |
| WUI4 | Galveston, TX |
| WUI5 | Alburquerque, NM |
| WUI6 | Fort Worth, TX |
| WUJ2 | Seattle, WA |
| WUJ4 | Walla Walla, WA |
| WUJ5 | Anchorage, AK |
| WUK2 | San Francisco, CA |
| WUK3 | Sacramento, CA |
| WUK4 | Los Angeles, CA |
| WUO | Washington, DC |

## U.S. Army Corps of Engineers
## Saudi Arabia

The Corps of Engineers run a COE HF Net in Saudi Arabia which operates 24 hours a day and is used by COE personnel and aircraft, USDAQ Jeddah C-12 utilizes the net for flight and followings and relaying ETAs and assistance requests to the U.S. Embassy in Jeddah.

| | |
|---|---|
| 9130 | Primary USB/LSB |
| 11425 | Secondary LSB |

| | |
|---|---|
| Castle 1 | Riyadh Airport |
| Castle 2 | Khamis Mushait Airport |
| Castle 3 | Jeddah Airport |
| Castle 4 | Tabuk Airport |
| Castle 6 | Dhahran Airport |

# ARMY

| Castle 7 | Jubail Airport |
| Castle 8 | Al Batin Airport |

The U.S. military runs a training mission from Sat. to Wed. 0430-1330 GMT using 7300 USB and 12112 USB.

| Hotel 1 | NCS Dhahran Int'l Airport |
| Hotel 2 | Riyadh Airport |
| Hotel 3 | Khamis Mushait Airport |
| Hotel 4 | Taif Airport |
| Hotel 8 | Tabuk Airport |

*MO Nat'l Guard at Palmerola AFB, Honduras*

## U.S. Army MARS

**Frequencies:**

| | | |
|---|---|---|
| 2001.0 | 4446.5 | 9181.0 |
| 2220.0 | 5115.0 | 9305.0 |
| 2259.5 | 5208.0 | 9419.0 |
| 2308.0 | 5217.0 | 9810.0 |
| 2358.5 | 5396.0 | 9990.0 |
| 2813.5 | 5401.0 | 10165.0 |
| 3235.5 | 5760.0 | 10815.0 |
| 3237.0 | 6766.0 | 11990.0 |
| 3245.0 | 6825.0 | 12072.0 |
| 3257.0 | 6910.0 | 13479.0 |
| 3273.5 | 6988.0 | 13505.0 |
| 3276.5 | 6995.5 | 13508.0 |
| 3289.0 | 6997.5 | 13511.0 |
| 3347.0 | 7309.5 | 13514.0 |
| 4001.5 | 7311.0 | 13743.0 |
| 4004.0 | 7313.5 | 13505.0 |
| 4012.0 | 7360.0 | 13508.5 |
| 4020.0 | 7405.0 | 13997.5 |
| 4025.0 | 7590.0 | 14402.0 |
| 4030.0 | 7720.0 | 14406.5 |
| 4035.0 | 7849.5 | 14440.0 |

| | | |
|---|---|---|
| 14485.5 | 17520.0 | 20520.0 |
| 14487.0 | 17546.5 | 20560.0 |
| 14510.0 | 17594.0 | 20650.5 |
| 14513.0 | 18129.5 | 20655.0 |
| 14580.5 | 18212.5 | 20665.0 |
| 14665.0 | 19004.5 | 20812.0 |
| 14847.5 | 19007.5 | 20921.5 |
| 14855.5 | 19010.5 | 20941.5 |
| 14877.5 | 19013.5 | 20975.0 |
| 14930.0 | 19024.0 | 20992.5 |
| 14936.5 | 19029.0 | 20995.5 |
| 14938.0 | 19531.0 | 21825.5 |
| 14939.5 | 19534.0 | 24012.5 |
| 16441.5 | 19840.0 | 24050.0 |
| 17459.0 | 20078.5 | 24197.5 |
| 17498.5 | 20105.5 | 24560.0 |
| 17501.5 | 20221.0 | 24761.5 |
| | | 24860.0 |

## U.S. Army MARS Call Sign Identifications

| Area | Callsign | Location |
|---|---|---|
| Eastern | *AAA3USA | Ft. Meade, MD |
| Central | *AAA6USA | Ft. Sam Houston, TX |
| Western | *AAA9USA | Presidio of San Francisco |
| Westcom | *ABM6USA | Schofield Bks HI |
| Europe | *AEM1USA | Pirmasens Germany |
| Pacific | ABM1US | Camp Zama Japan |
| | ABM2US | Zukeran, Okinawa |
| | *ABM4USA | Camp Coiner (Seoul) Korea |

*\* Gateway/relay station*

**Eastern**

| | |
|---|---|
| AAR1USA | Ft Devins MA |
| AAM2USA | Ft Monmouth NJ |
| AAR2USA | Ft Dix NJ |
| AAR2USB | Ft Drum NY |
| AAR3USA | Ft Belvoir VA |
| AAR3USD | Carlisle Barracks PA |

*Pohakuloa Training Area, Hawaii*

| | |
|---|---|
| AAR3USE | Ft Monroe VA |
| AAR3USF | Ft Eustis VA |
| AAR3USG | Ft Story VA |
| AAR3USH | Ft Lee VA |
| AAR3WCQ | Tobyhanna PA |
| AAR4USB | Ft Benning GA |
| AAR4USC | Ft Bragg NC |
| AAR4USD | Ft Jackson SC |
| AAR4USE | Ft McClellan AL |
| AARUSG | Ft Knox KY |
| AAR4USH | Ft Campbell KY |
| AAR7USF | Ft Rucker AL |
| AAT4USA | Ft Gordon GA |

## Central

| | |
|---|---|
| AAR5USE | Rock Island Arsenal IL |
| AAR5USF | Ft Ben Harrison IN |
| AAR6USC | Corpus Cristi TX |
| AAT6USE | Ft Hood TX |
| AAT6USF | Ft Bliss TX |
| AAT6USG | Ft Polk LA |
| AAT6USH | Pine Bluff Arsenal AR |
| AAT7USC | Ft Leonard Wood MO |
| AAT7USD | Ft Riley KS |
| AAT7USS | Ft Leavenworth KS |

## Western

| | |
|---|---|
| AAR5USB | Ft Lewis WA |
| AAR0USC | Silverton OR |
| ABM6EPV | Army Reserve Hilo HI |
| ABM6EQO | Schofield Bks HI |
| ALM7USA | Ft Richardson OK |
| AAR8USB | Ft Carson CO |
| AAR8USC | Ogden UT |
| AAR8USD | Lakewood CO |
| AAR8USE | Golden CO |
| AAR8USF | Denver CO |
| AAR8USH | Dugway UT |
| AAR8USI | Cheyenne WY |
| AAR9USB | Ft. Huachuca AZ |
| AAR9USC | Yuma Proving Grounds AZ |
| AAR9USD | Fort Ord CA |
| AAR9USE | Mesa AZ |
| AAR9USF | Phoenix AZ |
| AAR9USG | Santa Rosa CA |
| AAR9USH | Alameda CA |
| AAR9USI | San Jose CA |
| AAR9USJ | San Mateo CA |
| AAR9USK | Camp Parks CA |
| AAR9USL | Sparks NV |
| AAR9USM | Carson City NV |
| AAR9USO | Genoa NV |
| AAR9USP | Sebastopol CA |
| AAR9USQ | Petaluma CA |
| AAR9USR | Sierra Ad CA |
| AAR9USS | Sacramento Ad CA |
| AAR9UST | Chico CA |
| AAR9USU | Santa Barbara CA |

| | |
|---|---|
| AAR9USV | Ft Irwin CA |

## Europe

| | |
|---|---|
| AEM1AB | Aschaffenburg Germany |
| AEM1AC | Wurzburg Germany |
| AEM1AGG | Ansbach Germany |
| AEM1AH | Wiesbaden Germany |
| AEM1ASA | Ausburg Germany |
| AEM1BPM | Baumholder Germany |
| AEM1CC | Worms Germany |
| AEM1CLA | Ludwigsburg Germany |
| AEM1DR | Stuttgart Germany |
| AEM1DZ | Mainz Germany |
| AEM1EO | Geissen Germany |
| AEM1EWA | Frankfurt Germany |
| AEM1FHZ | Friedburg Germany |
| AEM1HKE | Schwabisch Gmund Germany |
| AEM1JY | Hanau Germany |
| AEM1KFD | Darmstadt Germany |
| AEM1KP | Karlsruhe Germany |
| AEM1KZ | Neckarsulm Germany |
| AEM1LT | Landstuhl Germany |
| AEM1MAN | Nellingen Germany |
| AEM1OV | Pirmasens Germany |
| AEM1QF | Berlin Germany |
| AEM1TX | Garlstedt Germany |
| AEM1US | Heidelburg Germany |
| AEM1USA | Lohnsfeld Germany |
| AEM1XL | Mannheim Germany |

## Pacific

| | |
|---|---|
| ABM1AD | Sagami Japan |
| ABM1AF | Yokota AFB Japan |
| ABM1CC | Camp Zama Japan |
| ABM1FC | Atsugi Japan |
| ABM1YP | Yokohama Japan |
| ABM2QW | Tori (Sobi) Sta Okinawa |
| ABM4AC | USASG-JSA (Panmumion) Korea |

*U.S. Army M1 tanks and OH-58 helicopter at Hohenfels Training Area, West Germany*

# ARMY

| | | |
|---|---|---|
| ABM4BC | CP Page (Chunchon) Korea | |
| ABM4CA | CP Ames Tajon Korea | |
| ABM4CC | CP Carroll Waegwan Korea | |
| ABM4CG | CP Red Cloud (Uijongbou) Korea | |
| ABM4CH | CP Humphres (Pyongtek) Korea | |
| ABM4EB | CP Mercer (Kimpo) Korea | |
| ABM4DL | CP Long (Wongu) Korea | |
| ABM4GO | CP Gary Owens Korea | |
| ABM4IH | CP Casey (Tongduchon) Korea | |
| ABM4SI | Sihungni Korea | |
| ABM4ST | CP Walker Korea | |
| ABM4TE | 38th ART BDE Korea | |
| ABM4TW | CP Stanley Korea | |
| ABM4UF | Haileah (Pusan) Korea | |
| ABM4VF | CP Howze (Musauni) Korea | |
| ABM4WI | Seoul Korea | |
| ABM9AJ | Jusmag Bangkok Thailand | |

## U.S. Defense Mapping Agency
### SSB/CW

7726.5  7812.5  13550.0  17520.0

| Callsign | Location |
|---|---|
| ACI | Fort Clayton, Panama |
| ACI21 | San Jose, Costa Rica |
| ACI23 | Tegucigalpa, Honduras |
| ACI25 | San Salvador, El Salvador |
| ACI29 | Santiago, Chile |
| ACI30 | Bogota, Colombia |
| ACI32 | Lima, Peru |
| CPP61 | La Paz, Bolivia |
| HC1C2 | Quito, Ecuador |
| HIGS4 | Santo Domingo, Dom. Rep. |
| TDG | Guatemala City, Guatemala |
| YV2QY | Caracas, Venezuela |
| ZP | Ascunsion, Paraguay |

# Canada

## Canadian Forces Amateur Radio
### (CFAR) Network

Similar to US MARS, the Canadian military also provides point-to-point USB communications for service personnel.

| Freq | Call | Station |
|---|---|---|
| 6905 | | Al Ismailiyah |
| 13971(Alpha) | CIC9 | Golan Heights, Syria |
| | CIS9 | Petawawa, Ont. |
| | CIW91 | Lahr, W. Germany |
| | CIW2102 | Victoria, BC |
| | CIW601-628 | Quebec, PQ |
| 14385 (Bravo) | CRW210 | |

| | CRW610 | |
|---|---|---|
| 14458.5 | VXN9 | Nicosia, Cyprus |
| | CIW819 | |
| 20957 | VXN9 | Nicosia, Cyprus |
| | VXV9 | Golan Heights, Syria |
| | CIX9 | Petawawa, Ont. |
| 20962 (Golf) | CIW202 | Vancouver, BC |
| 20970 | CIW202 | Vancouver, BC |

## Other Callsigns

| | | |
|---|---|---|
| CIW802-806 | Naval vessels | |
| VEV9 | Valcartier, PQ | |

# Great Britain

Belize City, Belize (MSS)
11462  USB/CW

# Bahamas
## Bahamas Defense Force

With headquarters in Nassau this expansive police agency uses two HF channels on upper sideband to link fixed and mobile units throughout the island territory. Nassau may identify as "Nassau Control," "Coral Harbor" or "Police Control" on 3786 kHz (night) or 5217 kHz (daytime).

# U.N. Forces
## Golan Heights
Austrian Army USB/RTTY/FAX

OEY32 (Golan Hts) to OEY42 (Salzburg)
Mon, Wed, Fri

OEY31 (Cyprus)
Tues, Thur

| Freq +/- 2 kHz | Designator |
|---|---|
| 10800 | Julius |
| 13986 | Otto (pri) |
| 18590 | Ulrich |
| 19425 | Victor |
| 20945 | Wilhelm |

# Computer Aided Scanning

*a new dimension in communications from Datametrics*

Now you can enhance your ICOM communications receiver through a powerful computer controlled system by Datametrics, the leader in Computer Aided Scanning. The system is as significant as the digital scanner was five years ago and is changing the way people think about radio communications.

- The Datametrics Communications Manager provides computer control over the ICOM R7000 or R71A receiver.

- Powerful menu driven software includes full monitoring display, digital spectrum analyzer and system editor.

- Innovative hardware design requires no internal connections.

- Comprehensive manual includes step by step instructions, screen displays, and reference information.

- Extends ICOM capabilities including autolog recording facilities, 1000 channel capacity per file, and much more.

- Overcomes ICOM limitations such as ineffective scan delay.

# Datametrics, Inc

___ R7000 system $ 349
___ R71A system $ 349
___ Manual and demo disk $15

Requires ICOM receiver and IBM PC with 512K and serial port. The R71A version also requires an ICOM UX-14.

Send check or money order to Datametrics, Inc., 2575 South Bayshore Dr, Suite 8A, Coconut Grove, Fl, 33133. 30 day return privileges apply.

# ARMY

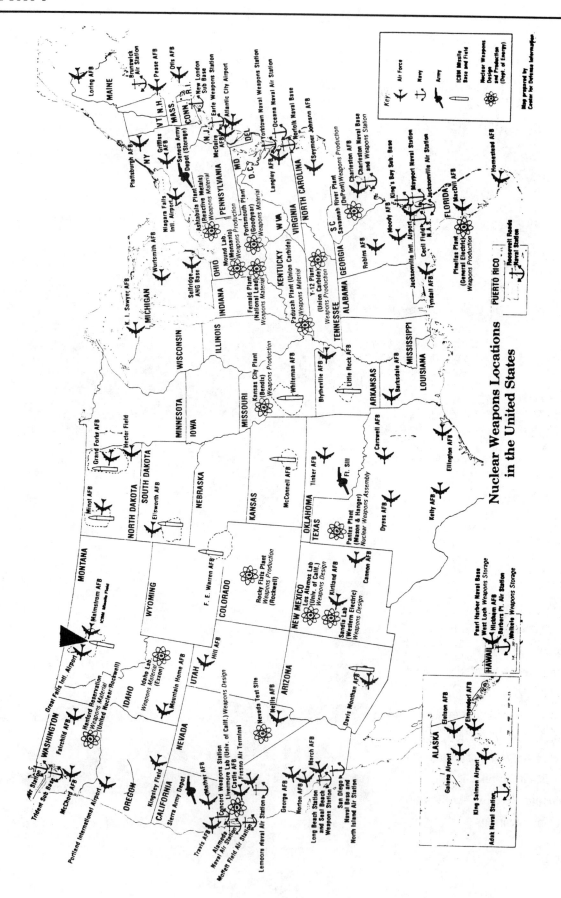

Nuclear Weapons Locations in the United States

64

**U.S. Coast Guard Stations**

| Callsign | Location |
|----------|----------|
| AOB50 | Estartit |
| DML | Sylt |
| JXL | Bo |
| JXP | Jan Mayen |
| NCF | Miami Beach, FL |
| NCI | Sellia Marina, Italy |
| NCI3 | Lampedusa, Greece |
| NCI4 | Kargaburun, Turkey |
| NMA | Miami, FL |
| NMA7 | Jupiter, FL |
| NMA10 | Mayport, FL |
| NMA21 | St. Petersburg, FL |
| NMB | Charleston, SC |
| NMB9 | Savannah, GA |
|  | Sitka, AK |
| NMC | San Francisco, CA |
| NMC6 | Monterey, CA |
| NMC11 | Humboldt Bay, CA |
| NMC17 | San Francisco, CA |
| NMD25 | Detroit, MI |
| NMD32 | Muskegon, MI |
| NMD35 | Alexandria Bay |
| NMD47 | Buffalo, NY |
| NMF | Boston, MA |
| NMF2 | Woods Hole, MA |
| NMF32 | Nantucket, RI |
| NMF33 | Caribou, ME |
| NMF44 | Southwest Harbor |
| NMG | New Orleans, LA |
| NMG8 | New Orleans, LA |
| NMG8 | North Bend, OR |
| NMG15 | Grand Isle, LA |
| NMG37 | Berwick |
| NMH1 | Washington, DC |
| NMJ1 | Juneau, AK |
| NMJ2 | Ketchikan, AK |
| NMJ3 | Valdez, AK |
| NMJ22 | Attu, AK |
| NMK | Cape May, NJ |
| NML5 | Buchanan |
| NML6 | Owensboro, KY |
| NML7 | Memphis, TN |
| NML21 | Keokuk |
| NMN | Portsmouth |
| NMN13 | Cape Hatteras, NC |
| NMN37 | Fort Macon |
| NMN70 | Chincoteague |
| NMN73 | Carolina Beach |
| NMN80 | Hampton Roads |
| NMO | Honolulu, HI |
| NMO12 | Barbers Point, HI |
|  | Boronquen, PR |
| NMO27 | Los Angeles, CA |
| NMP9 | Milwaukee, WI |
| NMQ8 | Channel Is., CA |
| NMQ27 | Los Angeles, CA |
| NMR | San Juan, PR |
| NMR1 | San Juan, PR |
| NMR26 | Borinquen, PR |
| NMW51 | Astoria, WA |
| NMW43 | Seattle, WA |
| NMW44 | Portland, OR |
| NMX | Baltimore, MD |
| NMY3 | New York, NY |
| NMY15 | Long Island Sound, NY |
| NMY42 | East Moriches, NY |
| NMY52 | Sandy Hook, NJ |
| NOB | San Francisco, CA |
| NOE | North Bend, OR |
| NOF | San Francisco, CA |
|  | Clearwater, FL |
| NOG | Sault St. Marie, MI |
| NOG14 | Duluth, MN |
| NOH | Chicago, IL |
| NOI | Detroit, MI |
| NOJ1 | Kodiak, AK |
| NOK | Key West, FL |
| NOL | Bermuda |
| NOM | Miami, FL |
| NOO | Sacramento, CA |
| NOP | Brooklyn, NY |
| NOQ | Mobile, AL |
| NOQ7 | Panama City, FL |
| NOR | San Diego, CA |
| NOS | Cape Cod, MA |
| NOT | Traverse City, MI |
| NOU | Annette, AK |

# COAST GUARD

## COMMUNICATION/RADIO STATIONS, ATLANTIC

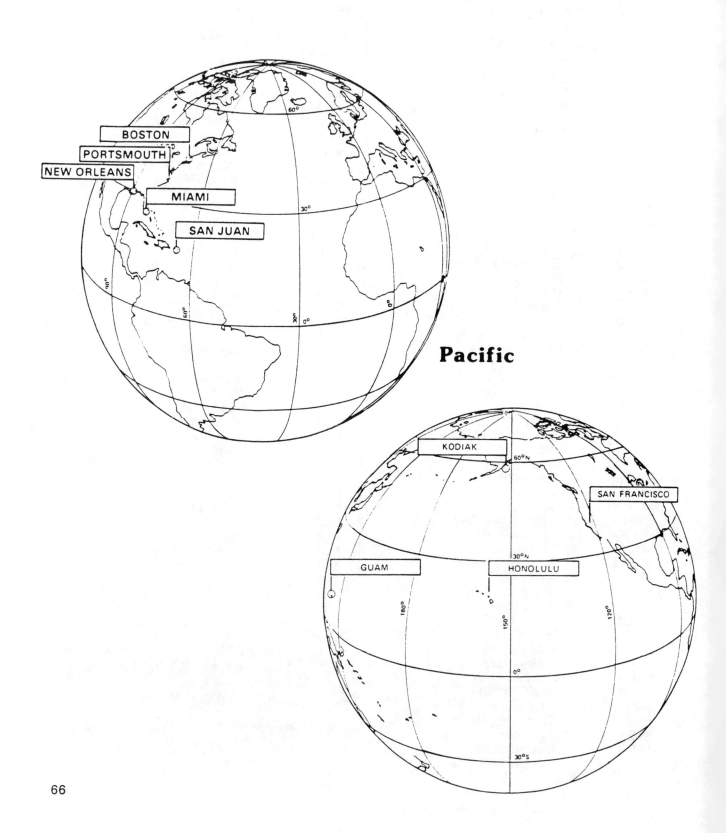

BOSTON
PORTSMOUTH
NEW ORLEANS
MIAMI
SAN JUAN

60°
30°
0°

**Pacific**

KODIAK
SAN FRANCISCO
GUAM
HONOLULU

60°N
30°N
0°
30°S

180°
150°
120°

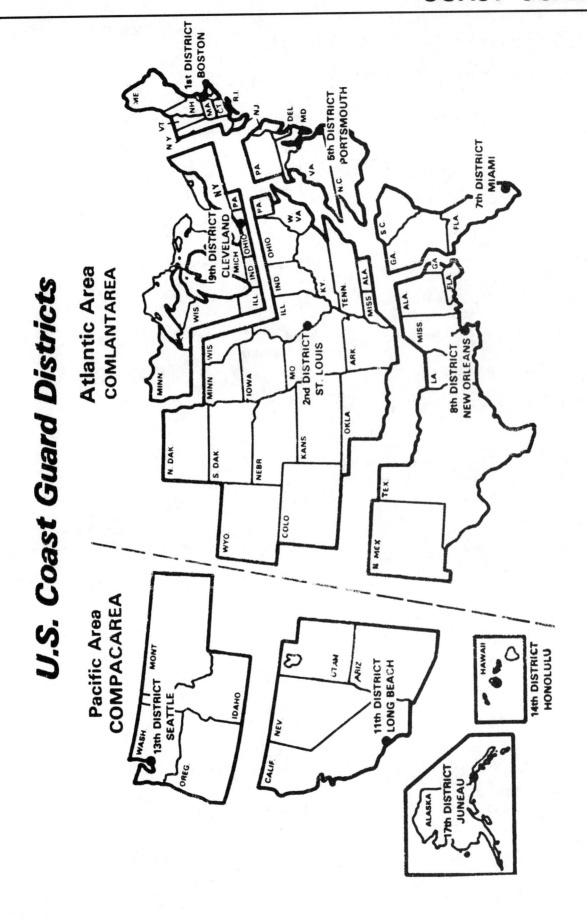

U.S. Coast Guard Districts

Atlantic Area COMLANTAREA

Pacific Area COMPACAREA

|        |                   |
|--------|-------------------|
|        | Arcata, WA        |
|        | Sitka, AK         |
| NOW    | Port Angeles, WA  |
| NOY    | Galveston, TX     |
| NOY8   | Corpus Christi, TX|
| NOY9   | Houston, TX       |
|        | Humboldt Bay, CA  |
| NOZ    | Elizabeth City, NC|
| NRO    | Johnston Island   |
| NRO5   | Upolo Point       |
| NRO7   | Kure Island       |
| NRT    | Yokota, Japan     |
| NRT2   | Gesashi, Japan    |
| NRT3   | Iwo Jima          |
| NRT9   | Hokkaido, Japan   |
| NRV    | Guam              |
| NRV6   | Marcus            |
| NRV9   | Barrigada         |
| NRW3   | Port Clarence     |
| OUN    | Ejde, Norway      |
| OVY    | Angissoq, Norway  |
| TFR    | Sandur            |
| TFR2   | Keflavik, Iceland |
| VDB    | Cape Race         |
| VDB2   | St. Anthony       |

## Marine Information Broadcasts

U.S. Coast Guard stations broadcast marine weather, hydrographic and safety information on a regular schedule. The mode is USB voice unless otherwise noted. NAVTEX is FEC/TOR mode.

| Freq. | Time |
|-------|------|

**Boston**

| | |
|---|---|
| 472 CW | 0000, 1450 |
| 518 NAVTEX | 0445, 1045, 1645, 2245 |
| 2670 | 0440, 1040, 1640, 2240 |
| 7530 FAX | |
| 8488 RTTY | |
| 8490 SITOR | 0200, 1700 |
| 12750 RTTY | |
| 13021 FEC | 0200, 1700 |
| 16969 FEC | 0200, 1700 |
| 3242 FAX | 0530, 1730 |

**East Moriches**

| | |
|---|---|
| 2670 | 0010, 1210 |

**Portsmouth**

| | |
|---|---|
| 448 CW | 0020, 1520 |
| 518 NAVTEX | 0130, 0730, 1330, 1930, |
| 4428.7, 6506.4, 8765.4 | |
| | 0400, 0530, 1000, 6506.4, |
| 8471 CW | |
| 8718 RTTY | |
| 8765.4, 13113.2 | |

| | |
|---|---|
| | 1130, 1600, 1730, 2200, 2330, |
| 8767.4, 13113.2, 17307.3 | |
| 13003 RTTY | |

**Ft. Macon**

| | |
|---|---|
| 2670 | 0103, 1233 |

**Chincoteague**

| | |
|---|---|
| 2670 | 0233, 1403 |

**Hampton Roads**

| | |
|---|---|
| 2670 | 0203, 1333 |

**Cape Hatteras**

| | |
|---|---|
| 2670 | 0133, 1303 |

**Cape May**

| | |
|---|---|
| 2670 | 1103, 2303 |

**Charleston**

| | |
|---|---|
| 2670 | 0420, 1620 |

**Mayport**

| | |
|---|---|
| 2670 | 0620, 1820 |

**Miami**

| | |
|---|---|
| 440 CW | 0050, 1500 |
| 518 NAVTEX | 0300, 0900, 1500, 2100 |
| 7843 RTTY | |
| 8455 CW | 1330, 2130 |

**Miami Beach**

| | |
|---|---|
| 2670 | 0350, 1550 |

**St. Petersburg**

| | |
|---|---|
| 2670 | 0320, 1420 |

**San Juan**

| | |
|---|---|
| 518 NAVTEX | 0415, 1015, 1615, 2215 |
| 2670 | 0305, 1505 |

**New Orleans**

| | |
|---|---|
| 432 CW | 0100, 1550 |
| 518 NAVTEX | 0000, 0600, 1200, 1800 |
| 2670 | 0550, 1035, 1235, 1635, 1750, 2235 |
| 6963 RTTY | |
| 7577 RTTY | |
| 8632 RTTY | |
| 9291 RTTY | |
| 12521 RTTY | |
| 18756 RTTY | |

**Galveston**

| | |
|---|---|
| 2670 | 1050, 1250, 1650, 2250 |

**Corpus Christi**

2670            1040, 1240, 1640, 2240

**Mobile**

2670            1020, 1220, 1620, 2220

**Channel Is.**

2670            0530, 1303, 2102

**San Francisco**

| | |
|---|---|
| 472 CW | 0300, 0400, 0500, 1600, 1700, 1830 |
| 2670 | 0203, 1403 |
| 4428.7 | 0430, 1030, 1630, 2230 |
| 8682 FAX | |
| 8765.4 | 0430, 1030, 1630, 2230 |
| 8774.7 | |
| 12730 FAX | |
| 13113.2 | 0430, 1030, 1630, 2230 |
| 17151.2 FAX | |
| 17205 FEC | |
| 17307.3 | 1630, 2230 |
| 8714.5 | 0000, 1800 |
| 17207 SITOR | 0000, 1800 |
| 4346 CW | 0630 |
| 8682, 12730 | 0300, 1900 |
| 17151.3 CW | 0030, 1900 |
| 8682, 12730 CW | 0630 |
| 4346, 8682 | 0145, 0300, 0500, 1500 |
| 12730 FAX | 0145, 0300, 0500, 1500 |
| 17151.3 FAX | 1500, 1715, 2015, 2330 |

**Monterey**

2670            0303, 1533

**Humboldt Bay**

2670            0303, 1503

**Astoria**

2670            0533, 1733

**Port Angeles**

2670            0615, 1815

**North Bend**

2670            0603, 1803

**Honolulu**

| | |
|---|---|
| 440 CW | 0500, 2100 |
| 2670, | 0545, 0903, 1145, 2103 |
| 6506.4, 8765.4 | 0545, 1145 |
| 8718 RTTY | |
| 8765.4, 2670 | 1745, 2345 |
| 12786 RTTY | |
| 12889.5 CW | |
| 13082.5 FEC | |
| 13113.2 | 1745, 2345 |

| | |
|---|---|
| 9050, 13655 | 0100, 0300, 0400, 0600, 0700, 1300, 1700, 2000, 2200 |
| 13655 CW | |
| 14731 RTTY | |
| 16457 | 0100, 0300, 0400, 0600, 0700, 1300, 1700, 2000, 2200 |
| 22476 CW | 0100, 0300, 0400, 0600, 0700, 1300, 1700, 2000, 2200 |
| 8716, 13082.5 | 0130, 0330, 0430, 0630, 0730, 1330, 2030, 2230 |
| 22203.5 SITOR | 0130, 0330, 0430, 0630, 0730, 1330, 2030, 2230 |
| 22572 CW | |

**Guam**

| | |
|---|---|
| 466 CW | 0100, 0800 |
| 2670 | 0705, 2205 |
| 6506.4 | 0930, 1530 |
| 8710.5 FEC | |
| 13077 FEC | |
| 13113.2 | 0330, 2130 |
| 13077, 17203 | 0001, 0200, 0400, 0500, 0900, 1100, 1400, 2300 |
| 22527 CW | |
| 22567 SITOR | 0001, 0200, 0400, 0500, 0900, 1100, 1400, 2300 |
| 8710.5, 13077 | 1400, 1500, 1900 |

# COAST GUARD

|  |  |
|---|---|
| 17203 SITOR | 1400, 1500, 1900 |
| 8150, 21760 CW | 0000, 0100, 0200, 0300, 0500, 0700, 0800, 1000, 1200, 1300, 2200 |

**Kodiak**

| | |
|---|---|
| 470 CW | 0530, 2000 |
| 6506.4 | 0203, 1645 |
| 4298, 8459 FAX | 0400, 1000, 1800, 2200 |

## International Ice Patrol

| Station | Time of Broadcast | Freq(kHz)/Mode |
|---|---|---|
| Boston/*NIK* | 0445, 1045 1645, 2245 | 518 FEC/TOR |
| Boston/*NIK* | 0018 1218 | 5320, 8502, 12750 85502, 12750 FEC/TOR |
| Boston/*NIK* | 0050 1250 | 5320, 8502, 12750 8502, 12750 CW |
| St. John's/*VON* | 0000, 1400 | 478 CW |
| Halifax/*CFH* | 0015, 1101 1301, 1401 2201, 2301 2230, 2239 if available | 122.5, 4271, 6330 10536 CW |
| Norfolk/*NMN/ NAM* | 0800-0900, 1500-1600, 1600-1700, 2100-2200 | 8090, 16180 CW |
| Thurso/*GXH* | 0800-0900, 1500-1600, 1600-1700, 2100-2200 | 7504.5, 12691 4001 CW |
| Keflavik/*NRK* | 0800-0900, 1500-1600, 1600-1700, 2100-2200 | 5167 |
| Key West/*NAR* | 0800-0900, 1500-1600, 1600-1700, 2100-2200 | 5870, 2675 |
| Rota/*AOK* | 0800-0900, 1500-1600, 1600-1700, 2100-2200 | 5917.5, 7705 |
| Nea Makri/*NGR* | 0800-0900 | 4623, 13372.5 |
| | 1500-1600, 1600-1700, 2100-2200 | |
| Boston/*NIK* | 1600 | 8502, 12750 FAX |
| Halifax/*CFH* | 0014, 1101 1301, 1401 2201, 2301 | 122.5, 4271, 6330 10536, 13510 |
| Bracknell/*GFE* | 1413 | 2618.5, 4782, 6330, 10536, 13510 |
| St. John's/*VON* | As required when new icebergs sighted | 2598 USB, 2182, 478 CW |
| Boston/*NIK* | As required when new icebergs sighted | 472 CW |
| Internat'l Ice Patrol Vessel/*NIDK* | When in the vicinity of ice in periods of darkness or fog | 2670 |

## Non-Scheduled Broadcasts

Urgent bulletins and safety broadcasts may be made at various times on CW on 466 and 500 kHz, and USB voice on 2003, 2638 and 2670 kHz.

## Polar Icebreakers (USB)

5384.5

## Notice to Atlantic Fishermen

CW transmissions involving the 200-mile Fisheries Conservation Zone, including fishing gear locations, are made on 472 and 8502 kHz at 1350 and 2150 UTC.

## New England Marine Facsimile

Surface analyses and weather prognoses are broadcast on 3242.5 and 7530 kHz at 0530 and 1730 UTC.

## U.S. COAST GUARD AIR STATIONS

The following air to surface frequencies utilize USB mode. Some frequencies are shared with the U.S. Navy.

| | | |
|---|---|---|
| 2141 | 5696 | 11201 |
| 2261 | 8980* | 15081 |
| 3120* | 8984 | 15084* |
| 3123 | 11195 | 15087 |
| 5692* | 11198* | |

*Helicopters

### Enforcement (USB)

| | | |
|---|---|---|
| 4376 | 4500 | 5480 |
| 6513 | 7527 | 8769 |
| 9802 | 11076 | 12220 |
| 13150 | 14371 | 14686 |
| 18666 | 23403 | |

### Air to Ground CW/RTTY/FAX

4335  6383
8648  12887.5

### Special Operations USB

| | |
|---|---|
| 3067 | 13244 |
| 4729 | 15015 |
| 6757 | 18002 |
| 8964 | 18019 |
| 11176 | 23217 |

### Tactical

5481

### Search and Rescue, Disaster

| | |
|---|---|
| 312 kHz CW | Distress and calling secondary |
| 500 CW | Distress and calling primary |
| 2182 USB | Distress and calling |
| 3023 SSB/CW | Search and rescue |
| 4040 USB | Joint Military Disaster primary |
| 4125 USB | Distress and safety |
| 4813.5 USB | Joint Military Disaster secondary |
| 6215.5 USB | Distress and safety |
| 5680 USB/CW | Search and rescue |
| 8364 CW | Survival craft |

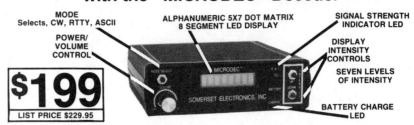
### Oceanographic Data Buoys

| | | |
|---|---|---|
| 4164.1 | 4164.7 | 4165.3 |
| 6246.1 | 6246.7 | 6247.3 |
| 8329.6 | 8330.2 | 8330.8 |
| 12481.1 | 12481.7 | 12482.3 |
| 16638.1 | 16638.7 | 16639.3 |
| 22162.1 | 22162.7 | 22163.5 |

### Intership Communications

| USB | RTTY |
|---|---|
| 2003 | 2328 |
| 2082.5 | 3387 |
| 2093 | 4177.5 |
| 2203 | 8295.6 |
| 2635 | 12520 |
| 2638 | |
| 2738 | |
| 2830 | |

### Air and Ship Intercommunications (USB)

2103.5  2667  2670  5422.5

### POINT-TO-POINT LAND NETWORKS

### Elizabeth City to Portsmouth (USB)

2675  5422.5

# COAST GUARD

## Miami to Frontier
## Guard Forces, Cuba (USB)

5763.5

## Great Lakes Emergency Net (USB)

3241    2678    5320

## Channel Harbor Island
## to Point Mugu (USB)

5080

## Turkey LORAN Net (USB/AM/CW)

5530      3663

*St. Louis Coast Guard
(courtesy Ron Seymour)*

## Hawaii and Central Pacific
## LORAN Nets (USB/CW/RTTY)

4050
5063
7473
9303
9630
12205

## Command Net
## All North American Regions
### (USB/RTTY)

27045
43357
6381
8648
12287.5
16907.7
22543
25378

## District Operations
### (USB/CW/RTTY)

| | |
|---|---|
| 2659 | 9th Dist. |
| 2662 | 1st Dist. |
| 2675 | 5th, 11th Dist. |
| 2678 | 7th, 9th, 17th Dist. |
| 2683 | 8th, 14th Dist. |
| 2686 | 5th, 11th Dist. |
| 2691 | 7th Dist. |
| 2694 | 1st, 11th Dist. |
| 2699 | 8th, 13th Dist. |
| 2702 | 5th, 14th Dist. |
| 2707 | 1st, 13th Dist. |
| 2710 | 11th Dist. |
| 2748 | 17th Dist. |
| 3241 | 9th Dist. |
| 3253 | 1st Dist. |
| 3382 | 11th Dist. |
| 4575 | 17 Dist. (daytime only) |
| 5320 | 5th, 7th, 8th,11th, 14th Dist. |

# COAST GUARD

## U.S. Coast Guard Cutter Frequencies and Modes

| | | | |
|---|---|---|---|
| 2003.0 | USB | 3253.0 | USB |
| 2082.5 | USB | 3382.0 | USB |
| 2093.0 | USB | 3387.0 | RTTY |
| 2182.0 | USB #1 | 4081.6/ | |
| 2203.0 | USB | 4376.0 | USB |
| 2234.0 | RTTY | 4106.4/ | |
| 2243.5 | RTTY | 4400.8 | USB |
| 2328.0 | RTTY | 4125.0 | USB |
| 2442.0 | RTTY | 4134.3/ | |
| 2512.0 | USB | 4428.7 | USB |
| 2606.0 | RTTY | 4170.0 to | |
| 2635.0 | USB | 4177.5 | RTTY |
| 2638.0 | USB | 4376.0 | USB |
| 2659.0 | USB | 4419.4 | USB |
| 2662.0 | USB | 5320.0 | USB |
| 2670.0 | USB | 5419.5 | RTTY |
| 2672.0 | USB | 5680.0 | USB #2 |
| 2675.0 | USB | 6200.0/ | |
| 2678.0 | USB | 6506.4 | USB |
| 2680.0 | RTTY | 6206.2/ | |
| 2683.0 | USB | 6512.6 | USB |
| 2686.0 | USB | 6218.6 | USB |
| 2690.0 | RTTY | 6221.6 | USB |
| 2691.0 | USB | 6260.0 to | |
| 2694.0 | USB | 6267.5 | RTTY |
| 2698.0 | RTTY | 6381.0 | RTTY |
| 2699.0 | USB | 6521.9 | USB |
| 2702.0 | USB | 7349.5 | RTTY #3 |
| 2707.0 | USB | 8195.0/ | |
| 2710.0 | USB | 8718.9 | USB |
| 2738.0 | USB | 8241.5/ | |
| 2748.0 | USB | 8765.4 | USB |
| 2830.0 | USB | 8245.0 | USB |
| 3023.0 | USB #2 | | |
| 3241.0 | USB | | |

| | | | |
|---|---|---|---|
| 8295.0 to | | 16475.5/ | |
| 8297.1 | RTTY | 17248.4 | USB |
| 8343.5 to | | 16497.2/ | |
| 8355.0 | RTTY | 17270.1 | USB |
| 8768.5 | USB | 16534.4/ | |
| 9108.0 | RTTY #3 | 17307.3 | USB |
| 9125.0 | RTTY #3 | 16574.7/ | |
| 10136.0 | RTTY #3 | 17347.3 | USB |
| 10166.0 | RTTY #3 | 16660.0 to | |
| 11434.0 | RTTY #3 | 16671.5 | RTTY |
| 12342.4/ | | 16693.0 to | |
| 13113.2 | USB | 16694.5 | RTTY |
| 12379.6/ | | 18195.5 | RTTY #3 |
| 13150.4 | USB | 22015.5/ | |
| 12426.1/ | | 22611.5 | USB |
| 13196.9 | USB | 22052.7/ | |
| 12491.0 to | | 22648.7 | USB |
| 12502.5 | RTTY | 22192.0 to | |
| 12518.5 | RTTY | 22203.5 | RTTY |
| 12519.5 | RTTY | 22224.5 | RTTY |
| 15654.5 | RTTY #3 | 27540.0 | RTTY #3 |

NOTES:
#1 - International distress & calling
#2 - International Search and Rescue
#3 - Arctic/Antarctic Icebreakers

(Freq/freq) are duplex USB ship/shore. Some cutters use USB simplex on shore duplex frequencies. (Freq to freq) are multiple duplex RTTY channels. RTTY is usually 75/170R or 75/850R.

# COAST GUARD

## What's in a Name?

With some exceptions, there is a method in assigning those colorful names to Coast Guard vessels. For example, 95' - 110' patrol boats utilize geographical capes (Cape Fox, Cape Horn), while 82' patrol boats will be named after points (Point Swift, Point Roberts); medium tugs are indian tribes (Chinook, Snohomish) and small tugs are ship rigging (Line, Shackle); buoy tenders are decidedly botanical (Red Birch, Sumac); high endurance cutters honor individuals (Sherman, Hamilton); and medium endurance cutters show their dedication (Steadfast, Durable).

## U.S. COAST GUARD CUTTERS
*(Includes new vessels not yet assigned to a port)*

| Name | Call | Hull No. | Home Port |
|------|------|----------|-----------|
| Acacia | NODY | WLB406 | Grand Haven, MI 49417 |
| Active | | WMBC618 | Port Angeles, WA 98362 |
| Acushnet | NNHA | WMEC167 | Gulfport, MS 39501 |
| Adak | | WPB1333 | Highlands, NJ 07732 |
| Alert | NZVE | WMEC630 | Cape May, NJ 08204 |
| Amelia Island | | WPB1338 (on order) | |
| Anacapa | | WPB1335 | |
| Annette Island | | WPB1339 (on order) | |
| Anvil | NMFN | WLIC75301 | Corpus Christi, TX 78401 |
| Apalachee | NRKC | WYTM711 | Baltimore, MD 21226 |
| Aquidneck | NBTC | WPD1309 | Portsmouth, VA 23703 |
| Assateague | | WPB1337 | |
| Attu | | WPB1317 | St. Thomas, VI 00802 |
| Axe | | WLIC85310 | Mobile, AL 36615 |
| Bainbridge Island | | WPB1340 (on order) | |
| Baranof | | WPB1318 | Miami, FL 33139 |
| Basswood | NODG | WLB388 | Guam,FPO Seattle,WA 96661 |
| Bayberry | NBKE | WLI65400 | Seattle, WA 98199 |
| Bear | NRKN | WMEC901 | Portsmouth, VA 23703 |
| Biscayne Bay | NRUS | WTGB104 | St Ignace, MI 49781 |
| Bittersweet | NODH | WLB389 | Woods Hole, MA 02543 |
| Blackberry | | WLI65303 | Southport, NC 28461 |
| Blackhaw | NODI | WLB390 | San Francisco, CA 94130 |
| Block Island | | WPB1341 | |
| Bluebell | NODD | WLI313 | Portland, OR 97217 |
| Bollard | | WYTL65614 | New Haven, CT 06512 |
| Boutwell | NYCQ | WHEC719 | FPO Seattle, WA 98799 |
| Bramble | NODK | WLB392 | Port Huron, MI 48060 |
| Bridle | | WYTL65607 | SW Harbor, ME 04679 |
| Bristol Bay | NRLY | WTG102 | Detroit, MI 48207 |
| Buckthorn | NADT | WLI642 | Sault Ste Marie, MI 49783 |
| Buttonwood | NRPX | WLB306 | Galveston, TX 77550 |
| Campbell | NRDC | WMEC909 | New Bedford, MA 02740 |
| Cape Carter | NJSJ | WPB95309 | Auke Bay, AK 99821 |
| Cape Corwin | NVUN | WPB95326 | Maui, HI 96793 |
| Cape Cross | NPDZ | WPB96321 | Crescent City, CA 955?? |
| Cape George | NYRW | WPB95306 | Guam,FPO San Fran,CA 96662 |
| Cape Gull | NRLZ | WPB95304 | Miami, FL 33139 |
| Cape Hatteras | NRDA | WPB95305 | Petersburg, AK 99833 |
| Cape Hedge | NZPO | WPB96311 | Morro Bay, CA 93442 |
| Cape Higgon | NRDE | WPB95302 | Glouchester, MA 01930 |
| Cape Horn | NPUF | WPB95322 | Ft Tilden, NJ 11695 |
| Cape Jellison | NJTB | WPB95317 | Seward, AK 99664 |
| Cape Knox | NGBO | WPB95312 | Charleston, SC 29401 |
| Cape Romain | NLBB | WPB95319 | Ketchikan, AK 99901 |
| Cape Small | NCFQ | WPB95300 | Hilo, HI 96720 |
| Cape Starr | NJVB | WPB95320 | Atlantic City, NJ 08401 |

| | | | |
|---|---|---|---|
| Cape Wash | NOBE | WPB95310 | Morro Bay, CA 93442 |
| Catenary | | WYTL65606 | Gloucester City,NJ 08030 |
| Capstan | NFKO | WYTL65601 | Ft Belvoir, VA |
| Chandeleur | | WPB1319 | Miami, FL 33139 |
| Chase | NLPM | WHEC718 | Governors Island, NY 10004 |
| Chena | | WLR75409 | Hickman, KY 42050 |
| Cherokee | NNGP | WMEC165 | Norfolk, VA 23520 |
| Cheyenne | NGXJ | WLR75405 | St Louis, MO 63111 |
| Chilula | NPIN | WMEC153 | Atlantic Beach, NC 28512 |
| Chincoteague | | WPB1320 | Mobile, AL 36615 |
| Chinook | NRKL | WYTM96 | Portsmouth, VA 23703 |
| Chippewa | | WLR75404 | Owensboro, KY 42301 |
| Chock | | WYTL65602 | Portsmouth, VA 23703 |
| Chokeberry | | WLI65304 | Crisfield, MD 21817 |
| Cimarron | | WLR65502 | Buchanan, TN 38222 |
| Citrus | NRPQ | WMEC300 | Coos Bay, OR 97420 |
| Clamp | | WLIC75306 | Galveston, TX 77550 |
| Cleat | | WYTL65615 | Gloucester, NJ 08030 |
| Clover | NRPK | WMEC292 | Eureka, CA 95501 |
| Confidence | NHKW | WMEC619 | New Bedford, MA 02740 |
| Courageous | NCRG | WMEC622 | Key West, FL 33040 |
| Cowslip | NAFO | WLB277 | Portsmouth, VA 23703 |
| Cushing | | WPB1321 | Mobile, AL 36615 |
| Cuttyhunk | | WPB1322 | Port Angeles, WA 98362 |
| Dallas | NPCR | WHEC716 | FPO Governors Is,NY 09567 |
| Dauntless | NDTS | WMEC624 | Miami Beach, FL 33139 |
| Decisive | NUHC | WMEC629 | St Petersburg, FL 33? |
| Dependable | NOWK | WMEC626 | Panama City, FL 32401 |
| Diligence | NMUD | WMEC616 | Port Canaveral, FL 32920 |
| Dogwood | NUNA | WLR259 | Pine Bluff, AR 71611 |
| Drummond | | WPB1323 | Port Canaveral, FL 32920 |
| Durable | NRUN | WMEC628 | Brownsville, TX 78521 |
| Eagle | NRCB | WIX327 | New London, CT 06320 |
| Edisto | NLKY | WPB1313 | Crescent City, CA 95531 |
| Elderberry | NUKD | WLI65401 | Petersburg, AK 99833 |
| Escanaba | NNAS | WMEC907 | Boston, MA 02109 |
| Escape | NBPG | WMEC6 | Charleston, SC 29408 |
| Evergreen | NRXD | WMEC295 | New London, CT 06320 |
| Farallon | NABK | WPB1301 | Miami Beach, FL 33139 |
| Fir | NRYR | WLB212 | Seattle, WA 98199 |
| Firebush | NODL | WLB393 | FPO Seattle, WA (Kodiak, AK) |
| Foreward | NICB | WMEC911 | Long Beach, CA 90822 |
| Gallatin | NJOR | WHEC721 | FPO GovernorsIs,NY 09570 |
| Galveston Island | | WPB1342 (on order) | |
| Gasconade | | WLR75401 | Omaha, NB 68112 |
| Gentian | NRPI | WLB290 | Atlantic City, NC 28512 |
| Glacier | NAAO | WAGB4 | Long Beach,FPO SF,CA 96666 |
| Grand Isle | | WPB1343 (on order) | |
| Hamilton | NMAG | WHEC715 | Boston, MA 02109 |
| Hammer | NNWM | WLIC75302 | Fort Pierce, FL 33450 |
| Harriet Lane | NHNC | WMEC903 | Portsmouth, VA 23703-2199 |
| Hatchet | | WLIC75309 | Galveston, TX 77550 |
| Hawser | | WYTL65610 | New York, NY 10004 |
| Hornbeam | NODM | WLB394 | Cape May, NJ 08204 |
| Hudson | NCWX | WLIC801 | Miami Beach, FL 33139 |
| Ingham | NRDL | WLEC35 | Portsmouth, VA 23703 |
| Iris | NODN | WLB395 | Astoria, OR 97103 |
| Ironwood | NRPN | WLB297 | FPO Seattle, WA 98799 |
| Isle Royale | | WPB1344 (on order) | |
| Jarvis | NAQD | WHEC725 | Honolulu,FPO SF,CA 96669 |
| Jefferson Island | | WPB1345 (on order) | |
| Kanawha | | WLR75407 | Memphis, TN 38173 |

| | | | |
|---|---|---|---|
| Katmai Bay | NRLX | WTGB101 | Sault Ste Marie,MI 49783 |
| Kennebec | NRDJ | WLIC802 | Portsmouth, VA 23703 |
| Key Biscayne | | WPB1346 (on order) | |
| Key Largo | | WPB1324 | Savannah, GA 31401 |
| Kickapoo | | WLR75406 | Vicksburg, MS 39180 |
| Kiska | | WPB1336 | |
| Knight Island | | WPB1347 (on order) | |
| Kodiak Island | | WPB1348 (on order) | |
| Lantana | NRQX | WLR80310 | Natchez, MS 34120 |
| Laurel | NRPJ | WLB291 | San Pedro, CA 90731 |
| Legare | NRPM | WMEC912 | Long Beach, CA 90822 |
| Liberty | | WPB1334 | |
| Line | | WYTL65611 | New York, NY 10004 |
| Lipan | NBOZ | WMEC85 | Fort George Isl, FL 32226 |
| Long Island | | WPB1349 (on order) | |
| Mackinaw | NRKP | WAGB83 | Cheboygan, MI 49721 |
| Mallet | NODO | WLIC75304 | Corpus Christi, TX 78401 |
| Mallow | | WLB396 | Honolulu,FPO SF,CA 96672 |
| Maninicus | NDIS | WPB1315 | Cape May, NJ 08204 |
| Manitou | NA5P | WPB1302 | Miami Beach, FL 33139 |
| Mariposa | NODP | WLB397 | Detroit, MI 48207 |
| Matagorda | NAYM | WPB1303 | Miami Beach, FL 33139 |
| Maui | NBEI | WPB1304 | Miami Beach, FL 33139 |
| Mellon | NDIT | WHEC717 | FPO Seattle, WA 98799 |
| Messenger | | WYTM85009 | Baltimore, MD 21226 |
| Mesquite | NRPW | WLB305 | Charlevoix, MI 49720 |
| Metomkin | | WPB1325 | Charleston, SC 29401 |
| Midgett | NHWR | WHEC726 | Alameda,FPO SF,CA 96672 |
| Mobile Bay | NRUR | WTGB103 | Sturgeon Bay, WI 54235 |
| Mohawk | NRUF | WMEC913 | Long Beach, CA 90822 |
| Mohican | NRKI | WYTM73 | Portsmouth, VA 23703 |
| Monhegan | NEGS | WPB1305 | |
| Monomoy | | WPB1326 | Woods Hole, MA 02543 |
| Morgenthau | NDWA | WHEC722 | Alameda,FPO SF,CA 96672 |
| Morro Bay | NRWY | WTGB106 | Yorktown, VA 23690-5000 |
| Munro | NGDF | WHEC724 | Honolulu,FPO SF,CA 96672 |
| Muskingum | | WLR75402 | Sallisaw, OK 74955 |
| Mustang | NJSH | WPB1310 | Seward, AK 99664 |
| Nantucket | | WPB1316 | San Juan, PR 00903 |
| Naushon | NEWR | WPB1311 | Ketchikan, AK 99901 |
| Northland | NLGF | WMEC904 | Norfolk, VA 23511 |
| Neah Bay | NRUU | WTGB105 | Cleveland, OH 44114 |
| Nunivac | NHPX | WPB1306 | Roosevelt Roads, PR (FPO Miami 34051) |
| Obion | | WLR65503 | St Louis, MO 63111 |
| Ocracoke | NGBL | WPB1307 | Roosevelt Roads, PR (FPO Miami 34051) |
| Orcas | | WPB1327 | Coos Bay, OR 97420 |
| Osage | | WLR65505 | Sewickly, PA 15143 |
| Ouachita | | WLR65501 | E Chattanooga, TN 37416 |
| Padre | | WPB1328 | Key West, FL 33040 |
| Pamlico | NAYE | WLIC800 | New Orleans, LA 70117 |
| Papaw | NRPZ | WLB308 | Charleston, SC 29401 |
| Patoka | | WLR75408 | Greenville, MS 38701 |
| Pendant | | WYTL65608 | Boston, MA 02109 |
| Penobscot Bay | NIGY | WTGB107 | Governors Island, NY 10004 |
| Petrel | NACN | WSES-4 | Key West, FL 33040 |
| Plantree | NRPY | WLB307 | FPO Seattle,WA 98799 |
| Point Arena | NJKT | WPB82346 | Norfolk, VA 23520 |
| Point Baker | NIQK | WPB82342 | Port Aransas, TX 78373 |
| Point Barnes | NLVA | WPB82371 | Ft Pierce,FL 33449 |
| Point Barrow | NUFN | WPB82348 | Monterey, CA 93940 |
| Point Batan | NAKH | WPB82340 | Point Pleasant,NJ 08742 |
| Point Bennett | NAVH | WPB82351 | Port Townsend, WA 98368 |

| | | | |
|---|---|---|---|
| Point Bonita | NSMF | WPB82347 | Woods Hole, MA 02543 |
| Point Bridge | NLDW | WPD82338 | Marina del Rey, CA 90291 |
| Point Brower | NMEX | WPB82372 | San Diego, CA 92106 |
| Point Brown | NKFW | WPB82362 | Norfolk, VA 27968 |
| Point Camden | NKIG | WPB82373 | San Pedro, CA 90731 |
| Point Carrew | NNZL | WPB82374 | San Pedro, CA 90731 |
| Point Charles | NJGJ | WPB82361 | Cape Canaveral, FL 32920 |
| Point Chico | NIOO | WPB82339 | Bodega Bay, CA 94923 |
| Point Countless | NVOA | WPB82335 | Port Angeles, WA 98362 |
| Point Divide | NRJG | WPB82337 | Corona del Mar, CA 92625 |
| Point Doran | NLLX | WPB82375 | Everett, CA 98201 |
| Point Estero | NZON | WPB82344 | Gulfport, MS 39501 |
| Point Evans | NCCE | WPB2354 | Long Beach, CA 90803 |
| Point Francis | NDCH | WPB82356 | Highland, NJ 07732 |
| Point Franklin | NYVC | WPB82350 | Cape May, NJ 08204 |
| Point Glass | NCNR | WPB82336 | Gig Harbor, WA 98335 |
| Point Hannon | NCNP | WPB82355 | West Jonesport, ME 04649 |
| Point Harris | NKEQ | WPB82376 | Honolulu, HI 96850 |
| Point Herron | NUEL | WPB82318 | Babylon, NY 11702 |
| Point Heyer | NVRC | WPB82369 | San Francisco, CA 94130 |
| Point Highland | NWNO | WPB82333 | Chincoteague, VA 23336 |
| Point Hobart | NSWA | WPB82377 | Oceanside, CA 92054 |
| Point Hope | NKJX | WPB82302 | Sabine, TX 77655 |
| Point Huron | NECA | WPB82357 | Norfolk, VA 23520 |
| Point Jackson | NSGB | WPB82378 | Woods Hole, MA 02543 |
| Point Judith | NJIL | WPB82345 | Santa Barbara, CA 93109 |
| Point Knoll | NTKD | WPB82367 | New London, CT 06230 |
| Point Ledge | NDLK | WPB82334 | Ft Bragg, CA 95437 |
| Point Lobos | NSDA | WPB82366 | Panama City, FL 32407 |
| Point Lookout | NELL | WPB82341 | Morgan City, LA 70380 |
| Point Martin | NRQJ | WPB82379 | Atlantic Beach, NC 28512 |
| Point Monroe | NBRM | WPB82353 | Freeport, TX 77541 |
| Point Nowell | NLEZ | WPB82363 | Port Isabel, TX 78578 |
| Point Richmond | NWGM | WPB82370 | Anacortes, WA 98221 |
| Point Roberts | NFDU | WPB82332 | Mayport, FL 32067 |
| Point Sal | NBEN | WPB82352 | Grand Isle,LA 70358 |
| Point Spencer | NXFN | WPB82349 | New Orleans, LA 70117 |
| Point Steele | NGYD | WPB82359 | Ft Myers Beach, FL 3393? |
| Point Stuart | NFIJ | WPB82358 | San Diego, CA 92106 |
| Point Swift | NLBI | WPB82312 | Clearwater, FL 33515 |
| Point Thatcher | NMDS | WPB82314 | Sarasota, FL 33578 |
| Point Turner | NICC | WPB82365 | Newport, RI 02840 |
| Point Verde | NZZF | WPB82311 | Pensacola, FL 32? |
| Point Warde | NUIH | WPB82368 | San Juan, PR 00903 |
| Point Wells | NTDG | WPB82343 | Montauk, NY 11954 |
| Point Whitehorn | NMEI | WPB82364 | St Thomas, VI 00802 |
| Point Winslow | NHMV | WPB82360 | Eureka, CA 45501 |
| Polar Sea | NRUO | WAGB11 | FPO Seattle, WA 98799 |
| Polar Star | NBTM | WAGB10 | FPO Seattle, WA 98799 |
| Primrose | NRZT | WLIC316 | Atlantic Beach, NC 285? |
| Rambler | NRPO | WLIC298 | Charleston, SC 29401 |
| Raritan | NRPS | WYTM93 | New York, NY 10004 |
| Red Beech | NJLE | WLM686 | Governors Is, NY 10004 |
| Red Birch | NGFH | WLM687 | Baltimore, MD 21226 |
| Red Cedar | NPDC | WLM688 | Portsmouth, VA 23703 |
| Red Oak | NPKF | WLM689 | Gloucester, NJ 08030 |
| Red Wood | NDTE | WLM685 | New London, CT 06320 |
| Reliance | NJPJ | WMEC615 | Cape Canaveral, FL 3293? |
| Resolute | NRLT | WMEC620 | Astoria, OR 97103 |
| Rush | NLVS | WHEC723 | Alameda,FPO SF,CA 96677 |
| Saginaw | NJOY | WLIC803 | Mobile, AL 36615 |
| Salvia | NODS | WLB400 | Mobile, AL 36690 |

# COAST GUARD

| | | | |
|---|---|---|---|
| Sangamon | | WLR65506 | Peoria, IL 61601 |
| Sanibel | NDCK | WPB1312 | Rockland, ME 04841 |
| Sapelo | | WPB1314 | Eureka, CA 95501 |
| Sassafras | NODT | WLB401 | Honolulu,FPO SF,CA 96678 |
| Scioto | | WLR65504 | Keokuk, IA 52632 |
| Sea Hawk | NEHM | WSES-2 | Key West, FL 33040 |
| Sedge | NODU | WLB402 | Homer, AK 99603 |
| Seneca | NFMK | WMEC906 | Boston, MA 02109 |
| Sepelo | NHKD | WPD1314 | |
| Shackle | | WYTL65609 | S Portland, ME 04106 |
| Shearwater | NKBT | WSES-3 | Key West, FL 33040 |
| Sherman | NMMJ | WHEC720 | Alameda,FPO SF,CA 96678 |
| Sitkinik | | WPB1329 | Key West, FL 33040 |
| Sledge | NSAD | WLIC75303 | Baltimore, MD 21226 |
| Smilax | NRYN | WLIC315 | Brunswick, GA 31521 |
| Snohomish | NRKR | WYTM72 | S Portland, ME 04106 |
| Sorrel | NRZI | WLB296 | New York, NY 10004 |
| Spar | NODV | WLB403 | S Portland, ME 04106 |
| Spencer | NWHE | WMEC905 | Boston, MA 02109 |
| Spike | | WLIC75308 | Mayport, FL 32267 |
| Steadfast | NSTF | WMEC623 | St Petersburg, FL 33701 |
| Storis | NRUC | WMEC38 | Kodiak, AK 99619 |
| Sturgeon Bay | | WTGB109 | South Portland, ME 04106 |
| Sumac | NTLZ | WLR311 | St Louis, MO 63111 |
| Sundew | NODW | WLB404 | Duluth, MN 55802 |
| Sweetbriar | NODX | WLB405 | FPO Seattle,WA 98799 |
| Sweetgum | NRQW | WLB309 | Mayport, FL 32267 |
| Swivel | | WYTL65603 | Rockland, ME 04841 |
| Tahoma | NCBE | WYTL65604 | New Bedford, MA 02740 |
| Tamaroa | NNGR | WMEC166 | New Castle, NH 03854 |
| Tampa | NIKL | WMEC902 | Portsmouth, VA 23703 |
| Tackle | | WYTL65504 | Crisfield, MD 21817 |
| Taney | NRDT | WHEC37 | Portsmouth, VA 23703 |
| Thetis | NYWL | WMEC910 | Long Beach, CA 90822 |
| Thunder Bay | NRTB | WTGB108 | South Portland, ME 04106 |
| Towline | | WYTL65605 | Bristol, RI 02809 |
| Tybee | | WPB1330 | San Diego, CA 92106 |
| Unimak | NBYG | WHEC379 | New Bedford, MA 02740 |
| Ute | NBXL | WMEC76 | Key West, FL 33040 |
| Valiant | NVAI | WHEC621 | Galveston, TX 77550 |
| Vashon | NJEH | WPB1308 | Roosevelt Roads, PR (FPO Miami 34051) |
| Venturous | NVES | WMEC625 | San Pedro, CA 90731 |
| Vise | | WLIC75305 | St Petersburg, FL 33701 |
| Vigilant (MARS NNNOCYU) NHIC | | WMEC617 | New Bedford, CT 33701 |
| Vigorous | NQSP | WMEC627 | New London, CT 06320 |
| Washington | | WPB1331 | Honolulu, HI 96850 |
| Wedge | | WLIC75307 | New Orleans, LA 70117 |
| Westwind | NLKL | WAGB281 | Mobile, AL 36633 |
| White Heath | NAQE | WLB545 | Boston, MA 02109 |
| White Holly | NAOY | WLM543 | New Orleans, LA 70117 |
| White Lupine | NARO | WLM546 | Rockland, ME 04841 |
| White Pine | NODE | WLM547 | Mobile, AL 36690 |
| White Sage | NAPC | WLM544 | Woods Hole, MA 02543 |
| White Sumac | NAED | WLM540 | St Petersburg, FL 33701 |
| Wire | | WYTL65612 | New York, NY 10004 |
| Woodrush | NODZ | WLB407 | Sitka,FPO Seattle,WA 98799 |
| Wrangell | | WPB1332 | New Castle, NH 03854 |
| Wyaconda | | WLR75403 | Dubuque, IA 52001 |
| Yocona | NNHB | WMEC168 | Kodiak, AK 99619 |

# Canadian Coast Guard

### 2182 USB    500CW

## Canadian CG Coast Stations

| | CW | | USB |
|---|---|---|---|
| VAF | | Alert Bay, NWT | 2054 2458 4125 |
| VAG | 484 | Bull Harbour, BC | 2054 |
| VFC | 6351.5 | Cambridge Bay, NWT | 2558 4363.6 5803 |
| VOK | 416 444 | Cartwright, Lab | 2514 2538 2582 2598 4376 |
| VCA | | Charlottetown, PEI | 2530 2582 |
| VAP | 420 6370 | Churchill, Man | 2582 4376 |
| VOO | | Comfort Cove, NFLD | 2514 2538 2582 2598 |
| VAC | | Comox, BC | 2054 |
| VFU 6 | | Coppermine, NWT | 4363.6 5803 |
| VFU | 484 | Coral Harbour, NWT | 2514 4376 6513.8 |
| VFF 3 | | Fort Chipewyan, Alta | 5803 |
| VFF 6 | | Fort Simpson, NWT | 5803 |
| VFF | 4236.5 430 484 6335.5 8443 12671 | Iqaluit, NWT | 2514 2582 4376 6512.6 8753 13100.8 |
| VFR | | Resolute, NWT | 6438(CW) |
| VFZ | | Goose Bay, Lab | 2514 2538 2582 2598 4379.1 |
| VCN | 440 | Grindstone, Magdelene Is, PQ | 2514 2582 2598 |
| VCS | 446 | Halifax, NS | 2514 2582 2598 4410.1 6518.8 |
| | 484 | | 8787.1 13138 17242.2 |
| | 4285, 6491.5 8440, 12874 16948.5, 22387 | | |
| VBJ 20 | | Halifax Traffic, NS | 2134 2237 |
| VFH 3 | | Hay River, NWT | 5803 |
| VAL | | Inoucdjouac, PQ | 2582 |
| VFA | 6351.5 | Inuvik, NWT | 2558 4363.6 5803 |
| VAW | 416 | Killineck, NWT | 2514 2582 4376 |
| VCF | 446 | Mont Joli, PQ | 2582 |
| VFN | 420 | Montreal, PQ | 2514 |
| VFN 8 | | Norman Wells, NWT | 5803 |
| VAJ | 420 | Prince Rupert, BC | 2054 4125 |
| VAV | | Poste della Balein,PQ | 2582 |
| VCC | 474 | Quebec, PQ | 2514 2582 |
| VFR | 474 4334 6438 | Quasuittuk, NWT | 2582 4376 8793.3 |
| VCG | 434 | Riviere au Renard,PQ | 2514 2582 2598 |
| VCM | 448 489 | St Anthony, NFLD | 2514 2582 2598 |
| VON | 460 478 | St John's, NFLD | 2538 2514 2582 2598 |
| VCC | 474 | Lauzon, PQ | 2582 |
| VCP | 420 | St Lawrence, NFLD | 2514 2538 2582 2598 |
| VAH | | Sandspit, BC | 2054 |
| VCK | 420 | Sept Iles, PQ | 2514 2582 2598 |
| VOJ | 416 | Stephenville, NFLD | 2514 2582 |
| VCO | 464 489 | Sydney, NS | 2598 2530 2582 2598 |
| VAE | 464 478 | Tofino, BC | 2054 2458 4125 |
| VYO 21 | | Tuktoyaktuk, NWT | 5803 |
| VAI | 440 4235 6493 8453 12876 17175.2 22368 | Vancouver, BC | 2054 4385.5 |
| VAK | 430 484 | Victoria, BC | 2054 2458 |
| VAU | 450 489 | Yarmouth, NS | 2538 2582 2598 |

### CW

| | | |
|---|---|---|
| VFF | Iqaluit, NWT | 4236.5 6493 8443 12671 |
| VCS | Halifax, NS | 4285 6491.5 8440 12874 16948.5 22387 |
| VAI | Vancouver, BC | 4235 6493 8453 12876 17175.2 22368 |

### Radiotelex

| | | |
|---|---|---|
| VCS | Halifax, NS | 4353 6497.5 8708 13090.5 17212.5 22590 |
| VAI | Vancouver, BC | 4354 6498.5 8717 13091.5 17213 22576.5 |

*Canadian Coast Guard Ship Alexander Henry in Lake Ontario*

# COAST GUARD

### Radiofacsimile

3251.1  4616  6915.1  7708.1  8113.1  10155.1
10169.1  12055.1  13440  14440  14624.6
14770  15642.1  17443.1  18168.1  18208.1
20168.1  20530.1

### Intership Simplex

| | |
|---|---|
| 2003 | Gulf of St. Lawrence & River |
| 2040 | Pacific |
| 2638 | |
| 2134 | |
| 2237 | |
| 2318 | Pacific |
| 2738 | |
| 2083.9 | |

### Private Ship - Shore

2066.5      2094.4

### Government Use

2010.4      2104.9

### Ice Breakers

St. Lawrence Seaway
6292.5

# FEDERAL GOVERNMENT

# Federal Emergency Management Agency
(FEMA)
## FEMA National Radio System
(FNRS)

Formerly the Defense Civil Preparedness Agency (DCPA), FEMA's prime responsibility is to assure continuity in government during an enemy attack, as well as to respond to natural disasters as declared by the President. Civilian employees man the facilities.

VIP relocation installations, such as those deep in Mount Weather, Berryville, VA (Blue Ridge Mountains, Rt 601 between Paris and Bluemont, VA), are part of this nationwide complex. A sophisticated USB/RTTY (100WPM/85Hz) network has been established to support this vital program.

Slow-speed, encrypted CW is also transmitted. ATS-3 satellite is also used. Communications security (COMSEC) is observed when discussing critical materials for stockpiling in case of aggression.

## Other Transmission Modes

ASCII, ? Hz Shift, 110 Baud, Reverse Polarity
ASCII, ? Hz Shift, 300 Baud, Reverse Polarity
Baudot, 850 Hz Shift, 75 Baud, Normal Polarity
Packet, 2400 baud

| Foxtrot | | Foxtrot | |
|---|---|---|---|
| 01 | 2320 | 19 | 6151 |
| 02 | 2360 | | |
| 03 | 2377 | 20 | 6176 |
| 04 | 2445 | | |
| 05 | 2658 | 21 | 6809 |
| 06 | 3341 | | |
| 07 | 3379 | 22 | 7348 |
| 08 | 3388 | | |
| 09 | 4603 | 23 | 7428 |
| 10 | 4780 | 24 | 9462 |
| 11 | 5211 | | |
| 12 | 5378 | 25 | 10194 |
| 13 | 5402 | | |
| 14 | 5821 | 26 | 10493 |
| 15 | 5961 | 27 | 10194 |
| 16 | 6049 | | |
| 17 | 6106 | 28 | 11721 |
| 18 | 6108 | 29 | 11801 |

*FEMA Network Control station console*

*State Emergency Operation Center*

# FEDERAL GOVERNMENT

| Foxtrot | | Foxtrot | |
|---|---|---|---|
| 30 | 11957 | 51 | 16201 |
| 31 | 11994 | 52 | 16430 |
| 32 | 12009 | 53 | 17519 |
| 33 | 12129 | 54 | 17649 |
| 34 | 12216 | 55 | 18744 |
| 35 | 12219 | 56 | 19757 |
| 36 | 13446 | 57 | 19969 |
| 37 | 13633 | 58 | 20027 |
| 38 | 13744 | 59 | 20063 |
| 39 | 13780 | 60 | 21866 |
| 40 | 13783 | 61 | 21919 |
| 41 | 14450 | 62 | 22983 |
| 42 | 14776 | 63 | 23028 |
| 43 | 14836 | 64 | 23390 |
| 44 | 14885 | 65 | 23451 |
| 45 | 14899 | 66 | 23550 |
| 46 | 14908 | 67 | 23814 |
| 47 | 15464 | 68 | 24008 |
| 48 | 15509 | 69 | 24282 |
| 49 | 15532 | 70 | 24526 |
| 50 | 15708 | 71 | 24819 |

## Test Net

All tests are conducted at 1600 UTC (1500 during Daylight Savings Time)

Working hours only, day-to-day all states monitor and work 10493. During any disaster, the affected state is the authority.

Mon,Thurs,Fri:
   Olney net control to all regions
   (USB)/10493
1st Tues in January, April, July, October:
   net control WGY903
1st Tues in February, May, August, November:
   net control WGY906
1st Tues in March, June, September, December:
   net control WGY904
2nd Tues in March, June, September, December:
   net control WGY905
Wed: all regions work each other
   (no net control station)

## FEMA Regions

**Reg Callsign ID/Location**

|  | WGY900 | FEMA IECC |
|---|---|---|
|  |  | Washington, DC |
|  | WGY911 | Telecommunications |
|  |  | Branch |
| 1 | WGY901 | MAYNARD, MASSACHUSETTS |
|  | WGY921 | Concord, NH |

WGY931 Montpelier, VT
WGY941 Augusta, ME
WGY951 Hartford, CT
WGY961 Framingham, MA
WGY971 Providence, RI

2  WGY902  NEW YORK CITY, NEW YORK
WGY942 Albany, NY
WGY982 North Trenton, NJ
WGY992 San Juan, Puerto
   Rico

3  WGY903  OLNEY, MARYLAND
(Net Control Station)
WGY923 Harrisburg, PA
WGY933 Pikesville, MD
WGY943 Charlestown, WV
WGY953 Delaware City,
   Delaware
WGY963 Richmond, VA
WGY983 Washington, D.C.

4  WGY904  THOMASVILLE, GEORGIA
(3rd alternate NCS)
WGY914 Canal Zone,
   (Balboa Hts)
WGY924 Nashville, TN
WGY934 Columbia, SC
WGY944 Atlanta, GA
WGY954 Montgomery, AL
WGY964 Jackson, MS
WGY974 Tallahassee, FL
WGY984 Raleigh, NC
WGY994 Frankfort, KY

5  WGY905  BATTLE CREEK, MICHIGAN
(4th alternate NCS)
WGY925 Madison, WI
WGY935 St Paul, MN
WGY945 Columbus, OH
WGY955 Springfield, IL
WGY965 Indianapolis, IN
WGY975 Lansing, MI

6  WGY906  DENTON, TEXAS
(2nd alternate NCS)
WGY926 Oklahoma City, OK
WGY936 Santa Fe, NM
WGY946 Baton Rouge, LA
WGY956 Austin, TX
WGY966 Conway, AR

7  WGY907  KANSAS CITY, MISSOURI
WGY947 Des Moines, IA
WGY957 Lincoln, NB
WGY977 Jefferson City, MO
WGY997 Topeka,KS

# FEDERAL GOVERNMENT

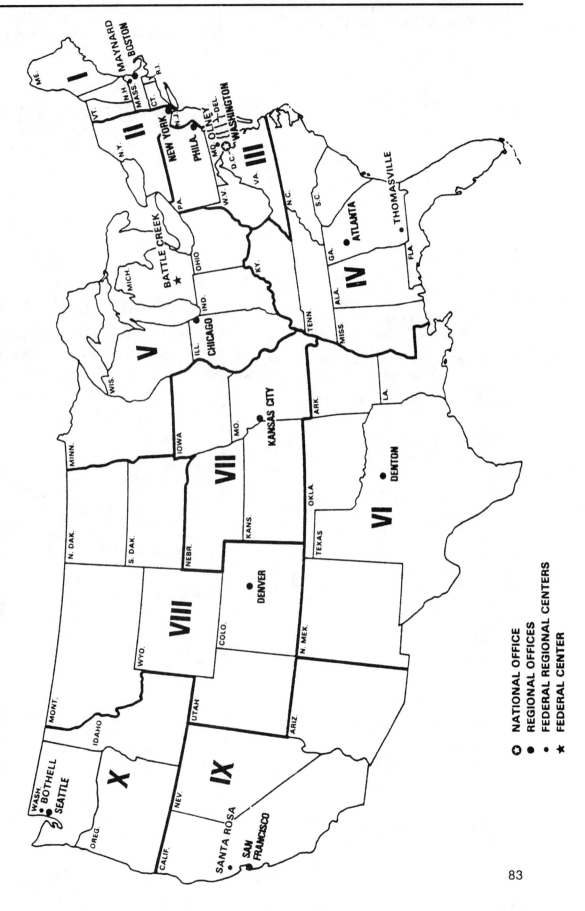

## FEDERAL EMERGENCY MANAGEMENT AGENCY

### REGIONAL BOUNDARIES WITHIN THE CONTERMINOUS UNITED STATES

✪ NATIONAL OFFICE
● REGIONAL OFFICES
· FEDERAL REGIONAL CENTERS
★ FEDERAL CENTER

83

# FEDERAL GOVERNMENT

| 8 | WGY908 | DENVER, COLORADO |
|---|---|---|
| | | (1st alternate NCS) |
| | WGY928 | Pierre, SD |
| | WGY938 | Cheyenne, WY |
| | WGY948 | Bismark, ND |
| | WGY958 | Helena, MT |
| | WGY968 | Golden, CO |
| | WGY998 | Salt Lake City, UT |

| 9 | WGY909 | SAN FRANCISCO, CALIFORNIA |
|---|---|---|
| | WGY929 | Carson City, NV |
| | WGY939 | Sacramento, CA |
| | WGY949 | Phoenix, AZ |
| | WGY959 | Honolulu, HI |
| | WGY979 | American Samoa |

| 10 | WGY910 | BOTHELL, WASHINGTON |
|---|---|---|
| | WGY920 | Boise, ID |
| | WGY930 | Olympia, WA |
| | WGY940 | Salem, OR |
| | WGY960 | Soldotna, AK |
| | WGY970 | Juneau, AK |
| | WGY980 | Alcantra, AK |

| | WGY911 | Telecommunications Management; FEMA HQ, Wash DC |
|---|---|---|
| | WGY912 | FEMA Special Facility Berryville, VA |
| | WGY915 | National Communications System HQ, Washington, DC |

### VIP Support Facilities

| Ft Myer, VA | WAR21 |
|---|---|
| Ft Belvoir | WAR22 |
| Hagerstown, MD | WAR30 |
| Mt Weather, VA | WAR42 |
| Mercersburg, PA | WAR45 |
| Ft Richie, MD | WAR46 |
| Boonsboro, MD | WAR47 |

### DOE Installations on this net

| KAL22 | Savannah River, South Carolina |
|---|---|
| KAL23 | Oak Ridge, Tennessee |
| KAL24 | Washington, DC (Headquarters) |
| KGO45 | Rocky Flats, Colorado |

### Other Installations on this net

| KGD34 | NCC, Arlington, VA |
|---|---|
| Tactical | USAF (SAC) |

## Nuclear Regulatory Commission (NRC)

Many FEMA channels are shared with other agencies. WGY912 has been heard testing in USB on USAF/SAC channels "Bravo Uniform" (3113 kHz). The NRC has access to FEMA channels for point-to-point administrative communications among the following NRC regional headquarters.

| Reg. | Location |
|---|---|
| 1 | Bethesda, MD |
| | King of Prussia, PA |
| 2 | Atlanta, GA |
| 3 | Glen Ellyn, IL |
| 4 | Arlington, TX |
| 5 | Walnut Creek, CA |

# Federal Highway Administration
## Emergency Communications System
(FHWA ECS)

Linking regional and field offices nationwide, the FHWA net is intended to inform DOT officials of major occurences, accidents and catastrophes involving the nation's highways.

A typical radio installation utilizes a Sunair GSB-900DX 100W transceiver with GSL-1900A 1kW linear amplifier on USB into a CGU-1935 antenna coupler and a 37 foot whip antenna, and Dovetron MPC-1000R terminal with Extel printer (DOT system) or Model 43 printer (FHWA system) for 110/300 baud (850 Hz shift) ASCII. 300 baud packet and FEC/TOR are also used.

Practice drills are held March, June, September and December from 1700-2131 UTC, Wednesday and Thursday, with participation from SHARES and other DOT agencies including FAA and Office of Emergency Transportation (OET). Occasional weekly tests are heard Wednesdays at 1800 UTC on 10891 (F5).

By reciprocal agreement, FHWA stations have been heard checking into USAF MARS and FEMA networks. Channels F1 through F9 are the most commonly used.

## Designators and Frequency Assignments (kHz)

| | | | |
|---|---|---|---|
| F1 | 3199.5 | F32 | 9930 |
| F2 | 5255 | F33 | 10225 |
| F3 | 7419 | F34 | 10918 |
| F4 | 9197 | F35 | 11045 |
| F5 | 10891 | F36 | 11518.7 |
| F6 | 12158 | F37 | 11605 |
| F7 | 14461 | F38 | 12064.5 |
| F8 | 16211.5 | F39 | 12094.5 |
| F9 | 19223 | F40 | 12171 |
| F10 | 3304.5 | F41 | 12178.7 |
| F11 | 3329.5 | F42 | 13434 |
| F12 | 3395 | F43 | 13493 |
| F13 | 4572.5 | F44 | 14953 |
| F14 | 4821 | F45 | 15910 |
| F15 | 4965 | F46 | 15969 |
| F16 | 4902 | F47 | 15981 |
| F17 | 5024.5 | F48 | 16330 |
| F18 | 5031 | F49 | 17525 |
| F19 | 5330 | F50 | 18403 |
| F20 | 5350 | F51 | 18716 |
| F21 | 5424 | F52 | 19934 |
| F22 | 5749 | F53 | 20095 |
| F23 | 5755.5 | F54 | 20330 |
| F24 | 5885 | F55 | 20843 |
| F25 | 7669.5 | F56 | 22926 |
| F26 | 7726.5 | F57 | 22975 |
| F27 | 7736 | F58 | 24040 |
| F28 | 7743 | F59 | 24793 |
| F29 | 7821 | F60 | 25490 |
| F30 | 9169 | F61 | 26703 |
| F31 | 9185 | F62 | 26905 |
| ? | 9918 | | |

Additional: 5009.5 7375 7583.5 9076 9918
11029.5 17422.5
FAA frequencies for FHWA check-in:
4056.5 7476.5 8126.5

### Regions and Call Signs

The FHWA ECS system is broken into 10 regions following the FEMA plan as follows (*primary):

| Call | Location | Assigned Channels |
|---|---|---|

#### Region HQ 1/2

WWJ40 Washington, DC (NCS)    1 2 3 4 5 6 7 8 9
WWJ41 Cairo, NY (NCS)    10 13* 20 25 30 33

WWJ42 Providence, RI
WWJ51 Montpelier, VT
WWJ52 Albany, NY
WWJ53 Trenton, NJ
WWJ54 Concord, NH
WWJ55 Taunton, MA
WWJ56 Augusta, ME
WWJ57 Wethersfield, CT

#### Region 3

WWJ43 Berkeley Springs, WV (NCS)    11 15 17 27 31*
WWJ58 Staunton, VA
WWJ59 Montoursville, PA
WWJ60 Beckley, WV
WWJ61 Georgetown, DE
WWJ62 Lavale, MD

#### Region 4

WWJ44 Dahlonega, GA (NCS)    7 12 14* 18 26 32* 34 35 48 50

WWJ63 Nashville, TN
WWJ64 Orangeburg, SC
WWJ65 Raleigh, NC
WWJ66 Greenwood, MS
WWJ67 Frankfort, KY
WWJ68 Avondale, GA
WWJ69 Defuniak Springs, FL
WWJ70 Montgomery, AL

#### Region 5

WWJ45 Homewood, IL (NCS)    10 16 21 29* 30 36 41 45 49

WWJ71 Madison, WI
WWJ72 Dundas, OH
WWJ73 St. Paul, MN
WWJ74 Cadillac, MI
WWJ75 Indianapolis, IN
WWJ76 Effingham, IL

#### Region 6

WWJ46 Worth, TX (NCS)    4 11 13 22 25* 37 42 43 46 51 53

WWJ77 Brownwood, TX
WWJ78 Oklahoma City, OK
WWJ79 Santa Fe, NM
WWJ80 Baton Rouge, LA
WWJ81 Arkadelphia, AR

# FEDERAL GOVERNMENT

## Region 7

WWJ47  Kansas City,MO(NCS)  12 14 23* 28 31
                            35 41 47  49  52
WWJ82  Grand Isle, NE
WWJ83  Jefferson City, MO
WWJ84  Topeka, KS
WWJ85  Ames, IA

## Region 8

WWJ48  Lakewood, CO (NCS)  11 13 24 25 30*
                           34 42 43 48 50 54
WWJ86  Cheyenne, WY
WWJ87  Cedar City, UT
WWJ88  Pierre, SD
WWJ89  Bismarck, ND
WWJ90  Helena, MT
WWJ91  Denver, CO

## Region 9

WWJ49  Petaluma, CA (NCS)  6 7 10* 15 19 27*
                           32 35 38* 46 50
                           55 57 58 59 61
WWJ92  Redding, CA
WWJ93  Carson City, CA
WWJ94  Phoenix, AZ
WWJ95  Lihue, HI

## Region 10

WWJ50  Newport, OR (NCS)   12 16 18 26 32
                           33* 39 44 45 49
                           52 53 56 59 60 62
WWJ50A Newport, OR (remote)
WWJ96  Olympia, WA
WWJ97  Salem, OR
WWJ98  Shoshone, ID
WWJ99  Juneau, AK

Command Net (net control stations) utilize channels 1-9

## Additional Call Signs

KWB406 DOT    Ames, IA
KWB407 DOT    Durango, CO
KWB408 DOT    Two Rock, CA
KTX20  DOT    Washington, DC
KCP63  FAA    Longmont, CO
KKU40  FAA    Kansas City, MO
KDM50  FAA    Hampton, GA
KGD34  SHARES Arlington, VA

Other check-ins have included WWH7, WWC9, WWJ6, WWJ7.

# Department of Transportation
## Emergency Radio System

| Freq. | Ch.* | Freq. | Ch.* |
|-------|------|---------|-------|
| 3303 | F1 | 11028 | F6 |
| 5008 | F2 | 13432.5 | F7 |
| 7373.5 | F3 | 17421 | F8 |
| 7582 | F4 | 4055 | (FAA) |
| 9074.5 | F5 | 7475 | (FAA) |
| | | 8125 | (FAA) |

*Channel numbers were being reassigned at this writing

| Call Sign | Location |
|-----------|----------|
| KIT88 | FAA, Martinsburg, WV |
| KTX20 | DOT headquarters, Washington, DC |
| KWB236 | Nashua, NH |
| KWB238 | Leesburg, VA |
| KWB239 | Hilliard, FL |
| KWB240 | Rensselaer, IN |
| KWB405 | Bryan, TX |
| KWB406 | Ames, IA |
| KWB407 | Laramie, CO |
| KWB408 | Petaluma, CA |
| KWB409 | Ephrata, WA |
| WSX70 | Kenai, AK |

# National Communications System (NCS)
## Shared Resources (SHARES)

Established in 1961 by Presidential Order, The National Communications System (NCS) is a confederation of 23 federal departments and agencies. Their shared resources (SHARES) HF radio communications plan provides for a nationwide network using the combined radio assets of the federal government on a shared basis, enabling agencies of the federal government to exchange critical information.

USB/LSB voice communications and even occasional RTTY have been heard on the folowing frequencies: 7475, 7540, 7552.1, 7635, 8125, 9197, 10195, 10588, 10891, 11028, 13630, 13993, 14383.5, 14450, 14905, and 16348 kHz. Common check-ins are heard from CAP, DOT, FAA, FBI, FEMA, FHWA and MARS.

**SHARES HF Radio Stations**

|  |  |
|---|---|
| ARNG | 54 Stations |
| USCG | 7 Stations |
| Bellcore | 1 Station |
| GSA | 1 Station |
| USN/MC MARS | 56 Stations |
| DMA | 13 Stations |
| DLA | 19 Stations |
| USNR | 40 Stations |
| NCC | 1 Station |
| DEA | 2 Stations |
| INS | 1 Station |
| 7th Sig Cmd | 17 Stations |
| VA | 7 Stations |
| NASA | 2 Stations |
| CAP | 64 Stations |
| FCC | 14 Stations |
| USAF MARS | 30 Stations |
| USCS | 1 Station |
| DOT/OET | 11 Stations |
| USDA | 4 Stations |
| USA MARS | 22 Stations |
| FEMA | 3 Stations |
| USA COE | 41 Stations |
| FHWA | 59 Stations |
| FBI | 63 Stations |
| DOI | 31 Stations |
| FAA | 37 Stations |

**SUMMARY**

| | |
|---|---|
| SHARES STATIONS | 601 |
| ENTITIES | 27 |
| OVERSEAS STATIONS | 28 |
| FREQUENCIES | 110 |

87

# FEDERAL GOVERNMENT

## Department of Energy (DOE)

### Nuclear Transport Safeguard Network

Unmarked convoys carrying fissionable materials and nuclear weapons along Interstate 40 take elaborate precautions to thwart any attempt to gain access to their cargo. In addition, a shortwave network ties the mobile units to nuclear facilities across the nation.

Hourly check-ins may be heard using secure communications (SECOM) rolling-code speech inversion accompanied by data transmissions to the Eclipse computer system.

| Channel | Frequency |
|---------|-----------|
| *1 | 3335 |
| 2 | 5308 |
| *3 | 5751 |
| 4 | 5947 |
| *5 | 7700 |
| 6 | 8013.5 |
| 7 | 9918 |
| *8 | 11555 |
| 9 | 14657 |
| 10 | 17397 |

\* *Only channels presently in use; Simulkeyed*

*Pantex Weapons Assembly Plant, Amarillo, TX*

### Control Points

| | |
|---|---|
| KRF263 | Albuquerque, NM (NCS) |
| KRF264 | Idaho Falls, ID |
| KRF265 | Belton, MO |
| KRF266 | Cheltenham, MD (HQ) |
| KRF267 | Savannah River, SC |
| KRF268 | Los Alamos, NM |
| KRF269 | Amarillo, TX |

## Nuclear Materials Production and Storage Facilities

### Point to Point Communications Networks

The Department of Energy utilizes USB radio communications to interlink many of their facilities. Some of the installations are for the production of radioactive materials for the industry, medicine and defense, while others in more remote regions are for the storage of nuclear warheads. Not all frequencies shown are in use, but all are assigned to the DOE.

| Call | Location |
|------|----------|
| KRR41 | Fernald, OH |
| KAL22 | Savannah River Plant, SC; Dupont Weapons Lab |
| KAL23 | Oak Ridge, TN; Y-12 Plant (produces plutonium triggers for bombs); Union Carbide |
| KAL24 | Washington, DC; BDF Facility |
| KAS22 | Kansas City, MO |
| KKR39 | Paducah, KY |
| KKR40 | Piketon, OH |
| KMF43 | Albuquerque, NM |
| KSJ87 | Lemont, IL; Argonne National Lab |
| KIV25 | St Petersburg/Brooksville, FL; G.E. Labs (makes neutron generators) Pinellas Plant |
| KBA63 | Upton, NY; Brookhaven National Institute |
| KMU77 | Livermore, CA; University of California, Lawrence Radiation Lab |
| KAY20 | Richland, WA; Hanford Reservation - Rockwell Internat'l |
| KBW48 | Idaho Falls,IA; Exxon - Idaho Weapons Lab |
| KBW49 | Nevada; Weapons Test Site |
| KLJ87 | New Mexico; Los Alamos Scientific Lab, University of Chicago |
| KGO45 | Estes Park, CO; Rocky Flats Plant, Rockwell Internat'l |
| KBA62 | Miamisburg, OH; Monsanto Chemical, Mound Lab |
| KLE39 | Amarillo, TX; Pantex - Weapons Assembly Plant |
| KAD71 | Hickam AFB, HI - Nuclear Defense Agency |
| KAN71 | Mauna Loa, HI |

# FEDERAL GOVERNMENT

## Frequencies

| | | | |
|---|---|---|---|
| 2287.5 | 6930.5 | 10555.5 | 17902.5 |
| 2446.5 | 6982.5* | 10871.5 | 18417.5 |
| 2610.5 | 7349.5 | 11126.5 | 19791.5 |
| 2624 | 7355.5 | 11397.5 | 20405.5 |
| 4480.5 | 7429.5 | 12020.5 | |
| 4600 | 7465.5 | 12233.5 | |
| 4669.5 | 7690.5 | 13261.5 | |
| 4776.5 | 7723.5 | 13803.5 | |
| 4946.5 | 7767.5 | 14400.5 | |
| 5212.5 | 7840.5 | 15454.5 | |
| 5377.5 | 9115.5 | 16060.0 | |
| 6646.5 | 9359.5 | 16065.6 | |

*NASA coordination*

# U.S. Department of State
## Diplomatic Telecommunications System

With transmitters located at Station C of the National Communications System (NCS), Warrenton Training Center, Remington, Virginia, the State Department uses secure H.F. as backup for their normal satellite links with American embassies worldwide.

This is also the primary site of the US-originated "spy numbers" transmissions. While occasional SSB signals are reported by listeners, the majority of diplomatic transmissions are CW or RTTY.

QRA marker beacons generally identify with the collective callsign KKN50, although discrete callsigns are actually assigned for individual frequencies ("SD," "RUES" and others) and relay sites. KKN39, for example, is in Miami, Florida. Underlined frequencies were recently reported.

**Callsign Freq**

| Callsign | Freq | Callsign | Freq |
|---|---|---|---|
| | 4880 | KKN36 | 15540 |
| | 4886 | | 16355 |
| KKN32 | 6925.5 | | 16363 |
| | 6928 | KKN39 | 17390 |
| | 7470 | | 17413.5 |
| | 10365 | | 17605 |
| | 10470 | | 18460 |
| KKN41 | 10637 | | 18700 |
| KKN42 | 11095 | KKN45 | 18525 |
| KKN43 | 12022.5 | | 18700 |
| | 12023 | KKN47 | 18972 |
| | 12111.5 | KKN48 | 20365 |
| | 12112 | | 20720 |
| KKN39 | 13387 | | 23862.5 |
| | 13646 | KKN51 | 23863 |
| | 14638 | | (RTTY) |
| | 14800 USB(SS) | KKN49 | 23975 |
| KKN35 | 15492 | KKN52 | 23982.5 |

**Probable Target Areas:**

| | |
|---|---|
| KKN44 | Africa |
| KKN39/46/50 | Central/S.America |
| KRH50/51 | Europe |
| KWL90 | Asia |
| KWS78 | Middle East/N.Africa |

*Entrance to the KKN39/NCS transmit facility outside of Miami*

## U.S. Embassies Worldwide Regional Relay Facilities

### Brussels, Belgium
Callsign: ONN31

| | | | | |
|---|---|---|---|---|
| 4548 | 4808 | 4903 | 5110 | 5732 |
| 7726.5 | 10522 | 10680 | 11474 | 12126 |
| 13718 | 14367 | 14548 | 14824 | 16150 |
| 16420 | 17426 | 18348 | 20929 | |

### La Paz, Bolivia
Callsign: KLA24 & KWR86

| | | | | |
|---|---|---|---|---|
| 5270 | 6925 | 7517 | 8035 | 14945 |
| 20123 | | | | |

### Croughton AFB, England (RTTY)
Callsign: KRH51

| | | | | | |
|---|---|---|---|---|---|
| 7315 | 7570 | 7867 | 9147 | 9437 | 9463 |
| 9785 | 9973 | 10740 | 11072 | 11635 | 13673 |
| 13788 | 13842 | 13942 | 14655 | 14930 | 15596 |
| 17512 | 19443 | 19445 | 19942 | 19945 | 23446 |

### Croughton AFB, England (CW)
Callsign: KRH50

| | | | |
|---|---|---|---|
| 3310 | 4589 | 4626 | 5379 |
| 5426 | 6788.5 | 7520 | 7724 |
| 7727 | 7865 | 7875 | 10627 |
| 10680 | 10770 | 11070 | 11142 |
| 13813 | 14450 | 16150 | 16349 |
| 16458 | 20569 | 20571 | |

89

# FEDERAL GOVERNMENT

### Accra, Ghana
Callsign: 9GV

14360

### Athens, Greece (CW/RTTY)
Callsign: KWS78

| | | | | |
|---|---|---|---|---|
| 3186 | 3822 | 3830 | 4435 | 4463 |
| 4468 | 4600 | 4910 | 5113 | 5270.5 |
| 5426 | 6978 | 7434 | 7620 | 7627 |
| 7640 | 7645 | 7652 | 7821 | 7816 |
| 7821 | 7822 | 8127 | 10255 | 10285 |
| 10310 | 11170 | 11220 | 11255 | 11653 |
| 12211.5 | 12268.5 | 12272 | 13776 | 14360 |
| 15704 | 15711 | 15780 | 16149.5 | 18143 |
| 18148 | 18152 | 18351 | 18400 | 18460.5 |
| 18480 | 18545 | 19571 | 23642 | |

### Tehran, Iran (Closed)
Callsign: KWK50

| | | | | |
|---|---|---|---|---|
| 7474 | 7555 | 7645 | 10790 | 15957 |

### Clark AB, Philippines
Callsign: KWL90

| | | | | |
|---|---|---|---|---|
| 4048 | 5443 | 5822.5 | 6866.5 | 7662 |
| 9224 | 10464 | 10900 | 12210 | 13210 |
| 13485 | 13700 | 14616 | 14782 | 16623 |
| 17552 | | | | |

### Monrovia, Liberia
(Note: May be closed)
Callsign: KKN44

| | | | | |
|---|---|---|---|---|
| 4886 | 5110 | 7634 | 7652 | 7830 |
| 11434 | 11474 | 11520 | 11635 | 11995 |
| 15917 | 16158.5 | 17404 | 17426 | 18043 |
| 18348 | 20353 | 20929 | 20950 | 21300 |
| 23425 | | | | |

### Warsaw, Poland
Callsign: KWK97

| | | | | |
|---|---|---|---|---|
| 9229 | 10621 | 11142 | 12105 | 14713 | 15664 |

### Cairo, Egypt
Callsign: KWK95 (CW)

11142

### Unidentified Locations

| Callsign | Freq. |
|---|---|
| KWN90 (CW) | 14360.5 |
| KWM24 (RTTY) | 19270 |
| KWK94 | 14757.5 |
| KWR94 | 18531 |
| KRH70 (CW) | 17426 |
| KRM32 (RTTY) | 4596 |
| KRZ66 (CW) | 14360 |
| KWK94 | 14757.5 |
| KWM24 (RTTY) | 19270 |
| KWN90 (CW) | 14360.5 |
| KWR94 | 18531 |

# Embassies

## Foreign Embassies in Washington, DC

The following frequencies are authorized for use by embassies and diplomatic offices of foreign governments in Washington, DC, for communications with their home offices. Virtually all transmissions will be CW or RTTY. It is common for users to move one or two kilohertz from their assigned frequencies.

Many Washington, D.C., embassies are moving toward satellite communications.

| | | | |
|---|---|---|---|
| 5793.5 | 11448 | 15804 | 18430 |
| 5856 | 12067.5 | 16065 | 18808 |
| 6928* | 12205 | 16125 | 18810 |
| 7321.5 | 13377.5 | 16242 | 18964 |
| 7719 | 13379 | 16255 | 19014.5 |
| 7887.5 | 13605 | 16275 | 19146 |
| 8135 | 13606.5 | 16392 | 19210 |
| 9040.5 | 13626* | 16395 | 19458 |
| 9297 | 14353.5 | 17368 | 19900 |
| 9410 | 14355* | 17375 | 20011 |
| 9440 | 14373 | 17450 | 20021 |
| 9862.5 | 14410 | 17570 | 20527.5 |
| 10640* | 14500 | 17580 | 20920 |
| 10642.5 | 14649 | 17625 | 20958 |
| 10922.5 | 14912.5 | 18055 | 21764 |
| 11090 | 14989 | 18169 | 21773 |
| 11105 | 15519 | 18248.5 | 21811.5 |
| 11166 | 15542 | 18250 | 22960 |
| 11423 | 15704 | 18306.5 | 22966.5 |
| | | | 23995 |

*Embassy pool worldwide

*Antennas atop Russian's lofty embassy in Wash DC*

# FEDERAL GOVERNMENT

## Embassies Worldwide

Diplomatic relations among governments of the world are maintained on a daily basis through embassies, consulates and missions around the globe. Radiocommunications provide a vital link in this chain of peace.

The following listings have been compiled from off-the-air monitoring, private correspondence and official ITU records. Frequencies often vary 2-3 kilohertz from registered listings. Paired frequencies are indicated where known.

### Algeria

Mode: CW/RTTY (50 baud)
Frequencies:

10100 10390 10703 10720 10730/11418
10096/11475 11005 13655 11463 14989
15918 16634 17368 18145/20150 18306
20190

| Callsign | Location |
|----------|----------|
| 7RP20 | Paris, France |
| 7RP30 | London, England |
| 7RP40 | Brussels, Belgium |
| 7RP50 | Geneva, Switzerland |
| 7RP70 | Rome, Italy |
| 7RP80 | Bonn, Fed.Rep.Germany |
| 7RP90 | Madrid, Spain |
| 7RQ20 | Min of Foreign Affairs Algeria |
| 7RQ30 | Beijing, Peoples Rep.China |
| 7RQ40 | Pyongyang,Dem.P.Rep.Korea |
| 7RQ70 | Islamabad |
| 7RQ80 | Tehran, Iran |
| 7RQ90 | Ankara, Turkey |
| 7RV20 | Moscow,USSR |
| 7RV30 | Stockholm, Sweden |
| 7RV40 | Prague, Czechoslovakia |
| 7RV50 | Sofia, Bulgaria |
| 7RV60 | Belgrade, Yugoslavia |
| 7RV70 | Warsaw, Poland |
| 7RV80 | Berlin, German Dem.Rep. |
| 7RX40 | Tripoli, Libya |
| 7RX60 | Khartoum, Sudan |
| 7RX70 | Niamey, Niger |
| 7RP50 | Geneva, Switzerland |
| 7RW20 | Cairo, Egypt |
| 7RW40 | Amman, Jordan |
| 7RW60 | Baghdad, Iraq |
| 7RW70 | Kuwait |
| 7RW90 | Jeddah, Saudi Arabia |
| 7RY20 | Accra, Ghana |
| 7RY40 | Conakry, Guinea |
| 7RY50 | Dakar, Senegal |
| 7RY60 | Brazzaville, Congo |
| 7RY80 | Bamako, Mali |
| 7RY90 | Lagos, Nigeria |
| 7RZ20 | Washington, USA (also uses KNY35) |
| 7RZ50 | Abu Dhabi,U.Arab Emirates |
| 7RZ90 | Presidential Delegation |
| CME356 | Havana, Cuba |
| ETN52 | Addis Ababa (14525) |

### Angola

Location: Brussels

9942 17458 22848

### Argentina

Frequency: 19330

### Austria

Mode: RTTY/FEC (75 and 100 baud)
Frequencies:

7882 7894 10298 10423 13617 13805 14485
16133 16162 17556 19728 20480 20496
20750 20973

| Callsign | Location |
|----------|----------|
| OEC | Vienna |
| OEC35 | New Delhi, India |
| OEC36 | Beijing, China |
| OEC44 | Tel Aviv, Israel |
| OEC57 | Pretoria, South Africa |
| OEC61 | Rome, Italy |
| OEC64 | Lagos, Nigeria |
| OEC72 | Lisbon, Portugal |

### Belgium

| Callsign | Location | | |
|----------|----------|---|---|
| KNY33 | Washington, DC | 11105 | 16275 |
| | | 18810 | |
| | Ankara, Turkey | 5546 | |
| | Warsaw, Poland | 10322 | |

### Benin

Mode: RTTY (50 baud)

| Freq | Call | Location |
|------|------|----------|
| 23964 | | Cotonou |
| 21854.5 | | Brussels, Belgium |
| 23964 | CME388 | Havana, Cuba |

### Bulgaria

LZG7 -MFA, Sofia, Bulgaria
Mode:RTTY/FEC (50 and 75 baud)

Sofia:

5924 9055 11122 11505 11531 12282 12306
13375-13382 13395 13916 14192 14382
16132 16608 16628 17507 18235 18362
18723 20808 20955 20929.5 23014

| Location | Call | Freq |
|---|---|---|
| Addis Ababa, Ethiopia | ETR45 | 16450 |
| Athens, Greece | | 9055 |
| Baghdad, Iraq | | 19007 |
| Beijing, China | | 23030 |
| | | 18365 |
| Berlin, GDR | | 12420 |
| Berne, Switzerland | | 14433 |
| Bogota, Colombia | | 13955 |
| Bonn, Fed.Rep.Germany | | 10804 |
| Brussels, Belgium | ONN29 | 13380 |
| Lisbon, Portugal | | 16066 |
| Luanda, Angola | | 19042 |
| | | 20948 |
| Madrid, Spain | | 20035 |
| Prague, Czechoslovakia | | 9982 |
| Rabat, Morocco | | 18941 |
| Stockholm, Sweden | | 14374 |
| | | 13523 |
| Tripoli, Libya | | 14482 |
| | | 16602 |
| Tunis, Tunisia | | 13414 |
| Washington, U.S.A. | KNY32 | 13605 |
| | | 14728 |
| | | 16256 |
| | | 18055 |
| | | 20164 |
| | | 22996 |

## China

Mode:RTTY (75 baud)
Beijing (BIM)

| Location | Call | Frequencies | |
|---|---|---|---|
| Aden, Yemen | 70B469 | 16300 | 18320 |
| Brussels, Belgium | ONN28 | 18037 | 21751 |
| Gaborone, Botswana | A2D500 | 8185 | |
| Havana, Cuba | CME360-369 | 10480 | 10760 |
| | | 11550 | 12213 |
| | | 18320 | 18750 |
| | | 19920 | 22818 |
| Harare, Zimbabwe | ZEF81 | 16125 | 17385 |
| | | 19250 | 20970 |
| New Delhi, India | BXT7 | 18969 | |
| Warsaw, Poland | BXO44 | 5731 | 9955 |
| | | 11455 | |
| Unidentified: | BXT1 | 14402 | 20492 |
| | | 14393 | |
| | BXT2 | 18805 | |
| | BXT3 | 19878 | |

| | | |
|---|---|---|
| BXT5 | 10060 | 13310 |
| | 10096 | |
| BXT7 | 18969 | |
| BXT9 | 8917 | 15049 |
| | 23952 | |
| BXT11 | 11362 | 13295 |
| | 18682 | |
| BXT15 | 20012 | 20060 |
| BXT16 | 8334 | |
| BXT18 | 14652 | 16595 |
| | 18685 | 16451 |
| BXT23 | 14562 | 19803 |
| BXT32 | 19501 | |
| BXT39 | 10520 | |
| BXT42 | 16542 | |
| BXT44 | 19821 | |
| BXT49 | 13314 | 13905 |
| | 19662 | |
| BXT58 | 20247 | |
| BXT71 | 13302 | |
| BXM62 | 19542 | |
| BXM72 | 13702 | |
| BXM87 | 19808 | |
| BXR75 | 10078 | |
| BXP51 | 16551 | |

## Congo

Mode: CW and RTTY (50 baud/425 Hz; 75 baud/425 Hz)

| Location | Call | Freq |
|---|---|---|
| Havana, Cuba | CME405 | 18054 |

## Cuba

Mode:CW/RTTY (50 baud/425 Hz)
Frequencies:

8815 9130 10040 10045 0507 10620
10675 11230 12115 12225 12230 12235
12290 12295 12344 12980 13115 13275
13284 13290 13295 13300 13311 13316
13322 13333 13340 13345 13350 13371
13375 13380 13382 13385 13388 13390
13401 13856 13878 13920 13925 13930
13935 13940 13945 13950 13955 13965
13970 13975 13981 13985 13989 13991
14461 14472 14625 14730 14755 14818
14812 14823 14920 14945 14960 14989
14995 15040 15935 16010 16040 16145
16165 16170 16175 16316 16325 16355
16355 16358 16360 16370 16380 16405
16420 16425 16430 16470 17255 17485
17517 17720 17894 18030 18035 18040
18050 18110 18112 18115 18160 18185
18236 18420 18450 18454 18612 18628
18630 18638 18640 18645 18650 18785

# FEDERAL GOVERNMENT

| | | | | | |
|---|---|---|---|---|---|
| 19060 | 19065 | 19160 | 19250 | 19420 | 19425 |
| 19429 | 19585 | 19715 | 19765 | 19785 | 19810 |
| 19815 | 19850 | 19882 | 19945 | 19977 | 19985 |
| 19990 | 20044 | 20050 | 20130 | 20148 | 20155 |
| 20165 | 20419 | 20450 | 20419 | 20455 | 20610 |
| 20695 | 20700 | 20720 | 20735 | 20740 | 20765 |
| 20770 | 20775 | 20805 | 20810 | 20825 | 20830 |
| 20840 | 20850 | 20917 | 20940 | 20945 | 20868 |
| 20874 | 20880 | 20955 | 20990 | 21890 | 21965 |
| 21980 | 22030 | 22740 | 22830 | 23000 | 23181 |
| 23635 | 23860 | 23865 | 24806 | 25041 | |

| Call | Location | Frequencies | |
|---|---|---|---|
| CLP1 | Havana, Cuba | 13284 | 13340 |
| | | 13350 | 13385 |
| | | 13390 | 13925 |
| | | 13945 | 13955 |
| | | 13975 | 13981 |
| | | 13989 | 14920 |
| | | 14989 | 16010 |
| | | 16470 | 17894* |
| | | 18236 | 18454 |
| | | 18653.5 | 19060 |
| | | 19945* | |
| CLP2 | Panama | 8815 | 12344 |
| CLP4 | Bissau, Guinea | | |
| CLP5 | Algiers, Algeria | | |
| CLP6 | Damascus, Syria | | |
| CLP7 | Brazzaville, Congo | | |
| CLP8 | Conakry, Guinea | 20043.7 | |
| CLP9 | Aden, Yemen | | |
| CLP13 | | 12290 | |
| CLP15 | Cotonou, Benin | | |
| CLP16 | Malabo | 13315.9 | |
| CLP18 | Dar es Salaam, Tanzania | | |
| CLP23 | Lagos | | |
| CLP25 | Maputo, Mozambique | 12115 | |
| CLP33 | Addis Ababa, Ethiopia | | |
| CLP38 | Paramaribo, Surinam | | |
| CLP45 | Luanda, Angola | 13375 | |
| CLP55 | Georgetown, Guyana | 12290 | 12230 |
| | | 13115 | 13345 |
| | | 14989 | 14995 |
| | | 18040 | |
| CLP65 | Managua, Nicaragua | 13940 | |
| CLP67 | Baghdad, Iraq | | |

\* CLP1 Listening frequencies

## Czechoslovakia

Mode:RTTY (75 baud)
Prague (OMZ):

| | | | | | | |
|---|---|---|---|---|---|---|
| 4900 | 5204 | 5250 | 6839.5 | 6871 | 7622.5 | 7690 |
| 7960 | 8129 | 9268 | 9278 | 10101 | 10705 | 11400 |
| 11500 | 13397.5 | 14650 | 16100 | 16180 | 16368 | |
| 16395 | 16450 | 18300 | 18320 | 18400 | 19320 | |
| 19330 | 20620 | 21810 | 22800 | 22820 | | |

Embassy Frequencies:

| | | | | | | |
|---|---|---|---|---|---|---|
| 5305 | 5755 | 6860 | 7605 | 7620 | 7628 8130 9104 | |
| 10567 | 10607 | 10618.5 | 14350 | 14355 | 14360 | |
| 14370 | 14390 | 14400 | 14478 | 14498 | 14550 | |
| 14560 | 14628 | 14638 | 14650 | 14959 | 16378 | |
| 18323 | 18362 | 18392 | 18500 | 18902 | 18909 | |
| 19074 | 19458 | 19465 | 20622 | 20632 | 20707 | |
| 20752 | 20778 | 20781 | 20788 | 20810 | 20880 | |
| 20895 | 20900 | 20960 | 20970 | 20980 | 20990 | |
| 22880 | 22886 | | | | | |

| Location | Call |
|---|---|
| Addis Ababa, Ethiopia | ETN40/ ETT90 |
| Algiers, Algeria | 3G5 |
| Ankara, Turkey | 3Q5 |
| Baghdad, Iraq | 7A1 |
| Belgrade, Yugoslavia | |
| Beirut, Lebanon | 3K5 |
| Beijing, China | |
| Bonn, Fed.Rep.Germany | |
| Bucharest, Hungary | 5F3 |
| Cairo, Egypt | 2C6 |
| Conakry, Guinea | 5Q3 |
| Damascus, Syria | 7T1 |
| Hanoi, N.Vietnam; Nairobi, Kenya; Pyongyang | 2F6 |
| Havana, Cuba | 7L1 |
| Jakarta, Indonesia | |
| Kabul, Afghanistan | |
| London, England | 3P5 |
| Lagos, Nigeria | 6C2 |
| Luanda, Angola | |
| Maputo, Mozambique | |
| Moscow, USSR | 5W3 |
| Montreal, Canada | |
| New Delhi, India | |
| Paris, France | 1S7 |
| Phnom Penh, Cambodia | 6M2 |
| Sofia, Bulgaria | |
| Tehran, Iran | 2D6 |
| Tirana, Albania | |
| Ulan Bator, Mongolia | |
| Warsaw, Poland | 0MZ9 |
| Washington, USA | KNY23 |
| 7719 13377 14649 15704 15804 18430 19458 19468 | |
| Unidentified | 2M8 |
| | 5Y3 |
| | 6M3 |

## Denmark

Mode:SSB/RTTY (TOR-ARQ)
Frequencies:

7452 7462 13491 16401 18586 18592 26131

| Call | Location |
|------|----------|
| OZU25 | Copenhagen |
| OZU30 | Cairo, Egypt |
| OZU33 | Tel Aviv, Israel |
| OZU34 | Jeddah, Saudi Arabia |
| OZU35 | Beijing, China |
| OZU37 | Ankara, Turkey |
| OZU38 | Lagos, Nigeria |
| OZU39 | Nairobi, Kenya |

## Egypt

Mode:CW/TOR-ARQ

CW: 9997 15400 15600 16201
ARQ: 6726 6754 8217 8302 8932 10402
11240 11250 13342 13352 13502 14652
14902 16302 16352 16562 16802 18014
18050 18062 18802 19002 19452 20002
20022 20038 20054 20102 20402 20792
23106

| Location | Call | Freq |
|----------|------|------|
| Amman | | 14652 |
| Bucharest, Hungary | | 8302 |
| Jeddah | | 13342 |
| Kuwait | | 15014 |
| Madrid, Spain | | 13507 |
| Nairobi, Kenya | | 23001 |
| Paris, France | | 20007 |
| Prague, CSSR | OMZ26 | 16298.5 |
| Washington,USA | KNY29 | 9297 11106 14353.5 18808 |

## Ethiopia

13377 19425 20375 20458
ADL MFA, Addis Ababa
KNY44    Washington, DC

## Finland

Mode:SSB/RTTY
Frequencies:
7778 10737 10748 10886 13434 13914
14662 14694 14817 14926 14955 15727
15871 16135 17419 17444 17562 17576
17603 18762 18842 20125 20187 20259
20264 20177

| Call | Location |
|------|----------|
| OHU20 | Helsinki |
| OHU21 | Paris, France |
| OHU22 | Bucharest, Roumania |
| OHU23 | Tel Aviv, Israel |
| OHU24 | Cairo, Egypt |
| OHU25 | Beijing, China |
| OHU26 | Belgrade, Yugoslavia |
| OHU27 | Addis Ababa, Ethiopia (also ETN66) |
| OHU28 | Warsaw, Poland |
| OHU30 | Lagos, Nigeria |
| OHU31 | Dar es Salaam |
| OHU32 | Jeddah, Saudi Arabia |
| OHU33 | Hanoi, Vietnam |
| OHU34 | Lusaka, Zambia |
| OHU35 | Tehran, Iran |
| OHU36 | Baghdad, Iraq |
| OHU38 | Tripoli, Libya |
| CME390-394 | Havana, Cuba |

## France

Mode:RTTY (50 baud)

Paris (P6Z):
5125 9811 10725 10714 10732 10792 11572
13951 14485 15873 16260 18760 19632

| Location | Call | Frequencies | |
|----------|------|------|------|
| Algiers, Algeria | H6L | 10426 | 11046 |
| | | 11042 | 11406 |
| | | 12215 | 14548 |
| | | 16078 | 16082 |
| | | 16088 | |
| Brussels, Belgium | ONN27 | 8176 | 15996 |
| | | 21840 | |
| Belgrade,Yugoslavia | G8T | 11573 | 14438 |
| | | 14442 | |
| Bucharest, Roumania | A9C | 13946 | 19636 |
| Budapest, Hungary | D6Z | 10560 | 10755 |
| Fort de France | | 16107 | |
| Jeddah | O6F | | |
| Madrid, Spain | AND3 | 9052 | |
| Rabat, Morocco | J5W | 14403 | |
| Tel Aviv, Israel | W5E | | |
| Tunis, Tunisia | K4X | 14400 | 14405 |
| | | 14417 | 14443 |
| Warsaw, Poland | H7K/SRZ944 | 10707 | 10714 |
| | | 10730 | 11042 |
| | | 14548 | |

# FEDERAL GOVERNMENT

## East Germany (GDR)

Mode:RTTY (50, 75, 100 baud)
Berlin:

| | | | |
|------|-------|------|-------|
| Y7A22 | 4020 | Y7A59 | 14619 |
| Y7A23 | 4480 | Y7A60 | 14818 |
| Y7A24 | 4840 | Y7A61 | 15911 |
| Y7A25 | 4880 | Y7A62 | 15963 |
| Y7A26 | 5830 | Y7A63 | 16199 |
| Y7A27 | 5868 | Y7A64 | 16243 |
| Y7A28 | 6706 | Y7A65 | 16268 |
| Y7A29 | 6803 | Y7A66 | 16308 |
| Y7A30 | 6974 | Y7A67 | 16358 |
| Y7A31 | 6987 | Y7A68 | 17126 |
| Y7A32 | 7538 | Y7A69 | 17128 |
| Y7A33 | 7647 | Y7A70 | 17161 |
| Y7A34 | 7812 | Y7A71 | 17367 |
| Y7A35 | 8008 | Y7A72 | 17390 |
| Y7A36 | 9062 | Y7A73 | 17395 |
| Y7A37 | 9078 | Y7A74 | 17670 |
| Y7A38 | 9086 | Y7A75 | 19242 |
| Y7A39 | 9140 | Y7A76 | 19390 |
| Y7A40 | 9784 | Y7A77 | 19445 |
| Y7A41 | 9910 | Y7A78 | 19700 |
| Y7A42 | 9920 | Y7A79 | 19885 |
| Y7A43 | 10210 | Y7A80 | 20020 |
| Y7A44 | 10430 | Y7A81 | 20170 |
| Y7A45 | 10464 | Y7A82 | 20840 |
| Y7A46 | 10554 | Y7A83 | 22722 |
| Y7A47 | 11000 | Y7A84 | 22732 |
| Y7A48 | 11448 | Y7A85 | 22793 |
| Y7A49 | 11459 | Y7A86 | 22950 |
| Y7A50 | 11541 | Y7A87 | 22955 |
| Y7A51 | 11574 | Y7A88 | 23460 |
| Y7A52 | 12130 | Y7A89 | 24000 |
| Y7A53 | 13436 | Y7A90 | 24300 |
| Y7A54 | 13538 | Y7A91 | 24800 |
| Y7A55 | 13946 | Y7A92 | 26450 |
| Y7A56 | 14410 | Y7A93 | 26800 |
| Y7A57 | 14460 | Y7A94 | 27400 |
| Y7A58 | 14560 | | |

Embassy Frequencies:

| | | | | | |
|------|------|------|------|------|------|
| 5793.5 | 6787 | 6828 | 6982 | 9065 | 9145 | 11441 |
| 11676 | 14880 | 16355 | 17362 | 17375 | 19385 |
| 19393 | 19395 | 19241 | 19495 | 19500 | 20004 |
| 20023 | 20506 | 20728 | 21842 | 22729 | 23342 |
| 24005 | | | | | |

| Call | Location |
|------|----------|
| AND20 | Madrid, Spain |
| H6A21 | Managua, Nicaragua |
| Y7B23 | Moscow, USSR |
| Y7B25 | Bucharest, Hungary |
| Y7B26 | Sofia, Bulgaria |
| Y7B27 | Tirana, Albania |
| Y7B32 | Belgrade, Yugoslavia |
| Y7B33 | Helsinki, Finland |
| Y7B45 | Nicosia, Cyprus |
| Y7B72 | Paris, France |
| Y7B73 | Rome, Italy |
| Y7B84 | Ankara, Turkey |
| Y7C24 | Prague, Czechoslovakia |
| Y7C31 | Warsaw, Poland |
| Y7C65 | Vienna, Austria |
| Y7C66 | Berne, Switzerland |
| Y7C69 | Stockholm, Sweden |
| Y7C71 | London, England |
| Y7C81 | Copenhagen, Denmark |
| Y7D35 | Conakry, Guinea |
| Y7D39 | New Delhi, India |
| Y7D42 | Luanda, Angola |
| Y7D49 | Dar es Salaam, Tanzania |
| Y7F44 | Damascus, Syria |
| Y7F47 | Baghdad, Iraq |
| Y7F54 | Algiers, Algeria |
| Y7F59 | Tripoli, Lybia |
| Y7F68 | Beirut, Lebanon |
| Y7G22 | Beijing, PRC |
| Y7G29(45) | Pyongyang |
| Y7G34 | Hanoi, Vietnam |
| Y7G41 | Ulan Bator (Tokyo relay) |
| Y7G59 | New Delhi, India |
| Y7L36 | Havana, Cuba |
| KNY37 | Washington, DC |
| | 8135 11166 19210 |
| ONN30 | Brussels, Belgium |
| CUV523 | Lisbon, Portugal |

## West Germany

Mode:CW/RTTY

| Freq | Call | Location |
|------|------|----------|
| 16432 | DMK | Bonn |
| 20417 20422 | DFY | Tel Aviv, Israel |
| 5220 7983 13894 | DMK260 | Warsaw, Poland |

## Ghana

Mode:RTTY (50 baud)

| Location | Frequencies | |
|----------|-------------|---|
| Accra | 18890 | 20948 |
| Prague, Czechoslovakia | 18890 | 20948 |
| London, England | 18890 | 20948 |
| Washington, USA (KNY22) | 17580 | |

## Great Britain

Mode: RTTY

| Location | Frequencies |
|----------|-------------|
| Washington, USA (KNY 29/30) | 11106 18808 |
| | 18964 |

## Hungary

Mode: CW/RTTY (50 baud)

Budapest:
2450 4500 5450 5470 5475 5725 6762 6765
6795 6799 6800 6802 7937 8174 9046 9055
9125 10318 10325 10393 10400 11460
11462 11464 12257 12605 13374 13385
13396 14377 14490 14809 14814 14820
14829 15487 15720 15728 15988 16398
16400 16430 16457 17503 18045 18051
18060 18065 18068 18392 18402 18585
19230 19514 20001 20011 20073 20398
20802 21755 23363 23370 23385

Embassy Frequencies:
5475 7442 8160 9051 9301 10398
10853 12254 12260 12265 16432 16443
16452 16460 18435 18500 19512 20014
22809 23360

| Call | Location |
|------|----------|
| HGX20 | Hanoi, Vietnam |
| HGX21 | Budapest |
| HGX22 | Athens, Greece |
| HGX23 | Rome, Italy |
| HGX27 | Berne, Switzerland |
| HGX28 | Vientiane, Laos |
| HGX29 | Paris, France |
| HGX31 | Warsaw, Poland |
| HGX38 | London, England |
| HGX39 | New Delhi, India |
| HGX41 | Damascus, Syria |
| HGX42 | Tehran, Iran |
| HGX44 | Baghdad, Iraq |
| HGX45 | Islamabad, Pakistan |
| HGX49 | Beirut, Lebanon |
| HGX53 | Accra, Ghana |
| HGX54 | Conakry, Guinea |
| HGX55 | Algiers, Algeria |
| HGX56 | Cairo, Egypt |
| HGX57 | Rabat, Morocco |
| HGX60 | Khartoum, Sudan |
| ONN39 | Brussels, Belgium |
|  | 8160 9046 12257 |
| CME339-341/377 | Havana, Cuba |
|  | 14815 16458 17460 12260 |
| 8EH39 | Jakarta, Indonesia |
|  | 14809 19800-19805 |
| KNY26 | Washington, USA |
|  | 9040 10642.5 11090 15728 |
|  | 16065 16392 17573 20011 |

## India

Mode: CW/RTTY (50 baud)
Frequencies:
10024 10145 11987 13486 13496 13523
15482 16372 16397 16418 18320 19007
19016 19022 19050 19055 19435 19440
20525 20610 20888

| Location | Callsign |
|----------|----------|
| New Delhi | 8WD4 8WD6 8WD7 |
| Belgrade, Yugoslavia | 8WB1 |
| Moscow, USSR | 8WW3 |
| Port Louis, Mauritia | 8WB3 |
| Tehran, Iran | 8WB4 |
| Bangkok, Thailand | |
| Hanoi, Vietnam | |
| Rangoon, Burma | |
| Dhaka, Bangladesh | |

## Indonesia

Mode: SSB/RTTY (50/57 baud)
Frequencies:
10900 12168 15918 16399 17480 17850
18050 18317 18321 18403 18448 18488
18498 18505 18689 18701 18707 18808
19006 19020 19094 19101 19133 20398
20405 20908 20938 20977 20988 20996

Locations

| | |
|--|--|
| Algiers, Algeria | Lagos, Nigeria |
| Cairo, Egypt | Mexico City, Mexico |
| Hanoi, Vietnam | New Delhi, India |
| The Hague, Netherlands | Paramaribo, Surinam |
| Jeddah, Saudi Arabia | Prague, Czechoslovakia |
| Tehran, Iran | Tunis, Tunisia |

## Iran

| Location | Callsign | Frequencies |
|----------|----------|-------------|
| Brussels, Belgium | ONN34 | 12238.5 16108.5 |

## Ireland

| Location | Callsign | Frequency |
|----------|----------|-----------|
| Brussels, Belgium | ONN37 | 8105.5 |

# FEDERAL GOVERNMENT

## Israel

Mode:ARQ
Frequencies: 13690  24300   Jerusalem

## Italy

Mode:ARQ
Frequencies:
```
    4050  6771  7960  8115  9445  10230  10415
    10482 10547 11086 11181 11441 11449
    11461 11422 12311 13759 13781 13914
    13926 13936 14513 14522 14657 14695
    14710 14734 14738 14821 14841 15781
    15806 15811 15831 16100 16301 16331
    18052 18337 18402 18421 19321 19514
    19518 19816 20161 20321 20327 20712
    20722 22711 23272
```

| Location | Callsign | Frequency |
|---|---|---|
| Addis Ababa, Ethiopia | 9EU76 | 21755 |
| Algiers, Algeria | | 13761 |
| Baghdad, Iraq | | 20327 |
| Beirut, Lebanon | | 15831 18052 |
| Belgrade | IPG27 | |
| Islamabad | | 15784 |
| Jeddah, Saudi Arabia | | 18402 |
| Mogadishu, Somalia | | 19816 |
| Nicosia, Cyprus | | 18421 |
| Rabat, Morocco | | |
| Tripoli | IPG71 | |
| Tunis, Tunisia | | |
| Warsaw, Poland | IPG22 | |

## Korea

Mode: RTTY (50 baud)
Frequencies:
```
    13347 13416.8 14696 15768 19231 19635
    20095 20420
    21863.5 North Korean Embassy, Havana,
    Cuba
```

## Kuwait

| Location | Callsigns |
|---|---|
| Amman, Jordan | 9KE317 |
| Baghdad, Iraq | 9KE205 9KE217 9KE238 |
| | 9KE246 9KE261 9KE275 |
| Beirut, Lebanon | 9KE222 9KE251 9KE280 |
| | 9KE296 9KE302 9KE326 |
| Cairo, Egypt | 9KE247 9KE263 9KE293 |
| | 9KE316 9KE341 9KE361 |
| Damascus, Syria | 9KE221 9KE249 9KE279 |
| | 9KE292 9KE304 9KE325 |
| Jeddah, Saudi Arabia | 9KE295 (11416 kHz) |

| | |
|---|---|
| Khartoum, Sudan | 9KE305 9KE338 9KE387 (20722 kHz) |
| New Delhi, India | 9KE278 |
| Tehran, Iran | 9KE245 |

## Libya

| Location | Callsign | Frequencies |
|---|---|---|
| Aden, Yemen | 70B452 | 11125 11985 |
| Brussels, Belgium | ONN42 | 11508  20682 |
| | | 20732 20932 |
| Havana, Cuba | CME406 | 11508 |
| Prague, Czechoslvka | OMZ30 | 4500  11508 |
| | | 14387  18297 |
| | | 20682 |

## Mongolia

Mode: RTTY (50 baud)

| Location | Call | Frequencies |
|---|---|---|
| Ulan Bator JTF23 | | 17538 17542 |
| | | 17554-17585 |
| | | 18667 |
| Prague, Czechoslovakia | OMZ24 | 17541  18668 |
| | | 18933 |

## Morocco

Mode: CW and 50 bauds standard RTTY
Frequencies: 11433 20378
Stations:
```
    Cairo      L6E
    Rabat      J5W
    Unidentified:  E7U F9O H9N O4T S4K
```

## Netherlands

Mode: SSB/ARQ
Frequencies:
```
    7700  7714  7800  13412  13610  13620
    13770 15645 15727 15966 15987 18462
    18499 18643 18707 18715 18743 19012
    19124 19303 19730 21843 23549 23560
    23570 25599
```

| Call | Location |
|---|---|
| PCW1 | The Hague |
| PCW2 | Tel Aviv, Israel |
| PCW3 | Beijing, China |
| PCW4 | Jakarta, Indonesia |
| PCW5 | Cairo, Egypt |
| PCW6 | Jeddah, Saudi Arabia |
| PCW7 | Damascus, Syria |
| PCW8 | Tripoli, Libya |
| PCW9 | Lagos, Nigeria |

| | |
|---|---|
| PCW10 | Islamabad, Pakistan |
| PCW11 | Bangkok, Thailand |
| PCW12 | Kinshasa, Zaire |
| PCW13 | Tehran, Iran |
| PCW16 | Accra, Ghana |
| PCW20 | Warsaw, Poland |
| Z2H6 | Harare, Zimbabwe |
| - | Ankara, Turkey |
| - | Baghdad, Iraq |
| - | Beirut, Lebanon |
| - | Khartoum, Sudan |
| - | Moscow, USSR |
| - | Paramaribo, Surinam |
| - | Yaounde, Cameroon |

## Niger Republic

Mode:SSB

| Location | Frequency |
|---|---|
| Niamey | 15941.5 |
| Algiers, Algeria | 15941.5 |

## Nigeria

Mode: SSB/RTTY (TOR-ARQ)

| Location | Call | Frequency |
|---|---|---|
| Lagos | | 23104 |
| Addis Ababa, Ethiopia | ETS43,ETW21 | 19463 |
| | | 23104 |
| Warsaw, Poland | SRZ911 | 21846 |
| | | 23104 |
| Washington, USA | KNY36 | 8135 11166 |
| | | 19210 |

## Norway

Mode: SSB/ARQ
Frequencies: 13386 16251 18401 20129 21763 21953

| Call | Location |
|---|---|
| LJA20 | Oslo |
| LJA21 | Tel Aviv, Israel |
| LJA22 | Nairobi, Kenya |
| | Ankara, Turkey |
| | Dar es Salaam, Tanzania |

## Pakistan

Mode: SSB/RTTY (50 baud)
Frequencies:

```
13252 14365 14370 14377 14407 16202
16351 18973 19015 19022 19747 19905
20010 20028 20099 20369 20373 20381
21856 21876 21882
```

| Location | Callsign |
|---|---|
| Islamabad | ASP32 |
| Karachi | ASK21 |
| Abu Dhabi, United Arab Emirates | |
| Algiers, Algeria | |
| Amman | |
| Ankara, Turkey | ASS6 |
| Baghdad, Iraq | |
| Beijing, China | |
| Beirut, Lebanon | |
| Belgrade, Yugoslavia | |
| Bonn, Fed.Rep.of Germany | |
| Cairo, Egypt | |
| Damascus, Syria | |
| Dhaka, Bangladesh | ASP23 |
| Doh'a | |
| Dubai, United Arab Emirates | |
| Geneva, Switzerland | |
| Jeddah, Saudi Arabia | |
| Kathmandu, Nepal | |
| Kuala, Lumpur | |
| Kuwait | |
| London, England | |
| Mogadishu, Somalia | |
| Moscow, USSR | ASM41 |
| Muscat | |
| New Delhi, India | |
| Paris, France | |
| Pyongyang (Berlin relay) | 9047.5 |
| Riyadh | |
| Rome, Italy | ASZ57 |
| San'a | |
| Stockholm, Sweden | ASN69 |
| Tehran, Iran | |
| Tripoli, Libya | |

## Philippines

| Location | Frequencies |
|---|---|
| Tel Aviv, Israel | 14678.5 14694 17365 |

## Poland

Mode: CW/RTTY (50 and 75 baud)
Warsaw:

```
5431  6780  6991  7737  8154  9079
9112  9162  9881  10423 10538     11145
11464 12078 12124 13699 13722 13872
14448 14772 15701 16127 16240 16269
16350 16366 17430 18063 18572 19428
19617 19809 20319 20322 20473 20740
20939 20970 21847 23952
```

# FEDERAL GOVERNMENT

| Location | Callsign | Frequencies |
|---|---|---|
| Accra, Ghana | SPP51 | 17336 23157 |
| Addis Ababa, Ethiopia | ETK47/ | 11478 20375 |
| | ETT37 | 20624 |
| Algiers, Algeria | SPP312 | |
| Baghdad, Iraq | | 18572 |
| Bangkok, Thailand | SPP317 | |
| Beijing, PRC | | 10484 |
| Beirut, Lebanon | | 10832 11037 |
| Berlin, German Dem.Rep. | | 5300 |
| Berne, Switzerland | | 12131 |
| Brussels, Belgium | SPP322 | 10900 |
| Damascus, Syria | | 11123 14832 |
| Havana, Cuba | CME301- | 9481 10484 |
| | 304 | 24500 |
| Islamabad, Pakistan | SQL47-49 | 19668 19707 |
| | | 20245 |
| Jakarta, Indonesia | | 18662 20568 |
| Jeddah, Saudi Arabia | SQR61 | 16358 |
| Kinshasa | | 24330 |
| Lagos, Nigeria | SPP341 | 12300 19676 |
| | | 20915 |
| Luanda, Angola | SPP342 | 19973 20797 |
| Maputo, Mozambique | | 20205 |
| Montreal, Canada | | 14681 |
| Moscow, USSR | | 11447 |
| New Delhi, India | SPP5/ | 20473 24573 |
| | SQL45 | 18114 |
| Nairobi, Kenya | | 24612 |
| Ottawa, Canada | CYS22 | 15582 18092 |
| | | 23376 |
| (Paired freqs to | | 5194/5398 |
| | | 7791/8097 |
| Warsaw shown) | | 11688/10942 |
| | | 12060/18092 |
| | | 14684/14448 |
| | | 15990/18187 |
| | | 16024/16413 |
| | | 17532/18060 |
| | | 18355/18842 |
| | | 20030/21817 |
| Paris, France | | 13710 14731 |
| Phnom Penh | SPP351 | |
| Stockholm, Sweden | | 8123 10183 |
| Tehran, Iran | | 15595 19704 |
| Vienna, Austria | SRT52 | 6936 7933 |
| Washington, USA | KNY20 | 15804 19458 |

## Roumania

Mode: RTTY/CW

| Location | Callsign | Frequencies |
|---|---|---|
| Abidjan, Ivory Coast | TUR19 | 11590 15500 |
| | | 18150 |
| Brussels, Belgium | ONN33 | 13550 16250 |
| Havana, Cuba | CMR342- | 9966 10550 |
| | 348 | 11980 15960 |
| Washington, USA | KNY25 | 9040 11090 |
| | | 16065 16392 |
| | | 19950 |

## Saudi Arabia

Ottawa, Canada:
Callsign: VCR976
Frequencies: 5806 14745 15595 17448 17624
23100

## Senegal

Mode: SSB
            Frequencies: 18120  13295

## Spain

Mode:SSB & RTTY

| Freq | Location |
|---|---|
| 14537 14542 | Madrid |
| 17435 | Yaounde, Cameroon & Nouakchott, Mauretania |
| 18205 18210 | Malabo, Equatorial Guinea |
| 19295 | Madrid |

## Sweden

Mode: SSB/RTTY (75 and 100 baud)
Frequencies:
    6980  7605  7972 10150  10164  10958
   12101 12225 12304 12310 14962 14972
   17428 18758 18803 18810 20010 20942
   20954 20964 20986 23210 23505 23512
   23526 23532 23584 26660 25035

| Call | Location |
|---|---|
| SAM20 | Athens, Greece |
| SAM21 | Berlin, German D.R. |
| SAM23 | Helsinki, Finland |
| SAM24 | Copenhagen, Denmark |
| SAM25 | Lisbon, Portugal |
| SAM26 | London, England |
| SAM30 | Madrid, Spain |
| | (also AND5) |

# FEDERAL GOVERNMENT

| | |
|---|---|
| SAM34 | Harare, Zimbabwe |
| SAM35 | Belgrade, Yugoslavia |
| SAM36 | Budapest, Hungary |
| SAM37 | Bucharest, Roumania |
| SAM38 | Moscow, USSR |
| SAM39 | Prague, Czechoslovakia |
| SAM40 | Warsaw, Poland |
| SAM41 | Berlin, FRG |
| SAM45 | Ankara, Turkey |
| SAM46 | Baghdad, Iraq |
| SAM47 | Beirut, Lebanon |
| SAM48 | Damascus, Syria |
| SAM49 | Jeddah, Saudi Arabia |
| SAM50 | Cairo, Egypt |
| SAM51 | Tehran, Iran |
| SAM52 | Tel Aviv, Israel |
| SAM53 | Amman, Jordan |
| SAM54 | Vientiane, Laos |
| SAM55 | Bangkok, Thailand |
| SAM57 | Dhaka, Bangladesh |
| SAM58 | Jakarta, Indonesia |
| SAM59 | Hanoi, Vietnam |
| SAM60 | Islamabad, Pakistan |
| SAM61 | New Delhi, India |
| SAM62 | Beijing, China |
| SAM64 | Pyongyang, N.Korea |
| SAM65 | Seoul, S.Korea |
| SAM66 | Neutral Nations Supervisory Commission, S. Korea |
| SAM67 | Colombo, Sri Lanka |
| SAM70 | Addis Ababa, Ethiopia |
| SAM71 | Dar es Salaam, Tanzania |
| SAM72 | Kinshasa, Zaire |
| SAM73 | Lagos, Nigeria |
| SAM74 | Monrovia, Liberia |
| SAM75 | Nairobi, Kenya |
| SAM76 | Pretoria, South Africa |
| SAM77 | Tripoli, Libya |
| SAM78 | Tunis, Tunisia |
| SAM79 | Lusaka, Zambia |
| SAM80 | Gaborone, Botswana |
| SAM81 | Bissau, Guinea Bissau |
| SAM82 | Maputo, Mozambique |
| SAM83 | Santiago, Chile |
| SAM84 | United Nations, NY, USA |
| SAM86 | Mexico City, Mexico |
| SAM88 | Buenos Aires, Argentina |
| SAM89 | Havana, Cuba |
| KNY34 | Washington, USA 11106 14355 18808 The Hague, Netherlands |

## Switzerland

Mode: RTTY
Frequencies:
5768 7650 7657 7675 9166 10960
10962 10969 13580 13695 13982 16109
18248 18270 18285 22963 13605 18425
22977

| Call | Location |
|---|---|
| HBN25 | Berne |
| HBD20 | Berne |
| HBD21 | Rome, Italy |
| HBD22 | New Delhi, India |
| HBD49 | Budapest, Hungary |
| HBD57 | Bonn, GDR |
| HBD63 | London, England |
| HBM41 | Punmunjon, Korea |
| HBD41 | Pretoria, South Africa |
| HBD56 | Warsaw, Poland |
| ONN38 | Brussels, Belgium |
| KNY27 | Washington, USA 10642 13605 18250 22960 Paris, France Budapest, Hungary |

## Thailand

Mode: SSB/RTTY
Frequencies:
13300 13745 13865 13918 16352 18812
20103 20109 20735 20908 21876 21948

| Location | Callsign |
|---|---|
| Bangkok | HSF212BKK |
| Bonn, Fed.Rep.of Germany | HSF212BNN |
| New Delhi, India | HSF212NDH |
| Washington, USA 12205 15542 19015 21773 | HSF212WSN |

## Tunisia

Mode: RTTY (50 baud)
Tunis 6CV / 0MP
Frequencies:
2755 2855 5556 5566 5621 7676 7776 7979
8866 8888 8898 16285

## Turkey

Mode: CW/RTTY
Frequencies:
11123 13931 13942 13815 14600-14800
16455 18965 19965 19703-19705 20235

# FEDERAL GOVERNMENT

| Call | Location |
|------|----------|
| SFA/SFB2/ | |
| SFC | Ankara |
| YMK | Beirut, Lebanon |
| YMK1 | Damascus, Syria |
| YMK2 | Nicosia, Cyprus |
| YMK4 | Athens, Greece |
| YMK5 | Sofia, Bulgaria |
| YMK6 | Tel Aviv, Israel |
| YMK7 | Cairo, Egypt |
| YMK8 | Bucharest, Roumania |
| YMK9 | Amman, Jordan |
| YML1 | Rome, Italy |
| YML2 | Belgrade, Yugoslavia |
| YML3 | Baghdad, Iraq |
| YML4 | Vienna, Austria |
| YML5 | Prague, Czechoslovakia |
| YML6 | Brussels, Belgium |
| YML7 | Moscow, USSR |
| YML8 | Alger, Algeria |
| YML12 | Paris, France |
| YML13 | Kabul, Afghanistan |
| YME | Stockholm, Sweden |
| YMO | Kuwait |
| YMN | New Delhi, India |
| YMS | Warsaw, Poland |
| YMN2 | London, England |
| YMN4 | Copenhagen, Denmark |
| YMN5 | Karachi, Pakistan |
| YMN6 | Ministry of Foreign Affairs, Ankara |
| YMN8 | Tehran, Iran |
| YMN9 | Tirana, Albania |
| YMN12 | Madrid, Spain |
| ONN32 | Brussels, Belgium (15657) |

## U.S.S.R.

Mode:CW/RTTY

| Location | Call | Frequencies |
|----------|------|-------------|
| Madrid, Spain | AND8 | 13965 |
| Luanda, Angola | D3M74 | 20688 |
| Washington,USA | KNY31 | 5856  12065  14915 16242 10925 20525 |
| Dublin, Ireland | RLX | 4893  4903  5181 5744  6842  9217 9842 11148  12211  14426 17528 |
| Warsaw, Poland | RXO | 4530  5410  6972 11152 12030 |

## Vietnam

| Location | Call | Frequency |
|----------|------|-----------|
| Addis Ababa, Ethiopia | ETM85 | 13850 |
| Aden, Yemen | 70B455 | 11650 |
| Other frequencies | 8823.9 | 18161.2 |

## Yemen

| Location | Callsign | Frequencies |
|----------|----------|-------------|
| Addis Ababa, Ethiopia | ETM99 | 13997 |

## Yugoslavia

Mode: RTTY (50, 75 and 100 baud)

| Location | Callsign | Frequencies |
|----------|----------|-------------|
| Belgrade ("RGB") | | 6824 14913 |
| Addis Ababa, Ethiopia ("MGE") | ET183,ETT95 | 9840 20950 |
| Athens, Greece | | 9973 |
| Brasilia, Brazil | | 20413 |
| Dar es Salaam ("RDS") | | 21008 |
| Helsinki, Finland | | 11411 14662 |
| Kinshasa ("UCU") | | |
| Lusaka ("ZGA") | | 21003 21010 |
| Nairobi, Kenya ("WUZ") | | 20957 |
| Nicosia, Cyprus | | 13378 |
| Ottawa, Canada | VCS838 | 7445  8172 10780  12295 14515  15705 16459 18247 |
| Paris, France | | 14912 |
| Prague, Czechoslovakia | | 9982 |
| Tehran, Iran | | 18968 |
| Warsaw, Poland | SRZ922 | 7678  8120 9985 10110 |
| Washington, USA | KNY21 | 7719  11305 13375 14649 14875 15704 18430 19225 20001 20149 |
| Other diplomatic freqs | | 11145 18042 18972 10434 |

## Zaire

| Location | Callsign | Frequencies |
|----------|----------|-------------|
| Brussels, Belgium | ONN25 | 20348 |

## Pine Gap Monitoring Station

The United States government maintains an intelligence-gathering monitoring station at Pine Gap, Australia. Licensed as VL5TY, the station maintains an HF link with Clark Field, Philippines, on the following frequencies:

4048 5920 6766 6866 8093 8170 9224 9915 10587 12210 14782 15895 17523 17623 23355 23690

## Spy Numbers Stations

Many countries of the world utilize the shortwave spectrum to send encoded transmissions to foreign agents in adversarial countries. These are most often encountered by listeners as four- or five-digit numbers in female voice with computerized cadence since the recorded voice is accessed digit-by-digit via a keyboard. An optical sleeve with a recording of the appropriate spoken language is slid over a cylinder and read by a sensor each time the corresponding key is pressed on a keyboard. Any language can be originated in this manner. The voice is known among American agents as "Cynthia from Langley", a play on the letters C, I and A; CIA is headquartered at Langley, Virginia.

The format is usually a callup ("Attention") followed by a callsign trinome (three numbers repeated at the beginning of each transmission) which contains the majority of the message. The first two numbers identify the agent and the third is his basic instructions; for example, "240" would be directed to agent 24.

The third numeral, in this case a zero, indicates a dummy message. Other third numerals indicate (approximately): (1) report to pickup point; (2) read message and follow their instructions; (3) arrange meeting and read message for instructions; (4) prepare for trip but no instructions; (5) prepare for trip and read message for instructions; (6) meet contact at safe house; (7) caution, prepare to evacuate; (8) evacuate according to set procedures; (9) destroy all and leave immediately the best way.

The five-digit code groups are done on a trigraph matrix (three-way look-up table) and, whether five straight digits or so-called "3/2" sets, are interpreted the same: one group indicates the line in the decoding book.

After the header group, the first three digits of the next group apply to the matrix, either 3 x 12 or 6 x 6 columns and rows, followed by the last two digits of that group and first digit of the next group.

Many of these transmissions are practice or dummy and are used for propagation testing. The operative travels to two or three cities over a period of time and sends a reception report back to headquarters indicating best times and frequencies, noting overall signal quality and presence of interference.

More than one agent may monitor the same transmission at different locations to determine the best schedule for future messages. Frequencies may be paired to assure reception.

These test transmissions account for many of the repeat messages and simulcasts reported by listeners over various periods of time.

Transmissions usually begin on quarter hours, with on-the-hour, repeated on the half hour, the most common. The mode is normally reduced-carrier upper sideband. Morse code numbers transmissions are also heard; five-digit groups on the hour and four-digits groups on the half hour. They equate as: A=1, N=2, D=3, U=4, W=5, R=6, I=7, G=8, M=9 and T=0.

Four-digit cut-number transmissions are also heard and equate as: A=1, U=2, V=3, 4=4, E=5, 6=6, B=7, D=8, N=9 and T=0.

Four-digit Spanish language and 3/2 digit English language broadcasts originate from Remington, Virginia (National Communications System

*Dozens of towers and masts in Remington, VA, home of US-based "spynumbers" broadcasts and US State Department's KKN50 CW/RTTY transmissions.*

# FEDERAL GOVERNMENT

transmitter site which also houses the KKN50 State Department tranmitter), and the State Department's KKN39 transmitter site on Krome Avenue in Miami, Florida.

Frequencies used for these broadcasts include 4307, 5090, 5812, 6840, 11532 and 16450 kHz.

The Morse code stations that end their broadcasts with either three or five cut zeroes (long dashes) are KGB; they often originate from the GRU communications facility just outside Havana, Cuba, as do five-digit Spanish and English language transmissions.

Czech language transmission most likely originate from Vratislawice, Czechoslovakia. Most of the female-voice German language stations are run not by the East Germans, but by the Hungarian security services.

The most active five-digit Spanish language frequencies at this writing include: 3690, 4025, 4028, 4030, 4044, 4050, 4055, 4125, 4780, 4825, 5060, 5070, 5080, 6835, 6840, 7320, 7527, 7532, 8212, 9122, and 11467 kHz.

German language numbers transmission (beginning "Achtung"!) are commonly reported on 3228, 4543, 5015, 8173, 9450, 10177, 11108, 13775 and 15610 kHz.

For many years, the East German STASI (Staats-Sicherheit or state security) operated a five-digit numbers station at Willmersdorf near Berlin, transmitting on 3220 kHz. In January, 1990, the site was occupied and partially destroyed by demonstrators, ending the transmissions. Another numbers station is at Nauen, East Germany.

Quite probably the most overt of these broadcasts are made by the West German PTT (Deutches Bundespost) at Frankfurt. Prefaced by a voice ID ("Hier ist DFC37" or "Hier ist DFD21") on 3370 or 4010 kHz and followed by a sequence of 20 musical tones, the numbers transmission begins.

Variants on the "numbers" format include the phonetic transmissions like "kilo papa alpha (one or two)" preamble heard on 7445 kHz, probably from the Israeli Mossad intelligence organization. Besides KPA, ART has been reported on 3415 and 5437 kHz; CIO on 6790, 9325 and 9965 kHz; and YHF has been heard on 5820 and 7918 kHz. If the number 2 is heard in the preamble rather an 1, no message will be sent.

A list of recently-monitored numbers transmission schedules follows. Since these schedules are subject to frequent changes, no claim for present accuracy is made. We would appreciate corrections from our readers.

## Abbreviations and Symbols

CW  Morse Code
CZ   Czechoslovakian
EE   English
FF   French
GG  German
KPA2, MIW2, etc = Phonetic Stations
RR   Russian
SS   Spanish
3-2  Three digits, pause, two digits
4,5  Number of characters in a group
//   Two or more simultaneous transmissions

Su, Mn, etc = days of the week

NOTE:  Frequencies often vary as much as 2 kHz.

*Miniature "spy" radio*

| Time/<br>Freq | Language/<br>Group | Paired Freq/<br>Day Heard |
|---|---|---|
| **0000** | | |
| 4010 | SS/5 | |
| 4030 | SS/5 | |
| 4125 | SS/5 | |
| 4638 | EE/3-2 | |
| 4670 | VLB2 | //7605, 12950 |
| 5047 | EE/3-2 | |
| 5090 | CW/5 | //6840 WD |
| 6227 | SS/5 | |
| 6276 | SS/5 | //5015 Su Tu |
| 7404 | GG/5 | |
| 7445 | EE/5 | KPA2 Tu |
| 9074 | SS/4 | //11,532 Tu Wd Th Fr |
| 9972 | GG/5 | |
| 11110 | GG/5 | |

| | | |
|---|---|---|
| 11532 | SS/4 | //9074 |
| 13437 | SS/5 | |

## 0030

| | |
|---|---|
| 4030 | SS/5 |
| 4040 | SS/5 |
| 4050 | SS/5 |
| 4125 | SS/5 |

## 0100

| | | |
|---|---|---|
| 3040 | SS/5 | |
| 3445 | SS/5 | |
| 4027 | SS/5 | |
| 4080 | SS/5 | |
| 4125 | SS/5 | |
| 4320 | SS/5 | |
| 4640 | SS/5 | |
| 4670 | XLE2 | //7605, 12950 |
| 5812 | SS/4 | |
| 6861 | CW/5 | |
| 6890 | SS/4 | |
| 7000 | GG/5 | |
| 7415 | CW/5 | |
| 7445 | KPA2 | |
| 7605 | XLE2 | //4670, 12950 |
| 9222 | SS/4 | //11532 Fr |
| 11543 | SS/5 | |
| 11563 | SS/5 | |
| 12301 | SS/4 | Th |
| 12950 | XLE2 | //7605, 4670 |
| 14454 | SS/5 | |
| 14622 | GG/5 | |

## 0130

| | |
|---|---|
| 3665 | SS/5 |
| 4030 | SS/5 |
| 4444 | SS/5 |
| 5060 | SS/5 |
| 5070 | SS/5 |
| 5080 | SS/5 |
| 5135 | SS/5 |
| 5280 | SS/5 |
| 7605 | VLB2 |

## 0200

| | | |
|---|---|---|
| 3290 | SS/5 | |
| 3445 | SS/5 | |
| 4025 | SS/5 | |
| 4030 | SS/5 | |
| 4037 | SS/5 | |
| 4044 | SS/5 | |
| 4050 | GG/5 | |
| 4100 | SS/5 | Tu |
| 4125 | SS/5 | |
| 4210 | SS/5 | |
| 4300 | SS/5 | Wd Fr |

| | | |
|---|---|---|
| 4640 | SS/4 | Fr Su |
| 5060 | SS/5 | |
| 5070 | SS/5 | |
| 5080 | SS/5 | |
| 5135 | SS/5 | |
| 5413 | EE/5 | |
| 5642 | SYN2 | |
| 5812 | SS/4 | //8418 Fr |
| 5870 | CW/5 | |
| 5916 | CW/5 | Sa Th |
| 6790 | CIO2 | |
| 6840 | SS/4 | //9958 DAILY |
| 6865 | CW/5 | Sa |
| 6870 | SS/4 | Wd |
| 6963 | EE/3-2 | //8176 |
| 6980 | SS/4 | Mn |
| 6990 | SS/4 | Wd |
| 7376 | CW/5 | |
| 7445 | KPA2 | |
| 7828 | SS/5 | |
| 7887 | SS/5 | |
| 8175 | GG/5 | |
| 8872 | CW/5 | Sa |
| 9050 | GG/5 | |
| 9430 | GG/5 | Wd |
| 9435 | EE/5 | Su |
| 9972 | EE/5 | |
| 10247 | EE/5 | Sa |
| 10382 | SS/5 | Sa |
| 10990 | SS/5 | Sa |
| 11580 | SS/5 | |
| 11610 | SS/5 | SaT |
| 13485 | SS/5 | |

## 0230

| | | |
|---|---|---|
| 3444 | SS/5 | |
| 4030 | SS/5 | |
| 4040 | SS/5 | |
| 4044 | SS/5 | |
| 4780 | SS/5 | |
| 7606 | EE/VLB2 | |
| 10820 | EE | Su |
| 10860 | EE/5 | |

## 0300

| | | |
|---|---|---|
| 3227 | RR/5 | OM |
| 3270 | SS/5 | |
| 3290 | SS/5 | |
| 3445 | SS/5 | |
| 4025 | SS/5 | |
| 4030 | SS/5 | |
| 4054 | SS/5 | |
| 4200 | SS/5 | Wd |
| 4300 | SS/5 | Th |
| 4442 | SS/5 | |
| 4610 | RR/5 | |
| 4630 | EE/4 | |

# FEDERAL GOVERNMENT

| | | |
|---|---|---|
| 5182 | SS/5 | Mn |
| 5413 | EE/3-2 | //7740 |
| 5812 | SS/5 | |
| 6784 | EE/5 | Fr |
| 6790 | CIO2 | |
| 6812 | SS/5 | |
| 6835 | SS/5 | |
| 6840 | CW/4 | //11605 DAILY |
| 6963 | EE/5 | |
| 7000 | GG/4 | |
| 7025 | SS/5 | |
| 7430 | SS/5 | Mn |
| 7445 | KPA2 | |
| 7500 | SS/5 | Sa |
| 8173 | GG/5 | |
| 8430 | SS/5 | |
| 9052 | RR/5 | |
| 9074 | SS/4 | //11532 |
| 9075 | GG/5 | M T |
| 9265 | GG/5 | |
| 9265 | EE/5 | |
| 9380 | SS/5 | |
| 9445 | SS/5 | |
| 10135 | SS/5 | Sa |
| 11520 | SS/5 | |
| 11532 | SS/4 | |

## 0330

| | |
|---|---|
| 3075 | SS/5 |
| 3444 | SS/5 |
| 4030 | SS/5 |
| 4040 | SS/5 |
| 4445 | SS/5 |
| 6225 | SS/5 |
| 6835 | SS/5 |

## 0400

| | | |
|---|---|---|
| 3125 | SS/5 | |
| 3135 | SS/5 | |
| 4030 | SS/5 | |
| 4728 | SS/5 | |
| 6226 | SS/5 | |
| 6235 | GG/5 | Sa |
| 6250 | SS/5 | Mn Wd |
| 6270 | GNZ2 | |
| 6772 | SS/5 | |
| 6784 | SS/5 | Su |
| 6802 | SS/5 | |
| 6880 | SS/5 | |
| 6892 | SS/5 | |
| 7382 | SS/5 | |
| 7410 | RR/5 | |
| 7415 | SS/5 | |
| 7445 | KPA2 | |
| 7527 | SS/5 | |
| 8020 | SS/5 | |
| 8173 | GG/5 | |

| | | |
|---|---|---|
| 8418 | SS/4 | //11532 Su |
| 9074 | SS/4 | //11532 Th |
| 9222 | SS/4 | //11532 Wd |
| 9240 | SS/5 | |
| 9990 | RR | |

## 0430

| | |
|---|---|
| 3444 | SS/5 |
| 4080 | SS/5 |
| 4125 | SS/5 |
| 5190 | SS/5 |
| 5210 | SS/5 |
| 9440 | SS/5 |

## 0500

| | | |
|---|---|---|
| 3102 | SS/5 | |
| 3190 | SS/5 | |
| 3230 | GG/5 | |
| 3292 | SS/5 | |
| 4010 | SS/5 | |
| 4020 | SS/5 | |
| 4025 | SS/5 | |
| 4030 | SS/5 | |
| 4044 | SS/5 | |
| 4060 | SS/5 | |
| 4125 | SS/5 | |
| 4445 | SS/5 | |
| 4825 | SS/5 | |
| 5010 | SS/5 | |
| 5060 | SS/5 | |
| 5065 | SS/5 | |
| 5070 | SS/5 | |
| 5080 | SS/5 | |
| 5095 | SS/5 | |
| 6625 | SS/5 | Tu |
| 6785 | SS/5 | |
| 6825 | GG/5 | |
| 6860 | GG/3-2 | |
| 6925 | SS/5 | |
| 7375 | EE/4 | Sa |
| 7415 | CW/5 | |
| 7428 | SS/5 | Fr |
| 7627 | SS/5 | |
| 7537 | RR/5 | |
| 7918 | YHF2 | |
| 8046 | SS/5 | |
| 8056 | SS/5 | Su |
| 8100 | SS/5 | |
| 8124 | SS/5 | Wd |
| 8186 | SS/5 | |
| 8418 | SS/4 | //11532 Wd |
| 8873 | SS/5 | Mn |
| 9383 | EE/5 | |
| 9452 | SS/5 | |
| 10135 | SS/5 | Sa |
| 10180 | RR/5 | |
| 10344 | SS/5 | |

| | | |
|---|---|---|
| 10519 | SS/5 | |
| 10685 | EE/4 | Wd |
| 13923 | CIO | |

## 0530

| | | |
|---|---|---|
| 3060 | SS/5 | |
| 3125 | SS/5 | |
| 3290 | SS/5 | |
| 4025 | SS/5 | |
| 5070 | SS/5 | |
| 7210 | SS/5 | |
| 7415 | CW/5 | |
| 7862 | SS/5 | |
| 8156 | SS/5 | |

## 0600

| | | |
|---|---|---|
| 2707 | GG/5 | //5015, 11110 |
| 3225 | RR/5 | |
| 3240 | SS/5 | |
| 3444 | SS/5 | |
| 4125 | SS/5 | |
| 4825 | SS/5 | |
| 4881 | SS/5 | |
| 5010 | SS/5 | |
| 5060 | SS/5 | |
| 5135 | SS/5 | |
| 5692 | GG/5 | Sa |
| 5746 | GG/5 | |
| 6625 | SS/5 | Tu |
| 6733 | SS/5 | |
| 6772 | SS/5 | |
| 6872 | SS/5 | Tu Th |
| 6892 | SS/5 | Su |
| 6954 | SS/5 | Wd |
| 6996 | GG/5 | |
| 7375 | GG/5 | |
| 7404 | GG/5 | //5015 Su |
| 7887 | SS/5 | |
| 7908 | SS/5 | Sa |
| 7918 | YHF2 | |
| 8057 | SS/5 | Su |
| 11120 | EE/5 | |
| 11416 | SS/5 | Wd |
| 11532 | SS/4 | //8417 |
| 11634 | SS/5 | |

## 0630

| | | |
|---|---|---|
| 3135 | SS/5 | |
| 3444 | SS/5 | |
| 4010 | SS/5 | |
| 5061 | SS/5 | |
| 5974 | SS/5 | |
| 6853 | SS/5 | |
| 6895 | SS/5 | |
| 7345 | SS/5 | |
| 7688 | SS/5 | |

| | | |
|---|---|---|
| 8112 | SS/5 | |
| 8186 | SS/5 | |

## 0700

| | | |
|---|---|---|
| 3120 | SS/5 | |
| 3290 | SS/5 | |
| 4010 | SS/5 | |
| 4020 | SS/5 | |
| 4022 | SS/5 | |
| 4030 | SS/5 | |
| 4040 | SS/5 | |
| 4835 | SS/5 | Th |
| 4881 | CZ/5 | |
| 5692 | GG/5 | |
| 5746 | GG/5 | |
| 6294 | SS/5 | |
| 6554 | SS/5 | Wd Fr |
| 6625 | SS/5 | Tu |
| 6733 | SS/5 | |
| 6772 | SS/5 | |
| 6853 | SS/5 | |
| 6872 | SS/5 | Tu Th |
| 6892 | SS/5 | Su |
| 6954 | SS/5 | Wd |
| 7375 | GG/5 | |
| 7404 | GG/5 | //11532 |
| 7918 | YHF2 | |
| 8057 | SS/5 | Su |
| 8117 | SS/5 | |
| 8417 | SS/4 | //11532 |
| 11416 | SS/5 | Wd |
| 11532 | SS/4 | Wd |
| 11634 | SS/5 | |

## 0730

| | | |
|---|---|---|
| 3125 | SS/5 | |
| 3290 | SS/5 | |
| 4444 | SS/5 | |
| 6826 | SS/5 | |
| 7532 | GG/3-2 | |
| 8122 | SS/5 | |
| 10675 | SS/5 | |

## 0800

| | | |
|---|---|---|
| 3225 | RR/5 | |
| 5800 | SS/5 | |
| 5837 | SS/5 | |
| 6920 | SS/5 | |
| 6954 | SS/5 | |
| 7375 | GG/5 | |
| 7404 | GG/5 | //5015 Su |
| 7415 | SS/5 | |
| 7845 | SS/5 | |
| 8057 | SS/5 | |
| 11532 | SS/4 | //8417 |
| 11634 | SS/5 | |

# FEDERAL GOVERNMENT

| | | |
|---|---|---|
| 12905 | GG/5 | |
| 14603 | EE/5 | |

**0830**

| | |
|---|---|
| 6971 | SS/5 |
| 7345 | SS/5 |
| 7415 | SS/5 |
| 7445 | KPA2 |
| 7812 | SS/5 |

**0900**

| | | |
|---|---|---|
| 5135 | CW/5 | |
| 5182 | GG/5 | |
| 5410 | GG/5 | Sa |
| 5820 | GG/5 | |
| 6295 | SS/5 | |
| 6874 | SS/5 | |
| 7249 | SS/5 | |
| 7412 | GG | |
| 7415 | CW/5 | |
| 7725 | SS/4 | //10324 DAILY |
| 8418 | SS/4 | //11532 |
| 17170 | MIW2 | |
| 17410 | EZI2 | |

**0930**

| | |
|---|---|
| 6226 | SS/5 |
| 6925 | SS/5 |
| 7887 | SS/5 |

**1000**

| | | |
|---|---|---|
| 4025 | SS/5 | |
| 4030 | SS/5 | |
| 4035 | SS/5 | |
| 4040 | SS/5 | |
| 4044 | SS/5 | |
| 7341 | SS/5 | |
| 11272 | GG/5 | Sa |
| 12905 | GG/5 | |

**1030**

| | |
|---|---|
| 6825 | SS/5 |
| 7415 | SS/5 |
| 7872 | SS/5 |

**1100**

| | |
|---|---|
| 4080 | SS/5 |
| 5540 | CZ/5 |
| 7661 | EE/3-2 |
| 8118 | SS/5 |
| 17170 | MIW2 |

**1130**

| | |
|---|---|
| 4825 | SS/5 |

| | |
|---|---|
| 7661 | EE/3-2 |
| 8113 | SS/5 |
| 9376 | GG/3-2 |

**1200**

| | | |
|---|---|---|
| 4320 | SS/5 | |
| 5410 | GG/5 | |
| 6890 | SS/5 | Fr |
| 7415 | CW/5 | |
| 17170 | MIW2 | |

**1230**

| | |
|---|---|
| 4780 | SS/5 |
| 5080 | SS/5 |
| 5135 | SS/5 |
| 9085 | SS/5 |
| 10240 | RR |
| 15612 | SS/5 |

**1300**

| | | |
|---|---|---|
| 3290 | SS/5 | |
| 4028 | SS/5 | |
| 4030 | SS/5 | |
| 4040 | SS/5 | |
| 4044 | SS/5 | |
| 4125 | SS/5 | |
| 4825 | SS/5 | |
| 5410 | GG/5 | |
| 17170 | MIW2 | //12747 |

**1330**

| | |
|---|---|
| 3290 | SS/5 |
| 4835 | SS/5 |
| 5190 | SS/5 |

**1400**

| | |
|---|---|
| 4030 | SS/5 |
| 4825 | SS/5 |
| 8062 | GG/5 |
| 12800 | EE/5 |
| 17170 | MIW2 |

**1430**

| | |
|---|---|
| 3290 | SS/5 |
| 5135 | SS/5 |

**1500**

| | | |
|---|---|---|
| 4888 | GG/5 | |
| 7405 | GG/5 | |
| 13854 | SS/5 | |
| 14425 | SS/4 | //17436 |
| 18456 | SS/5 | |

**1530**

| | |
|---|---|
| 4725 | SS/5 |
| 5010 | SS/5 |
| 5080 | SS/5 |

**1600**

| | | |
|---|---|---|
| 5895 | SS/5 | |
| 7662 | GG/5 | |
| 11491 | SS/4 | //16450 |
| 16221 | SS/5 | |
| 20520 | SS/5 | |

**1630**

| | |
|---|---|
| 5080 | SS/5 |
| 5085 | SS/5 |
| 5210 | SS/5 |

**1700**

| | |
|---|---|
| 3371 | GG/5 |
| 4010 | SS/5 |
| 4028 | SS/5 |
| 4825 | SS/5 |
| 4950 | SS/5 |
| 5210 | SS/5 |

**1730**

| | |
|---|---|
| 8558 | GG/4 |
| 16856 | SS/5 |

**1800**

| | |
|---|---|
| 4030 | SS/5 |
| 4825 | SS/5 |
| 5440 | GG |
| 5750 | GG/5 |
| 6370 | GG |
| 6902 | GG |
| 7580 | GG/5 |
| 7692 | GG |
| 8174 | GG |
| 8186 | EE |
| 8312 | GG |
| 8314 | GG |
| 11462 | EE |
| 12158 | SS |
| 12237 | SS/5 |
| 12620 | EE/5 |

**1830**

| | |
|---|---|
| 5190 | SS/5 |
| 5280 | SS/5 |

**1900**

| | |
|---|---|
| 4030 | SS/5 |

**1930** (continued from left)

| | |
|---|---|
| 4078 | SS/5 |
| 5137 | EE |
| 5228 | GG |
| 5230 | GG |
| 5413 | EE |
| 6852 | GG/3-2 |
| 7408 | TGG |
| 7694 | GG |
| 8492 | GG/3-2 |
| 8562 | GG |
| 10134 | GG/3-2 |
| 10254 | GG/3-2 |
| 10734 | GG |
| 11540 | EE/F |
| 12950 | VLB2 //7605, 4670 |
| 16724 | CW/5 |

**1930**

| | |
|---|---|
| 3910 | EE/5 |
| 5135 | SS/5 |
| 5268 | CZ/5 |
| 5280 | SS/5 |
| 7530 | 3-2 |
| 11532 | SS/5 |

**2000**

| | |
|---|---|
| 3290 | SS/5 |
| 3410 | GG/5 |
| 4022 | GG/5 |
| 4547 | GG |
| 5175 | FF |
| 6708 | GG/3-2 |
| 6778 | RR |
| 6780 | CZ |
| 6820 | RR/5 |
| 6855 | GG/5G |
| 6860 | GG/3-2 |
| 6920 | GG |
| 7445 | KPA2 |
| 8120 | GG/4 |
| 8326 | RR/5 |
| 8490 | GG |
| 9120 | GG |
| 10123 | EE |
| 10135 | GG/4 |
| 10403 | GG |
| 11328 | RR/5 |
| 12301 | SS/4 Th |
| 13682 | SS/5 |

**2015**

| | |
|---|---|
| 3820 | GG/5 |

# FEDERAL GOVERNMENT

## 2030

| | | |
|---|---|---|
| 4820 | GG/3-2 | |
| 5212 | SS/5 | |
| 5338 | GG/5 | |
| 15614 | SS/5 | |

## 2100

| | | |
|---|---|---|
| 3230 | RR/5 | |
| 3833 | EE/5 | |
| 4010 | GG/5 | |
| 4022 | GG/5 | |
| 4926 | GG/5 | |
| 5088 | GG | |
| 5316 | GG/3-2 | |
| 5437 | EE/ART | |
| 6366 | GG | |
| 6784 | EE | |
| 7445 | KPA2 | |
| 7740 | RR/5 | |
| 7860 | EE | |
| 8216 | GG | |
| 8218 | GG | |
| 8310 | GG | |
| 8422 | EE | |
| 8425 | SYN2 | |
| 9450 | GG/5 | |
| 9964 | MIW2 | |
| 10128 | GG/5 | |
| 11467 | SS/5 | Sa |

## 2130

| | | |
|---|---|---|
| 9122 | SS/5 | |

## 2200

| | | |
|---|---|---|
| 4025 | SS/5 | |
| 4030 | SS/5 | |
| 5342 | GG/5 | |
| 11300 | EE/4, SS/4 | |
| 12301 | SS/4 | |
| 12325 | SS/5 | |
| 12950 | VLB2 | //17605 |

## 2230

| | | |
|---|---|---|
| 4030 | SS/5 | |
| 5080 | SS/5 | |
| 10666 | SS/4 | |
| 11615 | SS/5 | |

## 2300

| | | |
|---|---|---|
| 3808 | GG/5 | |
| 3820 | GG/5 | Sa |
| 4030 | SS/5 | |
| 4040 | SS/5 | |
| 4050 | SS/5 | |

| | | |
|---|---|---|
| 4055 | SS/5 | |
| 4060 | SS/5 | |
| 4066 | SS/5 | Sa |
| 4395 | GG/5 | |
| 4445 | SS/5 | Sa |
| 4669 | VLB2 | //7605, 12950 |
| 4780 | KPA2 | Sa |
| 4990 | GG/5 | |
| 5330 | EE/3-2 | //7740 |
| 6825 | GG/5 | Su Wd |
| 7445 | KPA2 | |
| 7605 | VLB2 | |
| 8425 | SYN2 | //5641 |
| 9040 | GG/5 | |
| 9074 | SS/4 | //11532 Tu Th |
| 11225 | SS/5 | |
| 11532 | SS/4 | //9074, 14418 |
| 11605 | EE/5 | |
| 12161 | CW/4 | Sa |
| 15651 | SS/4 | //16450 Fr |

## 2330

| | | |
|---|---|---|
| 4271 | PCD2 | |
| 5046 | EE/3-2 | |
| 6840 | SS/4 | |
| 7415 | SS/5 | |

# Alaskan Agencies

**Alaska Dept of Fish and Game**
*(Salmon Season Net)*

3201   3230   3411   4125   5195

| | | |
|---|---|---|
| Bethel | KJK678 | WZK44 |
| St Marys | KE6628 | KET156 |

**Alaska Fish and Wildlife Net** (SSB)

| | |
|---|---|
| 2512 | 17270 |
| 3215 | 22695 |
| 5925 | 22701.5 |
| 8753 | |

### Alaska Bureau of Land Management
(Forest Fire Net)

3223.5  5287.5  5242.5  4863.0

### Alaska Native Health Service Net
(SSB)

3277  3296  3343  3385  5150

### Alaska Emergency/Calling Net
(USB)

5167.5

## Department of the Interior Pacific Trust Territory Net
(USB)

Call sign prefixes KUP, KUS, KWQ
Island groups: Carolines (Micronesia), Marshalls, Marianas

| | | |
|---|---|---|
| 2724 | 5470 | 8115 |
| 3385 | 7875 | 15626 |
| 3990 | 7878 | 16412.5 |
| 5150 | 7935 | 16591.6 |
| 5205 | 8025 | |

## Immigration and Naturalization Service
(Border Patrol)

The once-expansive HF CW/SSB Border Patrol network has been replaced by a VHF/UHF repeater system. Some point-to-point traffic is still heard, however on the HF frequencies.

4617.5
5912.5
9435.0
1165.0
14585.0
14577.5

## International Boundary & Water Commission
(IBWC)

**Rio Grande Border, TX**

3238.5  5326  6783.5  6976  9946

## All Agencies:
### Land Mobile Common, Nationwide
27575     27585

### Disaster Communications
*OPERATION SECURE*
(State Emergency Capability
Using Radio Effectively)

All frequencies 1 kW maximum power, USB;
Contiguous 48 United States only

**Fixed/Base/Mobile**

| | | |
|---|---|---|
| 2326 | 2511 | 7477 |
| 2411 | 2535 | 7480 |
| 2414 | 2569 | 7802 |
| 2419 | 2587 | 7805 |
| 2422 | 2801 | 7932 |
| 2439 | 2804 | 7935 |
| 2463 | 2812 | |
| 2466 | 5135 | |
| 2471 | 5140 | |
| 2474 | 5192 | |
| 2487 | 5195 | |

## Federal Communications Commission (FCC)

The following frequencies are assigned to the FCC for RTTY (60 wpm/400 Hz normal shift) point to point communications and rarely SSB. No CW has been copied recently. Inversion of bits 2 or 3 provide encryption; the code is changed at 0000 UTC. HF repeaters are used.

### Fixed Station Network and Miscellaneous Units

| Chan | Freq | Use |
|---|---|---|
| 1 | 2220 | Mobiles (CW) |
| 1A | 2295 | Monitoring stations (RTTY) |
| 2 | 4483 | Mobiles (CW/SSB) |

# FEDERAL GOVERNMENT

| | | |
|---|---|---|
| 2A | 5133 | Monitoring stations (CW) |
| 2B | 5373 | Monitoring stations (RTTY) |
| 3 | 7790 | Mobiles (CW/SSB/RTTY) |
| 3A | 7604 | Monitoring stations (CW)(Repeater) |
| 4 | 10655 | Mobiles (CW) |
| 4A | 10902 | Monitoring stations (CW/SSB/RTTY) (Repeater) |
| 5 | 13830 | Monitoring stations (CW/SSB/RTTY) (Repeater) |
| 5A | 13493 | Monitoring stations (RTTY) |
| 5B | 13990 | Mob/Mon stations (CW/RTTY/SSB) |
| 5C | 13992 | Monitoring stations (RTTY) |
| 6 | 18050 | Mobiles (CW) |
| 6A | 19230 | Monitoring stations (RTTY) (Repeater) |
| 7 | 22964 | Mobiles (CW) |
| 7A | 23035 | Monitoring stations (RTTY) (Repeater) |

| Call | ID | Location |
|---|---|---|
| KAA60 | GI | Grand Isle, NE (repeater site) |
| KCA35 | BE | Belfast, ME |
| KGA91 | LR | Laurel, MD |
| KGA93 | WA | Washington, DC Headquarters |
| KIA84 | PS | Powder Springs, GA |
| KIA85 | FL | Vero Beach |
| KKA59 | DS | Douglas, AZ |
| KKA60 | KA | Kingsville, TX |
| KMB27 | LV | Livermore, CA |
| KOA56 | FE | Ferndale, WA |
| KQA62 | AL | Allegan, MI |
| KUN70 | WP | Waipahu, HI |
| KWC41 | AN | Anchorage, AK |
| WWQ20 | SS | Sabana Seca, PR |

## Department of Commerce

## National Oceanic and Atmospheric Administration (NOAA)

### Coast Radio Stations

These stations communicate with U.S. Government vessels in the Atlantic Ocean, Gulf of Mexico, Pacific Ocean, Gulf of Alaska, and Hawaii areas.

#### Atlantic Stations

| Stn | Agcy | Location |
|---|---|---|
| KVH | NOS | Norfolk, VA |
| KVK | NOS | Miami, FL |
| KAC | NMFS | Woods Hole, MA |
| KAF | " | Sandy Hook, NJ |

*One of two FCC monitoring positions in Powder Springs, Georgia*

# FEDERAL GOVERNMENT

| KBR | " | Beaufort, NC |
| KHW | " | Pascagoula, MS |
| KAI | " | Galveston, TX |
| KAG | AOML | Miami, FL |

### Pacific Stations

| Stn | Agcy | Location |
|---|---|---|
| KVJ | NOS | Seattle, WA |
| WWD | NMFS | La Jolla, CA |
| KAB | NOS | Honolulu, HI |

### Great Lakes Station

| Stn | Agcy | Location |
|---|---|---|
| KVR | NOS | Detroit, MI |

### Alaska Station

| Stn | Agcy | Location |
|---|---|---|
| KWL | NMFS | Auke Bay, AK |

### USCG Atlantic Area

| Station | Location |
|---|---|
| NMF | Boston, MA |
| NMN | Portsmouth, VA |
| NMA | Miami, FL |
| NMG | New Orleans, LA |

### USCG Pacific Area

| Station | Location |
|---|---|
| NMC | San Francisco, CA |
| NOX | Adak, AK |
| NOJ | Kodiak, AK |
| NMO | Honolulu, HI |
| NMT | Barrow, AK |
| NRV | Guam |
| NMJ | Ketchikan, AK |
| NMQ | Long Beach, CA |

## Agency Abbreviations

AOML Atlantic Oceanographic and Meteorological Laboratories
NMFS National Marine Fisheries Service
NOS National Ocean Survey
OOE Office Ocean Engineering
USCG United States Coast Guard

## Frequency Assignments

### CW Frequencies (Simplex)

| | |
|---|---|
| 4213.5 | 12640.5 |
| 6320.25 | 16641 |
| 8364 | 16854.5 |
| 8427 | 22252 |
| 8427.5 | 22251.5 |

### CW (Two Frequency Simplex)

| Coast Frequency | Ship Frequency |
|---|---|
| 6503.5 | 6265.6 |
| 8714 | 8353 |
| 13080.5 | 12500.5 |
| 17206.5 | 16669.5 |
| 22570.5 | 22201.5 |

### Duplex Radiotelephone (USB)

| Coast Frequency | Ship Frequency |
|---|---|
| 4379.1 | 4084.7 |
| 4431.8 | 4137.4 |
| 4434.9 | 4140.5 |
| 6509.5 | 6203.1 |
| 8753 | 8229.1 |
| 8802.6 | 8278.7 |
| 13141.1 | 12370.3 |
| 13156.6 | 12385.8 |
| 17267 | 16494.1 |
| 17270.1 | 15497.2 |
| 22695.2 | 22099.2 |
| 22701.4 | 22105.4 |

*(NOTE: NOAA vessels also use USB simplex on limited coastal channels 4A-22E - see index)*

## NOAA Callsigns

| Location | Callsign |
|---|---|
| Atlantic Marine Center | WZ3050-59 |
| Pacific Marine Center | WZ3060-70 |
| Office of Fleet Ops | KA8133,34 |
| Geodesy | KE8200-26 |
| | KF3712-3802 |
| Field 742 | WS2535-37 |
| Photogrammetry | WZ2531-34 |
| | KD7998,99 |
| | KE8227-34 |

# FEDERAL GOVERNMENT

|  |  |  |  |
|---|---|---|---|
|  | KF3700-11 | French Frigette Shoals | KAW65 |
| Auka Bay, AK | KWL21 | Pearl & Hermes Reef | KAW64 |
| Anchorage, AK | KWL59 |  | KAW63 |
| College, AK | KAW53, | New Orleans, LA | WZ2538-39 |
|  | KD7992-3 | New York, NY    (MESA) | KAW61 |
| Kodiak, AK | KWL38 |  | WZ2517-19 |
|  | KD8057 | Beaufort, NC | KD8051 |
| Juneau, AK | KWL36 | Rockville, MD | KCU721-27 |
| Sitka, AK | KAW57 | (Security) | KCU729 |
| St George, AK | KWL41 | (Motor Pool) | KCU730 |
| St Paul, AK | KWL44 |  | KA8132 |
| Pt Barrow, AK | KAW54 |  | KD7986-91 |
| Valdez, AK | WZ2520 | Boston, MA | WZ2513-16 |
| Tucson, AZ | KAW51 | Detroit, MI | WVW |
|  | KD7996,7 | (Portable Units) | KD8000-50 |
| Los Angeles, CA | WZ2550-52 | Portland, OR | WZ2570-71 |
| San Diego, CA | KDX | (Same as RAINIER) |  |
| San Francisco, CA | WZ2553-59 | Prescott, OR | KD8055,56 |
| San Miguel Is, CA | KF7424 |  | KD8063 |
| Tiburon | KD3977 | Galveston & Pt Arkansas,TX | KA8129-31 |
| Terminal Island | KHU |  | KD8052-54 |
| Guam |  | Fredericksburg, VA | KAW56 |
| Honolulu, HI | WZ2593-4 | Norfolk, VA | KAW60 |
|  | KD7994-5 | Seattle, WA | KAW56,66 |
|  | KAW52,68 |  | KAW67 |
|  |  |  | KD8058-62 |

*NOAA ship Fairweather*

# FEDERAL GOVERNMENT

## NOAA FLEET

| Vessel | Length Ft | *Port | *Hull Des | Ship Callsign | Callsign | Location |
|---|---|---|---|---|---|---|
| OCEANOGRAPHER | 303 | SE | R101 | WTEP | WZ2633-42 | |
| DISCOVERER | 303 | SE | R102 | WTEA | WZ2608-19 | S.E. Alaska |
| MALCOM BALDRIGE | 278 | MI | R103 | WTER | WZ2540-49 | E. Caribbean |
| SURVEYOR | 292 | SE | S132 | WTES | WZ2560-69 | Oregon Coast |
| FAIRWEATHER | 231 | SE | S220 | WTEB | WZ2645-50 | Gulf of Alaska |
| RAINIER | 231 | SE | S221 | WTEF | WZ2570-86 | Gulf of Alaska |
| MILLER FREEMAN | 215 | SE | R223 | WTDM | | Gulf of Alaska |
| MT MITCHELL | 231 | NK | S222 | WTEG | WZ3015-29 | N. Atlantic |
| PIERCE | 163 | NK | S328 | WTEQ | WZ2598-2602 | Delaware Bay |
| WHITING | 163 | NK | S329 | WTEW | WZ2603-07 | Honduras |
| MCARTHUR | 175 | SE | S330 | WTEJ | WZ3030-38 | Seattle, WA |
| DAVIDSON | 175 | SE | S331 | WTEK | WZ3039-49 | Oregon Coast |
| OREGON II | 170 | PA | R332 | WTDO | | Gulf of Mexico |
| GEORGE B. KELEZ | 177 | NK | R441 | KNBG | WZ2006 | |
| ALBATROSS IV | 187 | WH | R342 | WMVF | WZ2158 | N. Atlantic |
| TOWNSEND CROMWELL | 164 | HU | R443 | WTDF | | Hawaiian Is. |
| DAVID STARR JORDAN | 171 | SD | R444 | WTDK | | S. California |
| DELAWARE II | 156 | SH | R445 | KNBD | WZ2035 | N. Atlantic |
| FERREL | 133 | NK | S492 | WTEZ | WZ2628-30 | Norfolk, VA |
| RUDE | 90 | NK | S590 | WTET | WZ2625-27 | N. Atlantic |
| HECK | 90 | NK | S591 | WTEY | WZ2620-22 | N. Atlantic |
| OREGON | 100 | SE | R551 | KNBH | WZ2022 | |
| JOHN N COBB | 94 | SE | R552 | WMVC | WZ2023 | Gulf of Alaska |
| CHAPMAN | | | R446 | WTED | | Gulf of Mexico |
| MURRE II | | | S663 | KJLM | | Gulf of Alaska |

## Abbreviations

| | | | |
|---|---|---|---|
| R | Research | WH | Woods Hole |
| S | Survey | HU | Honolulu |
| SE | Seattle | SD | San Diego |
| MI | Miami | SH | Sandy Hook |
| NK | Norfolk | KK | Kodiak |
| PA | Pascagoula | | |

*NOTE: WCGS used as a radio callsign for broadcast to all NOAA ships; i.e. National Emergency.*

# FEDERAL GOVERNMENT

## Woods Hole (MA) Oceanographic Institute

KC2XGF to all areas
2398    6970

KXC713 to Atlantis II (KADC)
to Knorr (KCEJ)
to Oceanus (WYN4084)

| | | | | |
|---|---|---|---|---|
| 2613 | 4431.8 | 6509.5 | 8753 | 8802.6 |
| 13141.1 | 16590.2 | 17267 | 22695.2 | |

## National Marine Fisheries Service

Scripps Bucom Fisheries, La Jolla, CA

| Callsign | Frequency |
|---|---|
| WWD | 8789.7 |
| | 17105.0 |

## (NOAA) National Weather Service (NWS)

| | | | |
|---|---|---|---|
| 2463.5 | 6509.5 | 12894 | 17280.9 |
| 2614 | 6516.8 | 13065 | 17902 |
| 2638 | 7880* | 13142 | 22025.9 |
| 3333 | 8282.6 | 13180.4 | 22099.4 |
| 4140.9 | 8646 | 16485.9 | 22106.4 |
| 4380 | 8754 | 16566.4 | 22109.4 |
| 4404.4 | 8800.6 | 16573.4 | 22700 |
| 4433.2 | 12401.4 | 17084 | 22429.4 |
| 6211.8 | 12422.4 | 17175 | 23651.4 |
| 6379 | 12425.9 | 17268 | |
| 6393 | | | |

*\* Weather frequency*

### Coast Stations

| | |
|---|---|
| KAB | Honolulu Ocean Inst. |
| KAF | Atlantic Highlands, NJ |
| KAI | San Juan, PR |
| KBR | Beaumart, SC |
| KHW | Pensacola, FL |
| KJS | Kings Point, NY |
| KVD | Norfolk, VA |
| KVH | Norfolk Atlantic Marine Center |
| KVK | Miami SE Marine Support Facility |
| KVJ | Pacific Marine Ctr, Seattle, WA |
| KVR | Detroit, MI |
| KAW51 | Tucson, AZ |
| KAW52 | Honolulu, HI |
| KAW53 | College, AK |
| KAW54 | Pt Barrow, AK |
| KAW55 | Seattle, WA |
| KAW56 | Fredricksburg, VA |
| KAW57 | Sitka, AK |
| KAW58 | Guam |
| KAW60 | Norfolk, VA |
| HRK6A | Swan Island |

*Hurricane hunter technician preparing a dropsonde*

## National Hurricane Center
*Air to Ground Hurricane Reconnaisance*

Two aircraft are WP-3, variation of Navy P-3 Orion reconnaissance aircraft, a modification of the Lockheed Electra. Aircraft can stay aloft for 12 hours at altitudes up to 24,000 feet, and cruise at 300 knots.

### Frequencies USB

| | |
|---|---|
| 3407 | 10015 |
| 5562 sec. | 11398 |
| 6673 | 13267 (Pri daytime) |
| 8876 | 21937 |

| Callsign | Location |
|---|---|
| KJY74 | Miami, FL |

## NWS Emergency Backup Network
*(Also Regional Fire Net)*

| Ch | Freq |
|---|---|
| 1 | 2776 |
| 2 | 3363 |
| 3 | 5925 |
| 4 | 6977.5 (Pri) |
| 5 | 9947.5 |
| 6 | 14729 |

## TVA Construction Division

| Channel | Frequency |
|---------|-----------|
| 1 | 2094.4 |
| 2 | 4195.0 |
| 3 | 6223.0 |
| 4 | 8295.6 |
| 5 | 12433.7 |

## Federal Aviation Administration (FAA)

The FAA, a participant in the SHARES system, is undergoing a "Phase IV" upgrade of their emergency communications system, allowing full duplex voice and data communications with remote sites at Atlanta and Denver and transportable equipment as well.*

Frequencies are used for emergency, emergency readiness, maintenance and scene of accident investigation. Weekly call-ups keep participating stations honed.

Although USB is the dominant mode, some LSB, ASCII (170 Hz shift, 75/150/300 baud) and even CW (14, 25, 45 WPM) may be heard on the new NARACS (national Radio Communication System) Eastern and Western Region nets.

*The remote console is an AT&T 6300 computerized panel with VP-100 voice privacy and phone patch capability. The radio rack contains a Rockwell/Collins HF-8070A receiver/exciter, HF-8020 1 KW power amplifier with matching HF-8030 power supply, an automatic communications processor and an antenna switching matrix.

An accessory rack includes an AEA PK-232 data controller, dual-channel tape recorder, printer, AT&T 6300 computer, video monitor, telephone and keyboard.

HF antennas include a Granger 3002 Spira-Cone omnidirectional array and a TCI RLPA.

## NARACS Eastern Region Net

(Callup Wednesdays at 10:45 Eastern Time on 8125 kHz)

| Callsign | Location |
|----------|----------|
| KDM50 | Hampton, GA (Net Control) |
| KEM80 | Washington, DC (FAA HQ) |
| KIT88 | Martinsburg, WV |
| KIA21 | Oklahoma City, OK |
| KLN80 | Atlantic City, NJ |
| KJK82 | Jamaica, NY |
| KJK76 | Ronkonkoma, NY |
| KCD73 | Islip, NY |
| KJK80 | Leesburg, VA |
| WHX51 | Des Plaines, IL |
| KJB96 | Aurora, IL |
| KLA25 | Oberlin, OH |
| KLB48 | Indianapolis, IN |
| KCJ20 | Farmington, MN |
| WHX45 | Burlington, MA |
| KLD70 | Nashua, NH |
| KDM49 | Atlanta, GA |
| KJK79 | Hilliard, FL |

# FEDERAL GOVERNMENT

| | |
|---|---|
| KMA47 | Miami, FL |
| KDM52 | Memphis, TN |
| KDM45 | San Juan, Puerto Rico |

Other frequencies include 6870 (ch 4), 13630 (ch 11), 16348 (ch 13) and 20852 (ch 15).

## NARACS Western Region Net

(Callup Wednesdays at 10:45 Pacific Time on 13630 kHz)

| Callsign | Location |
|---|---|
| KCP63 | Longmont (Denver), CO (Net Control) |
| KKU40* | Kansas City, MO |
| KDM47 | Ft. Worth, TX |
| KKY51* | St. Louis, MO |
| KCC97 | Diamond Head, HI |
| KJK77 | Palmdale, CA |
| KJK73 | Los Angeles, CA |
| KMU31 | Houston, TX |
| KMR96 | Fremont, CA |
| WHX20 | Seattle, WA |
| WHX44 | Auburn, WA or Mt. Kaala, HI |
| KDM53 | Anchorage, AK |
| KBX44 | Anchorage, AK |
| WSX70 | Kenai, AK |
| KWB407* | Durango, CO (DOT) |
| KWB409* | Ephrata, WA (DOT) |
| WWJ47* | Kansas City, MO (FHWA) |
| WWJ50 Alpha* | Newport, OR (FHWA) |
| WWJ97* | Salem, OR (FHWA) |
| WWJ98* | Shoshone, ID (FHWA) |

*Occasional SHARES participants

## U.S. Nationwide (USB)

3331   3428   4055   5571   5860   8912
11288   13312   17952

## U.S. and Possessions

4675   7611   9914   11637   13457   13630
15851   16348   19410   27575   27585   27625
27665

Since many of the new frequencies -- and most of the callsigns -- are being drawn from the old southwest, southern, eastern and central nets, that obsolescent list is reprinted here for reference.

### New York/Santa Maria

| New York (WSY70) | Santa Maria, Azores (CSY) |
|---|---|
| 4475 | 3194 |
| 8130 | 5402 |
| 12196 | 8045 |
| 16280 | 12240 |
| 23211 | 16234 |
| 20130 | |

### San Juan - Local Hurricane Net
WRW70

2196   7508.5

### Balboa/Bogota (ISB)

| Balboa, PNZ (WHZ70) | Bogota, Colombia (HKG210) |
|---|---|
| 3161.5 | 2845 |
| 4572 | 4610 |
| 9796.5 | 9950 |

### Balboa/Guayaquil (ISB)

| Balboa, PNZ (WHZ70) | Guayaquil, Ecuador (HC2G) |
|---|---|
| 8119 | 4476 |
| 5162.5 | 6810 |
| 10437.5 | 10163 |
| 12188.5 | 14607.5 |

### Balboa/San Andres

| Balboa, PNZ to San Andres, Colombia | |
|---|---|
| (WHZY0) | (CSZ) |
| 3390   5805   7405 | |

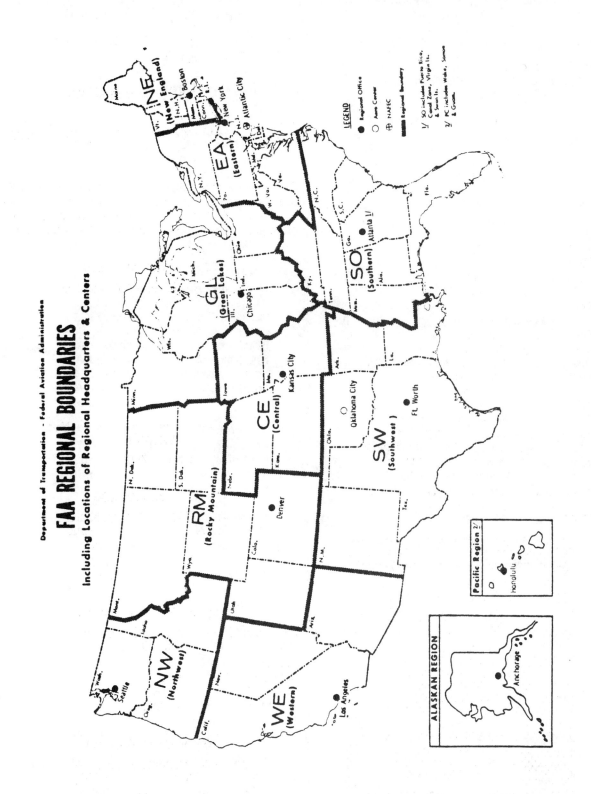

Department of Transportation - Federal Aviation Administration

## FAA REGIONAL BOUNDARIES

Including Locations of Regional Headquarters & Centers

# FEDERAL GOVERNMENT

## Southwest Region Emergency Net

| 3331 | 4055 | 5860 |
|------|------|------|
| 6870 | 8125 | 13626 |

| Station | Callsign |
|---------|----------|
| Albuquerque, NM | KGH23 |
| Albuquerque NM (ARTCC)* | KEA37 |
| Baton Rouge, LA | KAO31 |
| Beaumont, TX | KED67 |
| Brownsville, TX | KCA37 |
| Chickasha, OK | KOS41 |
| Corpus Christi, TX | KJB35 |
| Denton, TX | KKD50 |
| Farmington, NM | KGA46 |
| Fort Worth, TX (RCCC)* | KDM47 |
| Fort Worth, TX (ARTCC)* | KBQ25 |
| Houston, TX | KIC30 |
| Houston,TX(ARTCC)* | KMU31 |
| Lafayette, LA | KCK47 |
| Lake Charles, LA | KCA29 |
| Martinsburg, WV | KIT88 |
| Mesa Rica, NM | KDM44 |
| Midland, TX | KAP24 |
| Mineral Wells, TX | KPA54 |
| Mobile Units | |
| New Orleans, LA | WEK71 |
| Oklahoma City, OK (Center) | KIA21 |
| Palacios, TX | KGC36 |
| San Antonio, TX | KIA48 |
| Santa Fe, NM | KBY33 |
| Shreveport, LA | KOB92 |
| Tucumcari, NM | KAB34 |
| Wichita Falls,TX | KPA54 |

*Scene of accident investigation*

## Southern Region Emergency Net

(Check in Tuesdays 9:45 Eastern on 6870 LSB or 7475 USB)

| 4055 | 7475 |
|------|------|

| Station | Callsign |
|---------|----------|
| Anniston, AL | KLE78 |
| Asheville, NC | KKW51 |
| Atlanta, GA (S.O.) | KDM49 |
| Atlanta, GA | KUV62 |
| Augusta, GA | KUZ66 |
| Bimini, Bahamas | KLB79 |
| Birmingham, AL | KVA93 |
| Bristol, TN | KKW23 |
| Brunswick, GA | KUQ95 |
| Charleston, SC | KUR64 |
| Chattanooga, TN | KAN34 |
| Columbia, SC | KUZ31 |
| Columbus, GA | KVA92 |
| Covington, KY | KUX58 |
| Crestview, FL | KUU58 |
| Crossville, TN | KKX70 |
| Daytona Beach, FL | KXG85 |
| Dotham, AL | KUR68 |
| Elizabeth City,NC | KAO28 |
| Fayetteville, NC | KLJ61 |
| Florence, SC | KUS79 |
| Fort Lauderdale,FL | KKY53 |
| Fort Myers, FL | KWF53 |
| Fulton Co., GA | KLM44 |
| Gainesville, FL | KUA21 |
| Greenwood, MS | KKY54 |
| Greer, SC | KBZ80 |
| Gulfport, MS | KEG95 |
| Hampton, GA(ARTCC) | KDM50 |
| Hattiesburg, NS | KUR87 |
| Hilliard, FL (ARTCC) | KJK79 |
| Huntsville, AL | KVA95 |
| Jackson, MS | KVA94 |
| Jacksonville, FL | KGC34 |
| Jacksonville, FL | WHX49 |
| Key West, FL | KLD95 |
| Knoxville, TN | KFC87 |
| London, KY | KNH47 |
| Louisville, KY | KUW42 |
| Lynch, KY | KEF63 |
| Macon, GA | KVA27 |
| McComb, MS | KLA23 |
| Melbourne, FL | KXC70 |
| Memphis, TN(ARTCC) | KDM52 |
| Memphis, TN | KLE86 |
| Meridian, MS | KDM51 |
| Miami, FL (ARTCC) | KMA47 |
| Mobile,AL | KMJ20 |
| Montgomery, AL | KLC48 |
| Myrtle Beach, SC | KUR65 |
| Nashville, TN | KUV70 |
| New Bern, NC | KUR66 |
| Orlando, FL | KOR82 |
| Panama City, FL | KSE31 |
| Raleigh, NC | KCG58 |
| Raleigh, NC | KCA58 |
| San Juan,PR(ARTCC) | KDM45 |
| San Juan, PR | WRW70 |
| Sarasota, FL | KVH40 |
| Savannah, GA | KBN72 |
| St Croix, VI | KUU97 |
| St Petersburg, FL | WHX48 |
| St Thomas, VI | KSB53 |
| Tallahassee, FL | KKA36 |
| Tampa, FL | KEX63 |
| Tuscaloosa, AL | KLM37 |
| Valdosta, GA | KUR67 |
| Vero Beach, FL | WHX47 |
| Wilmington, NC | KDQ21 |

*A modern Air Traffic Control Center*

## Eastern Region Backup Net

| 3331 | 3353 | 3395 | 4055 |
|------|------|------|------|
| 5860 | 7475 | 8120 | 8125 |

| Station | Callsign |
|---------|----------|
| Albany, NY | KKY71 |
| Buffalo, NY | KKZ75 |
| Hagerstown, MD | KLM92 |
| National Airport, VA | KKW89 |
| Newark, NJ | KLC46 |
| Norfolk, VA | KLD46 |
| Pittsburgh, PA | KKW22 |
| Roanoke, VA | KLD57 |
| Starr Hill, NY | KLM94 |

## Central Region
### Emergency and Defense Readiness Net

| 3353 | 4055 | 7475 | 8125 | 13626 |
|------|------|------|------|-------|

| Station | Callsign |
|---------|----------|
| Chicago, IL | KLG27 |
| Chicago, IL | KLE72 |
| Indianapolis, IN (ARTCC) | KLB48 |
| Farmington, MN (ARTCC) | KCJ20 |
| Kansas City, MO (CE RO) | KKU40 |
| Kansas City, MO (RS) | KBC28 |
| Great Falls, MT (ARTCC) | KIN80 |
| Olathe, KS (ARTCC) | KKA82 |
| Aurora, IL (ARTCC) | KJB96 |

## Inter-Regional
### Emergency and Defense Readiness Net
*(And ARTCC Emergency Net)*

| 4055 | 7475 | 8125 |
|------|------|------|
| 13626 | 16280 | 20852 |

### *Regional Net*

| Station | Callsign |
|---------|----------|
| Washington, DC(HQ) | KEM80 |
| Atlantic City, NJ | KLN80 |
| Martinsburg, WV | KIT88 |
| Jamaica, NY | KJK82 |
| Atlanta, GA (SO) | KDM49 |
| Fort Worth, TX | KDM47 |
| Kansas City, MO | KKU40 |
| Oklahoma City, OK | KIA21 |
| Los Angeles, CA | KJK73 |
| Anchorage, AK | KDM53 |
| Honolulu, HI | KGA32 |
| Ronkonkoma, NY | KJK76 |
| Burlington, MA | WHX45 |
| Denver, CO | WHX46 |
| Des Plaines, IL | WHX51 |
| Seattle, WA | WHX20 |
| Kenai, AK | WSX70 |
| Jacksonville, FL | KJK79 |

# FEDERAL GOVERNMENT

## ARTCC Net

| Station | Callsign |
|---|---|
| Oberlin, OH | KLA25 |
| Nashua, NH | KLD70 |
| Islip, NY | KCD73 |
| Leesburg, VA | KJK80 |
| Fort Worth, TX | KBQ25 |
| Houston, TX | KMU31 |
| Albuquerque, NM | KEA37 |
| Miami, FL | KMA47 |
| Hilliard, FL | KJK79 |
| Hampton, GA | KDM50 |
| Memphis, TN | KDM52 |
| Aurora, IL | KJB96 |
| Indianapolis, IN | KLB48 |
| Olathe, KS | KKA82 |
| Farmington, MN | KCJ20 |
| Great Falls, MT | KIN80 |
| Auburn, WA | KCJ70 |
| Longmont, CO | KCP63 |
| Salt Lake City, UT | KDC20 |
| Palmdale, CA | KJK77 |
| Fremont, CA | KMR96 |

## New England Region Emergency and Defense Readiness Net

3331  3353  3395  4055  5860
7475  8120  8125  13626  16280

| Station | Callsign |
|---|---|
| Augusta, ME | KKU70 |
| Bangor, ME | KKU80 |
| Burlington, MA | WHX45 |
| Houlton, ME | KKU56 |
| Islip, NY (New York,ARTCC) | KCD73 |
| Jamaica, NY | KJK82 |
| Leesburg,VA (Washington,ARTCC) | KJK80 |
| Nashua, NY | KJK81 |
| Portland, ME | KKU54 |
| Ronkonkoma, NY | KJK76 |

## Great Lakes Region Emergency and Defense Readiness Net

4055  7475  8125  13626

| Station | Callsign |
|---|---|
| Aurora, IL | KJB96 |
| Battlecreek, MI | KLM43 |
| Bismarck, ND | KKU50 |
| Cleveland, OH | KLD69 |
| Columbus, OH | KLB79 |
| Des Plaines, IL | WHE51 |
| Detroit, MI | KLE80 |
| Indianapolis, IN | KLB48 |
| Farmington, MN | KCJ20 |
| Grand Rapids, MI | KLD27 |
| Green Bay, WI | KEB24 |
| Lansing, MI | KAE21 |
| Milwaukee, WI | KKY53 |
| Minneapolis, MN | KLD73 |
| North Canton, OH | KLF21 |
| Oberlin, OH | KLA25 |
| Pierre, SD | KKU51 |
| St. Paul, MN | KGJ42 |
| Vandalia, OH | KLA21 |

## Illinois Maintenance Net

| Station | Callsign |
|---|---|
| Moline, IL | KKZ74 |
| Springfield, IL | KBU29 |

## Indiana Maintenance Net

| Station | Callsign |
|---|---|
| Indianapolis, IN | KLD58 |
| South Bend, IN | KKY72 |

## Intra-Alaska Voice Backup Net
### (Weather Collection)

3253

| Station | Callsign |
|---|---|
| Cordova | KAC26 |
| Eklutna | KIS70 |
| Kotzebue | KEJ28 |

## Bettles/Chandalar Lake, AK Off-Airways Weather Collection Net

2758  4055  5357.5

| Station | Callsign |
|---|---|
| Bettles, AK | KBE43 |
| Chandalar Lake, AK | KEE67 |
| Cordova | KAC26 |

## FAA Southeastern Alaska Maritime Mobile Net

4139.5

| Station | Callsign |
|---|---|
| Sisters Island, AK | KAS24 |
| Level island, AK | KQA59 |
| Coghlan Islands,AK | KGC38 |
| Juneau, AK | KAJ28 |
| Gustavus, AK | KGG20 |
| Haines, AK | KAH94 |

# FEDERAL GOVERNMENT

## Alaska FAA Net
### 7475 LSB

| Station | Callsign |
|---|---|
| Anchorage | KBX44 |
| Barrow | KBA29 |
| Bethel | KBB54 |
| Big Delta | KGD82 |
| Cold Bay | KCD20 |
| Cordova | KAC26 |
| Deadhorse | KLK70 |
| Dillingham | KKE47 |
| Fairbanks | KBE94 |
| Farewell | KCF31 |
| Galena | KCG31 |
| Gulkana | KGA25 |
| Homer | KCH20 |
| Ilianna | KAI82 |
| Juneau | KAJ28 |
| Kenar | WSX70 |
| Ketchikan | KKY21 |
| King Salmon | KCK35 |
| Kodiak | KAK20 |
| Kotzebue | KEJ28 |
| McGrath | KCR60 |
| Middleton Is. | KCB20 |
| Murphy Dome | KLH70 |
| Nome | KBN58 |
| Northway Junction | KBL85 |
| Sitka | KAW27 |
| Talkeetna | KAV83 |
| Unalakleet | KAU51 |
| Valdez | KLL45 |
| Yakutat | KKW92 |

## Alaska Regional Scene of Accident
### 3253   5286   5357.5   5906   7375

## Alaska Defense Readiness Net
### 1790   3253   5357.5   7375

## Western Region Emergency Net
### 3353   4055   4060   7475   13903.5

| Station | Callsign |
|---|---|
| Fort Worth, TX | KDM47 |
| Fremont, CA (Oakland ARTCC) | KMR96 |
| Kansas City, KS | KKU40 |
| Los Angeles, CA | KJK73 |
| Los Angeles, CA | KOJ77 |
| Olathe, KS | KKA82 |
| Palmdale,CA(ARTCC) | KJK77 |
| Seattle, WA | WHX20 |

## Pacific Region Emergency, Administration, and Maintenance Net
### 3353   6870

| Station | Callsign |
|---|---|
| Honolulu, HI | KGA32 |
| Hilo, HI | KBH49 |
| Lanai, HI | KLI51 |
| Kahului, Maui, HI | KCU94 |
| Keahole, HI | KLI52 |
| Lihue, Kauai, HI | KAL38 |
| Mt Kaala, HI | WHX44 |
| Diamond Head, HI (ARTCC) | KCC97 |
| Molokai, HI | KLK48 |

## Northwest Region Emergency and Maintenance Net
### 4055   7475

| Station | Callsign |
|---|---|
| Auburn, WA | WHX44 |
| Boise, ID | WHX41 |
| Eugene, OR | WHX25 |
| Idaho Falls, ID | WH42 |
| Klamath Falls, OR | WHX36 |
| Medford, OR | WHX26 |
| Moses Lake, WA | WH44 |
| Olympia, WA | WHX30 |
| Pasco, WA | WHX33 |
| Pocatello, ID | WHX37 |
| Portland, OR | WHX31 |
| Redmond, OR | WHX27 |
| Seattle, WA(RCCC) | WHX20 |
| Spokane, WA | WHX30 |
| Spokane, WA | WHX42 |

## Rocky Mountain Region Emergency Net
### 4055   4060   7475   8125   13626

| Station | Callsign |
|---|---|
| Billings, MT | KIM33 |
| Bismarck, ND | KKU50 |
| Butte, MT | KKV33 |
| Casper, WY | KAI82 |
| Cedar City, UT | KAX36 |
| Denver, CO | KCP63 |
| Fargo, ND | KDM46 |
| Grand Forks, ND | KEA20 |
| Great Falls, MT | KIN80 |
| Great Falls, MT | KKW91 |
| Huron, SD | KBR27 |
| Jackson, WY | KKW27 |
| Longmont,CO(ARTCC) | KLP63 |
| Miles, MT | KKX67 |
| Minot, ND | KGC28 |
| Phoenix, AZ | KPS50 |

# FEDERAL GOVERNMENT

| Rapid City, SD | KCE24 |
|---|---|
| Rock Springs, WY | KKW90 |
| Salt Lake City, UT | KDC20 |
| Sioux Falls, SD | KBC28 |
| Trinidad, CO | KGF97 |

## Pacific Region/Scene of Accident Command and Control Network

6874    14464    20480    27625

| Station | Callsign |
|---|---|
| Honolulu | KGA32 |
| Diamond Head | KCC97 |
| Guam | KCD72 |
| Pago Pago | KDM48 |

## Honolulu to Kwajalein
(FACSIMILE)

Honolulu (KVM70)

9982.5    11090    16135    23331.5

## Samoa/Fiji (ISB)

| Pago Pago, Samoa (KDM48) | Nandi, Fiji (ZQD) |
|---|---|
| 3235.5 | 4027 |
| 5362.5 | 5010 |
| 8122 | 5085 |
| 10985 | 7956 |
| 12180 | 8084 |
| 15674 | 9410 |
| 18764 | 10563 |
| | 10985 |
| | 14950 |
| | 18615 |
| | 23240 |

## Honolulu/Samoa (ISB)

| Honolulu (KVM70) | Samoa (KDM48) |
|---|---|
| 3336 | 3394 |
| 5220 | 5925 |
| 5945 | 7770 |
| 6850 | 9785 |
| 9846 | 11121.5 |
| 9982.5 | 17669 |
| 11471 | 23345 |
| 14885 | |
| 16135.6 | |
| 17621 | |
| 23474 | |

## Samoa Airport Link

| Tafuna (KDM48) | Faleolo (ZKFA) |
|---|---|
| 2544 | 3158 |
| 6895 | 6776 |

## Intra-Alaska Air to Ground
(USB)

FSS service for Barrow, McGrath, Cold Bay, Cordova, Kotzebue, and Ketchikan.

2861    5631

## Tactical Identifiers

SNOWBALL   Tactical identifier for regional offices
TRADEWIND  Tactical identifier for Washington DC administrative offices

# Department of Justice

## FBI

### RTTY/USB Network

*(Tone-encoded Squelch)*

| 2332 | | 5913 | | 9238.5 | XE 13660 | 17603 |
|---|---|---|---|---|---|---|
| 2808.5 | XA | 6594 | | 9311.5 | 14460 | 18173 |
| 4030 | | 6952.5 | | 10550 | 14493.5 | 22345 |
| 4616 | | 7778.5 | XD | 10915 | 14534 | 23402 |
| 5014 | | 7903.5 | | 11075 | 15955 | 23675 |
| 5058.5 | XC | 9015 | | 11210 | 16376 | 23875 |
| 5388.5 | | 9183.5 | | 11488.5 | 17405 | 27740 |

### Callsigns and Locations
*(Test Monday 9-10 am local time)*

| KAG69 | Denver, CO |
|---|---|
| KAG81 | Minneapolis, MN |
| KAG82 | Rapid City, SD |
| KAH63 | St. Louis, MO |
| KCC61 | Boston, MA |
| KCC76 | New Haven, CT |
| KEC86 | Newark, NJ |
| KGD83 | Baltimore, MD |

# FEDERAL GOVERNMENT

| | | | | |
|---|---|---|---|---|
| KGE22 | Quantico, VA | | KKJ45 | Jackson, MS |
| KGE55 | Lorton, VA | | KKJ78 | Little Rock, AR |
| KGG64 | Philadelphia, PA | | KMI66 | Los Angeles, CA |
| KGG76 | Pittsburgh, PA | | KOG83 | Portland, OR |
| KGG83 | Washington, DC | | KQC67 | Cincinnati, OH |
| KGG85 | Washington, DC | | KQC77 | Cleveland, OH |
| KII50 | Columbia, SC | | KQC87 | Detroit, MI |
| KII66 | Norfolk, VA | | KSC81 | Springfield, IL |
| KII74 | Richmond, VA | | KSD73 | Sacramento, CA |
| KII88 | Savannah, GA | | KUR20 | Honolulu, HI |
| KII95 | Jacksonville, FL | | KWX20 | Anchorage, AK |
| KIJ44 | Tampa, FL | | WWR20 | San Juan, PR |
| KKI68 | Dallas, TX | | | |
| KKI73 | El Paso, TX | | | |
| KKI88 | Houston, TX | | | |

## Standard Time/Frequency Stations Worldwide

*(All 24 Hours Except When Shown)*

| Callsign | Location | Frequency and Schedule |
|---|---|---|
| ATA | New Delhi, India | 5000 (1230-0330)  10000  15000 (0330-1230) |
| BPM | Xian, China | 10000 15000 (Weekdays 1600-2200, 0300-0600) |
| BSF | Taiwan, Rep of China | 5000 15000 (0100-0900) |
| CBV | Valparaiso, Chile | 4298 8677 (1155-1200, 15555-1600, 1955-2000, 0055-0100) |
| CHU | Ottawa, Canada | 3330 7335 14670 |
| DCF77 | Mainflingen, FRG | 77.5 |
| DGI | Oranienburg, GDR | 177 (0559:30-0600, 1159:30-1200, 1759:30-1800) |
| EBC | Cadiz, Spain | 6840 (1025-1055) 12008 (0959-1025) |
| GBR | Rugby, UK | 16 (0255-0300, 0855-0900, 1455-1500, 2055-2100) |
| HBG | Prangins, Switzerlnd | 75 |
| HD2IOA | Guayaquil, Ecuador | 3810 (0500-1700); 5000 (1700-1800); 7600 (1800-0500) |
| HLA | Taejon, Korea | 5000 (Weekdays 0100-0800) |
| IAM | Rome, Italy | 5000 (Weekdays 0730-0830, 1030-1130. One hour earlier in summer.) |
| IBF | Turin, Italy | 5000 (Weekdays 0645-0700, 0845-0900, 0945-1000, repeated hourly to 1800. One hour earlier in summer.) |
| JG2AS | Chiba, Japan | 40 |
| JJY | Koganei, Japan | 2500 5000 8000 10000 15000 |
| LOL | Buenos Aires, AR | 5000 10000 15000 (1100-1200, 1400-1500, 1700-1800, 2000-2100, 2300-2400); 4856 8030 17180 (0055-0100, 1255-1300, 2055-2100) |
| LQB9 | Planta Gral Pacheco | 8167.5 (2200-2205, 2345-2350) |
| LQC20 | Planta Gral Pacheco | 17551.5 (1000-1005, 1145-1150) |
| NPO | Subic Bay, Philippines | 4445 10440.5 12804 (0555-0600, 1155-1200, 1755-1800, 2355-2400) |
| OBC | Callao, Peru | 485 8650 12307 (1555-1600, 1855-1900, 0055-0100) |
| OLB5 | Podebrady, Czech. | 3170 |
| OMA | Liblice, Czech. | 50 2500 (Ten minute intervals) |
| PKI | Jakarta, Indonesia | 8542 (0055-0100) |
| PLC | Jakarta, Indonesia | 11440 (0055-0100) |
| PPE | Rio de Janeiro, Brazil | 8721 (0025-0030, 1125-1130, 1325-1330, 1825-1830, 2025-2030, 2325-2330) |
| PPR | Rio de Janeiro, Brazil | 435 4244 8634 13105 17194.4 22603 (0125-0130, 1425-1430, 2125-2130) |
| RBU | Moscow, USSR | 66.666... (For five minutes after each hour, 000-2200) |
| RCH | Tashkent, USSR | 2500 (0530-0400) 5000 (0200-0400, 1400-1730, 1800-0130) 10000 (0530-0930, 1000-1330) |
| RID | Irkutsk, USSR | 5004 10004 15004 |
| RIM | Tashkent, USSR | 5000 (5 min intervals 0000-0130, 0215-0350, 1815-2400); 10000 (5 min intervals 0530-0930, 1015-1330, 1415-1730) |

# FEDERAL GOVERNMENT

| | | |
|---|---|---|
| RKM | Irkutsk, USSR | 10004 (5 min intervals 0000-0110, 1351-2400); 15004 (5 min intervals 0151-1310) |
| RTA | Moscow, USSR | 10000 (0200-0500, 1400-1730 1800-0130) 15000 (0630-0930, 1000-1330) |
| RTZ | Irkutsk, USSR | 50 (For 5 minutes after each hour, 0100-2400) |
| RWM | Moscow, USSR | 4996 9996 14996 |
| RW166 | Irkutsk, USSR | 200 (2200-2100) |
| RW76 | Novosibirsk, USSR | 272 (2200-2000) |
| VPS | Kowloon, Hong Kong | 500 (Even hours) |
| VPS8 | Kowloon, Hong Kong | 4232.5 (Odd hours 1100-2100) |
| VPS22 | Kowloon, Hong Kong | 22536 (Odd hours 0100-0900) |
| VPS35 | Kowloon, Hong Kong | 8539 (Odd hours) |
| VPS60 | Kowloon, Hong Kong | 13020.5 (Odd hours 0100-1500) |
| VPS80 | Kowloon, Hong Kong | 17096 (Odd hours 2100-1300) |
| VWC | Calcutta, India | 434 4286 (1625-1630); 12745 (0825-0830) |
| WWV | Ft Collins, CO | 2500 5000 10000 15000 20000 |
| WWVB | Ft. Collins, CO | 60 |
| WWVH | Kaui, HI | 2500 5000 10000 15000 |
| XSG | Shanghai, China | 522.5 6454 8487 12954 16938 (0255-0300, 0855-0900) |
| Y3S | East Berlin, DDR | 4525 |
| YVTO | Caracas, Venezuela | 6100 |
| ZLF | Wellington, New Zeal | 2500 F(Wednesday 0100-0400) |
| ZSC | Capetown, S.Africa | 418 4291 8461 12724 17018 22455 (0755-0800, 1655-1700) |
| ZUO | Olifantsfontein,S AF | 2500 5000 (1800-0400) |
| 4PB | Colombo, Sri Lanka | 482 8473 (0553-0600, 1323-1330) |
| --- | Belconnen, Australia | 6449 12982 |

## Standard Time Frequency Cross Reference

| Freq | Station Callsigns | |
|---|---|---|
| 16 | GBR | |
| 20.5 | UNW3 UPD8 UQC3 USB2 UTR3 | |
| 23 | UNW3 UPD8 UQC3 USB2 UTR3 | |
| 25 | UNW3 UPD8 UQC3 USB2 UTR3 | |
| 25.1 | UNW3 UPD8 UQC3 USB2 UTR3 | |
| 25.5 | UNW3 UPD8 UQC3 USB2 UTR3 | |
| 40 | JG2AS | |
| 50 | OMA, RTZ | |
| 60 | WWVB | |
| 66.666.. | RBU | |
| 75 | HBG | |
| 77.5 | DCF77 | |
| 177 | DGI | |
| 200 | RW166 | |
| 272 | RW76 | |
| 418 | ZSC | |
| 434 | VWC | |
| 435 | PPR | |
| 482 | 4PB | |
| 485 | OBC | |
| 500 | VPS | |
| 522.5 | XSG | |
| 2500 | BPM JJY OMA RCH WWV WWHV ZLF ZUO | |
| 3170 | OLB5 | |
| 3330 | CHU | |
| 3810 | HD210A | |
| 3842 | VPS8 | |

| Freq | Station Callsigns | |
|---|---|---|
| 4228 | CBV | |
| 4244 | PPR | |
| 4286 | VWC | |
| 4291 | ZSC | |
| 4298 | CBV | |
| 4445 | NPO | |
| 4525 | Y3S | |
| 4856 | LOL2 | |
| 4996 | RWM | |
| 5000 | ATA BSF HD210A HLA IAM IBF JJY LOL1 RCH RIM WWV WWVH ZUO | |
| 5004 | RID | |
| 5430 | BPM | |
| 6100 | YVTO | |
| 6449 | (Belconnen, Australia) | |
| 6454 | XSG | |
| 6840 | EBC | |
| 7335 | CHU | |
| 7500 | VNG | |
| 7600 | HD210A | |
| 8000 | JJY | |
| 8030 | LOL | |
| 8167.5 | LQB9 | |
| 8461 | ZSC | |
| 8473 | 4PB | |
| 8487 | XSG | |
| 8539 | VPS35 | |
| 8542 | PKI | |
| 8558 | CCV | |
| 8634 | PPR | |
| 8650 | OBC | |
| 8677 | CBV | |

| Freq | Station Callsigns | |
|---|---|---|
| 8721 | PPE | |
| 9351 | BPM | |
| 9996 | RWM | |
| 10000 | ATA BPM JJY LOL1 RCH RIM RTA WWV WWVH | |
| 10004 | RID RKM | |
| 10440.5 | NPO | |
| 11173.5 | EBC | |
| 11440 | PLC | |
| 12008 | EBC | |
| 12307 | OBC | |
| 12724 | ZSC | |
| 12745 | VWC | |
| 12804 | NPO | |
| 12954 | XSG | |
| 12982 | (Belconnen, Australia) | |
| 13020.5 | VPS60 | |
| 13105 | PPR | |
| 14670 | CHU | |
| 14996 | RWM | |
| 15000 | ATA BPM BSF JJY LOL RTA WWV WWVH | |
| 15004 | RKM | |
| 16938 | XSG | |
| 17018 | ZSC | |
| 17096 | VPS80 | |
| 17180 | LOL | |
| 17194.4 | PPR | |
| 17551.1 | LQC20 | |
| 20000 | WWV WWVH | |
| 22455 | ZSC | |
| 22536 | VPS22 | |
| 22603 | PPR | |

# International Criminal Police Organization (INTERPOL)

Virtually every western country and one Communist nation--Yugoslavia--belong to this mutual-assistance informational exchange.

While INTERPOL is forbidden by its charter to engage in religious, political, racial or espionage cases, nearly all other classes of criminal investigation are pursued.

Member countries provide National Central Bureaus for cooperative exchange of information relating to drugs, counterfeiting, fugitives and other illegal activities. In England, Scotland Yard provides the facilities and in the United States, the Treasury Department.

Virtually 100% of the HF communications are conducted by RTTY (TOR-ARQ and FEC). Underlined frequencies are most often reported. U.S. participation is via cable and satellite.

## Frequencies

| | | | |
|---|---|---|---|
| 2593 | 7401 | 13820 | 19405 |
| 2840 | 7532 | 14607.5 | 21785 |
| 3593 | 7832 | 14707 | 21807.5 |
| 3705 | 7906 | 14817.5 | 24072 |
| 3714 | 8038 | 14827 | 24110 |
| 4444 | 8045 | 15502.5 | 27845 |
| 4632.5 | 9105 | 15592 | |
| 4837.5 | 9200(Pri) | | |
| 4855.5 | 9285 | 15684 | |
| 5104 | 9821 | 15738 | |
| 5208 | 10295 | 18190 | |
| 5305.5 | 10390(Pri) | 18380(Pri) | |
| 5895 | 11538 | 18755.8 | |
| 6792 | 13520 | 19130 | |
| 6905 | 13747 | 19360 | |

## Alphabetic Callsign List - National Central Bureaus

| Callsign | Station-Bureau |
|---|---|
| AVD204 | Ranchi, Argentina |
| AYA47 | Buenos Aires, Arg. |
| CNP/CNT | Rabat, Morocco |
| CSJ26 | Lisbon, Portugal |
| DHA33/DEB | Wiesbaden, FDR |
| DUN356 | Manila, Phillipines |
| EEQ | Madrid, Spain |
| EEQ20 | Las Palmas, Canary Is |
| EP5X | Tehran, Iran |
| FSB57 | HQ Paris, France |
| GMP | West Wickam, Eng. |
| HEP39/58 | Zurich, Switzerland |
| HMA22 | Seoul, Korea |
| HK3M-TI | Bogota, Colombia |
| HSQ | Bangkok, Thailand |
| IUV81 | Rome, Italy |
| JPA21-27 | Tokyo, Japan |
| JPA56 | Nagoya, Japan |
| LJP20 | Oslo, Norway |
| LXF50 | Luxembourg |
| LZH7 | Sofia, Bulgaria |
| OAV84-87 | Lima, Peru |
| ODW22 | Beirut, Lebanon |
| OEO35 | Vienna, Austria |
| OGX | Helsinki, Finland |
| ONA20 | Brussels, Belgium |
| OWS4 | Copenhagen, Denmark |
| PDB2 | Utrecht, Holland |
| PPC55/PYZ2 | Brasilia, Brazil |
| SHX | Stockholm, Sweden |
| SUA81 | Cairo, Egypt |
| SXP | Athens, Greece |
| TCC2 | Ankara, Turkey |
| TTR103-148 | N'Djemena, Chad |
| TUW220 | Abidjan, Ivory Coast |
| VRD | Hong Kong |
| XJD48 | Ottawa, Canada |
| XJE57 | Almonte, Canada (RCMP) |
| YO99 | Bucharest, Rumania |
| YVZ32 | Caracas, Venezuela |
| ZPZ | Asuncion, Paraguay |
| 3VA | Tunis, Tunisia |
| 4NX7-25 | Belgrade, Yugoslavia |
| 4XP41 | Tel Aviv, Israel |
| 4XP63 | Jerusalem, Israel |
| 5BP6 | Nicosia, Cyprus |
| 50P25 | Lagos, Nigeria |
| 5TP25 | Nouakchott, Mauritania |
| 5YG | Nairobi, Kenya |
| 7RA20 | Algiers, Algeria |
| 8UF75 | New Delhi, India |
| 9TK21 | Kinshasa, Zaire |
| EEQ21 | Santa Cruz, Canary Is |

# FEDERAL GOVERNMENT

## CANADA

### Royal Canadian Mounted Police
(Mostly RTTY, some SSB; gradually being phased out)

| KHZ | PROVINCE |
|---|---|
| 2788 | Alberta NWT |
| 4765 | NWT |
| 4776.5 | NWT Alberta Nfld |
| 4785 | Yukon NWT BC |
| | Alta Man Ont PQ |
| | NS WashDC |
| 4798.5 | BC Manitoba |
| 4812.5 | Sask |
| 5445 | Alta Sask Man |
| | Ont PQ |
| 6792 | Paris |
| 7780 | Yukon NWT BC Alt |
| | Sask Man Ont PQ |
| | NB NS Nfld Paris |
| 9105 | NWT Alta Ont NB |
| | NS Nfld Paris |
| 9200 | Paris |
| 10390 | Paris |
| 14620 | Yukon,NWT,Alta, |
| | Man,Ont,PQ,NB, |
| | NS,Nfld |
| 14817.5 | Paris |
| 19130 | " |
| 19360 | " |
| 21785 | " |
| 21807.5 | " |
| 24110 | " |

### RCMP Call Signs: Eastern Arctic

| | |
|---|---|
| XJA 22 | Cambridge Bay |
| XJD 33 | Pond Inlet |
| XJD 34 | Pangnirtung |
| XJD 58 | Sachs Harbour |
| XJD 270 | Nanisivik |
| XJD 276 | Sanikilauq |
| XJD 503 | Snowdrift |
| XJD 559 | Franklin |
| XJD 563 | Holman Island |
| XJD 563 | Igloolik |
| XJD 597 | Wrigley |
| XJD 711 | Tungsten |
| XJD 717 | Paulatuk |
| XJE 74 | Spence Bay |
| XJE 78 | Grise Fiord |
| XJE 223 | Norman Wells |
| XJE 270 | Sanikilvaq |
| XJE 272 | Inuvik |
| XJE 273 | Rae |
| XJE 274 | Yellowknife |
| XJE 275 | Aklavik |
| XJE 330 | McPherson |
| XJE 331 | Tuktoyaktuk |
| XJE 332 | Coppermine |
| XJE 333 | Good Hope |
| XJE 335 | Norman |
| XJE 361 | Baker Lake |
| XJE 415 | Fort Smith |
| XJE 423 | Liard |
| XJE 441 | Rankin Inlet |
| XJE 441 | Resolute Bay |
| XJE 442 | Frobisher Bay |
| XJE 443 | Clyde River |
| XJE 459 | Lake Harbour |
| XJE 460 | Igloolik |
| XJE 473 | Cape Dorset |
| XJE 474 | Simpson |
| XJE 476 | Resolution |
| XJO 43 | Eskimo Point |
| Route 9 | Broughton Island |

### Ministry of Natural Resources
SSB

| | |
|---|---|
| 4460 | Northwest Region |
| 4520 | |
| 4535 | Northcentral Region |
| 4580 | |
| 4880 | Other Regions |
| 5170 | |
| 5240 | |
| 5540 | Air to Ground |
| 9172 | Air to Ground, Long Range |

### Quebec Dept of Natural Resources
3368.5    4982

### Canada Energy, Mines & Resources
SSB

| | |
|---|---|
| 4441 | NWT |
| 4474 | Yukon,NWT,BC,Alta,Sask, |
| | Man,Ont,PQ,Nfld |
| 4532 | Rankin Inlet Drilling |
| | Operation (NWT) |
| 4765 | Echo Bay Mines, |
| | Great Bear Lake (NWT) |
| 4982 | Yukon,BC,Ont,PQ,NB,NS, |
| | Nfld |
| 5341.5 | NWT,PQ |
| 7347.5 | NWT,BC,Alta,Sask,Man, |
| | Ont,PQ,Nfld |

## Environment Canada
### SSB

| | |
|---|---|
| 2221.5 | Atmospheric, PQ |
| 2727.5 | Atmospheric, PQ NWT |
| 2319.4 | Pacific Coast |
| 2508 | NWT,PQ |
| 2598.5 | Pacific coast |
| 2617.5 | Yukon,NWT,Pacific coast |
| 2846.5 | Pacific coast |
| 3191.5 | Atlantic Coast |
| 4554 | Yukon,BC,Nfld |
| 4861.5 | Atmospheric, PQ |
| 4982 | NS |
| 5124 | Pacific coast |
| 5811.5 | Atmospheric, PQ NWT |
| 9813 | Atlantic Coast |
| 12126.5 | NWT,Pacific coast |
| 27655 | Nationwide |

## Quebec Hydro

| SSB | RTTY |
|---|---|
| 4625.5 | 4496.5 |
| 4883.5 | 4625.5 |
| 5346.5 | 5013.5 |
| 5347.5 | 6771.5 |
| 6771.5 | |
| 7306.5 | |

## Dept of Fisheries

2752     4462     4510
Patrol boat "Cygnus"

## Hunters and Trappers Association
### SSB

5031     (Eastern Arctic; Inuit language)

## Wildlife Service (NWT)
### SSB

| | |
|---|---|
| 5053 | (4PM local call-in) |
| XLI 334 | Frobisher Bay |
| XLI 336 | Arctic Bay |
| XLK 201 | Coral Harbour |
| XLK 202 | Resolute Bay |
| XMC 808 | Regional Office |
| XMC 823 | Igloolik |
| XMD 315 | Clyde River |
| XMD 316 | Broughton Island |
| XMD 317 | Cape Dorset |
| XMD 318 | Pangnirtung |
| XMD 319 | Pond Inlet |

## Communications Canada (DOC)

| SSB Ontario & NS | RTTY (Hermes Electronics, Ltd) Ontario/NS | |
|---|---|---|
| 7555.5 | | |
| 8046.5 | 3174.5 | 9392.5 |
| 9391 | 3214 | 11623 |
| 13714 | 3253.5 | 11636 |
| 13717.5 | 3278 | 13714 |
| 19572.5 | 4081.5 | 13717.5 |
| 20529.5 | 4084 | 14645 |
| 22834.5 | 4573.5 | 14733.5 |
| 23068.5 | 5107.5 | 19572.5 |
| 25149 | 5181 | 20529.5 |
| | 5276.5 | 22834.5 |
| | 7555.5 | 23068.5 |
| | 7558.5 | 25147.5 |
| | 8046.5 | 25150.5 |
| | 9389.5 | |

Kenora, Ont.
USB Phone Patch

| | |
|---|---|
| 4666 | 4565 |
| 7546 | 7400 |

Val d'Or, Que.
USB Phone Patch

| | |
|---|---|
| 4971.1 | 4766.5 |
| 7326.5 | 7461.5 |

## Yukon Radiotelephone
### (NorthwesTel Inc.)

| Ch. | USB Base | Mobile | Call | Location |
|---|---|---|---|---|
| 1 | 5222.5 | 5009 | CFY79 | Ft. Nelson |
| 3 | 5396 | 5856 | CHB615 | Hay River |
| 4 | 5215 | 4950 | CFY82 | Whitehorse |

*Alcan's Shipshaw Dam in Quebec*

# FEDERAL GOVERNMENT

*Montreal headquarters of Bell Canada, gateway to bush country communications.*

### Bell Canada CGD206 (Quebec)
Bush radiotelephone

| Ch. | Freq. |
|-----|-------|
| 1 | 3166 |
| 2 | 7465 |
| 3 | 5390 (pri) |
| 4 | 5430 (sec) |

### Indian Affairs and Northern Development

| | |
|---|---|
| 2247.5 | Yukon, NWT, BC, Alta |
| 2508 | NWT, PQ |
| 5140 | Native Community of Trout Lake (NWT) |

### Dene Nation (NWT)

2240 4840 4898 4967
5031 5410 5730 6955 9114

### Raven Society of British Columbia
SSB

9115.5  (also Nowachant Band Council-XJL970)

## Ministry of Transport
### Point to Point
#### SSB

| | |
|---|---|
| 1792 | ONT,PQ,NB,NS,NFLD |
| 4885 | YUKON,NWT,BC,ALTA,SASK |

#### RTTY

| | |
|---|---|
| 2221.5 | PQ |
| 2292 | Man PQ |
| 2727.5 | PQ,NWT |
| 3273 | NWT |
| 4861.5 | PQ |
| 4886.5 | YUKON,BC |
| 4927.5 | PQ,NFLD |
| 5380 | NWT,NFLD |
| 5795 | NWT |
| 5811.5 | PQ,NWT |
| 5885 | NWT,LABRADOR |
| 7484 | NFLD,REYKJAVIK |
| 7630 | PQ,NFLD |
| 7675 | NWT,LABRADOR |
| 7742 | NWT |
| 7825 | NFLD,PQ |
| 8135 | NFLD,SHANNON IRELAND |
| 8151.5 | NWT |
| 9265 | NWT,LABRADOR |
| 10290 | NWT |
| 11055 | NFLD,REYKJAVIK |
| 11077.5 | NFLD,SHANNON IRELAND |
| 11470 | PQ,NFLD |
| 11505 | LABRADOR,NFLD |
| 11695 | NWT,LABRADOR |
| 13595 | NFLD,PQ |
| 13700 | PQ,NFLD |
| 16356 | NFLD,SHANNON IRELAND |

### Marine
#### SSB

| | |
|---|---|
| 1792 | ONT,PQ,NB,NS,NFLD |
| 2104.9 | NWT,BC,NB,NS,PEI,NFLD |
| 2201.5 | BC |

#### RTTY

| | |
|---|---|
| 4171 | BC |
| 4173.5/4353 | BC,NS |
| 4174.5/4354 | BC,NS |
| 4769 | NFLD,REYKJAVIK |
| 4990 | PQ,NFLD (Air-Nav) |
| 6259.5/6497.5 | BC,NS |
| 6260.5/6498.5 | BC,NS |
| 8340.5 | BC |
| 8355.5/8716.5 | BC,NS |
| 8356/8717 | BC,NS |
| 12501 | BC |
| 12510.5/13090.5 | BC,NS |
| 12511.5/13091.5 | BC,NS |

# FEDERAL GOVERNMENT

| | |
|---|---|
| 16675.5/17212.5 | BC,NS |
| 16676/17213 | BC,NS |
| 22207.5/22576.5 | BC,NS |
| 22221/22590 | BC,NS |

### Ships

| | |
|---|---|
| 2119.4 | ATLANTIC,GT LAKES |
| 2206/2583.4 | EAST COAST |
| 2207.4/2583.4 | EAST COAST |
| 2237 | EAST COAST |
| 2340/2458 | PACIFIC COAST |
| 2670 | GREAT LAKES,SHIPS TO CONTACT US COAST GUARD |

### Aviation
#### RTTY

| | |
|---|---|
| 2221.5 | PQ |
| 4861.5 | PQ |
| 5115.5 | PQ |

### Maintenance and Operations
#### RTTY

| | |
|---|---|
| 2377.5 | BC |
| 5115.5 | NWT,MAN,ONT,PQ |
| 5352.5 | BC |
| 7318 | BC |

| | |
|---|---|
| 11019.5 | BC |
| 16051.5 | BC |

### Northwest Territory Forestry Dept.
#### USB

| | |
|---|---|
| 4270 | 5730 |

# AUSTRALIA

## Australian Conservation and Emergency Services

| | |
|---|---|
| 2470 | Country fire authority(VIC) |
| 2488 | " |
| 2512 | " |
| 2580 | " |
| 2580 | Emergency fire service(SA) |
| 2600 | Country fire authority(VIC) |
| 2612 | Forestry commission (VIC) |
| 2620 | Country fire authority(VIC) |
| 2620 | Emergency fire service (SA) |
| 2668 | " |
| 2680 | " |
| 2692 | Country fire authority(VIC) |
| 2708 | Forestry commission (VIC) |
| 2720 | Emergency fire service(SA) |

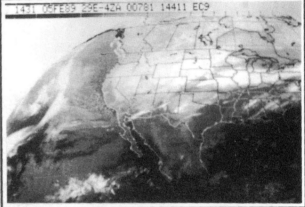

# FEDERAL GOVERNMENT

| | |
|---|---|
| 2728 | Forestry commission (VIC) |
| 2740 | Country fire authority(VIC) |
| 2752 | Emergency fire service (SA) |
| 2780 | Country fire authority(VIC) |
| 2780 | Emergency fire service (SA) |
| 2792 | Forestry commission (VIC) |
| 2808 | Emergency fire service (SA) |
| 2836 | " |
| 2840 | Forestry commission (VIC) |
| 2850 | Country fire authority(VIC) |
| 3152 | Emergency fire service (SA) |
| 3155 | St Johns search and rescue |
| 3158 | Emergency fire service (SA) |
| 3196 | Forestry commission (NSW) |
| 3260 | Country fire authority (VIC |
| 3290 | " |
| 3340 | " |
| 3360 | " |
| 3375 | State emergency services |
| 3393 | Country fire authority(VIC) |
| 3720 | State emergency service |
| 3836 | Country fire authority(VIC) |
| 3848 | " |
| 4428.7 | Coast stns working freq. |
| 4450 | Forestry commission(NSW) |
| 4510 | Emergency fire service(SA) |
| 4524 | Country fire authority(VIC) |
| 4571 | State emergency service |
| 4574 | " |
| 4880 | Forestry commission (VIC) |
| 5345 | " |
| 5370 | " |
| 5450 | " |
| 6940 | " |

## Australian Customs Department

2149.5 5286.5 7691.5 7961.5 10436.5 13781.5

## Australian National Conservation Agencies

2270 3176 3372 3373.5 3717.5
4616.5 5945 5946.5 6821.5 9221.5 9986.5

## Australian National Railways

| | |
|---|---|
| 4475 | South Australia, West Australia, Northern Territory |
| 5900 | South Australia |
| 5901.5 | |
| 9783 | |
| 9785 | South Australia |

## Australian Mines Department

3717.5  6821.5  9986.5

## Australian Main Roads Department

4461.5  4481.5

## Australian Transport Department

2601.4  4031.4  6806.4  9911.4

## Forestry Commission of New South Wales

3194.5  3196  4448.5  4450  6940

## Australian Telecommunications Commission

3724

## Australian Bureau of Meteorology

3315  5827  9072  13804

# Worldwide Operational Control

All worldwide HF voice communications are upper sideband and are authorized every 3 kilohertz in the following ranges:

| | |
|---|---|
| 2851-3019 | 8816-8960 |
| 3401-3497 | 10006-10096 |
| 4651-4696 | 11276-11396 |
| 5451-5475 | 13261-13357 |
| 5481-5676 | 17901-17967 |
| 6526-6682 | 21940-21997 |

## Worldwide Search and Rescue
*(Scene of Action):*
3023  5680

## Long Distance Operational Control (LDOC)

While HF-SSB communications are not generally used for regular operations by civilian airlines inside a country's borders (VHF and UHF AM is the authorized mode), it is the only means of communication between international aircraft and distant airports.

Tones preceding some transmissions comprise SELCAL (selective calling) which keeps an aircraft's receiver squelched until activated by a call directed to it. Tone frequencies are designated "Red A" through "Red S" (312.6-1479.1 Hz).

Worldwide air route traffic control centers (ARTCCs) are listed below with their allotted frequencies:

### AREA 1: EUROPE AND ASIA

Paris, France
3010 6637 11351 13351 17916 21940

London, England
3497 5535 5921 10072 13333 17922 21946

Portishead, England
3482 4807 5610 8170 8185 8960 11306 12133 17405 18210 19510 20065

Prague, Czechoslovakia
3497 5532 8924 10027 13351 17940 21967

Tel Aviv, Israel
3010 6637 8924 13351 17940 21952

Madrid, Spain
5529 8936 10027 13327 17940 21967

Amsterdam, Holland
3010 5532 8924 13336 17940 21973

Frankfurt, Germany
3010 4687 6637 10078 13327 17931 21979

Hamburg, Germany
10078 13327

Berne, Switzerland
3010 4654 6643 8936 10069 13205 13324 17931 1802321988

Dublin, Ireland
3010 5532 8924 13351 17916 21952

Rome, Italy
4687 5532 10027 13336 17940 21952

Stockholm, Sweden
5541 8930 11345 13342 17916 21997 23210

Lisbon, Portugal
3010 5532 8924 13336 17940 21952

Moscow, Tashkent, Khabarovsk, USSR
5529 8924 10030 13345 17940 21958

Brussels, Belgium
5529 8924 10078 13351 17940

Warsaw, Poland
4687 8924 11351 13345 17940 21967

Budapest, Hungary
5532 8930 11351 13336 17940 21973

Las Palmas, Conakry
5529 8936 10027 13327 17940 21967

Athens, Greece
3010 6637 10078 13327 17916 21979

Berlin, Germany
3010 5529 8930 10030 13351 17916 21973

Belgrade, Yugoslavia
3010 5532 11351 13336 17916 21940

# AIRCRAFT

Bucharest, Roumania
3497 4687 10027 13345 17931 21952

Valetta, Malta
3497 4687 10078 13327 17931 21967

Larnaca, Cyprus
6637 10078 13351 17931 21979

AREA 2:
NORTH AMERICA AND CENTRAL AMERICA

New York, US / San Juan, PR
3494 6640 11342 13330 17925 21964

Houston, TX
5529 10075 13330 17940 21964

San Francisco, Honolulu, USA
3013 6640 11342 13348 17925 21964

Toronto, Can.
3007 6646 10027 13339 17919 21985

Vancouver, Canada
3007 5544 8927 13339 17934 21985

Santo Domingo, Dominican Republic
5529 8936 10027 13348 17934 21985

Havana, Cuba
3007 5544 8927 13339 17934 21985

Mexico City/Matamoros/Tijuana, Mexico
3007 5544 8927 13339 17919

Merida, Mexico
3007 5544 8927 11354 13339 17919

Curacao, Netherlands Antilles
3007 5544 8936 13339 17919 21985

Cayman, Cayman Islands
6646 8936 13339

AREA 3: PACIFIC

Auckland, New Zealand
3007 6637 10072 13333 17940 21970

Hong Kong, HK
3007 6637 8921 13333 17940 21970

Sydney, Australia
3007 6637 8921 10078 13342 17922 21970

Bombay, Indonesia
3007 6637 8930 10072 13351 17916 21949

Manila, Philippines
3007 6637 11351 13351 17916 21949

Tokyo, Japan
4687 6637 11342 13324 17928 21949

Singapore, Singapore
6637 8930 10078 13333 17934 21949

Beijing, China
3007 6637 8930 11351 13351 17934

Jakarta, Indonesia
3007 6637 8921 10072 13351 17916 21970

Karachi, Pakistan
4687 6637 8930 11351 13342 17940 21970

Suva, Fiji
3007 6637 10072 13333 17916

Nauru, Nauru
6637 11351 17916 21949

Colombo, Sri Lanka
4687 6637 8921 13324 17922 21949

Seoul, Korea
3007 4687 10072 13333 17916 21970

Bangkok, Thailand
3007 4687 8930 13351 17922 21970

Rangoon, Burma
3007 4687 11351 13342 17934 21949

Kathmandu, Nepal
6637 13342

AREA 4: SOUTH AMERICA

Buenos Aires, Argentina
3010 6643 10030 13327 17919 17928 21955

Lima, Peru
3010 5535 8924 11345 17937 21976

Santiago, Chile
3010 6643 10030 13345 17937 21991

Rio de Jan., Brazil
3010 5541 8924 10069 17919 21991

Caracas, Venezuela
3010 6643 8924 11345 17937 21976

Pt. of Spain, Trinidad
3010 5535 8924 13336 17928 21955

Montevideo, Uruguay
3010 5535 8924 13336 17919 21976

Bogota, Colombia
3010 5535 11345 13336 17928 21955

San Salvador, El Salvador
3010 6643 10030 13345 17937 21976

Easter Island
3010 13345

AREA 5: MIDDLE EAST

Blantyre, Malawi
3013 5544 8933 13339 17931 21982

Kinshasa, Zaire
3013 5538 10033 13339 17931 21982

Nairobi, Kenya / Entebbe, Uganda / Dar Es Salaam, Tanzania
3013 6640 8927 13339 17931 21982

Addis Ababa, Ethiopia
3013 5538 8927 13339 17931 21982

Asmara, Asm.
3013 5538 8927 13339 17931 21982

Bahrain, Bah.
3013 5538 11354 13339 17931 21943

Kuwait, Kuw.
3013 6040 10033 13330 17931 21994

Beirut, Lebanon
3013 5538 10075 13330 17931 21943

Lagos, Nigeria
3013 5532 8927 13339 17931 21982

Monrovia. Liberia
3013 6646 11348 13348 17937 21961

Jeddah/Dhahran/Riyadh, Saudi Arabia
3013 5544 8927 13339 17925 21994

Mahe, Seychelles
3013 5544 8933 13339 17931 21982

Tunis, Tunisia
3013 6640 10033 13339 17925 21994

Cairo, Egypt
3013 6640 8933 17931 21982

Kabul, Afghanistan
3013 6640 10075 13330 17925 21982

Djibouti, Dji.
3013 5544 8933 13339 17925 21994

Muscat, Oman
3013 5538 10033 13330 17931 21943

Tehran, Iran
3013 5532 11348 13348 17937 21961

Harare, Zimbabwe
3013 5532 10075 13330 17937

Luanda, Angola
5544 8927 11354

AREA 5: AFRICA

Kigali, Rwanda
3013 6640 10075 13339 17931 21982

Lusaka, Zambia
3013 6640 10075 13330 17937 21982

Antananarivo, Madagascar
3013 5538 8927 13330 17931 21982

Dubai, UAE
10075 13348

Abu Dhabi, UAE
5532 10033 13348 17931 21943

Mauritius, Maur.
3013 5538 8933 13348 17937 21982

Algiers, Algeria
3013 5532 6646 8933 10075 11354 13348 17937 21943

Ouagadougou, Nigeria
3013 5544 6640 10033 21943

Abidjan, Ivory Coast
5538 8927 13330

Johannesburg, South Africa
3013 5532 8933 11354 13330 17925 21943

# International Flight Support Services

## European Sector
### Rainbow Station

| | |
|---|---|
| 3458 | 13286 |
| 5604 | 17910 |
| 8819 | |

### U.K. Station

| | |
|---|---|
| 3482 | 12133 |
| 4807 | 13865 |
| 5610 | 14890 |
| 6634 | 14890 |
| 8170 | 17405 |
| 8960 | 19510 |
| 10291 | 20065 |
| 11306 | 21765 |

## African Sector
### U.K. Station

| | |
|---|---|
| 3482 | 12133 |
| 4807 | 13865 |
| 5610 | 14890 |
| 6634 | 16370 |
| 8170 | 17405 |
| 8185 | 18210 |
| 8960 | 19510 |
| 10291 | 20065 |
| 11306 | 21765 |

# AIRCRAFT

## Atlantic Sector
### U.K. Station

| | |
|---|---|
| 3482 | 12133 |
| 4807 | 13865 |
| 5610 | 14890 |
| 6634 | 16370 |
| 8170 | 17405 |
| 8185 | 18210 |
| 8960 | 19510 |
| 10291 | 20065 |
| 11306 | 21765 |

### Rainbow Station

| | |
|---|---|
| 3458 | 13285 |
| 5604 | 13420 |
| 8819 | 17910 |

## Far East Sector
### U.K. Station

| | |
|---|---|
| 6634 | 14890 |
| 8170 | 16370 |
| 8185 | 17405 |
| 8960 | 18210 |
| 10291 | 19510 |
| 12133 | 20065 |
| 13865 | 21765 |

## South American Sector
### Lima Station

| | |
|---|---|
| 5535 | 11306 |
| 8885 | 17937 |

### U.K. Station

| | |
|---|---|
| 6634 | 14890 |
| 8170 | 16370 |
| 8185 | 17405 |
| 8960 | 18210 |
| 10291 | 19510 |
| 12133 | 20065 |
| 13865 | 21765 |

## South Pacific Sector
### U.K. Station

| | |
|---|---|
| 6634 | 16370 |
| 8960 | 17405 |
| 10291 | 18210 |
| 12133 | 18510 |
| 14890 | 20065 |
| 13865 | 21765 |

### Lima Station

| | |
|---|---|
| 5535 | 11306 |
| 8885 | |

## Sydney Sky Comms
### (Phone patch to Australia)

| | |
|---|---|
| 3007 | 10072 |
| 4666 | 11342 |
| 8903 | 13300 |
| 8930 | 17934 |
| 8936 | 17940 |

# Regional and Domestic Air Route Areas
## (RADARA)

*(authorized to governments, subcontracted to air carriers)*

## Airline Company Frequencies

In order to maintain communications with their international flights, many airline companies operate HF networks. These may be company-owned, or leased from contractors like Aeronautical Radio Incorporated (ARINC).

International callsigns are registered in accordance with the Table of International Callsign Prefixes in the Glossary.

## Company-Owned Network Frequencies

| kHz | User |
|---|---|
| 3007 | Cubana-Havana |
| | Quantas-Sydney |
| | KAL-Seoul |
| | Thai Airways-Bangkok |
| 3010 | KLM-Amsterdam |
| | Lufthansa-Frankfurt |
| | Olympic Airways-Athens |
| | Air France-Paris Radio |
| | Avianca-Bogota |
| | El Al-Tel Aviv |
| | Varig-Rio De Janeiro |
| | Sabena-Brussels |
| | Aerolineas Argentinas-Buenos Aires |
| 3458 | Rainbow Radio - Canada |
| 3497 | British Airways-London |
| 4654 | Swissair-Berne Radio |
| 4687 | KLM-Amsterdam |
| | Lufthansa-Frankfurt |
| | Alitalia-Rome |
| | Thai Airways-Bangkok |
| 5526 | Singapore |
| 5529 | Universal Aviation-Houston |
| | Iberia-Madrid/Las Palmas |
| | Sabena-Brussels |
| | Aeroflot-Moscow |

| | |
|---|---|
| 5532 | KLM-Amsterdam |
| | South African Airways-Johannesburg "Springbok" |
| | Alitalia-Rome |
| 5535 | British Airways-London "Speedbird" |
| | Eastern-Bermuda |
| 5536.5 | Ethiopian Airlines |
| 5538 | Gulfair-Bahrain |
| | MEA-Beirut |
| 5541 | SAS-Stockholm Radio |
| | Varig-Rio de Janeiro |
| 5544 | Cubana-Havana |
| | Saudia Airlines-Jeddah |
| 5553 | Varig-Rio de Janeiro |
| 5604 | Rainbow Radio - Canada |
| 5645 | KLM-Amsterdam |
| | Lufthansa-Frankfurt |
| 5710 | Aeroflot - Moscow |
| 5745 | Interflug-Berlin |
| 6526 | Air France-Paris Radio; |
| | Quantas-Sydney |
| 6637 | Air France-Paris Radio |
| | Lufthansa-Frankfurt |
| | Olympic Airlines-Athens |
| | Quantas-Sydney |
| | El Al-Tel Aviv |
| | JAL-Tokyo |
| 6640 | Kenya |
| 6643 | Swissair-Berne Radio |
| | Aerolineas Argentineas |
| 7541 | Ethiopian Airlines |
| 8819 | Rainbow Radio - Canada |
| 8837 | El Al-Tel Aviv |
| 8893 | South African |
| 8895 | Air Algerie-Algiers |

| | |
|---|---|
| 8921 | British Airways-London "Speedbird" |
| | LTU-Dusseldorf Radio |
| | German Airlines |
| | Quantas-Sydney |
| | Lufthansa-Frankfurt |
| | Saudi Arabia Airlines |
| 8924 | KLM-Amsterdam |
| | El Al-Tel Aviv |
| | Varig-Rio de Janeiro |
| | Sabena-Brussels |
| | Aeroflot-Moscow |
| | CSA-Prague |
| | LOT-Warsaw |
| | Pluma-Montevideo |
| 8927 | Cubana-Havana |
| | Saudia Airlines-Jeddah |
| | Ethiopian Airlines-Addis Ababa |
| | Kenya |
| 8930 | SAS Stockholm R.-Stockholm |
| | Thai Airways-Bangkok |
| | Malev-Budapest |
| | El Al-Tel Aviv |
| | Pakistan |
| 8933 | South African Airways-Johannesburg "Springbok" |
| 8936 | Swissair-Berne Radio |
| | Iberia-Madrid |
| | Aero Condor-Cartagena |
| 8952 | KLM (maintenance)-Amsterdam |
| 8963 | Nigeria |
| 9003 | Alia-Amman Amman |
| 10027 | Iberia-Madrid/Las Palmes |
| | Alitalia-Rome |
| 10030 | Aerolineas Argentinas |
| | Aeroflot-Moscow |

*Honolulu ARINC Communications Center*

# AIRCRAFT

| | | | |
|---|---|---|---|
| | Interflug-Berlin | 13351 | Air France-Paris Radio |
| 10069 | Swissair-Berne Radio | | Interflug-Berlin |
| | Varig-Rio de Janeiro | | Sabina-Brussels |
| 10072 | British Airways-London "Speedbird" | | El Al-Tel Aviv |
| | KAL-Seoul | 13420 | Rainbow Radio-Canada (phone patches) |
| | Air India-Bombay | | |
| | Air New Zealand-Auckland | 17910 | Rainbow Radio - Canada |
| 10075 | Universal Aviation-Houston | 17916 | Air France-Paris Radio |
| | MEA-Beirut | | Olympic Airways-Athens |
| 10078 | Lufthansa-Frankfurt | | KAL-Seoul |
| | Olympic Airways-Athens | 17919 | Varig-Rio de Janeiro |
| | Sabena-Brussels | | Aerolineas Argentinas |
| | Quantas-Sydney | 17922 | Quantas-Sydney |
| | CSA-Prague | | British Airways-London |
| 10093 | Air France-Paris Radio | 17928 | Avianca-Bogota |
| | Quantas-Sydney | | Aerolineas Argentinas- |
| 11197 | Aeroflot-Moscow | | Buenos Aires |
| 11256 | Ethiopian Airlines "Holloway" | 17931 | Swissair-Berne Radio |
| 11312 | Aeroflot-Moscow | | Gulfair-Bahrain |
| 11345 | SAS-Stockholm Radio | 17934 | Cubana Dispatcho Havana |
| 11351 | Air France-Paris Radio | 17937 | Lima Radio-Peru |
| | Malev-Budapest | 17940 | KLM-Amsterdam |
| | Pakistan Int'l Airlines-Karachi | | Universal Aviation-Houston |
| 11354 | Gulfair-Bahrain | | Sabina-Brussels |
| | South African Airways- | | Iberia-Madrid/Las Palmes |
| | Johannesburg "Springbok" | | Alitalia-Rome |
| 11384 | Balkan | | El Al-Tel Aviv |
| 11401 | Continental, Caroline & Marshall Is. | | Aeroflot-Moscow |
| 13205 | Aeroflot-Moscow | 17975 | Interflug-Berlin |
| 13220 | Aeroflot-Moscow | 19904 | Aeroflot-Moscow |
| 13225 | Alia-Amman Amman | 21940 | Air France-Paris Radio |
| 13285 | Rainbow Radio - Canada | 21943 | South African Airways- |
| 13324 | LTU German Airlines- | | Johannesburg |
| | Dusseldorf Radio | | Gulfair-Bahrain |
| | Swissair-Berne Radio | | MEA-Beirut |
| | JAL-Tokyo | 21946 | British Airways-London |
| 13327 | Lufthansa-Frankfurt | 21949 | KAL-Seoul |
| | South African Airways-Johannesburg | 21952 | El Al-Tel Aviv |
| | Olympic Airways-Athens | | Alitalia-Rome |
| | Aerolineas Argentinas- | 21955 | Varig-Rio de Janeiro |
| | Buenos Aires | | Aerolineas Argentinas- |
| | Iberia-Madria | | Buenos Aires |
| 13330 | Middle East Airlines-Beirut | 21958 | Aeroflot-Moscow |
| | Universal-Houston | 21967 | Iberia-Madrid/Las Palmes |
| | Cubana-Havana | | Lufthansa-Frankfurt |
| 13333 | British Airways-London "Speedbird" | | LOT-Warsaw |
| | KAL-Seoul | 21970 | Quantas-Sydney |
| 13336 | KLM-Amsterdam | | KAL-Seoul |
| | Malev-Budapest | 21973 | KLM-Amsterdam |
| | Avianca-Bogota | | Interflug-Berlin |
| | Alitalia-Rome | | Malev-Budapest |
| 13339 | Cubana-Havana | 21979 | Lufthansa-Frankfurt |
| | Gulfair-Bahrain | | Olympic Airways-Athens |
| | Saudia Airways-Jeddah | 21985 | Cubana-Havana |
| 13342 | SAS-Stockholm | 21988 | Swissair-Berne Radio |
| | Saudia Airways-Jeddah | | Nigerian Airways-Lagos |
| | Quantas-Sydney | 21991 | Varig-Rio de Janeiro |
| 13345 | Aeroflot-Moscow | 21994 | Saudia Airways-Jeddah |

## Aircraft Identifiers
### (includes VHF domestic)

| | |
|---|---|
| ABEX | Airborne Express (US) |
| ACOM | Southern Jersey (US) |
| AEROMAR | Aeromaritime (France) |
| AEROSERVICE | Europe Aero Service (France) |
| AIR AM | Air America (US) |
| AIR DELTA | Delta Air (Germany) |
| AIR FERRY | British Air Ferries (UK) |
| AIR FREIGHTER | Aeron International (US) |
| ALIA BASE | Royal Jordanian Airlines (Jordan) |
| AMFLIGHT | Ameriflight (US) |
| AMSTERDAM | KLM Base (Netherlands) |
| AMTRAN | American Trans Air (US) |
| ANZA | Ansett New Zealand (New Zealand) |
| ARPA | Air Panama (Panama) |
| ASEA | Atlantic Southeast (US) |
| ASIA | Japan Asia Airways (Japan) |
| ASPRO | Inter European (UK) |
| AYLINE | Aurigny Air Services (UK) |
| BATMAN | Ratioflug (Germany) |
| BEATOURS | Caledonian Airways (UK) |
| BELGAIR | TEA (Belgium) |
| BERNA RADIO | Swissair Base (Switzerland) |
| BIG A | Arrow Air (US) |
| BLUE STREAK | Jetstream International (US) |
| BLUEBIRD | Finnaviation (Finland) |
| BOOMERANG | Air 500 (Canada) |
| BRITISLAND | British Island Airways (UK) |
| BROWNAIR | Capital Airlines (UK) |
| CARIB-X | Caribbean Express (US) |
| CACTUS | American West (US) |
| CAROLINA | CCAir (US) |
| CEDAR JET | MEA (Lebanon) |
| CITY | NLM (Netherlands) |
| CLIPPER | Pan Am (US) |
| CUBANA DISPATCHO | Cubana Airlines-Habana |
| DTA | TAAG (Angola) |
| ELITE | Canada 3000 (Canada) |
| EUROCITY | London City Airways (UK) |
| EXPRESS | Federal Express (US) |
| EXPRESSAIR | Channel Express (UK) |
| FALCON CTL | Gulfair Base |
| FIREFLY | Jetall (Canada) |
| FLAG SHIP | Express Airlines I (US) |
| FLAMINGO | NFD (Germany) |
| FRANKFURT CTL | Lufthansa (Germany) |
| FREIGHTER | Saber Aviation (US) |
| GRAND AIR | MGM Grand (US) |
| HOTEL INDIA | Hispania (Spain) |
| INTERALAS | Markair (US) |
| JEDDAH | SAUDI Base |
| JET SET | Air 2000 (UK) |

| | |
|---|---|
| LAKES AIR | Great Lakes Commuter (US) |
| LATE NIGHT | CCAir Cargo (US) |
| LLOYDAEREO | LAB (Bolivia) |
| MAROCAIR | Royal Air Maroc (Morocco) |
| NEWSOUTH | Air NSW (Australia) |
| NORSEMAN | Norskair (Norway) |
| ORANGE | Air Holland (Netherlands) |
| OVERNIGHT | Russow Aviation (Germany) |
| PANAM | Panam Express (US) |
| PARADISE | Aero Virgin Islands (US) |
| PARTNER | Onterio Express (Canada) |
| PEGASUS | Air Cargo America (US) |
| PIARCO RADIO | Port of Spain, Trinidad |
| PRINCESS | Minerve Canada |
| RAINBOW RADIO | Sea Link Ltd., St. Johns, NFLD |
| REGAL | Crownair (Canada) |
| ST CLAIR | South West Air (Canada) |
| SCANWINGS | Malmo Aviation (Sweden) |
| SHAMROCK | Aer Lingus (UK) |
| SHOW-ME | Baron Aviation (US) |
| SHUTTLE | British Airways (Shuttle) |
| SKY BUS | Skyfreighters (US) |
| SKY COURIER | Business Air (US) |
| SKY EXPRESS | Salair (Sweden) |
| SKY TRUCK | Trans-Air-Link (US) |
| SOUNDAIR | Air Toronto/Odyssey (Canada) |
| SPEEDBIRD | British Airways (UK) |
| SPRINGBOK ("JoBerg") | South African Airways |
| STOCKHOLM RADIO | SAS Base (Sweden) |
| SUNLINE | Sun Country (US) |
| TAGGE | Orion Air (US) |
| TEE AIR | Tower Air (US) |
| TIGER | Flying Tigers (US) |
| TITAN AIR | Viking Express (US) |
| TOADOMES | Japan Air System (Japan) |
| TRANCON | Trans Continental (US) |
| UKAY | Air UK |
| UNIVERSAL | Universal Services-Houston |
| WASHINGTON EAGLE | Presidential (US) |
| WATERBIRD | Virgin Is. Seaplane Shuttle |
| WATERSKI | Resort Air (US) |
| WESTERN | Ansett WA (Australia) |
| WHITESTAR | Star Air (Denmark) |

## Additional RADARA Frequencies

| South Pacific | Africa | | Indian Ocean |
|---|---|---|---|
| 3425 | 2851 | 6673 | 3425 |
| 6553 | 5652 | 8870 | 4657 |
| 8846 | 6559 | 8873 | 8879 |
| 11339 | 6574 | 8888 | |

# Major World Air Route Areas
## (MWARA)
### ARINC Frequencies and Regions

| CEP-1 | CEP-2 | SEA-1 | SEA-2 | SEA-3 | CWP-1/2 | NP-3/4 | SP-6/7 | SW/SAM-1/8 |
|-------|-------|-------|-------|-------|---------|--------|--------|------------|
| 3413  | 5547  | 3470  | 3485  | 3470  | 2998    | 2932   | 3467   | 2944       |
| 5574  | 11282 | 6556  | 5655  | 6556  | 4666    | 5628   | 5643   | 4669       |
| 8843  | 13288 | 10066 | 8942  | 11396 | 6532    | 6665   | 8867   | 6649       |
| 13354 | 17904 | 13318 | 11396 | 13318 | 6562    | 10048  | 13273  | 10024      |
| 17904 |       | 17907 | 13309 | 13297 | 8903    | 11330  | 13300  | 11360      |
|       |       |       |       |       | 11384   | 13273  | 17904  | 17907      |
|       |       |       |       |       | 13300   | 13294  | 21964  |            |
|       |       |       |       |       | 17904   | 17904  |        |            |

| NCA-1 | NCA-2 | NCA-3 | SAM-2 | SAT-1 | SAT-2 | CAR-A | CAR-B | NAT-A | NAT-B |
|-------|-------|-------|-------|-------|-------|-------|-------|-------|-------|
| 3019  | 2851  | 3004  | 3479  | 3452  | 2854  | 2887  | 3455  | 3016  | 2899  |
| 5646  | 4678  | 5664  | 5526  | 6535  | 5565  | 5550  | 5520  | 5598  | 5616  |
| 13315 | 6592  | 10039 | 8855  | 8861  | 11291 | 6577  | 6586  | 8825  | 8864  |
| 17958 | 10096 | 13303 | 10096 | 13357 | 13315 | 8846  | 8846  | 13306 | 13291 |
|       | 17958 | 17958 | 13297 | 17955 | 17955 | 8918  | 11387 | 17946 | 17946 |
|       |       |       | 17907 |       | 11396 | 17907 |       |       |       |
|       |       |       |       |       | 13297 |       |       |       |       |
|       |       |       |       |       | 17907 |       |       |       |       |

| NAT-C | NAT-D | NAT-E | AFI-1 | AFI-2 | AFI-3 | AFI-4 | EUR-A | MID-1 | MID-2 |
|-------|-------|-------|-------|-------|-------|-------|-------|-------|-------|
| 2872  | 2971  | 3476  | 3452  | 3419  | 3467  | 2878  | 3479  | 2992  | 3467  |
| 2962  | 4675  | 6628  | 6535  | 5652  | 5658  | 5493  | 5661  | 5667  | 5601  |
| 5649  | 8891  | 8906  | 6673  | 8894  | 11300 | 6586  | 6598  | 8918  | 5658  |
| 8879  | 11279 | 11309 | 8861  | 13273 | 13288 | 8903  | 10084 | 13312 | 10018 |
| 13306 | 13291 |       | 13357 | 13294 | 17961 | 13294 |       |       | 13288 |
|       |       |       | 17955 | 17961 |       | 17961 |       |       |       |

| MID-3 | AFI-5/ INO-1 | SW PACIFIC ISLANDS | CENTRAL AMERICA COMMON | MELANESIA COMMON |
|-------|--------------|--------------------|------------------------|------------------|
| 2944  | 3476         | 3008               | 5568                   | 2861             |
| 4669  | 5634         | 3460               | 10017                  | 5498             |
| 6631  | 8879         | 6575               |                        | 5666             |
| 8951  | 13306        | 8924               |                        | 8917             |
| 11375 | 17961        | 11319              |                        |                  |
| 17961 |              |                    |                        |                  |

# AIRCRAFT

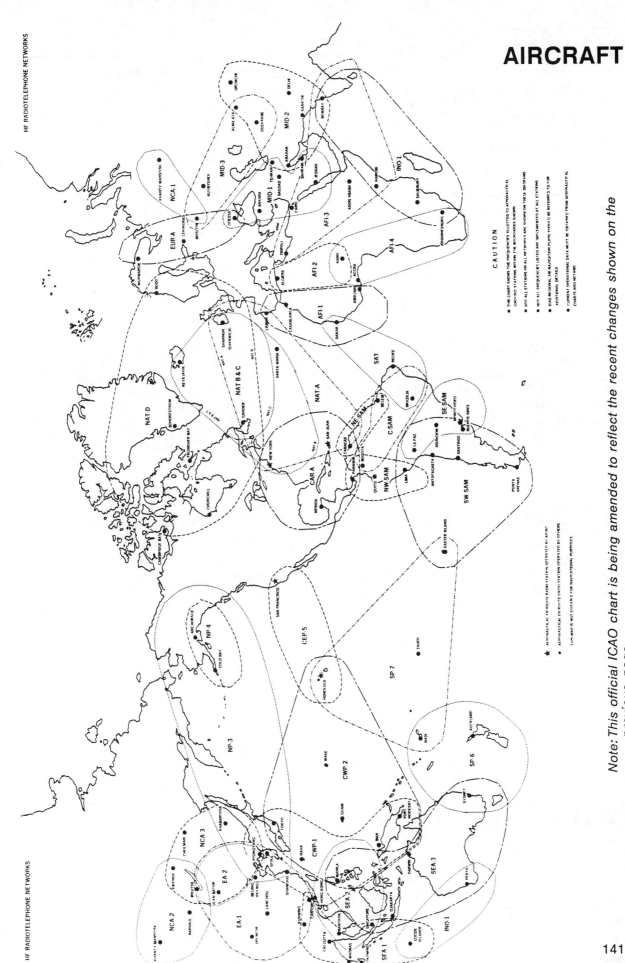

C A U T I O N

● THIS CHART SHOWS THE FREQUENCIES ALLOTTED TO AERONAUTICAL GROUND STATIONS WITHIN THE BOUNDARIES SHOWN

● NOT ALL FREQUENCIES ON ALL NETWORKS ARE SHOWN ON THESE DIAGRAMS

● NOT ALL FREQUENCIES LISTED ARE IMPLEMENTED AT ALL STATIONS

● ICAO REGIONAL AIR NAVIGATION PLANS SHOULD BE REFERRED TO FOR ADDITIONAL DETAILS

● CURRENT OPERATIONAL DATA MUST BE OBTAINED FROM AERONAUTICAL CHARTS AND NOTAMS

★ AERONAUTICAL EN ROUTE RADIO STATION OPERATED BY ADSIC

● AERONAUTICAL EN ROUTE RADIO STATION OPERATED BY OTHERS

THIS MAP IS NOT SUITABLE FOR NAVIGATIONAL PURPOSES

*Note: This official ICAO chart is being amended to reflect the recent changes shown on the previous page.*

141

# AIRCRAFT

## Foreign Aircraft Communications

### Regional Operational Control

Many countries authorize discrete frequencies in the international aircraft bands to resident airlines. Some frequencies are shared or regulated by their military agencies.

Afghanistan
6624

Algeria
| 3411 | 6638 | 8896 |
| 13273 | 13304 | |

Angola
| 2851 | 5493 |

Argentina
| 2346 | 2972 | 3008 |
| 4667 | 6554 | 6666 |
| 8842 | | |

Azores
8886

Australia (RAAF domestic net)
| 3008 | 3460 | 5498 |
| 6533 | 6633 | 6675 |
| 8896 | 8936 | 8952 |

Bahamas
| 6540 | 8959 |

Bahrain
2992

Bolivia
| 5469 | 6638 | 6659 |
| 6666 | | |

Botswana
6603

Brazil
| 6680 | 6704.5 | 8816 |
| 11195 | | |

Bulgaria
11245

*An international flight being readied for passenger loading*

Burma
| 2945 | 2973 | 2987 |
| 3526 | 5596 | 6659 |
| 6725 | | |

Canada
6540

Chile
| 4694 | 4745.5 | 5454 |
| 5710.5 | 5725.5 | 6666 |
| 7465.5 | 8842 | 8926 |
| 13304 | | |

China
| 6624 | 8931 | 13320 |

Colombia
| 5710 | 6666 | 8875 |
| 8938.4 | | |

Costa Rica
8993

Cuba
| 2681 | 2854 | 5461 |
| 8907 | 9001 | 13280 |
| 13296 | | |

Czechoslovakia
13296

Denmark
| 4742 | 6582 | 8871 |
| 8913 | | |

Easter Island
| 8667 | 19040 |

Ecuador
| 4694 | 5575 | 6533 |
| 6666 | | |

Egypt
4689

England
| 2838 | 3453 | 4550.7 |
| 5484 | 5645 | 6725 |

Fiji/Tonga
| 3460 | 6533 | 8846 |

France
| 4739 | 4742 | 9014 |

Ghana
6638

Greece
| 3015 | 4638 | 5600 |
| 5638 | | |

Greenland
4745.5

Guinea
| 5474 CW | 5567 | 5736 |
| 6537 CW | 7626 CW | 7979 |
| 13304.5 | | |

Honduras
| 6540 | 6927 |

Iceland
3330

India
| | | |
|---|---|---|
| 2913 | 2931 | 2948 |
| 3497.2 | 3499 | 5509.5 CW |
| 5514 | 5580 | 5584 |
| 5585 | 5668 | 6533 |
| 6760.5 | 6833 | 8917 |
| 8938 | 8963 | 8983 |

Indonesia
| | | |
|---|---|---|
| 3474 | 3756 | 3815 |
| 3855 | 4555 | 5188 |
| 5430 | 5536.5 | 5596 |
| 5659 | 6552 | 6554 |
| 6617 | 6659 | 7587.5 |
| 7589 | 7824 | 7825 |
| 8082.5 | 8820 | 8833 |
| 8871 | 13280 | 13320 |
| 17966.5 | | |

Iran
| | | |
|---|---|---|
| 3158 | 3650 | 5579 |
| 5960 | 6547 | 6624 |
| 7480 | | |

Iraq
| | | |
|---|---|---|
| 2972 | 3800 | 5710.5 |
| 9925 | | |

Italy
| |
|---|
| 5682 |

Jordan
| | |
|---|---|
| 2972 | 3453 |

Kuwait
| | | |
|---|---|---|
| 2992 | 6624 | 17920 |

Lebanon
| | |
|---|---|
| 2992 | 6634.5 |

Libya
| | | |
|---|---|---|
| 6574.5 | 6693 | 6697 |
| 6813.5 | 13273 | |

Madagascar
| | |
|---|---|
| 5484 | 8854 |

Mexico
| | |
|---|---|
| 3116 | 5668 |

Nepal
| |
|---|
| 5512 |

Netherlands
| | | |
|---|---|---|
| 3453 | 3474 | 5645 |

New Zealand
| |
|---|
| 3481.5 |

Nigeria
| | | |
|---|---|---|
| 3411 | 5493 | 6638 |
| 8896 | 8962 | 9960 |
| 13273 | | |

Norway
| | | |
|---|---|---|
| 3474 | 5645 | 6544 |

Pakistan
| | | |
|---|---|---|
| 2931 | 5512 | 6624 |

Papua New Guinea
| | | |
|---|---|---|
| 3008 | 5498 | 5666 |
| 6617 | 8833 | 8875 |
| 8917 | | |

Paraguay
| | |
|---|---|
| 2660 | 8842 |

Peru
| | | |
|---|---|---|
| 4694 | 5454 | 6617 |
| 6666 | 8842 | |

Philippines
| | |
|---|---|
| 6617 | 6700 |

Saudi Arabia
| | | |
|---|---|---|
| 3095 | 3453 | 5564 |
| 6701 | 8967 | 8990 |
| 17983 | | |

Somalia
| |
|---|
| 6552 |

Spain
| | | |
|---|---|---|
| 3137 | 4738.5 | 6737 |

Sudan
| | |
|---|---|
| 6537 | 8956 |

Surinam
| |
|---|
| 6533 |

Sweden
| | | |
|---|---|---|
| 11222 | 11345 | 13342 |
| 13942 | 15685 | 17916 |
| 23210 | 25035 | |

Switzerland
| | | |
|---|---|---|
| 3010 | 4654 | 6643 |
| 8936 | 10069 | 13205 |
| 15046 | 18023 | 21988 |
| 23285 | 25500 | |

Syria
| | |
|---|---|
| 2992 | 3939 |

Thailand
| | | |
|---|---|---|
| 4668 | 5491.5 | 5533 |
| 6533 | 6582 | 6617 |
| 6761 | 6768 | |

Tunisia
| | |
|---|---|
| 3411 | 4689 |

Turkey
| | | |
|---|---|---|
| 2972 | 6634.5 | 8896 |

Uruguay
| | | |
|---|---|---|
| 2445.5 | 2793 | 5659 |
| 8315 | | |

USSR
| | | |
|---|---|---|
| 2325 | 2868 | 2876 |
| 2882 | 2888 | 2895 |
| 2904 | 2924 | 2952 |
| 2966 | 2972 | 2976 |
| 2987 | 3024 | 3046 |
| 3102 | 3152 | 3439 |
| 3460 | 3462 | 3495 |
| 3905 | 3906 | 3918 |
| 3925 | 3930 | 3940 |
| 4694 | 4712 | 4736 |
| 4738 | 5250 | 5445 |
| 5491 | 5533 | 5557 |
| 5596 | 5642 | 5645 |
| 6536 | 6554 | 6603 |
| 6624 | 6704 | 6724 |
| 6743 | 6745 | 6748 |

# AIRCRAFT

| | | |
|---|---|---|
| 10089 | 11192 | 13220 |
| 13555 | | |

Venezuela

| | | |
|---|---|---|
| 3725 | 4045 | 5695.5 |
| 6540 | 6737 | 7710 |

Viet Nam

6557

Yemen

| | |
|---|---|
| 5935 | 6624 |

Yugoslavia

5688

Zaire

| | |
|---|---|
| 2851 | 5493 |

## Canada

Air/Ground

4675 Cambridge Bay, NWT air radio
5245 NWT bush camps

## Brazil

### Flight Information Region (FIR)

3452 3479 5526 8843 8855 8861
10042 10096 13357

## Honduras

Air/Ground

4665 Missionary Aviation Fellowship (1530 daily)
5120 SAHSA Airlines
5490 Missionary Aviation Fellowship Siguatepeque - 1530 daily)
7400 SOSA Airlines (La Ceiba)

## Flight Test

Since communications are a vital part of aerospace technology, adequate radio contact is necessary during many phases of aircraft development.

### Callsign Identification

*The flight of the Voyager in 1987 utilized many flight test frequencies.*

| Company | Call | Location |
|---|---|---|
| Boeing Aerospace | KOX3 | Seattle, WA |
| | KPS6 | Reston, WA |
| | KTF9 | Wichita, KS |
| | KRU7 | Everett, WA |
| Dayton Aircraft | WCT8 | Ft Lauderdale, FL |
| Grumman | KEC4 | Calverton, NY |
| | KEC6 | Bethpage, NY |
| | KEO8 | Bethpage, NY |
| | KFZ6 | Stuart, FL |
| | KOR3 | Stuart, FL |
| | WMS5 | Bethpage, NY |
| Lockheed | KMA6 | Los Angeles, CA |
| | KMN2 | Palmdale, CA |
| | WSO9 | Mountain View, CA |
| | WDN7 | California |
| McDonnell Douglas | KHR6 | USA |
| | KLP4 | Tulsa, OK |
| | KMF9 | Long Beach, CA |
| | KNU5 | Palmdale, CA |
| | WNS8 | St Louis, MO |
| | WJG3 | Yuma, AZ |
| | WDR3 | Yuma, AZ |
| Rockwell Int'l | KBU6 | Cedar Rapids, IA |
| | KLK2 | Newport Beach,CA |
| | KND9 | Los Angeles, CA |
| | KNI2 | Edwards AFB, CA |
| | KNV8 | Los Angeles, CA |
| | KQS3 | Palmdale, CA |
| Sunair Electronics | KVQ2 | Ft Lauderdale,FL |
| United Technologies | WNY6 | Stratford, CT |
| | KDN5 | Stratford, CT |
| E Systems, Inc. | KRY9 | Texas |
| | WKH2 | Greenville, TX |
| Aero Corporation | KNB3 | Lake City, FL |

### Flight Test Frequencies (USB)

| 2851 | 5469 | 10045 | 17964 |
|------|------|-------|-------|
| 3004 | 5571 | 11288 | 21931 |
| 3443 | 6550 | 11306 |       |
| 5451 | 8822 | 13312 |       |

# Offshore Drilling Support Aircraft
### (USB)

| 2878 | 3019 | 3434 |
|------|------|------|
| 4672 | 5463 | 5508 |

# Rescue

3023 3488 3939 5420 5670 5680 5695
6760 8364 8893 9025 18271

# Russian Aeroflot CW Communications

Aeroflot is the official airline of the Soviet Union. In flight between the USSR and Cuba, several Long Distance Operational Communications (LDOC) channels may be heard in USB (voice) as well as Morse code (CW). Frequently reported on CW are Havana (COL) and Moscow Kupavna Aero (RFNV).

| 6748 | |
| 8842 | CW |
| 11193 | |
| 11312 | CW |
| 11348 | CW (Commercial Flights) |
| 15024 | CW |
| 17936 | CW |

All 74,000 series tail numbers belong to IL-18 aircraft; 76,000 series are TU-114; 86,000 series are IL-62; 11000 series are AN-12; and 42303 is a YAK-42.

# VOLMET Aeronautical Weather Stations

A contraction of the French "flying weather," VOLMET stations broadcast recorded weather information on a fixed schedule to pilots. In nearly every case, the transmission mode is USB. Schedule times are shown as minutes past the hour.

## VOLMET-Aviation Weather Forecasting Stations

### PACIFIC REGION (PAC)

Frequencies

| 2863 | 10048 |
| 6679 | 13282 |
| 8828 | |

**H+00    H+30**
HONOLULU RADIO
Hickam AFB/Honolulu Intl
Gen Lyman Field/Hilo
Agana, Guam
Kahuli

**H+05    H+35**
HONOLULU RADIO
San Francisco
Los Angeles
Seattle/Tacoma
Portland
Sacramento Metro
Ontario
Las Vegas

**H+25    H+55**
HONOLULU RADIO
Anchorage
Fairbanks
Cold Bay
King Salmon
Sheyma
Anchorage
Vancouver

**H+10   H+40**
TOKYO RADIO
Tokyo-Narita Intl
Tokyo-Haneda Intl
Chitose
Nagoya
Osaka Intl
Fukuoka
Kimpo Intl (Korea)

**H+15   H+45**
HONG KONG RADIO
Kai Tak
Guang Zhou/Baiyan
Chiang Kai Shek Intl
Manila Intl
Naha
Kaohsiumg Intl
Mactan Intl
Hong Kong-Kai Tak

# AIRCRAFT

H+20      H+50
**AUCKLAND VOLMET**
Auckland
Christchurch
Wellington
Nandi
Noumean/La Tontouta
Pago Pago
Tahiti
Hong Kong-Kai Tak


H+25      H+55
**ANCHORAGE**
    2863  6679  8828  13282


## NORTH ATLANTIC REGION (NAT)

**Frequencies**
3485  6604  10051  13270

H+20      H+50
**GANDER AERADIO**
Calgary
Edmonton
Gander
Halifax
Montreal
Ottawa
Sidney
Skagway
Toronto
Winnipeg

H+00      H+30
**NEW YORK RADIO**
Detroit
Chicago
Cleveland
Niagara Falls
Milwaukee
Indianapolis

H+15    H+45
**NEW YORK RADIO**
Atlanta
Bermuda
Miami
Nassau
Freeport
Tampa
West Palm Beach

H+05    H+35
**NEW YORK RADIO**
Bangor
Pittsburgh

Windsor Locks
St Louis
Syracuse
Minneapolis

H+10     H+40
**NEW YORK RADIO**
New York
Newark
Boston
Baltimore
Philadelphia
Washington DC


## SHANNON AERADIO (EUR)

**Frequencies**
3413  5640  8957  13264

H+00 H+25  &  H+30 H+55
**SHANNON**
Madrid
Prestwick
London-Heathrow
Amsterdam
Oslo
Copenhagen
Athens
Paris


## EUROPE REGION (EUR)
## (Tel Aviv/Praha)

**Frequencies**
2980(night) 5575  6580  8957
11391(day) 13264

H+05     H+35
**BEN GURION RADIO**
Tel Aviv/Ben Gurion
Haifa/Ramat David
Elat
Jerusalem
Larnaca
Athens
Ankara/Esenboga
Istambul/Yesilkoy

H+15     H+45
**PRAGUE VOLMET**
Prague
Bratislava
Brno
Wien
Munich
Frankfurt Main
Berlin

*Shannon Aeradio*

Warsaw
Moscow
Budapest

H+25       H+55
SOFIA, BULGARIA
(also GORKI, USSR)
   11384

## SOUTH AMERICAN REGION (SAM)

Frequencies
2881   5601   10087   13279

| | |
|---|---|
| Lima | H+10 - H+40 |
| Brasilia | H+15 - H+45 |
| Buenos Aires | H+25 - H+55 |

### Additional South American VOLMET

Antofagasta, Chile
          H+20  3167.5  5280
                  7465.5
Asuncion Aeradio,   (0905-2395)
  Paraguay      H+05  5601  10067
Brasilia Metro,
  Brazil            6603  10057 13352
Carrasco Aeradio, Montevideo,
  Uruguay      H+15  5803
Comodoro Rivadavia Aeradio,

Argentina      H+30  4668  4675  8938
Cordoba Aeradio,
  Argentina    H+25  5475  8952
Ezeiza Aeradio, Buenos Aires,
  Argentina    H+15  2881  5601  11369
Galeao Metro, Rio de Janeiro,
  Argentina    H+00  10057
Puerto Montt     H+20  5280
        (Sun 1020-2320; Wed 1120-2320)
Resistencia Aeradio,
  Argentina    H+50  4667  4675
Salta, Argentina
             H+15  5475  5498
Tegucigalpa Aeradio,
  Honduras     H+50 (1200-2400) 4710

### AFRICAN/INDIAN OCEAN REGION (AFI)

Frequencies
2860  3404  5499  6538  8852  10057

H+00   H+30
JOHANNESBURG
Bloemfontein
Jan Smuts
Kimberley
Pietersburg
Durban
Matsapu

# AIRCRAFT

Phalaborwia
Welkom
Nelspriut
Newcastle
Skukuza
0400-2100 or
3047 6716 9026

H+00  H+25
BRAZZAVILLE(Eng)
N'Djamena
Bangui
Libreville
San Tome
Mogadiscio

H+30  H+55
BRAZZAVILLE(Fr)
Kinshasa
Kano
Lagos
Luanda

H+05 H+35
NAIROBI
Lusaka
Ndola
Kogali
Bujumbura
Mogadiscio

H+10 H+40
NAIROBI
Entebbe
Dar es Salem
Mombasa
Kilimanjaro

H+09
JEDDAH, SAUDI ARABIA
6570 (0001-0330
10215 (0400-2030)
H+00  H+30
6617  10037
H+25  H+55
5499  10057
ANTANANARIVO
(0225-1930)
Ivato
Majunga
Tamatave
Mahebourg
St Denis
Diego Suarez
Moroni Comores

## MIDDLE EAST REGION (MID)

Frequencies
3001  5561  8819

H+00  H+30
BAGHDAD RADIO
Baghdad Intl
Basrah/Magal

H+15  H+45
BEIRUT RADIO
Beirut Intl
Damascus Intl
Larnaca
Cairo Intl

H+05  H+35
TEHRAN RADIO
Tehran
Abadan
Tehran

H+20  H+50
CAIRO RADIO
Cairo
Damascus Intl
Beirut
Luxor

H+10  H+40
BAHRAIN RADIO
Bahrain
Dhahran
Kuwait

H+30
BASRAH RADIO (also 11336)
Basrah/Megal
Baghdad Intl

H+25  H+55
ISTANBUL RADIO
Istanbul
Yesilkoy

## NORTH CENTRAL ASIA (NCA-USSR)

ENGLISH LANGUAGE
Frequencies
3461  4663  5676  10090  13279

| | |
|---|---|
| Tashkent | H+05/H+35 |
| Novosibirsk | H+10/H+40 |
| Khabarovsk | H+15/H+45 |
| Kiev | H+20/H+25 |
| Moscow | H+25/H+55 |

RUSSIAN LANGUAGE
Network A
Frequencies
2941 3116 6617 8939 11297

| | |
|---|---|
| Riga | H+00/H+30 |
| Leningrad | H+05/H+35 |
| Moscow | H+10/H+40 |
| Kiev | H+20/H+50 |
| Rostov | H+25/H+55 |

RUSSIAN LANGUAGE
Network B
Frequencies
3407 6730 8819 11279

| | |
|---|---|
| Tbilisi | H+00/H+35 |
| Aktyubinsk | H+05/H+35 |
| Krasnodar | H+10/H+40 |
| Alma Ata | H+15/H+45 |
| Tashkent | H+20/H+50 |
| Baku | H+25/H+55 |

RUSSIAN LANGUAGE
Network C
Frequencies
6693 8888 11318

| | |
|---|---|
| Syktyvkar | H+00/H+30 |
| Sverdlovsk | H+05/H+35 |
| Novosibirsk | H+10/H+40 |
| Kuybyshev | H+15/H+45 |
| Tyuman | H+20/H+50 |

RUSSIAN LANGUAGE
Network D
Frequencies
5691 8852 13267

| | |
|---|---|
| Kirensk | H+00/H+30 |
| Yakutsk | H+10/H+40 |
| Khabarovsk | H+15/H+45 |
| Magadan | H+20/H+50 |
| Irkutsk | H+25/H+55 |

RUSSIAN LANGUAGE
Network E
Frequency
8861

| | |
|---|---|
| Pevek | H+10/H+40 |
| Khatanga | H+15/H+45 |

## Additional Russian VOLMET

| | |
|---|---|
| 4645 | Tallin |
| 4654 | Moscow/Sheremetievo H+15/H+45 |
| 5536 | Pechora H+25 |

| | |
|---|---|
| 6638 | Novosibirsk |
| 8990 | Novosibirsk |

## SOUTHEAST ASIA REGION (SEA)

Frequencies (except Karachi)
2965 6676 11387

H+00          H+30
SYDNEY VOLMET
Sydney/Kingsford Smith
Brisbane
Melbourne
Adelaide
Alice Springs
Perth

H+05          H+35
CALCUTTA RADIO
Calcutta
Bombay
Delhi
Dhaka/Tezgaon
Mingaladon

H+20          H+50
SINGAPORE RADIO
Singapore/Changi
Singapore/Paya
Lebar
Kuala Lumpur
Jakarta
Brueni
Kota Kinabalu
Bali
Penang

H+10   H+40
BANGKOK
Bangkok Intl
Mingaladon
Tan son Nhut
Kuala Lumpur Intl
Singapore Changi/Paya
Lebar
U-Tapeo

H+00 H+30 H+55
BOMBAY
Colombo/Katunayake Intl
Madras
Karachi civil
Ahmadabad

H+15          H+45
KARACHI
(1500-0130) 3432
(0130-1500) 6680 10017

# AIRCRAFT

Karachi Civil
Nawabshah
Lahore

## CARIBBEAN VOLMET NET (CAR)

Frequencies

2950  5580  11315

| | |
|---|---|
| Port of Spain | H+05/H+35 |
| Merida | H+10/H+40 |
| Miami | H+25/H+55 |

# Non-Directional Beacons

200-540 kHz non-directional beacons (NDBs) are heard all over the western hemisphere and many foreign locations as well. Used by airports to send homing signals to aircraft, the beacon band contains some of the oldest transmitters still in operation.

The transmissions are full carrier AM, identifying by Morse code every few seconds. Transmitting power ranges from 25 to 4000 watts. While the following list is nowhere near complete, it does contain hundreds of strong signals heard by low frequency listeners.

| kHz | ID | Location |
|---|---|---|
| 194 | TUK | Nantucket, MA |
| 198 | DIW | Dixon, NC |
| 200 | YAQ | Kasabonika, Ontario |
| 205 | XZ | Wawa, Ontario |
| 206 | QI | Yarmouth, Nova Scotia |
| 206 | GLS | Galveston, TX |
| 207 | YNE | Norway House, Manitoba |
| 208 | YSK | Sanikiluaq, N.W.T. |
| 212 | TS | Timmins, Ontario |
| 212 | UCF | Cienfuegos, Cuba |
| 212 | YGX | Gillam, Manitoba |
| 214 | CHX | Choix, Mexico |
| 215 | YSX | San Salvador, El Salvador |
| 216 | CLB | Wilmington, NC |
| 218 | RL | Red Lake, Ontario |
| 221 | SMS | San Marcos, Mexico |
| 223 | YYW | Armstrong, Ontario |
| 224 | MO | Moosonee, Ontario |
| 227 | YAC | Cat Lake, Ontario |
| 227 | LCE | La Ceiba, Honduras |
| 230 | ILT | Albuquerque, NM |
| 230 | QB | Quebec, Quebec |
| 230 | SH | Shreveport, LA |
| 230 | ZUC | Ignace, Ontario |

| kHz | ID | Location |
|---|---|---|
| 232 | GT | Grand Turk, Turks & Cacaios |
| 233 | QN | Nakina, Ontario |
| 233 | ZTC | Treasure Cay, Bahamas |
| 236 | GNI | Grand Isle, LA |
| 236 | ZRJ | Round Lake, Ontario |
| 242 | XC | Cranbrook, British Columbia |
| 244 | DG | Chute des Passes, Quebec |
| 244 | TH | Thompson, Manitoba |
| 245 | CB | Cambridge Bay, N.W.T. |
| 245 | YZE | Gore Bay, Ontario |
| 247 | YLH | Landsdowne House, Ontario |
| 248 | QL | Lethbridge, Alberta |
| 248 | UL | Montreal, Quebec |
| 248 | WG | Winnipeg, Manitoba |
| 248 | MBJ | Montego Bay, Jamaica |
| 250 | FO | Flin Flon, Manitoba |
| 251 | ZQA | Nassau, Bahamas |
| 257 | SQT | Melbourne, FL |
| 257 | XE | Saskatoon, Saskatchewan |
| 257 | YXR | Earlton, Ontario |
| 258 | ZSJ | Sandy Lake, Ontario |
| 260 | UFX | St. Fekix de Valois, Quebec |
| 260 | YAT | Attawapiskat, Ontario |
| 260 | MTR | Monteria, Colombia |
| 260 | TOY | Tongoy, Chile |
| 260 | HOR | San Jose, Costa Rica |
| 262 | CTM | Chetumal, Mexico |
| 262 | F | Iqaluit, N.W.T. |
| 263 | QY | Sydney, Nova Scotia |
| 263 | YGK | Kingston, Ontario |
| 266 | BR | Atlanta, GA |
| 266 | BY | Beechy, Saskatchewan |
| 266 | XD | Edmonton, Alberta |
| 267 | MAR | Maracaibo, Venezuela |
| 270 | OAX | Oaxaca, Mexico |
| 270 | HHP | Port au Prince, Haiti |
| 272 | YQA | Muskoka, Ontario |
| 272 | UVR | Varadero, Cuba |
| 273 | ZV | Sept Iles, Quebec |
| 276 | YEL | Elliot Lake, Ontario |
| 276 | TWT | Sturgis, KY |
| 278 | NM | Matagami, Quebec |
| 280 | ZWC | Walker Cay, Bahamas |
| 280 | IPA | Isla de Pascua, Easter Island |
| 281 | SGK | Knoxville, TN |
| 283 | PT | Pelee Island |
| 283 | UZG | Zaragoza, Cuba |
| 284 | QD | The Pas, Manitoba |
| 284 | RT | Rankin Inlet, N.W.T. |
| 284 | TEH | Techo, Colombia |
| 285 | UHA | Quaqtaq, Quebec |
| 286 | GD | Goderich, Ontario |
| 286 | OE | Dry Tortungas, Florida |
| 287 | SMR | Santa Marta, Colombia |
| 289 | YLQ | La Tuque, Quebec |
| 290 | CO | Chesapeake L.S., VA |
| 290 | FP | Frying Pan Shoals, L.S., NC |
| 290 | Y | Yankeetown, FL |

| | | |
|---|---|---|
| 290 | YYF | Penticton, British Columbia |
| 290 | CCG | Caia Cayman Grande, Cuba |
| 290 | YNP | Managua, Nicaragua |
| 292 | MIQ | Maiquetia, Venezuela |
| 294 | J | Jupiter Inlet L.S., FL |
| 295 | BDA | Gibbs Hill, Bermuda |
| 295 | RHC | Riohaca, Colombia |
| 296 | G | Galveston, TX |
| 296 | UBO | Batabano, Cuba |
| 300 | C | Mobile Point L.S., AL |
| 300 | LAP | La Paz, Mexico |
| 300 | YIV | Island Lake, Manitoba |
| 300 | FXG | Point-A-Pitre, Guadeloupe |
| 300 | SFM | San Francisco d'Marcoris, Dominican Republic |
| 300 | TFE | Tefe, Brazil |
| 301 | AB | Cabo San Antonio, Cuba |
| 302 | QW | North Battleford, Saskatchewan |
| 302 | MBY | Moberly, MO |
| 302 | ZCC | Chub Cay, Bahamas |
| 304 | BN | Nashville, TN |
| 304 | Z | Aransas Pass, TX |
| 305 | RO | Roswell, NM |
| 305 | YQ | Churchill, Manitoba |
| 305 | UIS | Isabella, Cuba |
| 306 | R | St. Johns L.S., FL |
| 307 | OT | Southwest Pass Ent. L.S., LA |
| 307 | EJA | Barrancabermeja, Colombia |
| 308 | TCG | Tulacingo, Mexico |
| 308 | PJM | St. Maarten, Netherland Antilles |
| 310 | H | Egmont Key L.S., FL |
| 311 | TBG | Panama City, Panama |
| 313 | PIL | Brazos Santiago L.S., TX |
| 313 | Z | Cape Canaveral, FL |
| 314 | YN | Swift Current, Saskatchewan |
| 314 | ZN | Portage, Manitoba |
| 314 | FXF | Forte de France, Martinique |
| 315 | USR | Simon Reyes, Cuba |
| 316 | M | South Pass West Jetty L.S., LA |
| 317 | CVP | Helena, MT |
| 317 | TB | Tybee L.S., GA |
| 317 | VC | La Ronge, Saskatchewan |
| 317 | ZWE | West End, Bahamas |
| 318 | L | San Juan, Puerto Rico |
| 320 | HTN | Miles City, MT |
| 320 | TY | Tyler, TX |
| 320 | W | Cape San Blas, FL |
| 320 | YQF | Red Deer, Alberta |
| 323 | UT | Calcasieu Pass, LA |
| 323 | UWP | Argentia, Newfoundland |
| 323 | BSD | St. Davids Head, Bermuda |
| 325 | LN | Lumsden, Saskatchewan |
| 325 | LRN | La Romano, Dominican Republic |
| 325 | VUP | Valledupar, Colombia |
| 326 | FC | Fredericton, New Brunswick |
| 326 | JY | Houston, TX |
| 326 | MA | Midland, TX |
| 326 | W | Wiarton, Ontario |
| 326 | YQK | kenora, Ontario |
| 327 | FXC | Cayenne, French Guiana |
| 328 | YTL | Big Trout Lake, Ontario |
| 328 | HB | Holsteinborg, Greenland |
| 329 | CH | Charleston, SC |
| 329 | TAD | Trinidad, CO |
| 329 | YEK | Eskimo Point, N.W.T. |
| 329 | YHN | Hornepayne, Ontario |
| 331 | LAN | San Salvador, El Salvador |
| 332 | FIX | Key West, FL |
| 332 | YFM | La Grande 4, Quebec |
| 332 | QT | Thunder Bay, Ontario |
| 332 | VT | Buffalo Narrows, Saskatchewan |
| 332 | XH | Medicine Hat, Alberta |
| 333 | SFD | San Fernando, Venezuela |
| 334 | YER | Fort Severn, Ontario |
| 335 | HP | Heath Point, Quebec |
| 335 | TTQ | Toccoa, GA |
| 335 | YLD | Chapleau, Ontario |
| 335 | MQU | Mariquita, Colombia |
| 336 | BNA | Barcelona, Venequela |
| 338 | ZU | Whitecourt, Alberta |
| 339 | UCU | Santiago, Cuba |
| 339 | A | Havana, Cuba |
| 340 | YY | Mont Joli, Quebec |
| 341 | TIZ | Tizayuca, Mexico |
| 341 | YCS | Chesterfield Inlet, N.W.T. |
| 341 | YFN | Cree Lake, Saskatchewan |
| 341 | YYU | Kapuskasing, Ontario |
| 343 | PJG | Willemstad, Curacao |
| 344 | CL | Cleveland, OH |
| 344 | JA | Jacksonville, FL |
| 344 | XX | Abbotsford, British Columbia |
| 344 | YGV | Havre-St. Pierre, Quebec |
| 344 | ZIY | Grand Cayman, Bahamas |
| 345 | BGI | Bridgetown, Barbados |
| 347 | YG | Charlottetown, Prince Edward Is. |
| 348 | RSD | Rock Sound, Bahamas |
| 348 | UHA | Havana, Cuba |
| 350 | ME | Chicago, IL |
| 350 | NY | Enderby, British Columbia |
| 350 | RB | Resolute Bay, N.W.T. |
| 350 | RG | Oklahoma City, OK |
| 350 | DAV | David, Panama |
| 351 | YKQ | Fort Rupert, Quebec |
| 351 | ANU | Antigua Island, British West Indies |
| 352 | RG | Rarotonga, Cook Island |
| 352 | TUR | Turbo, Colombia |
| 353 | IN | International Falls, MN |
| 353 | LI | Little Rock, AR |
| 353 | QG | Windsor, Ontario |
| 353 | HOT | Higuerote, Venezuela |
| 353 | UHG | Holquin, Cuba |
| 355 | YWP | Webequie, Ontario |
| 355 | SCO | Pisco, Peru |
| 355 | TGU | Tegucigalpa, Honduras |
| 356 | PB | West Palm Beach, FL |
| 356 | TIM | Georgetown, Guyana |

# AIRCRAFT

| | | |
|---|---|---|
| 359 | SDY | Sidney, MT |
| 359 | TPX | Tepexpan, Mexico |
| 360 | PN | Port Menier, Quebec |
| 360 | KIN | Kingston, Jamaica |
| 360 | ASN | Ascension Island |
| 361 | ZMH | Marsh Harbor, Bahamas |
| 362 | SB | Sudbury, Ontario |
| 362 | SC | Sherbrooke, Quebec |
| 362 | ZS | Coral Harbour, N.W.T. |
| 362 | C7 | Geraldton, Ontario |
| 362 | PSA | Point Salinas, Grenada |
| 363 | RNB | Millville, NJ |
| 364 | SG | Springdale, Newfoundland |
| 365 | AA | Fargo, ND |
| 365 | FT | Ft. Worth, TX |
| 365 | CKK | Miami, FL |
| 365 | LEN | Leon, Mexico |
| 365 | CGW | Cartago, Colombia |
| 365 | PAL | Palma, Ecuador |
| 366 | YMW | Maniwaki, Quebec |
| 367 | HA | Hao Atoll, French Polynesia |
| 368 | GYM | Guaymas, Mexico |
| 368 | L | Toronto, Ontario |
| 368 | SIR | Rawlings, WY |
| 368 | VX | Dafoe, Saskatchewan |
| 369 | ZDX | St. John, Antigua |
| 370 | GR | Iles de la Madeleine, Quebec |
| 370 | UCM | Camaguey, Cuba |
| 370 | VVC | Villaviencio, Colombia |
| 371 | ITU | Great Falls, MT |
| 371 | GW | Kuujjuarapik, Quebec |
| 371 | TS | Memphis, TN |
| 374 | SA | Sable Island |
| 375 | SPH | Spring Hill, LA |
| 375 | BM | Balmoral, Manitoba |
| 375 | DW | Tulsa, OK |
| 375 | ELM | Elmira, NY |
| 375 | TGE | Guatemala City, Guatemala |
| 376 | ZIN | Great Inagua, Bermuda |
| 378 | RJ | Roberval, Quebec |
| 378 | UX | Hall Beach, N.W.T. |
| 379 | CM | Channel Head, Newfoundland |
| 379 | GKQ | Newark, NJ |
| 379 | TL | Tallahassee, FL |
| 379 | YPQ | Peterborough, Ontario |
| 380 | LIO | Puerto Limon, Costa Rica |
| 380 | ASH | Aishalton, Guayana |
| 380 | UCY | Cayojabo, Cuba |
| 382 | EA | Empress, Alberta |
| 382 | AW | Arlington, VA |
| 382 | XU | London, Ontario |
| 382 | YPL | Pickle Lake, Ontario |
| 382 | UPA | Punta Alegra, Cuba |
| 382 | POS | Piarco, Trinidad |
| 385 | EMR | Augusta, GA |
| 385 | NA | Natashquan, Quebec |
| 385 | FLS | Flores, Guatemala |
| 385 | AUC | Arauca, Colombia |

| | | |
|---|---|---|
| 387 | PV | Providenciales, Turks & Caicos |
| 387 | PP | San Andres Island, Colombia |
| 388 | AM | Tampa, FL |
| 388 | MM | Fort McMurray, Alberta |
| 388 | BOG | Bogota, Colombia |
| 390 | JT | Stephenville, Newfoundland |
| 390 | VP | Kuujjuaq, Quebec |
| 390 | UCA | Ciego de Avila, Cuba |
| 391 | DDP | San Juan, Puerto Rico |
| 392 | ML | Charlevoix, Quebec |
| 392 | NAU | Nautla, Mexico |
| 392 | BEZ | Belize, Belize |
| 394 | YB | North Bay, Ontario |
| 395 | KM | Sept Iles, Quebec |
| 395 | YL | Lynn Lake, Manitoba |
| 396 | PH | Inukjuak, Quebec |
| 396 | ZBB | Bimini, Bahamas |
| 398 | G | Windsor, Ontario |
| 399 | UP | Upernavik, Greenland |
| 400 | QQ | Comox, British Columbia |
| 400 | BGA | Bucaramanga, Colombia |
| 400 | HIV | Santo Domingo, Dom Republic |
| 400 | IZP | Iztapa, Guatemala |
| 401 | YPO | Peawanuck, Ontario |
| 402 | USJ | San Julian, Cuba |
| 402 | C | Camaguey, Cuba |
| 403 | BK | Baker Lake, N.W.T. |
| 404 | OLF | Wolf Point, MT |
| 404 | YSL | St. Leonard, New Brunswick |
| 404 | ZR | Sarnia, Ontario |
| 404 | FNB | Falls City, NE |
| 405 | YXL | Sioux Lookout, Ontario |
| 407 | H | Montreal, Quebec |
| 407 | SWA | Swan Island |
| 407 | LET | Leticia, Colombia |
| 408 | SN | St. Catherines, Ontario |
| 409 | YTA | Pembroke, Ontario |
| 410 | GDV | Glendive, MT |
| 410 | HMM | Hamilton, MT |
| 411 | RM | Rocky Mountain House, Alberta |
| 412 | ALE | Alejandria, Colombia |
| 413 | TAM | Tampico, Mexico |
| 413 | YHD | Dryden, Ontario |
| 414 | AZE | Hazlehurst, GA |
| 414 | BC | Baie Comeau, Quebec |
| 414 | MSD | Mansfield, LA |
| 414 | OOA | Oskaloosa, IA |
| 414 | RPB | Belleville, KS |
| 414 | SKX | Taos, NM |
| 414 | SU | Sioux City, IA |
| 415 | ASJ | Ahoskie, NC |
| 415 | HJM | Bonham, TX |
| 415 | IEE | Offshore Platform, CA |
| 415 | SLS | Salinas, Ecuador |
| 416 | LB | North Platte, NE |
| 417 | EOG | Greensboro, AL |
| 417 | EVB | New Smyrna Beach, FL |
| 417 | HHG | Huntington, IN |

# AIRCRAFT

| | | |
|---|---|---|
| 417 | RGB | Rifle, CO |
| 418 | CW | Lake Charles, LA |
| 418 | PYF | Fairfield, TX |
| 419 | GB | Great Bend, KS |
| 419 | RYS | Grosse Ile, MI |
| 419 | UEW | Lawrenceville, GA |
| 420 | TU | Tupelo, MS |
| 423 | CKP | Cherokee, IA |
| 426 | FTP | Fort Payne, AL |
| 426 | IZS | Montezuma, GA |
| 428 | EEJ | Sanford, NC |
| 429 | JNM | Monroe, GA |
| 432 | IZN | Lincolnton, NC |
| 432 | MHP | Metter, GA |
| 435 | IIY | Washington, GA |
| 513 | PP | Omaha, NE |
| 513 | VIS | La Vista, Peru |
| 515 | CL | Port Angeles, WA |
| 515 | OS | Columbus, OH |
| 515 | PKV | Port Lavaca, TX |
| 515 | RRQ | Rock Rapids, IA |
| 515 | SAK | Kalispell, MT |
| 516 | VPX | Pineville, WV |
| 516 | YWA | Petawawa, Ontario |

| | | |
|---|---|---|
| 517 | FN | Clinton, IA |
| 519 | EAA | Eagle, AK |
| 520 | BHZ | Belo Horizonte, Brazil |
| 520 | LOR | El Valor, Peru |
| 521 | ADS | Dallas, TX |
| 521 | DWH | Houston, TX |
| 521 | GF | Cleveland, OH |
| 521 | GM | Greenville, SC |
| 521 | INE | Missoula, MT |
| 521 | ORC | Orange City, IA |
| 521 | TO | Topeka, KS |
| 524 | AJG | Mt. Carmel, IL |
| 524 | HEH | Newark, OH |
| 524 | HRD | Hardin County, TX |
| 524 | MNL | Valdez, AK |
| 524 | UOC | Iowa City, IA |
| 526 | OJ | Olathe, KS |
| 529 | FDV | Nome, AK |
| 529 | SQM | Sumner Strait, AK |
| 530 | AAW | Talkeetna Mountains, AK |
| 530 | NB | North Bay, Ontario |
| 530 | YCH | Chatham, New Brunswick |
| 534 | SZM | Salmon No. 1, AK |

*An aerial view of a major airport reveals the need for constant air to ground communications.*

| | | | | |
|---|---|---|---|---|
| 11252 | 13600 | 18022 | 19966 | 23840 |
| 11407 | 13676 | 18051 | 20186 | 23940 |
| 11414 | 13735 | 18196 | 20192 | 24240 |
| 11548 | 13742 | 18310 | 20198 | 24512 |
| 11621 | 13878 | 18331 | 20266 | 24530 |
| 11634 | 14497 | 18354 | 20272 | 24780 |
| 11984 | 14585 | 18434 | 20390 | 24914 |
| 11988 | 14615 | 18700 | 20475 | 25130 |
| 12107 | 14650 | 18769 | 20690 | 25161 |
| 12160 | 14896 | 18801 | 21810 | 25198 |
| 12277 | 14937 | 18990 | 22683 | 25245 |
| 12287 | 14967 | 19126 | 22755 | 25597 |
| 12876 | 15698 | 19143 | 22990 | 26356 |
| 13210 | 16216 | 19149.5 | 23035 | 26389 |
| 13227 | 16246 | 19303 | 23281 | 26515 |
| 13237 | 17470 | 19371 | 23325 | 26684 |
| 13244 | 17490 | 19390 | 23413 | 27065 |
| 13380 | 17554 | 19640 | 23479 | 27720 |
| 13468 | 17668 | 19928 | 23485 | |
| 13495 | 18009 | 19963 | 23661 | |

## Space Shuttle Audio Amateur Rebroadcast Frequencies:

**W3NAN   Goddard Space Flight Center Greenbelt, MD**

3860 LSB  7185 LSB  14295 USB  21395 USB 28650 USB

**W5RRR   Johnson Space Center Houston, TX**

3840 LSB  14280 USB

**W6VIO   Jet Propulsion Laboratories Pasadena, CA**

3840 LSB  21280 USB

*(Actual frequencies may vary somewhat to avoid interference)*

## Western Space and Missile Center (WSMC)

### USAF Pacific Missile Range
*(Formerly SAMTEC)*
*Edwards AFB/Vandenberg AFB, CA*

| | | | |
|---|---|---|---|
| 3162 | 6889 | 10510 | 14987 |
| 4486 | 7705 | 10804 | 15021 |
| 4760 | 9029 | 11510 | 15793 Air |
| 5700 sec. | 9212 | 13211 | 17428 |
| 5733 | 10272 | 13218 pri. | 19303 |
| 5822 | 10352 | 13756 | 19640 |
| 6820 | 10660 | 13900 | 20261 |

*(Primary frequencies underlined)*

# National Aeronautics and Space Administration (NASA)

## Eastern Space and Missile Center (ESMC)
### Cape Canaveral, Florida

Although the vast majority of NASA communications are conducted via satellite and wire line, some shortwave backup nets are still authorized. These become active during Space Shuttle missions. Frequencies recently monitored are underlined; they commonly link the Cape with Ascension Island and Antigua tracking stations. Voice and/or data may be transmitted on adjacent subchannels.

Most NASA HF communications are actually maintained by the U.S. Air Force because of air support during space flights.

Frequencies

| | | | | |
|---|---|---|---|---|
| 2405 | 4704 | 5775 | 7697 | 9170 |
| 2505 | 4714 | 5810 | 7742 | 9910 |
| 2622 | 4755 | 5822 | 7765 | 10159 |
| 2664 | 4825 | 6720 | 7804 | 10215 |
| 2678 | 4856 | 6750 | 7833 | 10230 |
| 2716 | 4860 | 6753 | 7860 | 10270 |
| 2744 | 4900 | 6810 | 7910 | 10301 |
| 2764 | 4992 | 6880 | 7919 | 10305 |
| 2800 | 5060 | 6896 | 7985 | 10310 |
| 2836 | 5180 | 6919 | 8077 | 10327 |
| 3024 | 5187 | 6937 | 8993 | 10475 |
| 3120 | 5190 | 7313 | 9018 | 10780 |
| 3187 | 5235 | 7412 | 9022 | 10850 |
| 3365 | 5246 | 7461 | 9043 | 10880 |
| 4376 | 5350 | 7525 | 9115 | 10905 |
| 4500 | 5436 | 7605 | 9132 | 10949 |
| 4510 | 5680 | 7676 | 9138 | 11104 |

# SPACE

## NASA Identifiers

| Tactical Callsign | Identification |
|---|---|
| 0 | Patrick AFB, FL |
| 1 | Cape Canaveral, FL |
| 3 | Grand Bahama Is. |
| 12 | Ascension Is. (AFE 83) |
| 28 | Jupiter, FL |
| 91 | Antigua (AFE 86) |
| Abnormal One Zero | Vandenburg AFB, CA (AGD) |
| Abnormal Two Zero | Wheeler AFB, HI (AGD25) |
| Abnormal Four Zero | Kwajalein (ABK) |
| ARIA (AGAR) | EC-135 Advanced Range Instrumentation Aircraft 4050th Test Wing, Wright Patterson AFB |
| Ashley 10 | Vandenberg AFB, CA |
| Ashley 12 | Wheeler AFB, HI |
| Canaveral Control | Port Canaveral Harbor Control (USN dockmaster) |
| Cape Leader | Contingency Emergency Base |
| Cape Radio | Cape Canaveral Communications (USAF) |
| Chase 1,2 | T-38 Chase Aircraft |
| Dishpan | USS Redstone (Tracking Ship) |
| Eyesite | US Navy P-3 Orion Aircraft |

| | |
|---|---|
| Fisher | Cape Radio to Ships |
| Freedom | Booster Recovery Vessel (civilian contractor) |
| Gull Photo | USAF WC-130 Aircraft |
| Independence | Booster recovery vehicle (civilian contractor) |
| Jolly | USAF/ARRS SH-3 Helicopters |
| King | USAF/ARRS Aircraft |
| Liberty | Booster Recovery Vessel (civilian contractor) |
| Lovelorn | Contingency (USN) director |
| Peapod | USAF AC-130 aircraft |
| Thinker 1 | Senior Range Operations Officer |
| Variety | Range safety officer |

### Additional NASA Frequency Allocations (kHz)

| | |
|---|---|
| 3380 | Nationwide tracking net |
| 3385 | " |
| 3395 | " |
| 6982.5 | " |
| 14455 | " |

## Russian Spacecraft

| Freq. | Spacecraft | Mode |
|---|---|---|
| 18000 | Soyuz | AM Voice |
| 18060 | Soyuz/Salyut | |
| 19542 | Cosmos | |
| 19954 | Cosmos | RTTY/Data |
| 19989 | Cosmos | RTTY/Data |
| 19994 | Cosmos | RTTY/Data |
| 20008 | Cosmos/Salyut/Soyuz | CW/Data |
| 29.357-29.403 | RS10 (ham) | CW |
| 29.403-29.453 | RS11 (ham) | CW |

## Russian Tracking Network

Ships    8417  8418  16836  16837 (CW)

| | |
|---|---|
| UISZ | Akademik Sergei Korolev |
| UTTX | Kosmonaut Pavel Belyayev |
| UUVO | Kosmonaut Vladamir Komarov |
| UZZV | Kosmonaut Georgi Dobrovolski |

Land Station    16697.5 (RTTY)  16835 (CW)

CMU967  Santiago, Cuba (Soviet Naval Radio Station)

# Radiolocation Frequency Assignments

The 1610-1800 kHz band is filled with unusual sounds at night. Radio location frequency assignments include those used for survey and position fixing with equipment such as Raydist, Hydrotract, Cubic Argo and Decca Hi-Fix.

Many of these 1600-1700 kHz systems are being deactivated or moved to the 200-400 kHz range in preparation for the expansion of the 540-1600 kHz medium wave broadcast band.

Recognizing 1610-1800 kHz signal patterns:
- Decca Hi-Fix (one short, three long dashes per second)
- Cubic Argo (2-4 second burst of uneven chirps)
- Central American Aeronautical beacons (2-3 letter CW)
- Fishing beacons (one letter, three numbers CW)

## Raydist Assignments
*(Average center frequencies shown)*

| Ch | Mobile | Interrogator |
|----|--------|--------------|
| A | 3288.5 | 1643.5-1644.9 |
| B | 3290.5 | 1644.5-1645.9 |
| C | 3294.5 | 1646.5-1647.9 |
| D | 3296.5 | 1647.5-1648.9 |
| E | 3300.5 | 1649.5-1650.9 |
| F | 3306.5 | 1652.5-1653.9 |

### Additional Raydist Frequencies

| Mobile | | Relay |
|--------|--|-------|
| 3281 | Alaska | 1640.3 |
| | | 1640.315/.725 |
| | | 1648 Delaware Bay |
| 3320.4 | Coastline | 1658.425 |
| 2398 | USA except Alaska | 1660.015 |
| 2456 | Alaska and Great Lakes | |
| 2510 | USA except Alaska | |
| 2848 | Alaska | |

## Hydrotrac/Argo Radiolocation Assignments

| Freq | Comment |
|------|---------|
| 1618.5/1798.5 | For lane ID |
| 1619.64/1799.6 | For lane ID |
| 1643 | US Coastline |
| 1649 | US Coastline |
| 1718.59 | US Coastline |

## Argo Assignments
### Frequency Pairs

| | |
|---|---|
| 1643.0/1798.5 | 1643.7/1798.5 |
| 1649.0/1798.5 | 1644.7/1798.5 |
| 1653.0/1798.5 | 1646.7/1798.5 |
| 1648.0/1798.5 | 1647.7/1798.5 |
| 1643.0/1799.6 | 1649.7/1799.6 |
| 1649.0/1799.6 | 1652.7/1799.6 |
| 1653.0/1799.6 | 1658.4/1799.6 |
| 1648.0/1799.6 | 1660.0/1799.6 |

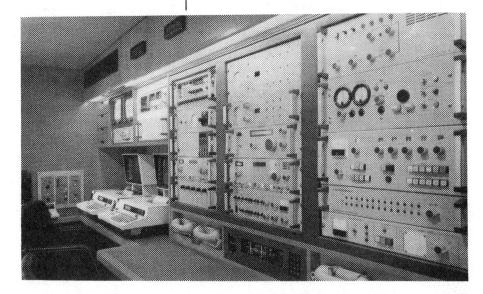

*A well-equipped ship's radio room.*

# MARITIME

## Canada

### Cubic Argo

| | | | | | |
|---|---|---|---|---|---|
| 1610 | 1627 | 1644 | 1715.5 | 1762.5 | 1767.5 |
| 1628.5 | 1645 | 1716 | 1763 | 1769 | |
| 1616 | 1630 | 1646.7 | 1746.8 | 1764.5 | 1770 |
| 1618.6 | 1632 | 1648 | 1750 | 1764.8 | 1771 |
| 1620 | 1632.5 | 1673 | 1753 | 1765 | 1772 |
| 1622.9 | 1638.9 | 1674 | 1757 | 1765.6 | 1785 |
| 1624 | 1639 | 1705 | 1759 | 1766 | 1788 |
| 1626.9 | 1640 | 1714 | 1761.5 | 1766.5 | |

## Limited Coastal Stations Nationwide

*(Private correspondence for business and shipping; see following pages for frequencies)*

| Callsign | Location |
|---|---|
| KBQ | Miami, FL |
| KCQ | Galveston, TX |
| KDL | Houston, TX |
| KED | New Orleans, LA |
| KJG | New Orleans, LA |
| KKP | Seattle, WA |
| KKV | Piney Point, MD |
| KLB | Houston, TX |
| KMB | Houston, TX |
| KNB 304 | Morgan City,LA |
| KEH | Houma, LA |
| KDR | St Louis, MO |
| KEC | New Orleans, LA |
| KFC | Morgan City, LA |
| KFN 540 | Pt Arthur, TX |
| KHT | Cedar Rapids, IA |
| KFV | Palm Beach, FL |
| KHU | San Diego, CA |
| KRM | Pt Arthur, TX |
| KSW | Oakland, CA |
| KUR | Pt Arthur, TX |
| KUY | Pt Arthur, TX |
| KWU | Golden Meadows, LA |
| KXQ | Houston, TX |
| KXE | Des Allemands, LA |
| KYZ | Galveston, TX |
| KZN | Miami, FL |
| KZU | Harvey, LA |
| WBK | Pittsburgh, PA |
| WBQ | Bowling Green, KY |
| WDJ | Morgan City, LA |
| WDM | New Orleans, LA |
| WEC | Norfolk, VA |
| WFE | Houston, TX |
| WFX 605 | Atlantic City, NJ |
| WGK | San Juan, PR |
| WIX | Lakin, WV |
| WJD | Tampa, FL |
| WJK | Jacksonville, FL |
| WJL | Narragansett, RI |
| WJU | Dravosburg, PA |
| WKC | St Petersburg, FL |
| WLN | San Juan, PR |
| WMP | W Palm Beach, FL |
| WNT | Greenville, MS |
| WPE | Jacksonville, FL |
| WRK | Morgan City, LA |
| WWB | St Louis, MO |

### Alaska

| | Call | Coast | Ship |
|---|---|---|---|
| Ft.Nelson | | 5222.5 | 5009.0 |
| | | 4792.5 | 5375.0 |
| Hay River | | 5396.0 | 5856.0 |
| Cold Bay | WGG53 | 2312.0 | 2134.0 |
| Cordova | WDU26 | 2397.0 | 2237.0 |
| Juneau | WGG58 | 2400.0 | 2240.0 |
| Ketchikan | WGG56 | 2397.0 | 2237.0 |
| Kodiak | WDU23 | 2309.0 | 2131.0 |
| Nome | WGG55 | 2400.0 | 2240.0 |
| Sitka | WDU29 | 2312.0 | 2134.0 |

## Radio Beacons
### 1600-1800 kHz

| kHz | ID | Location |
|---|---|---|
| 1600 | NHG | Palmer Station, Antarctica |
| 1602 | LML | Lomalinda, Colombia |
| 1610 | TDA | Trinidad, Colombia |
| 1615 | MIL | Quincemil, Peru |
| 1615 | NZ | Nadzab, Papua New Guinea |
| 1615 | OR | Ohura, New Zealand |
| 1620 | CEP | Concepcion, Bolivia |
| 1623 | FAM | Fazenda Amalia, Brazil |
| 1623 | GNY | Gurney,Papua New Guinea |
| 1623 | HUM | Humaita, Brazil |
| 1623 | PP | Oriximina, Brazil |
| 1625 | PAT | Pastaza, Ecuador |
| 1627 | NPA | Novo Paraiso, Brazil |
| 1627 | SAC | Sao Carlos, Brazil |
| 1630 | TM | Taumarunui, New Zealand |
| 1632 | OKT | Tabubil, Papua New Guinea |
| 1635 | LMC | Limoncocha, Ecuador |
| 1635 | ORI | Orito, Colombia |
| 1638 | URC | Urcos, Peru |
| 1640 | SJV | San Javier, Bolivia |
| 1645 | YPI | Yaupi, Ecuador |
| 1650 | SOT | Reyes, Bolivia |
| 1655 | RIO | Riobamba, Ecuador |
| 1662 | KUB | Kubuna, Papua New Guinea |
| 1665 | CBR | Cabo Norte, Brazil |
| 1670 | FU | Auckland, New Zealand |

| | | | |
|---|---|---|---|
| 1670 | PAH | Prainha, Brazil | |
| 1670 | TIPM | Palmar Sur, Costa Rica | |
| 1675 | TSL | Tsile Tsile, Papua New Guinea | |
| 1682 | QQ | Canoas, Brazil | |
| 1685 | MER | Mercaderes, Colombia | |
| 1689 | MH | Mt Hagen, Papua New Guinea | |
| 1690 | CXS | Campo Dos Bugres, Brazil | |
| 1692 | KIU | Kiunga, Papua New Guinea | |
| 1715 | BAN | Banos, Ecuador | |
| 1715 | PP | Vitoria, Brazil | |
| 1745 | CAN | Canar, Ecuador | |

*(Many of these are moving to the 200-400 kHz band)*

# Ship to Shore CW

**Public Correspondence Stations**
(TELEX)

WCC Chatham, MA
4238 4268 4331 6333.5 6337 6376 8586 8630
12926.5 12961.5 13033.5 16933.2 16972
16973.45 22347.5 22348.5 22366.5 22518
22521

KPH San Francisco, CA
4247 6477.5 8618 8642 12808.5 13002
17016.5 17016.8 22479 22557

WPD Tampa, FL
4274 6365.5 8615.5 13051.5 17170.4

KLC Galveston, TX
4256 6369 8508 8666 13038 16871.3 22467

WNU Slidell, LA
4294 4310 6326.5 6389.65 8525 8570
12826.5 13011 16861.7 17117.6 22431 22458

WLO Mobile, AL
4342.65 6414.5 6416 8514 8658 12660
12997.5 13024.9 16997.6 17021.6 22485
22487

**Marine Telex and Press**

| WCC Chatham, MA | | KPH San Francisco, CA | |
|---|---|---|---|
| Coast | Ship | Coast | Ship |
| 4356.5 | 4177 | 4356 | 4176.5 |
| 6504.5 | 6266.5 | 6500.5 | 6262.5 |
| 8715 | 8354 | 8711 | 8350 |
| 13081.5 | 12501.5 | 13077.5 | 12497.5 |
| 17207.5 | 16670.5 | 17203.5 | 16666.5 |
| 22571.5 | 22202.5 | 22567.5 | 22198.5 |

| KFS Palo Alto, CA | | WOM Ft. Lauderdale, FL | |
|---|---|---|---|
| Frequencies | | Coast | Ship |
| 2037.5 | 8558.4 | 4538 | 4048 |
| 2061.5 | 12695.5 | 4535 | 4015 |
| 4228 | 12844.5 | 4532 | 4057 |
| 4274 | 17026 | 4529 | 4051 |
| 6348 | 17184.8 | 7398 | 7980.5 |

| | | | |
|---|---|---|---|
| 6365.5 | 22425 | 7395 | 7977.5 |
| 8444.5 | 22515 | 7392.2 | 7974.5 |
| 8445 | | | |

Scheveningen, Netherlands

| Callsign | Coast | Ship |
|---|---|---|
| PCH52 | 4351.5 | 4172.0 |
| PCH95 | 4250.0 | 4178.0 |
| PCH53 | 6496.5 | 6258.5 |
| PCH96 | 6404.0 | 6268.5 |
| PCH4 | 8654.4 | 8298.1 |
| PCH54 | 8713.0 | 8352.0 |
| PCH5 | 12768.0 | 12521.5 |
| PCH55 | 13077.0 | 12497.0 |
| PCH92 | 17007.2 | 16698.0 |
| PCH56 | 17217.5 | 16680.5 |
| PCH98 | 22324.5 | 22226.0 |
| CH H | 1919.5 | 1972.5 |

# Medium Frequency Coastal Stations
(2 MHz)

*NOTE: Many coastal stations are abandoning their 2 MHz services. The following list is an example of this transition and many frequencies are now inactive.*

| Callsign/Location | Coast | Ship |
|---|---|---|
| KOW Seattle, WA | 2522 | 2126 |
| KFX Astoria, OR | 2442 | 2009 |
| | 2598 | 2206 |
| KTJ Coos Bay, OR | 2566 | 2031.5 |
| KOE Eureka, CA | 2506 | 2406 |
| | 2450 | 2366 |
| KLH San Fran, CA | 2506 | 2406 |
| | 2450 | 2003 |
| KOU San Pedro, CA | 2566 | 2009 |
| | 2598 | 2206 |
| | 2522 | 2126 |
| | 2466 | 2382 |
| KBP Kahuku, Hawaii | 2530 | 2134 |
| KMV Agana, Guam | 2506 | 2009 |
| WOU Boston, MA | 2506 | 2406 |
| | 2450 | 2366 |
| | 2566 | 2390 |
| WOX New York, NY | 2590 | 2198 |
| | 2522 | 2126 |
| | 2482 | 2382 |
| WOO Ocean Gate, NJ | 2558 | 2166 |
| WAE Pt Harbor, NC | 2538 | 2142 |
| WGB Norfolk, VA | 2450 | 2366 |
| WJO Charleston, SC | 2566 | 2390 |
| WNJ Jacksonvlle, FL | 2566 | 2390 |
| WFA Tampa, FL | 2550 | 2158 |
| | 2466 | 2009 |

# MARITIME

| | | | |
|---|---|---|---|
| WDR | Miami, FL | 2514 | 2118 |
| | | 2490 | 2031.5 |
| | | 2442 | 2406 |
| WLO | Mobile, AL | 2572 | 2430 |
| WAK | New Orleans,LA | 2598 | 2206 |
| | | 2482 | 2382 |
| WLF | Wilmington, DE | 2558 | 2166 |
| KGN | Delcambre, TX | 2506 | 2458 |
| KQP | Galveston, TX | 2530 | 2134 |
| | | 2450 | 2366 |
| KCC | Corpus Christi, TX | 2538 | 2142 |
| WCT | San Juan, PR | 2530 | 2134 |
| WAH | St Thomas, VI | 2506 | 2009 |

## Alaska

| Callsign/Location | | Coast | Ship |
|---|---|---|---|
| WGG53 | Cold Bay | 2312 | 2134 |
| WDU26 | Cordova | 2397 | 2237 |
| WGG58 | Juneau | 2400 | 2240 |
| WGG56 | Ketchikan | 2397 | 2237 |
| WDU23 | Kodiak | 2309 | 2131 |
| WGG55 | Nome | 2400 | 2240 |
| WDU29 | Sitka | 2312 | 2134 |

## Great Lakes

| Callsign/Location | | Coast | Ship |
|---|---|---|---|
| WBL | Buffalo, NY | 2514.0 | 2118.0 |
| | | 2550.0 | 2158.0 |
| | | 2582.0 | 2206.0 |
| | | 4415.8 | 4415.8 |
| | | 4428.6 | 4428.6 |
| | | 8783.2 | 8783.2 |
| WMI | Lorain, OH | 4369.8 | 4369.8 |
| | | 8796.4 | 8796.4 |
| WLC | Rogers City,MI | 2514.0 | 2118.0 |
| | | 2550.0 | 2158.0 |
| | | 2582.0 | 2206.0 |
| | | 4369.8 | 4075.4 |

## Additional Public Paired Channels

| Coast | Ship |
|---|---|
| 2582 | 2206 |
| 4072.4 | 4072.4 |
| 4371 | 4371 |
| 4415.8 | 4117.2 |
| 4419 | 4120.4 |
| 4428.6 | 4130 |
| 6147.5 | 6147.5 |
| 6455 | 6455 |
| 8210.8 | 8210.8 |
| 8793.2 | 8249.2 |

## 2 MHZ Simplex

| Freq | Use | Area |
|---|---|---|
| 2003 | Intership Safety | Great Lakes |
| 2082.5 | Intership Safety | All Areas |
| 2142 | Intership Safety | Pacific coast daytime |
| 2203 | Intership Safety | Gulf of Mexico |
| 2638 | Intership Safety | All Areas |
| 2670 | U.S.C.G. only | All Areas |
| 2738 | Intership Safety | All Areas except Great Lakes and Gulf of Mexico |
| 2782 | Intership Safety | All Areas |
| 2830 | Intership Safety | Gulf of Mexico |

### WOM Norwegian Steam Lines (USB)

| Coast | Ship |
|---|---|
| 4529.0 | 4051.0 |
| 4532.0 | 4057.0 |
| 4535.0 | 4015.0 |
| 4538.0 | 4048.0 |
| 7392.2 | 7974.5 |
| 7395.0 | 7977.5 |
| 7398.0 | 7980.5 |

## Mississippi Valley Simplex

| Callsign/Location | Frequencies | | |
|---|---|---|---|
| WCM Cincinnati | 2086.0 | 2782.0 | 4063.0 |
| | 6515.7 | 8213.6 | 12333.1 |
| | 16518.9 | | |
| WFN Jeffersonville/ Louisville | 2086.0 | 2782.0 | 4115.7 |
| | 6518.8 | 8725.1 | 13103.9 |
| | 17291.8 | | |
| WJG Memphis | 2086.0 | 2782.0 | 4087.8 |
| | 6209.3 | 8201.2 | 12333.1 |
| | 16518.9 | | |
| WGK St Louis | 2086.0 | 2782.0 | 4410.1 |
| | 6212.4 | 8737.5 | 13109.9 |
| | 17291.8 | | |

*WJG Memphis (photo by J.T. Pogue)*

# Ship to Shore Worldwide Duplex Voice
## *"High Seas" - (USB)*

The following frequency band plan has been adopted internationally to standardize frequency pairs used by ship and coastal radiotelephone stations.

Number/letter channel designators (4A through 22E) indicate simplex and are used by limited coastal stations and ships for private correspondence.

Channels 421, 606, 821, 1221, 1621 and 2221 are common international calling channels.

A superscript following a channel number (i.e., 401[1]) refers to the table of transmission times which follows this list.

| Channel # | Common User | Coast Freq | Ship Freq | USCG Use |
|---|---|---|---|---|
| 401[1] | KMI/WAH | 4357.4 | 4063.0 | |
| 402 | | 4360.5 | 4066.1 | |
| 403[3] | WOM | 4363.6 | 4069.2 | |
| 404 | | 4366.7 | 4072.3 | |
| 405 | PCH | 4369.8 | 4075.4 | |
| 406 | GKT | 4372.9 | 4078.5 | |
| 407 | | 4376.0 | 4081.6 | ALL |
| 408 | NOAA | 4379.1 | 4084.7 | |
| 409 | | 4382.2 | 4087.8 | |
| 410[7] | WOO | 4385.3 | 4090.9 | |
| 411[8] | WOO | 4388.4 | 4094.0 | |
| 412[4] | WOM | 4391.5 | 4097.1 | |
| 413 | | 4394.6 | 4100.2 | |
| 414 | DAN | 4397.7 | 4103.3 | |
| 415 | | 4400.8 | 4106.4 | ALL |
| 416[2] | KMI | 4403.9 | 4109.5 | |
| 416[9] | WOO | | | |
| 417[5] | WOM/KMI | 4407.0 | 4112.6 | |
| 418 | OXZ | 4410.1 | 4115.7 | |
| 419 | WLO | 4413.2 | 4118.8 | |
| 420 | SAG | 4416.3 | 4121.9 | |
| 421 | | 4419.4 | 4125.0 | |
| 4A | | | 4125.0 | |
| 4C | | | 4419.4 | |
| 422[10] | WOO | 4422.5 | 4128.1 | |
| 423[6] | WOM | 4425.6 | 4131.2 | |
| 424 | *2 | 4428.7 | 4134.3 | ALL |
| 425 | OST,OXZ | 4431.8 | 4137.4 | |
| 425 | NOAA | 4434.9 | 4140.5 | |
| 427 | (4B) | | 4143.6 | |
| 601 | DAN | 6506.4 | 6200.0 | ALL |
| 602 | OST,NOAA | 6509.5 | 6203.1 | |
| 603 | | 6512.6 | 6206.2 | A,G,R |
| 604 | LPL,WAH | 6515.7 | 6209.3 | |

*The radio room aboard the exploration vessel, "Discovery"*

| Channel # | Common User | Coast Freq | Ship Freq | USCG Use |
|---|---|---|---|---|
| 605 | WAH | 6518.8 | 6212.4 | ALL |
| 606 | (6C) | 6521.9 | 6215.5 | |
| 607 | (6A) | | 6218.6 | |
| 608 | (6B) | | 6221.6 | |
| 801 | SAG | 8718.9 | 8195.0 | ALL-C/G |
| 802[3] | WOM | 8722.0 | 8198.1 | |
| 803 | SAG | 8725.1 | 8201.2 | |
| 804[1] | KMI/WAH | 8728.2 | 8204.3 | |
| 805[4] | WOM | 8731.3 | 8207.4 | |
| 806 | PCH | 8734.4 | 8210.5 | |
| 807 | Vancouvr | 8737.5 | 8213.6 | |
| 808[9] | WOO | 8740.6 | 8216.7 | |
| 809[2] | KMI | 8743.7 | 8219.8 | |
| 810[4] | WOM | 8746.8 | 8222.9 | |
| 811[10] | WOO | 8749.9 | 8226.0 | |
| 812 | NOAA | 8753.0 | 8229.1 | |
| 813 | OST | 8756.1 | 8232.2 | |
| 814 | WOM | 8759.2 | 8235.3 | |
| 815[8] | WOO | 8762.3 | 8238.4 | |
| 816 | GKU | 8765.4 | 8241.5 | ALL |
| 817 | DAN | 8768.5 | 8244.6 | A,G,R |
| 818 | OXZ | 8771.6 | 8247.7 | |
| 819 | | 8774.7 | 8250.8 | ALL-C/G |
| 820 | | 8777.8 | 8253.9 | |
| 821 | | 8780.9 | 8257.0 | |

# MARITIME

| Ch | Station | Freq 1 | Freq 2 | Notes |
|---|---|---|---|---|
| 822[1] | KMI | 8784.0 | 8260.1 | |
| 823 | | 8787.1 | 8263.2 | |
| 824 | WLO | 8790.2 | 8266.3 | |
| 825[6] | WOM | 8793.3 | 8269.4 | |
| 826[7] | WOO | 8796.4 | 8272.5 | |
| 827 | | 8799.5 | 8275.6 | |
| 828 | *1,NOAA | 8802.6 | 8278.7 | |
| 829 | | 8805.7 | 8281.8 | |
| 830 | WLO | 8808.8 | 8284.9 | |
| 831 | WOM | 8811.9 | 8288.0 | |
| 832 | (8A) | | 8291.1 | |
| 833 | (8B) | | 8294.2 | |
| 1201[1] | KMI | 13100.8 | 12330.0 | |
| 1202[2] | KMI | 13103.9 | 12333.1 | |
| 1203[2] | KMI | 13107.0 | 12336.2 | |
| 1203[10] | WOO | | | |
| 1204 | WOO | 13110.1 | 12339.3 | |
| 1205 | | 13113.2 | 12342.4 | ALL |
| 1206[3] | WOM | 13116.3 | 12345.5 | |
| 1207 | PCH | 13119.4 | 12348.6 | |
| 1208[4] | WOM | 13122.5 | 12351.7 | |
| 1209[5] | WOM/ZSC | 13125.6 | 12354.8 | |
| 1210[7] | WOO | 13128.7 | 12357.9 | |
| 1211[8] | WOO | 13131.8 | 12361.0 | |
| 1212 | WLO | 13134.9 | 12364.1 | |
| 1213 | PCH,8PO | 13138 | 12367.2 | |
| 1214 | OXZ,NOAA | 13141.1 | 12370.3 | |
| 1215[6] | WOM | 13144.2 | 12373.4 | |
| 1216 | | 13147.3 | 12376.5 | |
| 1217 | | 13150.4 | 12379.6 | ALL |
| 1218 | | 13153.5 | 12382.7 | |
| 1219 | PCH,NOAA | 13156.6 | 12385.8 | A,G,R |
| 1220 | | 13159.7 | 12388.9 | |
| 1221 | | 13162.8 | 12392.0 | |
| 1222 | *1 | 13165.9 | 12395.1 | |
| 1223 | WOM | 13169.0 | 12398.2 | |
| 1224 | DAN | 13172.1 | 12401.3 | |
| 1225 | LGQ | 13175.2 | 12404.4 | |
| 1226 | WLO*1 | 13178.3 | 12407.5 | |
| 1227 | | 13181.4 | 12410.6 | |
| 1228[9] | WOO | 13184.5 | 12413.7 | |
| 1229[1] | KMI | 13187.6 | 12416.8 | |
| 1230 | WOM | 13190.7 | 12419.9 | |
| 1231 | *1 | 13193.8 | 12423.0 | |
| 1232 | *2 | 13196.9 | 12426.1 | ALL-C/G |
| 1233 | (12A) | | 12429.2 | |
| 1234 | (12B) | | 12432.3 | |
| 1235 | (12C) | | 12435.4 | |
| 1601[3] | WOM | 17232.9 | 16460 | |
| 1602[1] | KMI | 17236.0 | 16463.1 | |
| 1603[2] | KMI/WAH | 17239.1 | 16466.2 | |
| 1604 | | 17242.2 | 16469.3 | |
| 1605[8] | WOO*2 | 17245.3 | 16472.4 | |
| 1606 | IAR | 17248.4 | 16475.5 | A,G,R |
| 1607 | | 17251.5 | 16478.6 | |
| 1608 | ZSC | 17254.6 | 16481.7 | |
| 1609[4] | WOM | 17257.7 | 16484.8 | |
| 1610[5] | WOM | 17260.8 | 16487.9 | |
| 1611[6] | WOM | 17263.9 | 16491.0 | |
| 1612 | NOAA | 17267.0 | 16494.1 | |
| 1613 | NOAA | 17270.1 | 16497.2 | ALL |
| 1614 | | 17273.2 | 16500.3 | |
| 1615 | GKU | 17276.3 | 16503.4 | |
| 1616 | WOM | 17279.4 | 16506.5 | |
| 1617 | OXZ | 17282.5 | 16509.6 | |
| 1618 | OXZ | 17285.6 | 16512.7 | |
| 1619 | *1 | 17288.7 | 16515.8 | |
| 1620[9] | WOO | 17291.8 | 16518.9 | |
| 1621 | | 17294.9 | 16522.0 | |
| 1622 | *1 | 17298.0 | 16525.1 | |
| 1623 | PCH | 17301.1 | 16528.2 | |
| 1624 | KMI | 17304.2 | 16531.3 | |
| 1625 | | 17307.3 | 16534.4 | ALL |
| 1626[10] | WOO | 17310.4 | 16537.5 | |
| 1627 | | 17313.5 | 16540.6 | |
| 1628 | *1 | 17316.6 | 16543.7 | |
| 1629 | | 17319.7 | 16546.8 | |
| 1630 | | 17322.8 | 16549.9 | |
| 1631[7] | WOO | 17325.9 | 16553.0 | |
| 1632 | | 17329.0 | 16556.1 | |
| 1633 | GKW | 17332.1 | 16559.2 | |
| 1634 | | 17335.2 | 16562.3 | |
| 1635 | OXZ | 17338.3 | 16565.4 | |
| 1636 | PCH | 17341.4 | 16568.5 | |
| 1637 | GKT | 17344.5 | 16571.6 | |
| 1638 | | 17347.6 | 16574.7 | ALL-C/G |
| 1639 | PCH | 17350.7 | 16577.8 | |
| 1640 | *2 | 17353.8 | 16580.9 | |
| 1641 | WLO | 17356.9 | 16584.0 | |
| 1642 | (16A) | | 16587.1 | |
| 1643 | (16B) | | 16590.2 | |
| 1644 | (16C) | | 16593.3 | |
| 2201[9] | WOO | 22596.0 | 22000.0 | |
| 2202 | IAR | 22599.1 | 22003.1 | |
| 2203 | SAG | 22602.2 | 22006.2 | |
| 2204 | *1 | 22605.3 | 22009.3 | |
| 2205[7] | WOO | 22608.4 | 22012.4 | |
| 2206 | GKT | 22611.5 | 22015.5 | ALL |
| 2207 | DAN | 22614.6 | 22018.6 | |
| 2208 | | 22617.7 | 22021.7 | |
| 2209 | | 22620.8 | 22024.8 | |
| 2210[8] | WOO | 22623.9 | 22027.9 | |
| 2211 | | 22627.0 | 22031.0 | |
| 2212 | | 22630.1 | 22034.1 | |
| 2213 | | 22633.2 | 22037.2 | |
| 2214[2] | KMI | 22636.3 | 22040.3 | |
| 2215[3] | WOM | 22639.4 | 22043.4 | |
| 2216[4] | WOM | 22642.5 | 22046.5 | |
| 2217 | *2 | 22645.6 | 22049.6 | |
| 2218 | OXZ | 22648.7 | 22052.7 | ALL |
| 2219 | | 22651.8 | 22044.8 | |
| 2220 | GKU | 22654.9 | 22058.9 | |
| 2221 | | 22658.0 | 22062.0 | |
| 2222 | WOM | 22661.1 | 22065.1 | |
| 2223 | KMI/WAH | 22664.2 | 22068.2 | |
| 2224 | | 22667.3 | 22071.3 | |

| | | | |
|---|---|---|---|
| 2225 | | 22670.4 | 22074.4 |
| 2226 | | 22673.5 | 22077.5 |
| 2227 | | 22676.6 | 22080.6 |
| 2228 | KMI | 22679.7 | 22083.7 |
| 2229 | GKU | 22682.8 | 22086.8 |
| 2230 | SAG | 22685.9 | 22089.9 |
| 2231 | | 22689.0 | 22093.0 |
| 2232 | PCH | 22692.1 | 22096.1 |
| 2233 | NOAA | 22695.2 | 22099.2 |
| 2234 | | 22698.3 | 22102.3 |
| 2235 | NOAA | 22701.4 | 22105.4 |
| 2236 | KMI | 22704.5 | 22108.5 |
| 2236¹⁰ | WOO | | |
| 2237 | WLO | 22707.6 | 22111.6 |
| 2238 | | 22710.7 | 22114.7 |
| 2239 | | 22713.8 | 22117.8 |
| 2240 | | 22716.9 | 22120.9 |
| 2241 | (22A) | | 22124 |
| 2242 | (22B) | | 22127.1 |
| 2243 | (22C) | | 22130.2 |
| 2244 | (22D) | | 22133.3 |
| 2245 | (22E) | | 22136.4 |

*1-(FFL/FFS/FFT) St. Lys Radio, France
*2-(SVA/SVB/SVN) Athens Radio, Greece

## USCG Key

ALL = All stations use
A = NMA Miami
F = NMF Boston
G = NMG New Orleans
N = NMN Portsmouth
R = NMR San Juan
ALL-C/G = All except Caribbean
& Gulf of Mexico

## User Locations:

| | |
|---|---|
| DAN | Norddeich, Fed Rep Germany |
| GKT | Portishead Radio, England |
| GKU | Portishead Radio, England |
| GKW | Portishead Radio, England |
| IAR | Rome, Italy |
| KMI | Dixon, CA |
| LGQ | Rogaland, Norway |
| LPL | Gen Pac Radio, Buenos Aires, ARG |
| OST | Oostende, Belgium |
| OXZ | Lyngby, Denmark |
| PCH | Scheveningen, Netherlands |
| SAG | Goteborg, Sweden |
| WAH | St. Thomas, Virgin Islands |
| WOM | Ft Lauderdale, FL |
| WOO | Manahawkin (NY) New Jersey ("Oceangate Radio") |
| WLO | Mobile, AL |
| ZSC | Capetown Radio, S. Africa |
| 8PO | Bridgetown, Barbados |

## Transmission Times
### (TFC = traffic, WX = weather)

**(1) KMI**
| | |
|---|---|
| 0000 | TRFC/WX |
| 0300 | TRFC |
| 0600 | TRFC/WX |
| 0900 | TRFC |
| 1200 | TRFC |
| 1500 | TRFC/WX |
| 1800 | TRFC |
| 2100 | TRFC |

**(2) KMI**
| | |
|---|---|
| 0100 | TRFC |
| 0400 | TRFC |
| 0700 | TRFC |
| 1000 | TRFC |
| 1300 | TRFC/WX |
| 1600 | TRFC |
| 1900 | TRFC/WX |
| 2200 | TRFC |

**(3) WOM**
| | |
|---|---|
| 0030 | TRFC |
| 0430 | TRFC |
| 0830 | TRFC |
| 1230 | TRFC/WX |
| 1630 | TRFC |
| 2030 | TRFC |

**(4) WOM**
| | |
|---|---|
| 0130 | TRFC |
| 0530 | TRFC |
| 0930 | TRFC |
| 1330 | TRFC/WX |
| 1730 | TRFC |
| 2130 | TRFC |

**(5) WOM**
| | |
|---|---|
| 0230 | TRFC |
| 0630 | TRFC |
| 1030 | TRFC |
| 1430 | TRFC |
| 1830 | TRFC |
| 2230 | TRFC/WX |

**(6) WOM**
| | |
|---|---|
| 0330 | TRFC |
| 0730 | TRFC |
| 1130 | TRFC |
| 1530 | TRFC |
| 1930 | TRFC |
| 2330 | TRFC/WX |

**(7) WOO**
| | |
|---|---|
| 0000 | TRFC |
| 0400 | TRFC |
| 0800 | TRFC |
| 1200 | TRFC/WX |
| 1600 | TRFC |
| 2000 | TRFC/WX |

**(8) WOO**
| | |
|---|---|
| 0100 | TRFC |
| 0500 | TRFC |
| 0900 | TRFC |
| 1300 | TRFC/WX |
| 1700 | TRFC |
| 2100 | TRFC/WX |

| (9) **WOO** | | (10) **WOO** | |
|---|---|---|---|
| 0200 | TRFC | 0300 | TRFC |
| 0600 | TRFC | 0700 | TRFC |
| 1000 | TRFC | 1100 | TRFC |
| 1400 | TRFC/WX | 1500 | TRFC/WX |
| 1800 | TRFC | 1900 | TRFC |
| 2200 | TRFC/WX | 2300 | TRFC/WX |

# Automated Mutual Assistance Vessel Rescue

## (AMVER) System

### Worldwide Ship to Shore Channels (USB)

In order to ensure safety on the high seas, coastal stations around the globe maintain a constant radio watch on guarded channels to assist vessels in distress.

In addition to the channels listed, all stations (except U.S. Coast Guard NMF, NMN, NMA and NMR) monitor the international distress and calling frequency 2182 kHz. This frequency is usually the calling channel when no ship frequency is shown in the list.

## Atlantic Communications

### Western North Atlantic

| Call/Location | | Coast | Ship |
|---|---|---|---|
| CANADA | | | |
| (See listings under Canadian section, Canadian Coast Guard) | | | |
| UNITED STATES | | | |
| NMA | Miami, FL | 6506.4 | 6200.0 |
| NMF | Boston, MA | 6506.4 | 6200.0 |
| NMG | New Orleans, LA | 4428.7 | 4134.3 |
| | | 6506.4 | 6200.0 |
| | | 8765.4 | 8241.5 |
| | | 13113.2 | 12342.4 |
| NMN | Portsmouth, VA | 8465.0 | 8 MHz |
| | | 12718.5 | 12 MHz |
| | | 16976.0 | 16 MHz |
| | | 4428.7 | 4134.3 |
| | | 6506.4 | 6200.0 |
| | | 8765.4 | 8241.5 |
| | | 13113.2 | 12342.4 |
| BERMUDA | | | |
| ZBM | St George | 2582 | 2182 |

### Eastern North Atlantic

| Call/Location | | Coast | Ship |
|---|---|---|---|
| NORWAY | | | |
| LGA | Alesund | 1722 | 2442 |
| LGN | Bergen | 1743 | 2463 |
| LGP | Bodo | 2656 | 2139 |
| LGZ | Farsund | 1750 | 2470 |
| LGL | Floro | 2649 | 2132 |
| LGI | Hammerfest | 1722 | 2442 |
| LGH | Harstad | 1736 | 2456 |
| LFO | Orlandet | 2635 | 2118 |
| LGQ | Rogaland | 1729 | 2449 |
| LFW | | 4325.0 | 4 MHz 4185 |
| LGU | | 6432.0 | 6 MHz 6467 6277.5 |
| LFU | | 6467.0 | 6 MHz |
| LFN | | 8527.5 | 8 MHz |
| LGB | | 8574.0 | 8 MHz 8370 |
| LFB | | 8678.0 | 8 MHz 8678 |
| LGJ | | 12727.5 | 12 MHz |
| LFJ | | 12876.0 | 12 MHz 12555 |
| LFI | | 12961.5 | 12 MHz |
| LFT | | 16952.4 | 16 MHz |
| LGX | | 17074.4 | 16 MHz 16740 |
| LFF | | 17165.6 | 16 MHz |
| LGG | | 22425.0 | 22 MHz |
| LFG | | 22473.0 | 22 MHz 22242 |
| LGT | Tjome | 1736 | 2456 |
| LGE | Tromso | 1750 | 2470 |
| LGV | Vardo | 1729 | 2449 |
| | | 2182 | |
| SWEDEN | | | |
| SAG2 | Gothenburg | 4262.0 | 4 MHz |
| SAG3 | | 6372.5 | 6 MHz |
| SAG4 | | 8498.0 | 8 MHz |
| SAB4 | | 8646.0 | 8 MHz |
| SAG6 | | 12880.5 | 12 MHz |
| SAB6 | | 12755.5 | 12 MHz |
| SAG8 | | 17079.4 | 16 MHz |
| SAG9 | | 22413.0 | 22 MHz |
| SAG25 | | 25461.0 | 25 MHz |
| NETHERLANDS | | | |
| PCH | Scheveningen | 1764 | 2030 |
| | | 2824 | 2520 |
| | | 4369.8 | 4075.4 |
| | | 4419.4 | 4125 |
| | | 6509.5 | 6203.1 |

| | | |
|---|---|---|
| | 6521.9 | 6215.5 |
| | 8780.9 | 8257 |
| | 13162.8 | 12392 |
| | 17294.9 | 16522 |
| | 22658.0 | 22062 |
| | 6406 | 4 MHz |
| | 12966 | 12 MHz |
| PCH20 | 4250.0 | 4 MHz |
| PCH40 | 8562.0 | 8 MHz |
| PCH41 | 8622.0 | 8 MHz |
| PCH42 | 8654.4 | 8 MHz |
| PCH50 | 12768.0 | 12 MHz |
| PCH51 | 12799.5 | 12 MHz |
| PCH52 | 12853.5 | 12 MHz |
| PCH60 | 16902.0 | 16 MHz |
| PCH61 | 17007.2 | 16 MHz |
| PCH62 | 17104.2 | 16 MHz |
| PCH70 | 22324.5 | 22 MHz |
| PCH71 | 22539.0 | 22 MHz |
| PCG41 | 8796.4 | 8272.5 |
| PCG51 | 13138.0 | 12367.2 |
| PCG61 | 17341.4 | 16568.5 |
| PCG71 | 22608.4 | 22012.4 |

## IRELAND

| | | | |
|---|---|---|---|
| EJM | Malin Head | 1841 | 2182 |
| EJK | Valentia Is | 1827 | 2182 |

## SPAIN

| | | | |
|---|---|---|---|
| EAC | Cadiz | 4275.5 | 4 MHz |
| | | 6505.5 | 6 MHz |
| | | 8726 | 8 MHz |
| | | 13056 | 12 MHz |
| | | 17175.2 | 16 MHz |
| | | 22384 | 22 MHz |
| EAD | Aranjuez | 4349 | 4 MHz |
| EAD2 | | 6382.22 | 6 MHz |
| EAD3 | | 8682 | 8 MHz |
| EAD44 | | 12887.5 | 12 MHz |
| EAD4 | | 13065 | 12 MHz |
| EAD5 | | 17184.8 | 16 MHz |
| EAD6 | | 22446 | 22 MHz |
| EAF | Vigo | 6498.5 | 6 MHz |
| | | 8473 | 8 MHz |
| | | 13092 | 12 MHz |
| | | 17280.8 | 16 MHz |
| EAT | Santa Cruz de Tenerife | 6498.5 | 6 MHz |
| | | 8473 | 8 MHz |
| | | 13092 | 12 MHz |
| | | 6942.8 | 16 MHz |
| EDZ | | 4269 | 4 MHz |
| EDZ2 | | 6400.5 | 6 MHz |
| EDZ4 | | 8618 | 8 MHz |
| EDZ5 | | 12934.5 | 12 MHz |
| EDZ6 | | 17064.8 | 16 MHz |
| EDZ7 | | 22533 | 22 MHz |

## GREAT BRITAIN

(Consult ITU listings
for frequencies)

| | |
|---|---|
| GKA | Portishead |
| GKR | Wick (Scotland) |
| GND | Stonehaven (Scotland) |
| GCC | Cullercoats |
| GKZ | Humber |
| GNF | Northforeland |
| GNI | Niton |
| GLD | Lands End |
| GIL | Ilfracombe |
| GLV | Anglesey |
| GPK | Portpatrick (Scotland) |
| GHD | Hebrides |

### South Atlantic

| Call/Location | | Coast | Ship |
|---|---|---|---|
| **ARGENTINA** | | | |
| LPD68 | Gen.Pacheco | 4268.0 | 4 MHz |
| LPD86 | | 8646.0 | 8 MHz |
| LPD88 | | 12988.5 | 12 MHz |
| LPD46 | | 17045.6 | 16 MHz |
| | | 18081.5 | |
| LPD91 | | 22419.0 | 22 MHz |
| | | 20520 | |
| LOL | | 17665 | |
| LPL | | 17285.7 | |
| LPL5 | | 17232.9 | |
| **SOUTH AFRICA** | | | |
| ZSJ2 | Navcomcen Capetown | 4145 | 4 MHz |
| | | 4283 | 4 MHz |
| ZSJ3 | | 6386.5 | 6 MHz |
| ZSJ4 | | 8566.0 | 8 MHz |
| ZSJ5 | | 12849.0 | 12 MHz |
| ZSJ6 | | 17132.0 | 16 MHz |

## Pacific Communications

### Eastern Pacific

| Call/Location | | Coast | Ship |
|---|---|---|---|
| **CANADA** | | | |
| (See listings under Canadian Coast Guard) | | | |
| **UNITED STATES** | | | |
| NOJ | Kodiak, AK | 2670 | 2182 |
| | | 6506.4 | 6200.0 |
| NMC | San Francisco CA | 2670 | 2182 |
| | | 6383.0 | 6 MHz |
| | | 8574.0 | 8 MHz |
| | | 16800.9 | 16 MHz |
| | | 4428.7 | 4134.3 |

# MARITIME

|  |  | Coast | Ship |
|---|---|---|---|
|  |  | 6506 | 6200 |
|  |  | 8765.4 | 8241.5 |
|  |  | 13113.2 | 12342.4 |
| NMO | Honolulu,HI | 2670 | 2182 |
|  |  | 8650.0 | 8 MHz |
|  |  | 12899.5 | 12 MHz |
|  |  | 22476.0 | 22 MHz |
|  |  | 4428.7 | 4134.3 |
|  |  | 6506.4 | 6200.0 |
|  |  | 8675.4 | 8241.5 |
| KUQ | Pago Pago, Am Samoa | 5475 | 4 MHz |
|  |  | 6361.0 | 6 MHz |
|  |  | 8585.0 | 8 MHz |
|  |  | 12871.5 | 12 MHz |

## CHILE

| CBV | Valparaiso | 4349.0 | 4 MHz |
|---|---|---|---|
|  |  | 8478.0 | 8 MHz |
|  |  | 12714.0 | 12 MHz |
|  |  | 16945.0 | 16 MHz |
|  |  | 22473.0 | 22 MHz |

## ECUADOR

| HCG | Guayaquil | 8476 | 8 MHz |
|---|---|---|---|
|  |  | 12711 | 12 MHz |
|  |  | 16948 | 16 MHz |

## FRENCH POLYNESIA

| FJA | Mahina Tahiti | 2620 | 2182 |
|---|---|---|---|
|  |  | 8674 | 8230 |
|  |  | 8805.7 | 8281.8 |

## FIJI

| 3DP | Suva | 2111 | 2182 |
|---|---|---|---|
|  |  | 6215.5 | 6215.5 |
|  |  | 8690.0 | 8 MHz |
|  |  | 12700.0 | 12 MHz |

## AUSTRALIA

| VIS | Sydney | 2201 | 2182 |
|---|---|---|---|
|  |  | 4428.7 | 4125.0 |
| VIS53 |  | 4245.0 | 4 MHz |
|  |  | 6512.6 | 6215.5 |
| VIS3 |  | 6464.0 | 6 MHz |
| VIS5 |  | 12952.5 | 12 MHz |
|  |  | 12979.5 |  |
| VIS6 |  | 17161.3 | 16 MHz |
|  |  | 17194.4 |  |
| VIS26 |  | 8421.0 | 8 MHz |
|  |  | 8452.0 |  |
| VIS42 |  | 22474.0 | 22 MHz |
| VIP | Perth | 2201 | 2182 |
| VIP |  | 4428.7 | 4125.0 |
|  |  | 6512.6 | 6215.5 |
| VIP3 |  | 8597.0 | 8 MHz |
| VIP4 |  | 12994.0 | 12 MHz |
| VIP5 |  | 16947.6 | 16 MHz |
| VIP6 |  | 22315.5 | 22 MHz |
| VIP7 |  | 4229.0 | 4 MHz |
| VIO | Broome | 2201 | 2182 |
|  |  | 6407.5 | 6 MHz |
|  |  | 4428.7 | 4125.0 |
|  |  | 6512.6 | 6215.5 |
| VIC | Carnarvon | 2201 | 2182 |
|  |  | 6407.5 | 6 MHz |
|  |  | 4428.7 | 4125.0 |
|  |  | 6512.6 | 6215.5 |
| VID | Darwin | 2201 | 2182 |
|  |  | 6463.5 | 8 MHz |
|  |  | 4428.7 | 4125.0 |
|  |  | 6512.6 | 6215.5 |
| VIR | Rockhampton | 2201 | 2182 |
|  |  | 4255.6 | 4 MHz |
|  |  | 4428.7 | 4125.0 |
|  |  | 6512.6 | 6215.5 |
| VII | Thursday Is | 2201 | 2182 |
|  |  | 6333.5 | 8 MHz |
|  |  | 4428.7 | 4125.0 |
|  |  | 6512.6 | 6215.5 |
| VIT | Townsville | 2201 | 2182 |
|  |  | 6463.5 | 8 MHz |
|  |  | 4428.7 | 4125.0 |
|  |  | 6512.6 | 6215.5 |
| Emergency/Calling |  | 2524 |  |
|  |  | 2182 |  |
| Customs |  | 7160 |  |

## NEW ZEALAND

| ZLD | Auckland | 2207 | 2182 |
|---|---|---|---|
|  |  | 4143.6 | 4125 |
|  |  | 6218.6 | 6215.5 |
| ZLB | Awarua | 2423 | 2182 |
|  |  | 4143.6 | 4125 |
| ZLB2 |  | 4277.0 | 4 MHz |
| ZLB3 |  | 6393.5 | 6 MHz |
| ZLB4 |  | 8504.0 | 8 MHz |
| ZLB5 |  | 12740.0 | 12 MHz |
| ZLB6 |  | 17170.4 | 16 MHz |
| ZLB7 |  | 22533.0 | 22 MHz |
| ZLW | Wellington | 2153 | 2182 |
|  |  | 4143.6 | 4125 |
| ZLC | Chatham Is | 2104 | 2182 |

### Western Pacific

| Call/Location |  | Coast | Ship |
|---|---|---|---|

## UNITED STATES

| NRV | Guam | 2670 | 2182 |
|---|---|---|---|
|  |  | 8570.0 | 8 MHz |
|  |  | 12743.0 | 12 MHz |
|  |  | 17146.4 | 16 MHz |
|  |  | 22567.0 | 22 MHz |
|  |  | 6200.0 |  |
|  |  | 6506.4 |  |

12342.4
13113.2

<u>PHILLIPINES</u>

| DZG Las Pinas | 6441.0 | 6 MHz |
|---|---|---|
| | 8632.0 | 8 MHz |
| | 12948.0 | 12 MHz |
| | 17176.0 | 16 MHz |
| | 22502.0 | 22 MHz |

# Canada
## Limited Coastal Stations

| Halifax, N.S. | T | R |
|---|---|---|
| | 4133.2 | 4431.8 |
| | 8236.4 | 8770.4 |
| | 12375.5 | 13154.5 |
| | 16505.5 | 17300.5 |
| Inuvik, NWT | | |
| | 2220 | 2484 |

# Netherlands Antilles

| Curacao Radio, USB | T | R |
|---|---|---|
| | 2158.0 | 2550.0 |
| | 4110.8 | 4409.4 |
| | 8236.4 | 8770.4 |
| | 12375.5 | 13154.5 |
| | 16505.5 | 17300.5 |
| Emergency/ | | |
| Calling | 2182.0 | 2524.0 |
| Customs | 7160.0 | |

# Argentina
## General Pacheco Radio
### LPL Buenos Aires, Arg.

2086 2182 4394.6 4419.4 6512.6 8759.2
8780.9 13159.7 13165.9 17232.9 17285.6
17249.9 22605.3

### LPD (CW)

444.5 524 4262 4268 8646 12763.5
12988.5 17045.6 22419 22513.5 25130.5

# Australia
## Marine Radio

### Radiolocation/Radionavigation
1606.5-1800
1624
1782

*The R.V. Farnella, an oceanographic research vessel (Photo courtesy J. Marr Ltd.)*

**Traffic**
Coast: 2201 4428.7
Ship: 4134.3

**Weather**
Coast: 6512.6
Ship: 6206.2

**Distress & Safety**
2182 4125 6215.5

**Coastal Radiotelephone Stations**

| <u>Call</u> | <u>Station</u> |
|---|---|
| VIA | Adelaide Radio |
| VIB | Brisbane Radio |
| VIO | Broome Radio |
| VIC | Carnarvon Radio |
| VID | Darwin Radio |
| VIE | Esperance Radio |
| VIH | Hobart Radio |
| VII | Thursday Is Radio |
| VIM | Melbourne Radio |
| VIP | Perth Radio |
| VIR | Rockhampton Radio |
| VIS | Sydney Radio |
| VIT | Townsville Radio |
| VJN | Norfolk Is Radio |

| CH # | Coastal Station |
|---|---|
| 404 | VIB, VIH, VIM, VIP, VIT |
| 405 | VIS |
| 412 | VIA, VIB, VIT |
| 415 | VID, VIP |
| 417 | VIM, VIR, VIS |
| 419 | VIA, VID, VIT |
| 802 | VIS |
| 806 | VIP |

| | |
|---|---|
| 811 | VIB, VID, VIM, VIP |
| 815 | VID, VIP |
| 817 | VIA, VIT |
| 822 | VIT |
| 829 | VIA, VIS |
| 1203 | VIS, VIT |
| 1226 | VIM, VIP |
| 1227 | VIA |
| 1229 | VIB, VID, VIP |
| 1231 | VIS, VIT |
| 1602 | VIS |
| 1604 | VIP |
| 1610 | VIS |
| 1622 | VIS |
| 2203 | VIS |
| 2212 | VIP |
| 2223 | VIS |

*Note: See p. 151 for frequency plan.*

*Note: See p. 151 for frequency plan.*

# JAPAN
## Coastal Stations

| | | | |
|---|---|---|---|
| JDT | Yokosuka | JNG | Okinawa |
| JFA | Miwa | JNJ | Kagoshima |
| JFH | Okinawa | JNK | Nagasaki |
| JGC | Kanagawa | JNL | Hokkaido |
| JGC | Yokohama | JNN | Miyagi |
| JGD | Hyogo | JNR | Fukuoka |
| JNA | Tokyo | JNT | Nagoya |
| JNB | Okinawa | JNV | Niigata |
| JNC | Kyoto | JNX | Hokkaido |
| JNE | Hiroshima | | |

### Maritime Safety Agency

4276    8492    12942

Emergency Channel
4630

# Centre International Radio Medical (CIRM)

Station IRM in Rome provides medical advice to ships worldwide. CW communications are conducted on nine frequencies with ships replying on nearby telegraphy frequencies: 4342.5 4350.5 6365 6420 8685 12748 12760 17105 22525. Alternatively, ships may contact IAR or Rome Radio:

| IAR | Rome Radio |
|---|---|
| 4292 | 8778.6 |
| 4320 | 8796.4 |
| 6409.5 | 13125.6 |
| 6418.2 | 17239.1 |

| | |
|---|---|
| 6435.5 | 17248.4 |
| 8530 | 17304.2 |
| 8669.9 | 22599.1 |
| 13015.3 | 22627 |
| 16895.3 | |
| 17005 | |
| 17160.8 | |
| 22372.4 | |
| 22378 | |

# Medical Advisory System (MAS)

Many (if not most) vessels at sea are poorly equipped to handle medical emergencies which often arise with no hope of immediate assistance. Whether an accident or illness, stricken victims may face a bleak prospect of rescue and recovery.

With physicians on hand round the clock and operating out of Owings, Maryland, Medical Advisory Systems, Inc. (MAS) utilizes the following frequencies (USB) for ship to shore voice conferencing.

| | |
|---|---|
| 4983 | 16450 |
| 7952 | 16590.2 |
| 12327 | 22722 |

# Global Maritime Distress and Safety System
## (GMDSS)

2187.5

Scheduled for implementation between 1992 and 1999, this automated system will alert ship and shore stations of a vessel in distress, as well as provide navigational and meteorological warnings.

# International Red Cross Emergency Network

Due to its humanitarian nature, the International Red Cross enjoys immunity during armed conflict. In turn, no military or strategic involvement is allowed. Language spoken on the airwaves must be French or English and in the clear (no encoding).

Emission will be CW or SSB with field portable stations transmitting 150 watts of power. Geneva base station power is 800 watts. Frequencies underlined are most commonly reported.

## Frequencies

| | | |
|---|---|---|
| 3801.5 | 13950 | 20753 (Pri) |
| 3815.5 | 13965 | 20800 |
| 6998.5 | 13973 | 20815 |
| 13569.5 | 13998.5 | 20942 |
| 13915 | 14375 | 20998 |
| | | 27998 |

| Callsign | Location |
|---|---|
| 5VR88 | Lome, Togo |
| 5XR88 | Kampala, Uganda |
| 5ZR88 | Nairobi, Kenya |
| 6TR88 | Khartoum, Sudan |

| | |
|---|---|
| 9JR88 | Lusaka, Zambia |
| 9TO20 | Kinshsa, Zaire |
| C8Z650 | Maputo, Mozambique |
| CPR88 | La Paz, Bolivia |
| DEK23 | Berlin, GFR |
| DEK88 | Bonn, GFR |
| DUML | Manila, Philippines |
| EPR88 | Tehran, Iran |
| ETC88 | Addis Abeba, Ethiopia |
| HBC88 | Versoix, Switzerland |
| HBC881 | Geneva, Switzerland |
| HCR88 | Quito, Ecuador |
| HRC88 | Tegucigalpa, Honduras |
| HTCR88 | Managua, Nicaragua |
| HUCRS | San Salvador, El Salvador |
| J3E25 | St. George, Granada |
| JYL88 | Amman, Jordan |
| ODR80 | Tiro, Lebanon |
| ODR81 | Baalbek, Lebanon |
| ODR8? | Ksara, Lebanon |
| ODR83 | Naqoura, Lebanon |
| ODR8? | Tripoli, Lebanon |
| ODR88 | Beirut, Lebanon |
| OJR88 | Helsinki, Finland |
| ONC88 | Brussels, Belgium |
| PGA88 | The Hague, Netherlands |
| S2C88 | Dakha, Bangladesh |
| SAR88 | Stockholm, Sweden |
| SUR88 | Cairo, Egypt |
| TDR88 | Guatemala City, Guatemala |
| TJT88 | Kousserin, Cameroun |

*The central operations console of the International Red Cross, Geneva, Switzerland*

# PUBLIC SAFETY

| | |
|---|---|
| TTR85 | Bardai, Chad |
| TTR88 | N'Djamena, Chad |
| YKR88 | Damascus, Syria |
| YNCR88 | Managua, Nicaragua |
| Z2C88 | Harare, Zimbabwe |
| ZRL230 | Windhoek, Namibia |
| ZRS88 | Pretoria, RSA |

## Florida Division of Emergency Management

**Freq:** 7934.0    **Mode:** Baudot, 170/45R

(Exercises at 1500 UTC weekdays)

| Callsign | City | County |
|---|---|---|
| KNFZ232 | Inverness | Citrus |
| KNFZ233 | Jacksonville | Duval |
| KNFZ234[1] | Tallahassee | Leon |
| KNFZ236 | West Palm Bch | Palm Beach |
| KNFZ237[2] | Defuniak Spgs | Walton |
| KNFZ240 | Field Unit | Statewide |
| WNGX404 | Daytona Bch | Volusia |
| WNGX405 | Pensacola | Escambia |
| WNGX406 | Key West | Monroe |
| WNGX407 | Fort Myers | Lee |
| KB26292[3] | MECC-1 | Mobile unit |
| KB26292[3] | MECC-2 | Mobile unit |

[1]State Warning Point-Net Control
[2]Alternate State Warning Point-Net Control
[3]Mobile Emergency Communications Center

## Avalanche Transceiver

Manufactured by Ortovox, these devices can provide fixes on a buried victim and operate on 2.275 and 457 kHz.

## Australian Outpost Communications

To serve the extensive outpost separation on the Australian continent, an HF network has been implemented for convenience and safety.

### Royal Flying Doctor Service (RFDS)

The extreme remoteness of outback cattle and sheep stations mandates this extensive medical network. Supported mostly by private contribution and some government subsidy, RFDS centers monitor these emergency channels 24 hours a day

and can dispatch a doctor and ambulance plane to any point within two hours. There is no charge to patients.

| | | |
|---|---|---|
| Alice Springs | VJD | 2020 4350 5410 6950 |
| Port Augusta | VNZ | 2020 4010 5145 6890 8165 |
| Broken Hill | VJC | 2020 4055 6920 |
| Cairns | VJN | 2020 2260 4880 Tx 4926 Rx 5145 5300 6785 Rx 6866 Tx 7465 |
| Charleville | VJJ | 2020 4980 6845 7307 Tx 7410 Rx |
| Mount Isa | VJI | 2020 4606 Rx 4935 Tx 5110 6965 7392 Tx 7475 Rx |
| Carnarvon | VJT | 2020 2280 4045 6890 |
| Meekatharra | VKJ | 2020 2280 4010 4880 Tx 4926 Rx 6880 7517 Tx 8147 Rx |
| Port Hedland | VKL | 2020 2280 4030 6960 |
| Derby | VJB | 2020 2792 5300 6925 |
| Wyndham | VKF | 2020 2805 5300 6945 |
| Kalgoorlie | VJQ | 2020 2656 6825 7550 Tx 8144 Rx |
| Broome | VIO | 2760 4940 |
| Hobart | VIH | 5355 |
| Darwin | VJY | 2020 4010 6840 6975 |

## "School of the Air" Service

The vastness and sparse population of the Australian outback prohibits the establishment of school districts. Education is by HF radio.

| | | |
|---|---|---|
| Alice Spgs | VZ8BZ | 8340 5370 8035 |
| Katherine | VL5SK | 4860 5731 7340 |
| Broken Hill | VJC | 4635 5130 5735 5895 6920 7565 8150 |
| Cairns | VJN | 5300 5865 4980 7357 |
| Mt Isa | VJI | 4800 5445 7803 |
| Carnarvon | VJT | 4045 5230 6890 |
| Meekatharra | VKJ | 4010 5260 6880 |
| Port Hedland | VKL | 4030 6960 |
| Derby | VJB | 5200 5300 5850 6925 |
| Kalgoorlie | VJQ | 5010 5360 5740 6825 |
| Port Augusta | VNZ | 4010 5145 5845 |
| Charleville | VJJ | 4045 4980 5227 6845 6945 |

## Australian Mines Rescue

3177.5  5441.5

## Australian Safety Council

3159.5  5441.5  8088.5

## New South Wales Police
## VKG

2632 3252 3752 4557 4560 4780 5180 5915 6905
6985 7657 7660 7800 8939.5 10295 10502 10505
13727 13730 17675

## State Emergency Services
## of New South Wales

2563 2566 2569 2572 2574 3729 3732 3743 4567
7330 9300

## Queensland Police
## VKR

3752  6905  6985  10295

## Queensland State
## Emergency Services

**HF Allocations:**

| | | |
|---|---|---|
| 2566 | Zone 1 | South East Queensland |
| 2572 | Zone 2 | South West Queensland |
| 2575 | Zone 3 | Far North Queensland |
| 3732 | Statewide | |
| 3735 | Zone 2 | South West Queensland |
| 3743 | Zone 3 | Far North Queensland |
| 3746 | Zone 1 | South East Queensland |
| 4576 | Statewide | |

**Freqs for use by regional offices where communications may exceed 500 kms:**

5833  9300  11435  14745

## South Australian Police
## VKA

2316  5180 7800  8939.5  10295

## Northern Territory Police

5915

# Spain

| | | |
|---|---|---|
| 4053 | CW/RTTY | Spanish Police Network |
| 9092 | CW/RTTY | Spanish Police Network |

# Honduras

| | |
|---|---|
| 6800 | World Relief (food to refugee camps) |
| 7775 | Christian missions (San Luis) |

# The Caribbean
# Emergency Network

At the present time 20 Caribbean states participate in this disaster preparedness network which ranges from Belize in the west to Barbados is the east, and from Guyana in the south to the Bahama Islands in the north. All communications are SSB voice.

The net control station (NCS) identifies as DISPREP ANTIGUA and is located in the office of the Pan-Caribbean Disaster Preparedness and Prevention Project (PCDPPP) at St. John's Antigua.

Net practice sessions are held Tuesdays and Fridays at 1330 UTC on 7850 kHz USB (channel 2) for the eastern Caribbean group, and at 1400 UTC on 7453.5 kHz LSB (channel 5) for the western Caribbean group. Both of these frequencies are also used Caribbean-wide for routine traffic.

Other frequencies used throughout the Caribbean include:

| kHz | |
|---|---|
| 2182 | USB (Marine emergency only) |
| 2527 | USB (Marine emergency only) |
| 6977.5 | USB (National Weather Service net; ch.6) |
| 7453.5 | LSB (Routine traffic) |
| 10100 | LSB (Emergency only) |
| 13965 | USB (PCDPPP/Red Cross communications) |
| 14303 | USB (Amateur; emergency only) |

Occasional relays are needed when propagation is poor; in these cases, Federal Emergency Management Agency (FEMA) station WGY932 in Puerto Rico provides assistance.

**The Eastern Caribbean Group**

Utilizing the calling frequency 7850 kHz USB (channel 2), a common language (English) and, with the exception of Guyana, the same time zone, the following states monitor mutually; Guyana, Trinidad and Tobago, Grenada, St. Vincent, Barbados, St. Lucia, Dominica, Montserrat, Nevis, Anguilla, St. Kitts, Antigua, and Tortolla.

# PUBLIC SAFETY

Additional frequencies available for this group include:

| kHz | |
|---|---|
| 13996.5 | USB (Red Cross communications) |
| 7220 | USB (Emergency only) |
| 7453.5 | LSB (Routine traffic) |
| 3815 | USB (Emergency only) |
| 3616 | USB (Emergency; amateur; inter-island police) |

### The Western Caribbean Group

Crossing time zones and languages (English, French and Spanish), members include Belize, Dominican Republic, Jamaica, Haiti, Turks and Caicos Islands, and the Bahama Islands. This group monitors 7453.5 kHz LSB. Amateurs may use 7150 kHz LSB for emergencies only.

*Bahamian police use HF for inter-island communications (photo by Harry Baughn)*

### Call Signs and Locations:

| | |
|---|---|
| DISPREP | Pan-Caribbean Disaster Preparedness and Prevention Project |
| HHP57 | Haiti Emergency Operations Center |
| J6P | St. Lucia Police HQ |
| J6L | St. Lucia National Coordinator's Office |
| J39AI, J39YK | Grenada |
| WGY901 | St. Thomas Civil Defense (FEMA) |
| WGY932 | Puerto Rico Civil Defense (FEMA) |
| ZJL89 | Tortola |
| ZOA | Antigua Police HQ and EOC |
| ZOB | St. Kitts Police HQ and EOC |
| ZOD | Dominica National Coordinator's Office |
| ZOG | Grenada |
| ZOM | Montserrat Police HQ |
| ZON | Nevis |
| ZOU | Anguilla |
| 6YODP | Jamaica |
| 6YX | Jamaica Coast Guard |
| 8PF | Barbados Coast Guard |
| 9YA | Trinidad Coast Guard |
| 9Y4ST | University of the West Indies seismic unit |

### Netherlands Antilles

An emergency and weather network has been established for the Netherlands Antilles as well; listen for their practice drills daily during storm season at 1030 and 2230 UTC on the amateur frequency 3815 kHz LSB.

# Astrophysical Observatories

**Smithsonian Institution**
Cambridge, MA (Call Sign: KCW21/KEP78)
Frequency: 20610
Mode: ASCII (110 Baud/850 Hz shift); some LSB voice

Remote Facilities
(Satellite Tracking):
Arequipa, Peru - OCK25
Natel, Brazil - NA4XK (20089 kHz)
Orroval Valley, Australia
Mount Hopkins, AZ
Recife, Brazil - PP7A

*The Mayall telescope and (in background) the McMath solar tunnel telescope (courtesy National Optical Astronomy Observatories/Kitt Peak.*

**Kitt Peak National Observatory**
Tucson, Arizona
(Call Sign: KFK92)
Frequencies: 10190 and 20875 MHz
Mode: LSB/ASCII

Remote Facility:
Inter American Observatory at Cerro Tololo (La Serena), Chile
Call Sign: XQ8AFI

**Arecibo Ionospheric Observatory**
Puerto Rico (Call Sign: WDR71)
Frequencies:
3156 4438 5060 5730 6765 7300 9040 9775 10100 10190 11400 11975 13360 14350 15450 17360 18030
Mode: ASCII

Remote Facilities:
Unknown site WDR72
Ithaca, NY (Cornell University)
Cleveland, OH (Case/Western Reserve University)

**California Institute of Technology**
Variegated Glacier, AK
Frequency: 5472

**University of Delaware (Delaware Research)**
Lewes, Delaware
Callsign: KTD423
Frequency: 6521.9
Works their WZC800 located in Bermuda

**Massachusetts Institute of Technology**
20876.3 LSB

**Utah State University** (KK2XJT)
to **Siple Station Antarctica** (KK2XJS)
Frequencies: 1000 2505 5005 10005 15010 20010 21870

**University of Alaska**
Fairbanks, AK (Call signs KM2XNZ KA2XUA)

Frequencies: 3192.5 5736.5 9941.5 12256.5

## Firestone Network
### U.S. - Liberia Radio Corporation

The Firestone Tire and Rubber Company has maintained a RTTY (now TOR/ARQ) network between its Akron headquarters and its Harbel, Liberia, rubber plantation since 1929. It is registered as a common carrier offering its services on a secondary basis to other commercial users as well. Underlined frequencies are primary.

### Akron, Ohio

| Freq-kHz | Call | Freq-kHz | Call |
|---|---|---|---|
| 7775 | WQA27 | 15940 | WQB25 |
| 10265 | WQB20 | 19335 | WQB39 |
| 13945 | WQB23 | 19780 | WQB29 |
| 15780 | WQB35 | | |

# BUSINESS, SCIENTIFIC, PRIVATE

### Harbel, Liberia

| Freq-kHz | Call | Freq-kHz | Call |
|----------|------|----------|------|
| 7580 | ELE7 | 15940 | ELE25 |
| 10310 | ELE10 | 18530 | ELE18 |
| 13945 | ELE12 | 19335 | ELE39 |
| 14605 | ELE14 | 19780 | ELE19 |
| | | 19940 | ELE29 |

## Petroleum Networks

| Location | | Frequency |
|----------|--|-----------|
| U.S. | | 4634.5  4637.5 |
| | | 5645 |
| | Duplex | 4419.4/4125 |
| North Sea | | 5374.7 |
| Libya | | 7971 |
| Australia | | 9040.6 |

### Phillips Petroleum Net

| Location | Frequency |
|----------|-----------|
| Jurigas, Mindinao | 10405 (RTTY/LSB) |
| Jakarta, Indonesia | 13310 (RTTY/USB) |
| (RTTY) | 14445 |

### Petroleum Helicopters, Inc (PHI)

Gulf of Mexico Net (USB)
New Orleans to Brownsville and
to Offshore Oil Platforms

| Ch 1 | 4550 |
|------|------|
| Ch 2 | 8070 (pri) |
| Ch 3 | 11470 |

### Offshore Navigation, Inc.
(KA2307)

| Location | Frequency |
|----------|-----------|
| Harahan, Louisiana | 4455 (USB) |

### Zapata Offshore Drilling Platform

Gulf of Mexico
2292  2398  4637.5

### Mobile Oil Network (Canada)

Offshore petroleum exploration and recovery communications utilizing USB and facsimile may be heard on a number of frequencies among boats, helicopters, portable land stations, and drilling platforms around Nova Scotia.

| CJR681 | Sable Island | portable |
|--------|--------------|----------|
| VCC43 | Dartmouth | portable |
| VCD643 | Sable Island | drill rig |

| | | |
|------|--------|--------|
| 2381.5 | 4573.5 | 5283 |
| 2730.5 | 4603 | 5916.5 |
| 3214 | 4621.5 | 7350 |
| 3354.5 | 5116.5 | 7862.5 |
| 3496.5 | | |

## Industrial/Scientific/Medical (ISM)

| Frequency | Use |
|-----------|-----|
| 6780  13560  27120 | Heating, welding |

## Industrial Radio Service

Utilizing the block of frequencies set aside by the FCC for emergency use, several major power, fuel and wireline telecommunications companies perform routine on-air tests of their radio equipment. Oil drilling platforms also utilize some of these frequencies.

Bell Telephone operating companies are probably the most frequently heard, conducting their National Security Emergency Preparedness (NSEP) tests on channelized frequencies as shown below. Regional networks and their participants are also listed.

### OLD NET

A nationwide Bell net meets the last Tuesday of the month at 1500 UTC on channel 45 (7552 kHz) using USB, RTTY (170 HZ shift, 45/50/75/100/110 baud, reverse polarity), ASCII (170 Hz shift, 110 baud, reverse polarity) and 300 baud packet. Additional frequencies have included 2194, 3155, 4438, 5005, 6763 and 7300 kHz.

| Ch/Freq kHz/Notes | Ch/Freq kHz/Notes |
|-------------------|-------------------|
| **Base/Mobile** | **Base Only** |
| 11 2289 | 24 5046.5 E. US |
| 12 2292 | 25 5052.5 E. US |
| 13 2395 | 26 5055.5 E. US |
| 14 2398 | 27 5061.5 W. US |
| 15 3170 | 28 5067.5 |
| 16 4538.5 Night only | 29 5074.5 E. US |
| 17 4548.5 Night only | 30 5099.1 |
| 18 4575 | 31 5102 |
| 19 4610.5 | 32 5313.5 |
| 20 4613.5 | 33 6800  Night only |
| 21 4634.5 | 34 6803.1 |
| 22 4637.5 | 35 6806  W. US |
| 23 4647 | |

| | | |
|---|---|---|
| 36 6855 | Night only | |
| | /W.US | |
| 37 6858 | Night only | |
| 38 6861 | W. US | |
| 39 6885 | Night only | |
| 40 6888 | Night only | |
| 41 7480 | | |
| 42 7483 | | |
| 43 7486 | E. US | |
| 44 7549.1 | Day only | |
| 45 7552.1 | | |
| 46 7555 | W. US | |
| 47 7558 | W. US | |
| 48 7559 | W. US | |
| 49 7562 | W. US | |
| 50 7697.1 | | |

## NEW NET

Now scheduling tests on a random basis, new frequencies are:

| | | |
|---|---|---|
| 2194 | 9044.4 | 13373.4 |
| 3155 | 9497.4 | 13397.4 |
| 4438 | 11404.4 | 15454.4 |
| 5005 | 11452.4 | 15604.4 |
| 6763 | 11698.4 | 16456.4 |
| 7300 | 13364.4 | 18034.4 |
| | | 18064.4 |

# BUSINESS, SCIENTIFIC, PRIVATE

## Bell Telephone NSEP Net

| Call Sign | Company | Location |
|-----------|---------|----------|
| WNFT417 | Bell Communications Research | Morristown, NJ |
| WNFT417 | Bell Communications Research | Red Bank, NJ |
| Unit 1,2 | Bell Communications Research | Mobile |
| WNGN547 | Indiana Bell | Indianapolis, IN |
| WNGQ469 | New York Telephone | Linlithgo, NY |
| KJL412 | New York Telephone | Linlithgo, NY |
| WNHK760 | Chesapeake and Potomac | Staunton, VA |
| WNHK761 | Chesapeake and Potomac | Culpepper, VA |
| WNHM961 | Chesapeake and Potomac | Fairmont, WV |
| WNHM965 | Chesapeake and Potomac | Martinsburg, WV |
| WNHM969 | Chesapeake and Potomac | Beckley, WV |
| WNHN755 | Michigan Bell | Southfield, MI |
| WNHP857 | South Central Bell | Gadsden, AL |
| WNHU799 | Southern Bell Telephone | Atlanta, GA |
| WNIC425 | Illinois Bell | Champaign, IL |
| WNID658 | Southern Bell Telephone | Jacksonville, FL |
| WNIM867 | Southwestern Bell Telephone | St. Louis, MO |
| WNAX939 | Mountain States Telephone | Salilda, CO |
| WNIM435 | U.S. West Communications | ID |
| WNKD998 | Pacific Bell Telephone | CA |
| WNKG722 | Pacific Bell Telephone | ? |
| WNKR875 | Pacific Bell Telephone | ? |
| WNLP942 | Pacific Bell Telephone | ? |
| WNME910 | ? | ? |
| WNJN982 | ? | ? |
| WNPK982 | ? | ? |

## Southwestern Bell Net

| Call Sign | Company | Location |
|-----------|---------|----------|
| WNCP450 | Southwestern Bell Telephone | ? |
| WNCP451 | Southwestern Bell Telephone | ? |
| WNIM867 | Southwestern Bell Telephone | St. Louis, MO |
| WNHI785 | Southwestern Bell Telephone | Little Rock, AR |
| WNHT324 | Southwestern Bell Telephone | Oklahoma City, OK |
| WNID675 | Southwestern Bell Telephone | Topeka, KS |
| WNIY791 | Southwestern Bell Telephone | Houston, TX |
| WNIY791-8 | Southwestern Bell Telephone | Mobile |
| WNJK493 | Southwestern Bell Telephone | Kansas City, MO |

## Bell Atlantic Net

| Call Sign | Company | Location |
|-----------|---------|----------|
| WNHK760 | Chesapeake and Potomac | Staunton, VA |
| WNHK761 | Chesapeake and Potomac | Culpepper, VA |
| WNHM961 | Chesapeake and Potomac | Fairmont, WV |
| WNHM965 | Chesapeake and Potomac | Martinsburg, WV |
| WNHM969 | Chesapeake and Potomac | Beckley, WV |

## Additional Participants

| Call Sign | Agency/Company | Location |
|-----------|----------------|----------|
| KGD34 | National Coordinating Center | Arlington, VA |
| KPA525 | ? | Chicago, IL |

# BUSINESS, SCIENTIFIC, PRIVATE

## QSL Addresses

WNFT417

Bell Communications Research
435 South Street
Room MRE 1K126
Morristown, NJ  07960

WNFT417

Bell Communications Research
Attn: WNFT417 Manager, 2X410
331 Newman Springs Road
Red Bank, NJ  07701-5699

WNIM867

Southwestern Bell Telephone
Attn: Area Manager-NSEP Operations
1010 Pine Street
St. Louis, MO  63101

# Experimental Service

A variety of modes (SSB, CW, pulse, FSK) are used by licensees to develop new technologies. Some of these are used to support government communications, often air-to-ground or ship-to-shore.

| Licensee | Callsign | Location | Frequencies (kHz) |
|---|---|---|---|
| Magnavox | KM2XBO | Nationwide | 2005 2204 2510 5005 15010 20010 21870 25005 |
| SRI Int'l | KM2XBR | Palo Alto, CA | 2422 3240 3365 4438 5005 5730 5950 5985 7300 9040 9500 10100 11700 11975 13410 14350 15100 15450 17360 17700 18030 20010 21450 21750 22720 23100 23350 25600 25800 27410 29700 |
| SRI Int'l | KM2XDH | Stanford, CA | |
| SRI Int'l | KM2XET | Towson, CA | |
| Metromedia | KM2XFL | St. Louis, MO | 6950 7411 14446 14455 |
| RCA | KM2XHQ | Camden, NJ; Indianapolis, IN; Burlington, MA; Rockledge, FL; West Windsor, NJ | 3102 3104 4040 5270 6845 11195 15056 |
| Cincinnati Elec. | KC2XIG | Cincinnati,OH | 2200 5300 6900 |
| ITT | KC2XJB | Raleigh, NC | 2006 2031 2065 2079 2206 2458 2599 2782 3258 4072 4100 4169 4190 4225 6220 6252 6285 6339 8201 8334 8380 8450 12373 12490 12570 12660 16511 16648 16760 16890 22043 22178 22280 22360 |
| Bell Tel | KM2XRA | FL, WA, CA | 2118 2132 2142 2158 2206 2514 2530 2550 2538 2598 |
| Navidyne | KM2XRQ/ XRR | Newpt Nws,VA | 1600 4063 6200 8195 12330 16460 22000 |
| Motorola | KM2XQJ | Ft.Lauderdl,FL | 2003 2225 2638 4125 4145 4176 4178 4419 5250 6220 6268 6522 8291 8296 8298 12429 12432 12520 12523 16587 16695 22124 22226 |
| GTE | KM2XOP | Needham, MA; Rome, NY; Washington, DC; Raleigh-Durham, NC | 3340 4438 5005 5730 7300 9040 10100 11975 13360 14350 15450 17360 18030 20010 21750 22720 23350 25600 27410 29700 |
| Eyring Resrch | KM2XLP | Provost, UT | 1600 1614 1700 2197 2200 2242 2398 2458 2510 2598 2600 3191 4040 4044 5100 6100 6171 7354 7731 8015 9500 10100 10120 10707 15100 15343 19049 22903 23100 24930 25100 27743 29933 |

**Rockwell-Collins Network:**

KA2XAH

2396.5 2399.8 4795.5 4798.5 6151.5 6154.5 6171 6174 9655.5 9668.5 11761.5 11764.5 15341.5 15344.5 15406 15409 21725.5 21728.5 23098.5 23101.5 27738.5 27741.5 29928.5 29931.5

# BUSINESS, SCIENTIFIC, PRIVATE

| | | |
|---|---|---|
| KA2XDI | | 2396.5 2398 2399.5 4795.5 4797 4798.5 6171 6172.5 6174 9655.5 9657 9668.5 15406 15407.5 15409 21725.5 21727 21728.5 23098.5 23100 23101.5 29928.5 29930 29931.5 |
| KA2XKG | | 2810 4992.5 5278.5 |
| KM2XHY/KM2XHZ | Newport Bch, CA | 2398 4797 6153 6171 6174.5 9657 11763 16343 15407.5 21727 27740 29930 |
| KM2XLB | Cedar Rapids, IA; and | 5760 5960 6060 6160 10200 10400 10600 10800 15760 |
| KM2XLC | Richardson, TX | 15860 15960 16060 16160 25230 27440 29720 |
| KC2XKG | Cedar Rapids, IA; and | 5841 5912 7300 7657 9238.5 9497 9802 11073.5 11076 12138.5 12222 13658.5 14350 14686 14690 16376 18171 18666 19131 20348.5 23403 |
| KC2XKJ | Newport Bch, CA | 11073.5 11076 12222 14350 14686 14690 15955 17601 18666 19131 23403 |
| KM2XLB | | 5760 5960 6060 6160 10200 10400 10600 10800 15760 15860 15960 16060 16160 25230 27440 29720 |
| KM2XLF | Nationwide | 2398 3281 4797 6153 6174 9657 11763 15343 21727 25343 27740 29930 |
| KM2XMN | Richardson, TX | 2851 3004 3443 5251 5469 5571 6550 8822 10045 11288 11306 13312 17964 21931 |

*View of large antenna farm of TRT and AT&T in Ft. Lauderdale, FL*

# BUSINESS, SCIENTIFIC, PRIVATE

## Amateur and Citizen's Band Radio

### Amateur Radio Bands (USA)

| | |
|---|---|
| 160M | 1800-2000 |
| 80/75M | 3500-4000 |
| 40M | 7000-7300 |
| 30M | 10100-10150 |
| 20M | 14000-14350 |
| 17M | 18068-18168 |
| 15M | 21000-21450 |
| 12M | 24890-24990 |
| 10M | 28000-29700 |

#### 10-M REPEATERS

| In | Out |
|---|---|
| 29520 | 29620 |
| 29540 | 29640 |
| 29560 | 29660 |
| 29580 | 29680 |

### Amateur Satellites

Downlink: 29300 - 29500
Orbital periods: Oscar 95-103 min.; RS 118-120 min.

| Satellite | Beacon Frequency |
|---|---|
| Oscar-8 (USA) | 29402 (transponder 29400-29500) |
| Oscar-9 (USA) | 7001 14001 21001 28001 |
| RS-3 (USSR) | 29321 |
| RS-4 (USSR) | 29361 |
| RS-5 (USSR) | 29452 (transponder 29331) |
| RS-6 (USSR) | 29453 (transponder malfunctioned; carrier 29352) |
| RS-7 (USSR) | 29501 (transponder 29341) |
| RS-8 (USSR) | 29502 (transponder 29460-29500) |

#### Amsat Worldwide Network

| | |
|---|---|
| 3850 | (Wed. 0200/0300/0400) |
| 14282 | (Sun. 1900 UTC) |
| 21280 | (Sun. 1800 UTC) |

### Radio Amateur Civil Emergency Service (RACES)

During wartime, amateur radio operations are prohibited; only authorized RACES stations are allowed on the air as an adjunct to defense communications.

| | |
|---|---|
| 1800-1825 | 14047-14053 |
| 1975-2000 | 14220-14230 |
| 3500-3550 | 14331-14350 |
| 3930-3980 | 21047-21053 |
| 3984-4000 | 21228-21267 |
| 7079-7125 | 28550-28750 |
| 7245-7255 | 29237-29273 |
| | 29450-29650 |
| Military Calling | 3997 |

### Hurricane Networks

Florida area    3935, 3940, 7268, 7247, 14320, 14325

Carolinas area    3907, 3917, 3923, 3927

Emergencies/Disasters 14275

W1AW frequencies for ARRL bulletins and code practice

### CB Channels (USA)

| Ch | Freq kHz | | |
|---|---|---|---|
| 1 | 26965 | 21 | 27215 |
| 2 | 26975 | 22 | 27225 |
| 3 | 26985 | 23 | 27255 |
| 4 | 27005 | 24 | 27235 |
| 5 | 27015 | 25 | 27245 |
| 6 | 27025 | 26 | 27265 |
| 7 | 27035 | 27 | 27275 |
| 8 | 27055 | 28 | 27285 |
| 9 | 27065 | 29 | 27295 |
| 10 | 27075 | 30 | 27305 |
| 11 | 27085 | 31 | 27315 |
| 12 | 27105 | 32 | 27325 |
| 13 | 27115 | 33 | 27335 |
| 14 | 27125 | 34 | 27345 |
| 15 | 27135 | 35 | 27355 |
| 16 | 27155 | 36 | 27365 |
| 17 | 27165 | 37 | 27375 |
| 18 | 27175 | 38 | 27385 |
| 19 | 27185 | 39 | 27395 |
| 20 | 27205 | 40 | 27405 |

Certain common usage agreements are generally observed on the CB channels, including AM mode on channels 1 through 35 and SSB from channels 36 through 40. Channel 6 is populated by black ethnic conversation; Channel 9 is used for emergency and calling; channel 10 has its truckers' support groups; channel 19 is occupied by east/west truckers; and channel 21 is used by north/south truckers.

# BUSINESS, SCIENTIFIC, PRIVATE

## Fishing Fleets

The ready availability of continuous-frequency-coverage transceivers is an irresistable, although unlawful, temptation to saltwater fisherman. Recently monitored frequencies include both English and Spanish speaking seafarers.

5680 5770 5775 6538 6543 6666.6 6808
7777.7 8827 8889 8907 13277 13425

## Freeband Radio

A deluge of unlawful Japanese and Taiwanese transceivers featuring as many as 200 or more channels in 10 kHz increments, 25.5-28.5 MHz, with AM, FM and SSB modes, has encouraged the chaos surrounding the legitimate CB radio service. Most operators, however, have the presence of mind not to intrude into the 28 MHz amateur band.

Still, a pattern has emerged, subscribed to by bands of operators who may call themselves--and one another--"renegades", "rogues", "outlaws", "outbanders", or "freebanders".

Generally speaking, 26.325-26.965 MHz is inhabited by AM operators. Channelization is usually in 10 kHz increments.

Within the frequency range of legitimate CB there are several channels reserved for industrial services including 26.995 (3A), 27.045, 27.095, 27.145, and 27.195 MHz. Frequently assigned here are radio controlled toys, rural electric power companies (REA power grid control) and water and gas utility companies (level control).

SSB is the dominant mode (usually LSB) between 27.405 and 27.995 MHz, with the most popular U.S. frequencies between 27.445 and 27.605 MHz. These sideband operators may sound more professional and often use ham-type alphanumeric call signs. Groups here include "April", "Xray", "Whiskey", "World Radio", "Kolo Papa" (Knights Patrol International), "Adam and Eve", "SSB", "USS", and "Tango".

The outlaws do, however, have some favorite hangouts even within assignable CB channels. Channel 5 is popular among high power groups like the Armadillo Operators, Bullshippers, and Tuckers. When the channel gets crowded, they switch to channel 3A.

## Terrorist Networks

A suspected international terrorist net in South America has been reported on several frequencies from 14400-14500 kHz USB. 14436 appears to be a primary frequency and callups have been noted on some frequencies at 2100 UTC. Language is a mix of Spanish and Arabic.

## Drug Smugglers

Several common frequency ranges are used by drug smugglers, the majority of which bring their illicit cargos to the coasts of Florida and the Carolinas from Central and South America. Spanish language dominates, although salty English dialogue is common.

Traditionally, amateur radio transceivers are used, operated just above the 40 and 20 meter ham bands. Additionally, frequencies near the 6 MHz ship-to-shore band are also utilized. Frequently heard ranges include 6500-6900, 7400-7600 and 14350-14600 kHz. 3185 has also been reported.

## Unknowns

There will always be signals heard throughout the spectrum which, for one reason or another, remain unidentified. These may be incidental radiations from electronic equipment, illicit communications links, unlicensed transmitters, diplomatic correspondence, military transmissions and other taunting transmissions as well.

4472.5 5212 6650 6835 6930 6933 6960

## Collins Radio Networks

For decades, Collins Radio (now a division of Rockwell) has supported development of communications systems for all branches of military, government and industry. From a sophisticated electronic complex at Cedar Rapids, Iowa, additional transmitter sites are remotely keyed up at Richardson, Texas and Newport Beach, California.

Most of these transmissions are single sideband (upper sideband dominating). They are used for equipment tests in the aeronautical, maritime mobile and fixed services. KHR and KHT are maritime limited coastal radio stations, heard on frequencies in that service.

| Freq kHz/Callsign | |
|---|---|
| 2398 | Data |
| 4797 | Data |
| 5559 | Air/Ground |
| 5760 | Fixed |
| 5960 | Fixed |
| 6060 | Fixed |

*The Collins HF-80 series*

| | |
|---|---|
| 6153 | Data |
| 6160 | Fixed |
| 6172.5 | Data |
| 6669 | Air/Ground |
| 9657 | Data |
| 10200 | Fixed |
| 10400 | Fixed |
| 10600 | Fixed |
| 10800 | Fixed |
| 11287 | Air/Ground |
| 11763 | Data |
| 15343 | Data |
| 15407.5 | Data |
| 15760 | Fixed |
| 15860 | Fixed |
| 15960 | Fixed |
| 16060 | Fixed |
| 16160 | Fixed |
| 21727 | Data |
| 21931 | Air/Ground |
| 25230 | Fixed |
| 27440 | Fixed |
| 27740 | Data |
| 29720 | Fixed |
| 29930 | Data |

| Callsign | Location |
|---|---|
| KA2XAH | Richardson, TX |
| KHB | Newport Beach, CA |
| KHT | Cedar Rapids, IA |
| KK2XHQ | Richardson, TX |
| KK2XHR | Richardson, TX |
| KLK2 | Cedar Rapids, IA |
| KM2XLB | Cedar Rapids, IA |
| KM2XLC | Cedar Rapids, IA |
| KM2XLF | Puerto Rico |
| KM2XHY | Newport Beach, CA |
| KM2XHZ | Newport Beach, CA |
| WMF6 | Richardson, TX |

*Collins' "Comm Central" communications system*

# COMMON CARRIER

KHT at Cedar Rapids (Collins/Rockwell head-quarters) pinpoints their commercial maritime clients (Exxon, Maritime Oversea Corporation of New York, others) with three giant log periodic beam antennas.

The frequencies used are shared by other similar limited coast stations which are allowed to communicate only with contracted companies and the majority of two-way communications will be ship to shore phone patches.

KHR at Newport Beach, California, shares the same frequencies and is remotely keyed from Cedar Rapids. It has one log periodic antenna. Both antenna installations are capable of 6-60 MHz continuous coverage but are used only up to 22 MHz on the maritime bands.

## Aircraft

Collins also supports flight test operations of radionavigational equipment on frequencies authorized for that purpose. Listen for 71CR (turboprop), 80CR and 82CR (jets) on aircraft flight test frequencies.

All of these frequencies are "pool;" that is, they are shared with other services. On one occasion the Department of Energy (KP6) came up on 13312 and asked Collins what they were doing there! The shortwave spectrum is a busy place.

In the past, the Collins/Rockwell facilities have even checked into U.S. Air Force and Navy nets using the callsign "Rasputin."

# Alaska Public Fixed Radio Service

### HF SSB Carrier Frequencies

NOTES:
1. Ketchikan Receive
2. Kodiak Receive
3. RTTY operation may be authorized
4. Ketchikan Transmit
5. Cordova Receive
6. Juneau Receive
7. Kodiak Transmit
8. Juneau Transmit
9. Fairbanks Transmit
10. Anchorage Receive
11. Fairbanks Receive
12. Anchorage Transmit
13. Alaska Div of Emergency Services
14. Alaska Calling and Emergency
15. Cordova Transmit
16. Bethel Transmit
17. Bethel Receive
18. King Salmon Transmit
19. King Salmon Receive
20. Kotzebue Transmit
21. Kotzebue Receive
22. Nome Transmit
23. Nome Receive
24. Unalaska Transmit
25. Unalaska Receive
26. Cold Bay Transmit
27. Cold Bay Receive

| Freq | Ch | Freq | Ch | Freq | Ch | Freq | Ch |
|------|----|------|----|------|----|------|----|
| 1643 | | 2430 | | 2632[5] | | 3362[12/25] | 1 |
| 1646 | | 2447 | | 2691[27] | | 3365[10] | |
| 1649 | | 2450 | | 2694[6] | 1 | 4035 | |
| 1652 | | 2463[21] | | 2773[17] | | 4791.5[9] | 2 |
| 1657 | | 2466[19] | | 2776[1] | 2 | 5134.5[10] | 2[25] |
| 1660 | | 2471[23] | | 2781[7] | | 5135.0[13] | |
| 1705 | | 2474[2] | | 2784[8/22] | 1 | 5137.5[10] | 3 |
| 1709 | | 2479 | | 3164.5[18] | | 5164.5 | |
| 1712 | | 2482 | | 3167.5[9] | 1 | 5167.5[14] | |
| 2003 | | 2506[3] | | 3180[4] | 2 | 5204.5[17] | |
| 2006 | | 2509 | | 3183[12] | | 5207.5[11] | 2[23] |
| 2115 | | 2512 | | 3198 | | 5370[12/22] | 2[24] |
| 2118 | | 2535 | | 3201 | | 6948.5[3] | |
| 2253[10] | | 2538 | | 3238[10/12] | 1 | 7368.5[3] | |
| 2256[1] | 2 | 2563 | | 3241[8/26] | 2 | 8067 | |
| 2312[15] | | 2566 | | 3258 | | 8070 | |
| 2400 | | 2601[20] | | 3261 | | 11437[3] | |
| 2419 | | 2604[4][1][16] | | 3303 | | 11601.5[3] | |
| 2422 | | 2616 | | 3354[11] | 1 | | |
| 2427 | | 2629[17] | | 3357[6] | 2 | | |

# Canadian Radio Common Carriers

## BC Telephone Coast Stations
### Sideband

| | | | | |
|------|-------------------|------|------|--------|
| CFW 4 | Prince Rupert, BC | 2060 | 2590 | |
| CFW | Vancouver, BC | 2538 | 2558 | 4410.1 |
| | | 6515.7 | 8759.2 | |

## Canada Hydrant Service, Ltd
### SSB
29920

## Canadian Nickel Co, Ltd
### SSB
2567.5      Yukon BC Ont PQ Nfld

**Shell Resources Canada, Ltd**

SSB

| | |
|---|---|
| 2621.5 | NWT Alta Sask Ont PQ Nfld |

**Bell Canada**

SSB

| | |
|---|---|
| 3299.5 | NWT PQ |
| 4630.5 | NWT PQ |
| 5186.6/4964.5 | Ont |
| 7401.5/7547.5 | Ont |

**Teleglobe Canada**

SSB

| | |
|---|---|
| 6830 | PQ to France |
| 13745 | PQ to Moscow |
| 14392.5 | PQ to France |
| 18180 | PQ to Moscow |
| 19295 | PQ to Moscow |

RTTY

| | |
|---|---|
| 4465 | PQ to France |
| 6830 | PQ to France |
| 7310 | PQ to Eagles Nest |
| 10520 | PQ to France |
| 14357.5 | PQ to Chacarilla |
| 14392.5 | PQ to Carrington & France |
| 14790 | PQ to Eschborn |

**Panarctic Oil, Ltd**

Rea Pt., NWT, (XNU802)

| RTTY | | USB |
|---|---|---|
| 3311.5 | NWT | 3310 |
| 4036.5 | NWT | 5281.5 |
| | | 5411 |

**Bema Industries**

RTTY

| | |
|---|---|
| 5134 | BC |
| 7508.5 | BC |
| 9148.5 | BC |

**Mobile Oil Canada**

RTTY

| | |
|---|---|
| 6874.5/8092 | NS |

**G D Bruno**

RTTY

| | |
|---|---|
| 3311.5 | Alta |

**Canadian Broadcasting Corp**

SSB

| | |
|---|---|
| 5925 | Sackville NB to CFB Lahr W Germany |
| 9445 | same |

**Austin Airways, Ltd**

SSB

| | |
|---|---|
| 9081.5 | Ont PQ |
| 11651.5 | Ont PQ |

**British Petroleum Exploration Canada**

RTTY

| | |
|---|---|
| 2381.5/2633.5 | Nfld |
| 3214/3397.5 | Nfld |
| 4573.5/5061.5 | Nfld |
| 5149/5441.5 | Nfld |

**Asbestos Corp**

RTTY

| | |
|---|---|
| 2689.5/3206.5 | NWT |
| 3206.5/2689.5 | PQ |
| 4009/4448 | NWT, PQ to France |

**Lynes United Service**

RTTY

| | |
|---|---|
| 3311.5 | Alta |
| 4603 | Alta |
| 6966.5 | Alta |

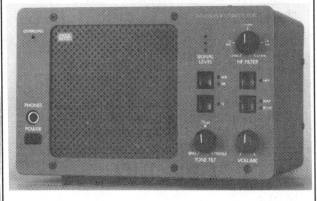

# COMMON CARRIER

**Lynes United Service**

RTTY

| | |
|---|---|
| 3311.5 | Alta |
| 4603 | Alta |
| 6966.5 | Alta |

**Hudson's Bay Co**

SSB

| | |
|---|---|
| 4837 | NWT Man Ont PQ |

**Gulf Minerals Canada, Ltd**

SSB

| | |
|---|---|
| 5289.5 | NWT Sask Man Ont |

**Imperial Oil, Ltd**

SSB

| | |
|---|---|
| 5436.5 | Yukon NWT BC Alta Sask PQ Nfld |

RTTY

| | |
|---|---|
| 4165 | Nfld |
| 6247 | Nfld |
| 8330.5 | Nfld |
| 12482 | Nfld |
| 16639 | Nfld |
| 22163 | Nfld |

**Remote Radiotelephone Networks**
*Remote Sites such as Logging Camps and Mines -- "bush telephones".*

Vancouver, British Columbia

| Ch | Receive | Transmit |
|---|---|---|
| VL1 | 6825 | 6790 |
| VL2 | 5405 | 5810 |
| VL3 | 4865 | 5405 |
| VL4 | 3270 | 3171 |
| VL5 | 3359 | 3213.5 |
| VL6 | 4820 | 4610 |
| VL7 | 4543.5 | 4573.5 |
| VL8 | 7953 | 7804 |
| VL9 | 5248.5 | 5313.5 |
| VL10 | 2160 | 2030 |
| VL11 | 9315 | 9462 |
| VL12 | 12080 | 12181 |
| VL13 | 7658.5 | 7818.5 |
| VL14 | 3300 | 3224 |

Red Deer, Alberta
3241.5    USB

Alberta
2621.5  5486.5  (simplex)

Manitoba
4837  5289.5  (simplex)

Newfoundland
2621.5  5486.5  (simplex)

Northwest Territory
2621.5  3299.5  3310  4630.5  5281.5  5289.5  5411  5436.5  (simplex)

Ontario
2621.5  4837  5289.5  9081.5  11651.5  (simplex)

5186.6/4964.5  7401.5/7547.5  (semi-duplex)

Quebec
2621.5  3299.5  4630.5  4837  5486.5  (simplex)

Saskatchewan
2621.5  5289.5  5486.5  (simplex)

Yukon
2567.5  5436.5  (simplex)

# Australia

**Outback Telephone**    12423

## Domestic Broadcasting

The 150-285 kHz portion of the spectrum is used for domestic (home) broadcasting in Europe and Asia. Rarely heard in the United States, this list is included for those listeners who are fortunate enough to pick up some rare DX. Power of these stations ranges from approximately 50 thousand to 2 million watts.

### World Longwave Broadcast Stations

| Freq | ID | Location | Kilowatts |
|------|------|----------|-----------|
| 153 | | Bechar, Algeria | 1000 |
| 153 | | Tromso, Norway | 10 |
| | DLF | Donebach, West Germany | 500 |
| | | Brasov, Romania | 1200 |
| | | Engels, USSR | 150 |
| | | Khabarovsk Kray (Far East), USSR | 1000 |
| | | Ufa, USSR | 500 |
| 162 | | Allouis, France | 2000 |
| 164 | | Ulan Bator, Mongolia | 150 |
| | | Agri, Turkey | 1000 |
| | | Tashkent (Uzbekistan), USSR | 150 |
| | | Norilsk, USSR | 50 |
| 171 | | Nador, Morocco | 1200 |
| 171 | | Minsk, USSR | 1000 |
| 171 | | Kransnodar, USSR | 1000 |
| 171 | | Tomsk, USSR | 1000 |
| | | Kaliningrad, USSR | 1000 |
| | | Lvov (Ukraine), USSR | 500 |
| | RW1315 | Moscow, USSR | 500 |
| | | Syktyvkar, USSR | 300 |
| | | Yakutsk (Siberia), USSR | 500 |
| | | Novosibirsk (Siberia), USSR | 150 |
| | | Rostov-Na-Donu. USSR | 1000 |
| | | Kyzyl (Siberia), USSR | 100 |
| 177 | | Oranienburg, East Germany | 750 |
| 180 | TAR | Polatli, Turkey | 1200 |
| 180 | | Aktyubinsk (Kazakhstan), USSR | 150 |
| 180 | | Alma Ata (Kazakhstan), USSR | 250 |
| 180 | | Petropavlovsk Kamchatka (Far East), USSR | 150 |
| 180 | | Tchita (Siberia), USSR | 150 |
| 183 | | Saarlouis (Felsberg), W.Ger. | 2000 |
| 189 | | Caltanissetta, Italy | 10 |
| | SBG | Motala (Orlunda), Sweden | 300 |
| | | Barnaul (Siberia), USSR | 50 |
| | | Blagoveshchensk(Siber.)USSR | 1000 |
| | (TB) | Tbilisi, USSR | 500 |
| 198 | | Burghead, England | 50 |
| 198 | | Droitwich, England | 400 |
| 198 | | Westerglen, England | 50 |
| 198 | | Warsaw, Poland | 200 |
| 198 | | Etimesgut (Ankara), Turkey | 200 |
| 198 | | Ovargla, Algeria | 1000 |
| 200 | | Aleksandrovsk-Sakhalinskiy (Far East), USSR | 50 |
| 200 | (AB) | Achkhabad (Turkemnia), USSR | 75 |
| | | Frunze (Kirghizia), USSR | 150 |
| | RW166 | Irkutsk (Siberia), USSR | 250 |
| | (KZ) | Kazan, USSR | 50 |
| | (LD) | Leningrad, USSR | 150 |
| | | Moscow, USSR | 100 |
| | | Ufa, USSR | 300 |
| 207 | | Muenchen-Erchling, W. Ger. | 500 |
| 209 | TFU | Eidar, Iceland | 10 |
| | TFU | Reykjavik (Vatnsendi), Iceland | 50 |
| | | Uglii, Mongolia | 60 |
| | | Dalanzadgad, Mongolia | 150 |
| | | Choibalsan, Mongolia | 75 |
| 209 | | Azilal, Morocco | 800 |
| | | Blagoveshchensk(Siberia),USSR | 30 |
| 209 | (KI) | Kiev I (Ukraine), USSR | 500 |
| | | Skovorodino (Siberia), USSR | 50 |
| | | Novosibirsk (Siberia), USSR | 150 |
| | | Nukus (Uzbekistan), USSR | 50 |
| 218 | LKO | Oslo, Norway | 200 |
| | (BK) | Baku, USSR | 500 |
| | (KN) | Krasnoyarsk kray(Siberia),USSR | 50 |
| | | Yeniseysk (Siberia), USSR | 300 |
| | | Birobidzhan (Far East), USSR | 30 |
| 225 | | Konstantynow (Warsaw),Poland | 2000 |
| 227 | | Altai, Mongolia | 150 |
| 227 | | Alma Ata (Kazakhstan), USSR | 50 |
| 227 | | Tyumen (Siberia), USSR | 50 |
| 234 | | Junglinster, Luxembourg | 2000 |
| 236 | | Arkhanghesk, USSR | 150 |
| | | Kichinev, USSR | 1000 |
| | (LD) | Leningrad I, USSR | 1000 |
| | | Magadan (Far East), USSR | 1000 |
| | | Mary (Turkmenia), USSR | 500 |
| | | Ulianovsk (Kubyshev), USSR | 1200 |
| | | Irkutsk (Siberia), USSR | 500 |
| 245 | | Belebey, USSR | 200 |
| | | Kalundborg, Denmark | 300 |
| | TCP | Erzurum, Turkey | 200 |
| | | Alma Ata (Kazakhstan), USSR | 500 |
| | | Karaganda (Kazakhstan), USSR | 100 |
| | | Vladivostok (Far East), USSR | |
| 254 | | Tipaza, Algeria | 1500/750 |
| 254 | | Clarkestown, Ireland | 500 |
| | | Lahti, Finland | 200 |
| | (FO) | Duchanbe (Tadzhikistan),USSR | 300 |
| | (KN) | Kazan, USSR | 150 |
| | (EW) | Erevan, USSR | 150 |
| 263 | | Plovdiv, Bulgaria | 500 |
| | | Karaganda (Kazakhstan), USSR | 150 |
| | (MK) | Moscow II, USSR | 2000 |
| | | Tchita (Siberia), USSR | 1000 |
| | | Alma Ata (Kazakhstan), USSR | 500 |

# BROADCASTING

| | | | |
|---|---|---|---:|
| 272 | | Topolna, Czechoslovakia | 1500 |
| | (NV) | Novosibirsk (Siberia), USSR | 150 |
| | | Orenburg, USSR | 15 |
| | | Tomsk (Siberia), USSR | 100 |
| | (HB) | Khabarovsk kray(Far East),USSR | 100 |
| | | Tashkent (Uzbekistan), USSR | 500 |
| 281 | | Achkhabad (Turkmenia), USSR | 150 |
| | (MS) | Minsk/Loshita (Byelorussia) | 500 |
| | | Sverdlovsk (Siberia), USSR | 50 |
| | | Ulan Ude (Siberia), USSR | 150 |
| | | Yuzno-Sakhalinsk(FarEast),USSR | 150 |
| 285 | | Roldal, Norway | 1 |

## International Broadcasting Allocations

| Frequency Band | Meter Band |
|---|---|
| 150-285 kHz | 1900 M |
| 140-1600 kHz | 200 M |
| 2300-2498 kHz | 120 M |
| 3200-3400 kHz | 90 M |
| 3900-4000 kHz | 75 M |
| 4750-4995 kHz | 60 M |
| 5950-6200 khz | 49 M |
| 7100-7300 kHz | 41 M |
| 9500-9775 kHz | 31 M |
| 11650-12050 kHz | 25 M |
| 13600-13800 kHz | 21 M |
| 15100-15600 khz | 19 M |
| 17550-17900 kHz | 16 M |
| 21450-21850 kHz | 13 M |
| 25670-26100 kHz | 11 M |

### International Broadcast Feeders and Relays

In order to secure program material from home offices, transmitter sites often receive transmissions in the shortwave spectrum for rebroadcast later. These transmissions may be in several sideband modes (see key).

While satellite relay is playing an increasingly greater role, a great deal of HF feeder activity is still monitored on these frequencies.

### Key

| | |
|---|---|
| AFRTS | American Forces Radio & TV Service |
| B | Bethany |
| BBC | British Broadcasting Corp |
| D | Dixon/Delano |
| DW | Deutsche Welle, Frankfurt |
| G | KWL61, Greenville |
| GWC | BBC Daventry |
| FS | Foreign Service |
| HS | Home Service |
| IS | Munich |
| ISB | Independent Side Band (uses both USB & LSB) |
| K | Kavala |
| LSB | Lower Side Band |
| M | Monrovia |
| R | Rhodes |
| RFE | Radio Free Europe, Holzkirchen |
| RM | Radio Moscow |
| RP | Radio Peking |
| SABS | Saudi Arabian Broadcasting Service, Riyadh |
| T | Tangier |
| UNID | Unidentified |
| USB | Upper Side Band |
| VOA | Voice of America |

| | |
|---|---|
| 4465 | RFE |
| 4475 | RFE |
| 4505 | RFE |
| 4555 | VOA-D |
| 4565 | RFE |
| 4565 | VOA-D |
| 5125 | RFE |
| 4195 | DW |
| 5230.8 | AFRTS-LSB |
| 5290 | RM |
| 5295 | RFE |
| 5390 | SABS |
| 5434 | VOA-T |
| 5455 | RM |
| 5460 | VOA-T |
| 5470 | RM-HS |
| 5742.5 | VOA-G-USB |
| 5760 | AFRTS-LSB |
| 5790 | RFE |
| 5810 | RM |
| 5815 | RM-HS |
| 5830 | DW-RM |
| 5845 | RFE |
| 5875 | SABS |
| 5880 | RIVADAVIA ARG |
| 5935 | VOA-USB |
| 5920 | RM |
| 6390 | RM-FS |
| 6470 | RM |
| 6550 | RP |
| 6770 | RM-HS |
| 6808 | RM |
| 6810 | RM |
| 6823 | RM-HS |
| 6825 | RM |
| 6825 | RP |
| 6838 | BBC-USB |

| | |
|---|---|
| 6840 | BBC |
| 6847.5 | VOA-K/R |
| 6873 | VOA-G-ISB/USB |
| 6888.5 | RM |
| 6902.5 | VOA-IS |
| 6910 | UNID |
| 6918.5 | RM-HS |
| 6942.5 | VOA-K/R |
| 6970 | RFE-LSB |
| 6975 | DW |
| 6980 | RM |
| 6987.5 | RM |
| 6993 | SABS |
| 6995 | RFE |
| 7408.5 | RM-HS |
| 7440 | UN GENEVA-RM |
| 7442.5 | VOA-M |
| 7470 | RM |
| 7478 | VOA-M |
| 7490 | DW |
| 7490 | RM-HS |
| 7530 | RM |
| 7541.5 | RM |
| 7548.5 | RM-FS |
| 7620 | VOA-D |
| 7648.5 | VOA-G-USB |
| 7651 | VOA-USB |
| 7653.5 | VOA-LSB |
| 7691 | BBC |
| 7724 | VOA-IS |
| 7768.5 | DW-LSB/ISB |
| 7768.5 | VOA-G-ISB/LSB |
| 7800 | RP |
| 7846 | BBC |
| 7893 | BBC |
| 7918 | DW |
| 7925 | RM-FS |
| 7955 | UNID |
| 7973 | BBC-USB |
| 7977 | UNID |
| 7991 | BBC |
| 8003.5 | RM-LSB |
| 8110 | VOA-D |
| 8125 | RM-FS |
| 8345 | RP |
| 8450 | RP |
| 8490 | RP |
| 8660 | RP |
| 9090 | UNID-LSB |
| 9100 | BBC |
| 9115 | RIVADAVIA ARG |
| 9140 | RM-HS |
| 9145 | RFE |
| 9145 | RM-HS |
| 9150 | RM-HS |
| 9170 | RFE |
| 9200 | RM-HS |
| 9210 | RM-HS |

| | |
|---|---|
| 9240 | RM |
| 9250 | RFE |
| 9290 | UNID |
| 9320 | BBC-USB/ISB |
| 9350 | VOA-D |
| 9858 | BBC |
| 9905 | RM |
| 9989 | SABS |
| 10120 | RM-FS |
| 10125 | RM |
| 10190 | RFE-LSB |
| 10235 | VOA-G-ISB |
| 10237.5 | VOA-LSB/ISB |
| 10315 | RFE-USB |
| 10338 | RM-HS |
| 10335.4 | UNID-USB |
| 10382.5 | VOA-G-LSB |
| 10385 | SABS |
| 10385 | VOA-D |
| 10402 | SABS |
| 10415 | R.NACIONAL, P-G-LSB |
| 10418 | UNID-LSB |
| 10420 | RFE |
| 10454 | VOA-G-ISB |
| 10456.5 | UN RADIO-LSB |
| 10476 | UNID |
| 10520 | VOA-IS |
| 10530 | SABS |
| 10538 | AFRTS-LSB |
| 10595 | RM |
| 10620 | RM-FS |
| 10660 | RM-HS |
| 10675 | RM |
| 10689 | VOA |
| 10690 | RM |
| 10690 | UNID-USB |
| 10740 | RM-FS |
| 10760 | VOA-IS |
| 10855 | RM-HS |
| 10860 | RFE |
| 10865 | RP |
| 10866.5 | VOA-USB/ISB |
| 10869 | UN RADIO |
| 10869 | VOA-G |
| 10872 | UNID |
| 10877.5 | VOA-USB/ISB |
| 10880 | VOA-K/R |
| 10922 | DW |
| 10935 | SABS |
| 10972 | VOA-T |
| 10990 | SABS |
| 11070 | BBC-LSB |
| 11090 | VOA-D-ISB/USB/LSB |
| 11092 | SABS |
| 11095 | SABS |
| 11152.5 | VOA-G-LSB |
| 11175 | SABS |

# BROADCASTING

| | | | | |
|---|---|---|---|---|
| 11445 | UNID | | 15752 | VOA-G |
| 11459 | BBC | | 15765 | VOA-G |
| 11495 | RM-FS | | 15770 | VOA-G |
| 11575 | RFE | | 15775 | RFE-USB |
| 11575 | RM-HS | | 15780 | RM-FS |
| 11585 | RM-HS | | 15790 | UNID |
| 12127 | BBC | | 15800 | VOA-IS |
| 12135 | RM | | 15849 | BBC-ISB/USB |
| 12165 | RM | | 15870 | R.AUSTRALIA |
| 12179 | BBC | | 15875 | VOA-M |
| 12205 | RM-HS | | 15877.5 | VOA-LSB |
| 12210 | VOA-G | | 15910 | BBC |
| 12223 | RM-HS | | 15913 | BBC |
| 12240 | RM-FS | | 15920 | VOA-M |
| 12246.5 | RFE | | 16030 | RM-HS |
| 12250 | RM | | 16065 | RFE |
| 12253 | RM-FS | | 16140 | RM |
| 12290 | R.AUSTRALIA-USB | | 16190 | RM |
| 12296 | R.MONTE CARLO | | 16222 | VOA-G |
| 13380 | RM-HS | | 16240 | RFE |
| 13395 | SABS | | 16247 | VOA-T |
| 13410 | RM | | 16250 | RM-HS |
| 13462 | SABS | | 16330 | RM-HS |
| 13486 | SABS | | 16330 | VOA-M |
| 13491.5 | VOA-G | | 16433 | R.BARIKI |
| 13512 | DW | | 17445 | RFE |
| 13590 | RM-HS | | 17490 | RP |
| 13690 | RFE | | 17560 | RM |
| 13710 | RM-HS | | 17580 | RM-HS |
| 13720 | UNID-USB/LSB | | 18137.5 | VOA-D |
| 13760 | RM-HS | | 18150 | UNID |
| 13770 | VOA-T | | 18157 | VOA-D |
| 13797 | UNID-USB | | 18172 | BBC-ISB |
| 13810 | SABS | | 18195 | RM-HS |
| 13818 | SABS | | 18210 | VOA-K-R-M |
| 13820 | RM-FS | | 18215 | VOA-G |
| 13855 | VOA-K/R | | 18265 | VOA-G-R-K |
| 13860 | VOA-K | | 18275 | VOA-G |
| 13865 | SABS | | 18285 | RM-HS |
| 13885 | VOA-K/R | | 18300 | RM |
| 13960 | RM-FS | | 18310 | RM |
| 13977 | RM-HS | | 18460 | RM |
| 13995 | VOA-M | | 18515 | VOA-K-R-M |
| 13997.5 | VOA-IS | | 18607.5 | VOA-LSB/ISB |
| 14365 | VOA-K/R | | 18653 | RM-FS |
| 14398 | VOA-D-LSB/USB/ISB | | 18782.5 | VOA-USB/ISB |
| 14456 | BBC | | 18830 | RM-HS |
| 14526 | VOA-G | | 19075 | VOA-T |
| 14713 | RFE | | 19155 | VOA-M |
| 14800 | VOA-IS | | 19210 | R.AUSTRALIA |
| 14850 | RM-HS | | 19261.5 | VOA-B-USB/LSB/ISB |
| 14985 | VOA-K/R | | 19291 | AFRTS-LSB |
| 15576 | SABS | | 19320 | RM |
| 15650 | VOA-G | | 19455 | BBC-ISB |
| 15670 | BBC-ISB | | 19480 | VOA-D-ISB |
| 15673 | VOA | | 19505 | VOA-G-USB/LSB/ISB |
| 15715 | VOA-G | | 19522 | RM |
| 15720 | RM | | 19532.5 | VOA-G |

| | |
|---|---|
| 19755 | BBC |
| 19845 | RM |
| 19912.5 | VOA-G-USB |
| 19915 | VOA-D |
| 19918 | VOA-G |
| 19922 | VOA |
| 20025 | UNID |
| 20060 | VOA-G-ISB/LSB |
| 20125 | VOA-G-LSB/ISB/USB |
| 20192 | AFRTS-Patrick to Ascension AFB-LSB Sunday |
| 20215 | RFE-LSB |
| 20280 | RM-HS |
| 20325 | UNID |
| 20345 | BBC-USB/ISB |
| 20455 | BBC-ISB |
| 20500 | RM |
| 20545 | RFE-ISB/LSB |
| 20605 | RM |
| 20650 | RFE |
| 20710 | RFE |
| 21455 | VOA-B |
| 21760 | VOA-M |
| 21800 | VOA-M |
| 22925 | VOA-D-M |
| 22930 | BBC |
| 22970 | RFE |
| 23191 | BBC |
| 23680 | RM-LSB/AM |
| 23826 | UNID |
| 23856 | RM |
| 25350 | RFE |
| 26640 | BBC |
| 26490 | BBC |
| 26680 | RFE |

# International Broadcasting

All broadcasts listed below are in English. In some cases, which are marked "ML" (multi-lingual), the broadcast is conducted in English and at least one other language.

Because this is a worldwide list -- that is, it includes transmissions intended for all parts of the world, not simply North America -- some stations may be difficult or impossible to hear from your location. Do not, however, assume anything to be impossible, for shortwave's most endearing charm is its unpredictability. If you don't hear it at first, try, try, try, try, try, try and try again.

The first number in the listing is the time at which the broadcast starts. It is followed, naturally, by the time that the broadcast ends. The next column, often left blank, is for special schedule notations. S=Sunday, M=Monday, T=Tuesday, W=Wednesday, H= Thursday, F=Friday, A=Saturday.

A letter followed by a comma indicated "and" as in "S,M" would mean that the broadcast is heard on Sundays and Mondays only. Two letters separated by a hyphen indicated "between." An example would be a broadcast listed as "M-F" which would be "Monday through Friday. See "Additional coding," below, for other notations that may appear in this column.

The next column is for the name of the station and its announced location. Often, a station will actually receive mail sent to it with this information alone. Keep in mind that the locations that are listed are not necessarily indicators of where the station's transmitters are actually located. The BBC, always listed as "London, England," has transmitters in places as far flung as the United States and Oman, to name a few.

The final column is the frequency at which the station is broadcasting. We recommend that you begin sampling at the low end and continue upwards in search of the best reception.

Remember that shortwave frequencies change often and without notice. To stay on top of the latest in international broadcasting, be sure to subscribe to Monitoring Times. Here, along with news from around the world of radio, you'll find the world's most comprehensive listing of English broadcasts to the world, updated and expanded every month.

Additional coding includes: IRR = Irregular  ML = Multi-lingual  MO = Music only  TEN = Tentative  TP = Temporary arrangement  V = Variable.

Frequencies written in subscript ($_{21595}$) indicate Single Side Band mode transmissions. A "/" between two frequencies indicates that they may alternate.

For updated versions of this list, check out Monitoring Times magazine, THE monthly source for all of you shortwave/scanner needs!

Radio New Zealand transmissions in this listing may not necessarily be on the air seven days a week, and are subject to change.

# BROADCASTING

| Time | Station | Frequencies |
|---|---|---|
| 0000-0015 | Radio Prague Int'l, Czechoslovakia | 7345 11680 11990 |
| 0000-0025 | Radio Finland, Helsinki | 11755 15185 |
| 0000-0030 | Radio Canada Int'l, Montreal | 5960 9755 |
| 0000-0030 | Kol Israel, Jerusalem | 9435 11605 12077 |
| 0000-0030 | Radio Australia, Melbourne | 11880 13605 15240 15380 15465 17600 17630 17750 17795 |
| 0000-0050 | Radio Pyongyang, North Korea | 11975 15115 |
| 0000-0100 | All India Radio, New Delhi | 9535 9910 11715 11745 15110 |
| 0000-0100 | Radio Thailand, Bangkok | 4830 9655 11905 |
| 0000-0100 | Radio New Zealand, Wellington | 17675 |
| 0000-0100 | BBC World Service, London, England | 5975 6005 6175 6195 7325 9590 9915 11750 12095 15260 17830 |
| 0000-0100 | Adventist World Radio, Costa Rica | 9725 11870 |
| 0000-0100 | Radio Moscow N.American Service | 11710 11730 11780 11850 11980 12040 15425 15580 15595 |
| 0000-0100 | Radio Sofia, Bulgaria | 11660 11720 |
| 0000-0100 | Radio Kiev, Ukraine | 7400 9860 11790 13645 15485 |
| 0000-0100 | Radio Moscow World Service | 15280 17670 17890 21690 21790 |
| 0000-0100 | CBC Northern Quebec Service, Can | 9625(ML) |
| 0000-0100 | CBN, St. John's, Nfld, Canada | 6160 |
| 0000-0100 | CBU, Vancouver, British Columbia | 6160 |
| 0000-0100 | CFCF, Montreal, Quebec, Canada | 6005 |
| 0000-0100 | CFCN, Calgary, Alberta, Canada | 6030 |
| 0000-0100 | CHNS, Halifax, Nova Scotia, Canada | 6130 |
| 0000-0100 | Christian Science World Svc, Boston | 7395 9850 13760 15225 15610 17555 (+17865 A,S) |
| 0000-0100 | CKWX, Vancouver, British Columbia | 6080 |
| 0000-0100 | CFRB, Toronto, Ontario, Canada | 6070 |
| 0000-0100 | FEBC Radio Int'l, Philippines | 15490 |
| 0000-0100 | KSDA, Guam | 15125 |
| 0000-0100 T-A | KUSW, Salt Lake City, Utah | 15590 |
| 0000-0100 | Radio Beijing, Beijing, China | 15100 17705 |
| 0000-0100 | Radio Luxembourg, Junglinster | 6090 |
| 0000-0100 | Spanish National Radio, Madrid | 9630 11880 |
| 0000-0100 | Voice of America-Americas Service | 5995 9775 9815 11580 15205 |
| 0000-0100 | Voice of America-Caribbean Service | 6130 9455 11695 |
| 0000-0100 | Voice of America-East Asia Service | 7120 9770 11760 15185 15290 17735 17820 |
| 0000-0100 | Radio for Peace Int'l, Costa Rica | 7375 13630 21566 |
| 0000-0100 | WHRI, Noblesville, Indiana | 7315 9495 |
| 0000-0100 | WINB, Red Lion, Pennsylvania | 15145 ML |
| 0000-0100 | WRNO Worldwide, Louisiana | 7355 |
| 0000-0100 | WWCR, Nashville, Tennessee | 7520 |
| 0000-0100 | WYFR, Okeechobee, Florida | 5985 13695 15170 |
| 0030-0100 T-S | Radio Canada Int'l, Montreal | 5960 9755 |
| 0030-0100 | Radio Australia, Melbourne | 11880 15160 15240 15465 15560 17630 17750 17795 21525 21740 21775 |

| | | |
|---|---|---|
| 0030-0100 | BRT Brussels, Belgium | 9925 13675 |
| 0030-0100 T-S | Radio Budapest, Hungary | 6110  9520  9585  9835 |
| | | 11910 15160 |
| 0030-0100 | Radio Netherlands Int'l, Hilversum | 6020  6165 11740 |
| 0035-0100 | HCJB, Quito, Ecuador | 15155 17875 25950ssb |
| 0050-0100 | Vatican Radio, Vatican City | 9605 11780 15180 |

| 0100 UTC | [9:00 PM EDT/6:00 PM PDT] |
|---|---|

| | | |
|---|---|---|
| 0100-0105 | Vatican Radio, Vatican City | 9605 11780 15180 |
| 0100-0115 | All India Radio, New Delhi | 9535  9910 |
| | | 11715 11745 15110 |
| 0100-0125 | RAI, Rome, Italy | 9575 11800 |
| 0100-0125 | Radio Netherlands Int'l, Hilversum | 6020  6165 11740 |
| 0100-0130 | Kol Israel, Jerusalem | 9435 11605 12077 |
| 0100-0130 S,M | Radio Norway, Oslo | 9615 11925 |
| 0100-0130 | Radio Australia, Melbourne | 11880 15160 15240 15465 |
| | | 15560 17630 17750 17795 |
| | | 21525 21740 21775 |
| 0100-0130 | Radio Canada International, Montreal | 5960  9755 |
| 0100-0130 | Radio Japan Americas Svc, Tokyo | 17755 |
| 0100-0130 | Radio Prague Int'l, Czechoslovakia | 5930  7345 11680 |
| 0100-0130 | CBC Northern Quebec Service, Can | 9625 (ML) |
| 0100-0130 | Radio Sweden, Stockholm | 15405 |
| 0100-0145 | Radio Yugoslavia, Belgrade | 5980  6005 11735 |
| 0100-0150 | Deutsche Welle, Koln, West Germany | 6040  6145  9565 |
| | | 15105 11865 |
| 0100-0200 | Radio Thailand, Bangkok | 4830  9655 11905 |
| 0100-0200 | WINB Red Lion, PA | 15145 ML |
| 0100-0200 | Radio Moscow North American Svc | 11690 11710 11730 11780 |
| | | 11850 11980 12040 15290 |
| | | 15425 15580 15595 |
| 0100-0200 | Radio Moscow World Service | 15280 17690 21690 21790 |
| 0100-0200 | BBC World Service, London, England | 5975  6005  6175  7325 |
| | | 9590  9915 11750 12095 |
| | | 15260 21715 |
| 0100-0200 S,M | Radio Canada Int'l, Montreal | 9535 11845 11940 |
| | | 13720 |
| 0100-0200 | Radio New Zealand, Wellington | 17675 |
| 0100-0200 | CBN, St John's, Newfoundland | 6160 |
| 0100-0200 | CBU, Vancouver, British Columbia | 6160 |
| 0100-0200 | CFCF, Montreal, Quebec, Canada | 6005 |
| 0100-0200 | CFCN, Calgary, Alberta, Canada | 6030 |
| 0100-0200 | CHNS, Halifax, Nova Scotia, Canada | 6130 |
| 0100-0200 | Christian Science World Svc, Boston | 7395  9850 13760 15225 |
| | | 15610 17555 (+17865 A,S) |
| 0100-0200 | CKWX, Vancouver, British Columbia | 6080 |
| 0100-0200 | CFRB, Toronto, Ontario, Canada | 6070 |
| 0100-0200 | FEBC Radio Int'l, Philippines | 15490 |
| 0100-0200 | HCJB, Quito, Ecuador | 17875 15155 |
| 0100-0200 T-A | KUSW, Salt Lake City, Utah | 15590 |
| 0100-0200 | Radio Havana Cuba | 11820 |
| 0100-0200 | Radio Japan General Svc, Tokyo | 5960 17765 17810 17835 |
| | | 17845 |
| 0100-0200 | Radio Luxembourg, Junglinster | 6090 |
| 0100-0200 | Radio for Peace Int'l, Costa Rica | 7375 (T-A add 13630) |
| 0100-0200 | Spanish National Radio, Madrid | 9630 11880 |

# BROADCASTING

| | | | | | |
|---|---|---|---|---|---|
| 0100-0200 | | Voice of America-Americas Service | 5995 | 9775 | 9815 11580 |
| | | | 15205 | | |
| 0100-0200 | | Voice of America-Caribbean Service | 6130 | 9455 | |
| 0100-0200 | | Voice of America-East Asia Service | 7115 | 7205 | 9740 11705 |
| | | | 15205 | 21525 | |
| 0100-0200 | | Voice of Indonesia, Jakarta | 11753 | 11785 | |
| 0100-0200 | | WHRI, Noblesville, Indiana | 7315 | 9495 | |
| 0100-0200 | | WRNO Worldwide, Louisiana | 7355 | | |
| 0100-0200 | | WWCR, Nashville, Tennessee | 7520 | | |
| 0100-0200 | | WYFR, Okeechobee, Florida | 5985 | 9505 | 11720 17612 |
| 0130-0200 | | Radio Budapest, Hungary | 6110 | 9520 | 9585 9835 |
| | | | 11910 | 15160 | |
| 0130-0200 | M-A | Voice of Greece, Athens | 11645 | 9395 | 9420 |
| 0130-0200 | | Radio Baghdad, Iraq | 11755 | 11810 | 11830 21585 |
| 0130-0200 | | Radio Austria International, Vienna | 9870 | 9875 | 13730 |
| 0130-0200 | S,M | Radio Canada Int'l, Montreal | 5960 | 9755 | |
| 0130-0200 | | Radio Australia, Melbourne | 11880 | 15160 | 15240 15465 |
| | | | 15560 | 17630 | 17750 17795 |
| | | | 21525 | 21740 | 21775 |
| 0145-0200 | | Radio Korea, Seoull | 6165 | 9640 | 15575 |
| 0145-0200 | | Vatican Radio, Vatican City | 9650 | 11750 | 15135 |

## 0200 UTC    [10:00 PM EDT/7:00 PM PDT]

| | | | | | |
|---|---|---|---|---|---|
| 0200-0220 | | Radio Veritas-Asia, Philippines | 15220 | 15360 | |
| 0200-0230 | | SLBC Domestic Service, Sri Lanka | 4940 | | |
| 0200-0230 | | Kol Israel, Jerusalem | 9435 | 11605 | 12077 |
| 0200-0230 | H,A | Radio Budapest, Hungary | 6110 | 9520 | 9585 9835 |
| | | | 11910 | 15160 | |
| 0200-0230 | | FEBC Radio Int'l, Philippines | 15490 | | |
| 0200-0230 | T-A | Voice of America | 5995 | 9775 | 9815 11580 |
| | | | 15205 | | |
| 0200-0230 | S.M | Radio Norway, Oslo | 9615 | 11735 | |
| 0200-0230 | | British Forces Radio, UK | 7125 | 9640 | 13745 |
| 0200-0230 | | Swiss Radio International, Berne | 6095 | 6135 | 9650 9885 |
| | | | 12035 | 17730 | |
| 0200-0250 | | Deutsche Welle, Koln, W. Germany | 7285 | 9615 | 9690 11835 |
| | | | 11945 | 15235 | 17770 |
| 0200-0300 | | Radio Thailand, Bangkok | 4830 | 9655 | 11905 |
| 0200-0300 | T-A | Radio Canada Int'l, Montreal | 9535 | 11845 | 11940 13720 |
| 0200-0300 | | BBC World Service, London, England | 5975 | 6005 | 6110 6175 |
| | | | 7135 | 7325 | 9410 9590 |
| | | | 9915 | 11750 | 12095 15260 |
| | | | 15390 | 21715 | |
| 0200-0300 | | RAE, Buenos Aires, Argentina | 11710 | | |
| 0200-0300 | | KSDA, Guam | 13720 | | |
| 0200-0300 | | Radio Moscow North American Svc | 11690 | 11710 | 11780 11850 |
| | | | 11980 | 12040 | 12050 13605 |
| | | | 13675 | 15290 | 15315 15425 |
| | | | 15435 | 15530 | 15580 15595 |
| 0200-0300 | | Radio Moscow World Service | 15280 | 17690 | 21690 21790 |
| 0200-0300 | | CBC Northern Quebec Service, Can | 9625 | (ML) | |
| 0200-0300 | | CBN, St. John's, Newfoundland, Can | 6160 | | |
| 0200-0300 | | CBU, Vancouver, British Columbia | 6160 | | |
| 0200-0300 | | CFCF, Montreal, Quebec, Canada | 6005 | | |
| 0200-0300 | | CFCN, Calgary, Alberta, Canada | 6030 | | |
| 0200-0300 | | CHNS, Halifax, Nova Scotia, Canada | 6130 | | |

| | | |
|---|---|---|
| 0200-0300 | Christian Science World Svc, Boston | 9455  9850 13720 13760 (+ 17865 & 17555 A,S) |
| 0200-0300 | CKWX, Vancouver, British Columbia | 6080 |
| 0200-0300 | CFRB, Toronto, Ontario, Canada | 6070 |
| 0200-0300 | HCJB, Quito, Ecuador | 15155 17875 25950ssb |
| 0200-0300 T-A | KUSW, Salt Lake City, Utah | 15590 |
| 0200-0300 | Radio Australia, Melbourne | 11880 15160 15240 15320 15465 15560 17630 17750 17795 21525 21740 21775 |
| 0200-0300 | Radio Baghdad, Iraq | 11755 11810 11830 21585 |
| 0200-0300 T-A | Radio For Peace Int'l, Costa Rica | 7375 USB (T-A add 13630) |
| 0200-0300 | Radio Romania Int'l, Bucharest | 5990  6155  9510  9570 11830 11940 15380 |
| 0200-0300 | Radio Cairo, Egypt | 9475  9675 |
| 0200-0300 | Radio Havana Cuba | 9710 11820 |
| 0200-0300 | Radio Luxembourg, Junglinster | 6090 |
| 0200-0300 | Voice of America-South Asia Service | 7115  7205  9740 11705 15160 15250 21525 |
| 0200-0300 | Radio Cultura, Guatemala | 3300 |
| 0200-0300 | Radio New Zealand, Wellington | 17675 |
| 0200-0300 | Voice of Free China, Taiwan | 5950  7445  9680 |
| 0200-0300 | WHRI, Noblesville, Indiana | 7315  9495 |
| 0200-0300 | WRNO Worldwide, Louisiana | 7355 |
| 0200-0300 | WWCR, Nashville, Tennessee | 7520 |
| 0200-0300 | WINB, Red Lion, Pennsylvania | 15145 |
| 0200-0300 | WYFR, Okeechobee, Florida | 6065  9505 11720 |
| 0230-0245 | Radio Pakistan (Slow speed news) | 9545 15115 17640 17690 17725 21730 |
| 0230-0300 T-A | Radio Portugal, Lisbon | 9600  9680  9705 11840 |
| 0230-0300 | Radio Sweden, Stockholm | 9695 11705 |
| 0230-0300 | Radio Tirana, Albania | 9500 11825 |

## 0300 UTC     [11:00 PM EDT/8:00 PM PDT]

| | | |
|---|---|---|
| 0300-0315 | Azad Kashmir Radio, Pakistan | 7286  4980  3665 |
| 0300-0330 | WINB Red Lion, PA | 15145 ML |
| 0300-0330 | Radio Australia, Melebourne | 11880 15160 15240 15320 15465 15560 17630 17750 17795 21525 21740 21775 |
| 0300-0330 | Radio Cairo, Egypt | 9475 9675 |
| 0300-0330 | Radio Japan , Tokyo | 15325 17825 21610 |
| 0300-0330 | Radio Prague Int'l, Czechoslovakia | 5930  7345 11680 |
| 0300-0330 | Radio Baghdad, Iraq | 11755 11810 11830 |
| 0300-0350 | Deutsche Welle, Koln, West Germany | 6085  6120  9545 15205 11810 |
| 0300-0355 | Radio Beijing, China | 9690 11715 15100 |
| 0300-0400 | Radio New Zealand, Wellington | 17675 |
| 0300-0400 | BBC World Service, London, England | 5975  6005  6175  6195 7135  7325  9410  9600 9915 11750 12095 15220 15260 15420 17705 21715 |
| 0300-0400 | CBC, Northern Quebec Service, Can | 9625 (ML) |
| 0300-0400 | Radio Moscow North American Svc | 9635 12050 13605 15180 15425 15455 15530 15580 15595 |
| 0300-0400 | Radio Moscow World Service | 15280 17690 21690 21790 |
| 0300-0400 | CBN, St. John's, Newfoundland, Can | 6160 |

# BROADCASTING

| | | |
|---|---|---|
| 0300-0400 | CBU, Vancouver, British Columbia | 6160 |
| 0300-0400 | CFCF, Montreal, Quebec, Canada | 6005 |
| 0300-0400 | CFCN, Calgary, Alberta, Canada | 6030 |
| 0300-0400 | CHNS, Halifax, Nova Scotia, Canada | 6130 |
| 0300-0400 | Christian Science World Svc, Boston | 9455  9850 13720 13760 15225 (+17865 & 17555 A,S) |
| 0300-0400 | CKWX, Vancouver, British Columbia | 6080 |
| 0300-0400 | CFRB, Toronto, Ontario, Canada | 6070 |
| 0300-0400 | Faro del Caribe,San Jose,Costa Rica | 5055  9645 |
| 0300-0400 | HCJB, Quito, Ecuador | 17875 15155 |
| 0300-0400 | Radio Cultural, Guatemala | 3300 |
| 0300-0400 | Radio Havana Cuba | 9710 11820 |
| 0300-0400 | Radio Thailand, Bangkokk | 4830  9655 11905 |
| 0300-0400 | Radio japan, Tokyo | 15195 17810 (+ 7125 to 0330) |
| 0300-0400 T-A | KUSW, Salt Lake City, Utah | 15590/11695 |
| 0300-0400 | Trans World Radio, Bonaire | 9535 11930 |
| 0300-0400 | Voice of America-Africa Service | 6035  7170  7280  9525 9575 11835 |
| 0300-0400 | Voice of Free China, Taiwan | 5950  7445  9680  9765 11745 15345 17845 |
| 0300-0400 | WHRI, Noblesville, Indiana | 7315  9495 |
| 0300-0400 | WRNO Worldwide, Louisiana | 7355 |
| 0300-0400 | WWCR, Nashville, Tennessee | 7520 |
| 0300-0400 | WYFR, Okeechobee, Florida | 6065  9505 15440 |
| 0310-0325 | Vatican Radio, Vatican City | 11725 |
| 0315-0330 | Radio for Peace Int'l, Costa Rica | 7375 USB |
| 0315-0345 | Radio France International, Paris | 3965  5990  7135  7280 9745  9790  9800 11705 11790 11995 15135 15155 15300 |
| 0330-0400 | Radio Netherlands Int'l, Hilversum | 9590 11720 |
| 0330-0400 | Radio Tirana, Albania | 9500 11825 |
| 0330-0400 | Radio Tanzania | 9684 |
| 0330-0400 | Radio Australia, Melbourne | 11880 15160 15240 15320 15465 15560 17795 21525 21740 21775 |
| 0330-0400 | United Arab Emirates Radio, Dubai | 11945 13675 15400 15435 |
| 0340-0350 M-A | Voice of Greece, Athens | 11645  9395 9420 |
| 0349-0357v | Radio Yerevan, Armenia | 11675 11790 15180 15455 15485 17555 |
| 0350-0400 | RAI, Rome, Italy | 11905 15330 17795 17690 17665 |

## 0400 UTC    [12:00 AM EDT/9:00 PM PDT]

| | | |
|---|---|---|
| 0400-0410 M-F | Radio Zambia, Lusaka | 4910 |
| 0400-0410 | RAI, Rome, Italy | 11905 15330 17795 |
| 0400-0415 | Radio Prague Int'l, Czechoslovakia | 5930  7345 11680 |
| 0400-0425 | Radio Cultural, Guatemala | 3300 |
| 0400-0425 | Radio Netherlands Int'l, Hilversum | 9590 11720 |
| 0400-0430 | Radio Tanzania | 9684 |
| 0400-0430 | Radio Thailand, Bangkok | 4830  9655 11905 |
| 0400-0430 | Radio Romania Int'l, Bucharest | 5990  6155  9510  9570 11830 11940 15380 |
| 0400-0430 | Radio Australia, Melbourne | 11880 15160 15240 15320 15465 15560 17795 21525 |

| | | |
|---|---|---|
| | | 21740 21775 |
| 0400-0430 | Swiss Radio International, Berne | 6135  9650  9885 12035 |
| 0400-0430 | Trans World Radio, Bonaire | 11930  9535 |
| 0400-0450 | Deutsche Welle, Koln, West Germany | 7225  7150  9765  9565 |
| | | 11765 15265 |
| 0400-0450 | Radio Pyongyang, North Korea | 15180 15230 17765 |
| 0400-0455 | Radio Beijing, China | 11685 11840 |
| 0400-0500 | Voice of America-Africa Service | 6025  6035  7280  9525 |
| | | 9575 11785 11835 |
| 0400-0500 T-A | KUSW Salt Lake City, Utah | 9815 IRR |
| 0400-0500 | Radio Canada Int'l, Montreal | 11925 |
| 0400-0500 | Radio Moscow North American Svc | 9635 11895 12050 13605 |
| | | 15180 15425 15455 15530 |
| | | 15595(+17605 from 0430) |
| 0400-0500 | Radio New Zealand, Wellington | 17675 |
| 0400-0500 | BBC World Service, London, England | 5975  6005  6195  7105 |
| | | 7120  9410  9580  9600 |
| | | 9610  9670  9915 12095 |
| | | 15070 15245 17885 21470 |
| | | 21715 |
| 0400-0500 | Radio Sofia, Bulgaria | 7115 11720 11735 11760 |
| 0400-0500 | Voice of Turkey, Ankara | 9445 17880 |
| 0400-0500 | Radio Moscow World Service | 15280 17690 21690 21790 |
| 0400-0500 | CBC, Northern Quebec Service | 9625 (ML) |
| 0400-0500 | Radio for Peace Int., Costa Rica | 7375 USB |
| 0400-0500 | Radio RSA, Johannesburg | 7270 11900 |
| 0400-0500 | CBN, St. John's, Newfoundland, Can | 6160 |
| 0400-0500 | CBU, Vancouver, British Columbia | 6160 |
| 0400-0500 | CFCF, Montreal, Quebec, Canada | 6005 |
| 0400-0500 | CFCN, Calgary, Alberta, Canada | 6030 |
| 0400-0500 | CHNS, Halifax, Nova Scotia, Canada | 6130 |
| 0400-0500 | Christian Science World Svc, Boston | 9455  9840 13720 13760 |
| | | 15225 17780 (+17555 A,S) |
| 0400-0500 | CKWX, Vancouver, British Columbia | 6080 |
| 0400-0500 | CFRB, Toronto, Ontario, Canada | 6070 |
| 0400-0500 | HCJB, Quito, Ecuador | 17875 15155 |
| 0400-0500 | KSDA, Guam | 15225 |
| 0400-0500 | Radio Havana Cuba | 9710  9750 11760 11820 |
| 0400-0500 S-F | WMLK Bethel, Pennsylvania | 9465 |
| 0400-0500 | Voice of America-Middle East Service | 3980  5995  6040  6140 |
| | | 7170  7200 11785 15205 |
| 0400-0500 | WHRI, Noblesville, Indiana | 7315  9495 |
| 0400-0500 | WRNO Worldwide, Louisiana | 6185 |
| 0400-0500 | WWCR, Nashville, Tennessee | 7520 |
| 0400-0500 | WYFR, Okeechobee, Florida | 6065  9505 |
| 0425-0440 | RAI, Rome, Italy | 5990  7275 |
| 0430-0500 M-F | NBC Windhoek, Namibia | 3270  3290 |
| 0430-0500 | Radio Australia, Melbourne | 11880 15160 15240 15320 |
| | | 15465 15560 17630 17750 |
| | | 17795 21525 21740 21775 |
| 0430-0500 IRR | Radio Truth | 5015 |
| (clandestine intended for Zimbabwe) | | |
| 0430-0500 | Radio Tirana, Albania | 9500 11835 |
| 0455-0500 | Voice of Nigeria, Lagos | 7255 |

# BROADCASTING

| Time | Station | Frequencies |
|---|---|---|
| 0500-0505 | Radio Lesotho | 4800 |
| 0500-0515 | Kol Israel, Jerusalem | 9435 11605 11655 12077 15640 17575 |
| 0500-0515 | Azad Kashmir Radio, Pakistan | 7268 4980 3665 |
| 0500-0520 | Vatican Radio | 6185 9645 |
| 0500-0530 | Vatican Radio African Service | 17710 17730 21650 |
| 0500-0530 M-F | NBC Windhoek, Namibia | 3270 3290 |
| 0500-0550 | Deutsche Welle, Koln, West Germany | 5960 6120 9670 11705 11845 |
| 0500-0600 | BBC World Service, London, England | 5975 6005 6195 7120 9410 9600 9640 9915 12095 15070 17740 17885 21470 21715 |
| 0500-0600 | CBU, Vancouver, British Columbia | 6160 |
| 0500-0600 | CFCF, Montreal, Quebec, Canada | 6005 |
| 0500-0600 | CFCN, Calgary, Alberta, Canada | 6030 |
| 0500-0600 | Radio Thailand, Bangkok | 4830 9655 11905 |
| 0500-0600 | WRNO New Orleans, Louisiana | 6185 |
| 0500-0600 | CHNS, Halifax, Nova Scotia, Canada | 6130 |
| 0500-0600 S-F | WMLK Bethel, Pennsylvania | 9465 |
| 0500-0600 | Christian Science World Svc, Boston | 9455 9840 13720 13760 15225 17780 (+ 17555 A,S) |
| 0500-0600 | Radio Moscow North American Svc | 9635 11895 12050 13605 15180 15425 15455 15530 15595 17605 |
| 0500-0600 | Radio Moscow World Service | 15280 17690 21690 21790 |
| 0500-0545 | Radio New Zealand, Wellington | 17675 |
| 0500-0600 | CKWX, Vancouver, British Columbia | 6080 |
| 0500-0600 | CFRB, Toronto, Ontario, Canada | 6070 |
| 0500-0600 | HCJB, Quito, Ecuador | 15155 17875 |
| 0500-0600 | Radio Australia, Melbourne | 11880 15160 15240 15320 15465 15560 17630 17750 17795 21525 21740 21775 |
| 0500-0600 | Radio Havana Cuba | 5965 9710 11760 11820 |
| 0500-0600 | Radio Japan General Service, Tokyo | 15195 17765 17810 17825 17890 |
| 0500-0600 | Radio for Peace Int., Costa Rica | 7375 USB |
| 0500-0600 | Spanish National Radio, Madrid | 9630 |
| 0500-0600 | Voice of America-Africa Service | 3990 6035 7280 9540 9575 |
| 0500-0600 | Voice of America-Middle East Service | 3980 5995 6140 7170 7200 11785 15205 |
| 0500-0600 | Voice of Nigeria, Lagos | 7255 |
| 0500-0600 | WHRI, Noblesville, Indiana | 7315 9495 |
| 0500-0600 | WWCR, Nashville, Tennessee | 7520 |
| 0500-0600 | WYFR, Okeechobee, Florida | 5985 11580 17640 15566 |
| 0510-0530 M-A | Radio Botswana | 3356 4830 7255 |
| 0530-0600 | Radio Austria International, Vienna | 6015 |
| 0530-0600 | Radio Romania Int'l, Bucharest | 15340 15380 17720 17745 17790 21665 |
| 0530-0600 M-F | NBC Windhoek, Namibia | 3270 |
| 0530-0600 | UAE Radio Dubai | 15435 17830 21700 |
| 0545-0600 | Radio New Zealand, Wellington | 9855/17675 |
| 0555-0600 | Voice of Malaysia, Kuala Lumpur | 6175 9750 15295 |

| 0600 UTC | [2:00 AM EDT/11:00 PM PDT] |
| --- | --- |

| | | |
| --- | --- | --- |
| 0610-0615 | Sierra Leone Brdcstng.Svc.,Freetown | 3316 |
| 0600-0645 | Radio For Peace, Int., Costa Rica | 7375 USB |
| 0600-0650 | Radio Pyongyang, North Korea | 15180 15230 |
| 0600-0650 | Deutsche Welle, Koln, W. Germany | 11765 13790 15185 17875 |
| 0600-0650 | CBU, Vancouver, British Columbia | 6160 |
| 0600-0700 | Radio Australia, Melbourne | 11880 13700 13705 15240 15465 17630 21525 21740 21775 |
| 0600-0700 | Radio Havana Cuba | 5965 11760 11820 |
| 0600-0700 M-A | Vatican Radio | 6248 9645 11740 ML |
| 0600-0700 | BBC World Service, London, England | 5975 6180 6195 7120 7150 9410 9580 9600 9640 12095 15070 15245 15280 15400 15420 17640 17710 17790 17885 21470 21715 |
| 0600-0700 M-F | NBC Windhoek, Namibia | 7165 7190 |
| 0600-0700 | CFCF, Montreal, Quebec, Canada | 6005 |
| 0600-0700 | SIBC Solomon Islands | 5020 9545 |
| 0600-0700 | Radio New Zealand, Wellington | 9855/17675 |
| 0600-0700 | WYFR, Okeechobee, Florida | 5985 6065 7355 13760 15566 17640 |
| 0600-0700 | ABC Domestic Network, Australia | 15425 |
| 0600-0700 S-F | WMLK Bethel, Pennsylvania | 9465 |
| 0600-0700 | CFCN, Calgary, Alberta, Canada | 6030 |
| 0600-0700 | CHNS, Halifax, Nova Scotia, Canada | 6130 |
| 0600-0700 | Christian Science World Svc, Boston | 9455 9840 11705 13720 15225 17780 |
| 0600-0700 | CKWX, Vancouver, British Columbia | 6080 |
| 0600-0700 | CFRB, Toronto, Ontario, Canada | 6070 |
| 0600-0700 | Radio Moscow North American Svc. | 9635 12050 13605 15180 15425 15530 15595 17605 |
| 0600-0700 | Radio Moscow World Service | 15280 17690 21690 21790 |
| 0600-0700 | Voice of the Mediterranean, Malta | 9765 |
| 0600-0700 | HCJB, Quito, Ecuador | 15155 17875 |
| 0600-0700 | ABC Brisbane, Australia | 9660 |
| 0600-0700 | Radio Tonga, Kingdom of Tonga | 5030v |
| 0600-0700 | Voice of America-Africa Service | 3990 6035 6080 6125 7280 9530 9540 9575 11915 |
| 0600-0700 | Voice of America-Middle East Serv | 3980 5965 5995 6060 6095 6140 7170 7200 7325 9715 11785 11805 11925 15195 15205 17715 |
| 0600-0700 | WHRI, South Bend, Indiana | 7315 9495 |
| 0600-0700 | Voice of Hope, Lebanon | 6280 |
| 0600-0700 | Voice of Malaysia, Kuala Lumpur | 6175 9750 15295 |
| 0618-0700 M-F | Radio Canada International, Montreal | 6050 6150 7155 9740 9760 11840 17840 |
| 0630-0645 | RTV Congolaise, Brazzaville* *(Experimental broadcasts) | 6115 7105 7175 9615 9715 |
| 0630-0700 | Radio Finland, Helsinki | 11755 9560 6120 |
| 0630-0700 | Vatican Radio African Service | 17710 17730 21650 |
| 0630-0700 | BRT. Brussells, Belgium | 13675 11695 |
| 0630-0700 | Radio Tirana, Albania | 9500 7205 |

# BROADCASTING

| | | | |
|---|---|---|---|
| 0630-0700 | Radio Polonia, Warsaw, Poland | 6135 7270 15120 9675 | |
| 0630-0700 | Swiss Radio International, Berne | 15430 17570 21770 | |
| 0645-0700 A | Radio for Peace Int., Costa Rica | 7375 USB | |
| 0645-0700 | GBC Radio, Accra, Ghana | 6130 | |
| 0645-0700 | HCJB, Quito, Ecuador | 9610 11835 (alt 6050) | |
| 0645-0700 | Radio Romania Int'l, Bucharest | 11810 11940 15335 17720 17805 21665 | |

## 0700 UTC　　[3:00 AM EDT/12:00 AM PDT]

| | | | |
|---|---|---|---|
| 0700-0710 | Sierra Leone Brdcstng.Svc.,Freetown | 3316 | |
| 0700-0715 | Radio Romania Int'l, Bucharest | 11810 11940 15335 17720 17805 21665 | |
| 0700-0725 | BRT Brussels, Belgium | 21815 11695 6035 | |
| 0700-0730 | Radio Australia, Melbourne | 11880 13705 15240 15465 17630 21525 21740 21775 | |
| 0700-0730 | Radio Tirana, Albania | 11835 9500 | |
| 0700-0750 | Radio Pyongyang, North Korea | 15340 17795 | |
| 0700-0800 A | Radio for Peace Int'l, Costa Rica | 7375 USB | |
| 0700-0800 | Voice of Hope, Lebanon | 6280 | |
| 0700-0800 | CBU, Vancouver, British Columbia | 6160 | |
| 0700-0800 | TWR Monte Carlo | 9480 | |
| 0700-0800 | Radio Havana Cuba | 11835 | |
| 0700-0800 | WYFR, Okeechobee, Florida | 6065 7355 13760 15566 | |
| 0700-0800 | Voice of the Mediterranean, Malta | 9725 | |
| 0700-0800 | ZBC-1, Zimbabwe | 7283 | |
| 0700-0800 | Radio New Zealand, Wellington | 9855 | |
| 0700-0800 | BBC World Service, London | 5975 7150 9410 9600 9640 9760 11940 12095 15070 15280 15360 15400 21715 | |
| 0700-0800 | Solomon Islands Broadcasting Co. | 5020 9545 | |
| 0700-0800 | Voice of Free China, Taiwan | 5950 | |
| 0700-0800 | WHRI Noblesville, Indiana | 7315 9495 | |
| 0700-0800 | ABC Brisbane, Australia | 9660 | |
| 0700-0800 | CFCF, Montreal, Quebec, Canada | 6005 | |
| 0700-0800 | CFCN, Calgary, Alberta, Canada | 6030 | |
| 0700-0800 | CHNS, Halifax, Nova Scotia, Canada | 6130 | |
| 0700-0800 | Christian Science World Svc, Boston | 9455 9840 11705 13720 15225 17780 | |
| 0700-0800 | Radio Moscow World Service | 15280 17690 21690 21790 | |
| 0700-0800 | CFRB, Toronto, Ontario, Canada | 6070 | |
| 0700-0800 | GBC Radio, Accra, Ghana | 6130 | |
| 0700-0800 | Radio Korea, Seoul | 7550 13670 | |
| 0700-0800 | HCJB, Quito, Ecuador | 9610 11835 15270 | |
| 0700-0800 | KNLS, Anchor Point, Alaska | 9785 | |
| 0700-0800 | Radio Japan, Tokyo | 17765 17810 17890 21590 21690 | |
| 0700-0800 | Voice of Malaysia, Kuala Lumpur | 6175 9750 15295 | |
| 0710-0800 | HCJB, Quito,Ecuador(S. Pacific Sv.) | 6130 9745 11925 | |
| 0730-0800 | Radio Prague Int'l, Czechoslovakia | 17840 21705 | |
| 0730-0800 | ABC, Alice Springs, Australia | 2310 (ML) | |
| 0730-0800 | ABC, Katherine, Australia | 2485 | |
| 0730-0800 | ABC, Tennant Creek, Australia | 2325 (ML) | |
| 0730-0800 | Radio Austria Int'l, Vienna | 21490 15410 13730 6155 | |
| 0730-0800 | HCJB Quito, Ecuador | 9745 11925 | |
| 0730-0800 | KTWR, Agana Guam | 15200 | |

| 0730-0800 | Radio Australia, Melbourne | 6035 11880 13705 15240 17630 21525 21775 |
| 0730-0800 | Radio Netherlands, Hilversum | 9630 15560 |
| 0730-0800 | Radio Sofia, Bulgaria | 11720 15160 17825 |
| 0730-0800 | Swiss Radio Int'l European Service | 3985  6165  9535 |

| 0800 UTC | [4:00 AM EDT/ 1:00 AM PDT] |
|---|---|

| 0800-0803 | | Radio Pakistan | 17555 21575 |
| 0800-0810 | | Sierra Leone Brdcstng Co., Freetown | 3316 |
| 0800-0825 | | Radio Netherlands Int'l, Hilversum | 9630 15560 |
| 0800-0825 | | Voice of Malaysia, Kuala Lumpur | 6175  9750 15295 |
| 0800-0825 | | Radio Finland, Helsinki | 17800 21550 |
| 0800-0830 | | Radio Australia, Melbourne | 13705 15160 15240 17630 17750 17795 21525 21775 |
| 0800-0830 | | Voice of Islam, Dacca, Bangladesh | 15195 11705 |
| 0800-0850 | | Radio Pyongyang, North Korea | 15180 15230 |
| 0800-0900 | | Radio Moscow World Service | 15280 17690 21690 21790 |
| 0800-0900 | | KTWR, Guam | 15200 |
| 0800-0900 | | Trans World Radio, Monte Carlo | 9480 |
| 0800-0900 | | ABC Brisbane, Australia | 9660 |
| 0800-0900 | | BBC, London | 15280  9640 12095 15070 15360 21715 15400  9410 21660 |
| 0800-0900 | | ABC, Alice Springs, Australia | 2310 (ML) |
| 0800-0900 | | ABC, Katherine, Australia | 2485 |
| 0800-0900 | | ABC, Perth, Australia | 15425 |
| 0800-0900 | | ABC, Tennant Creek, Australia | 2325 (ML) |
| 0800-0900 | A | Radio for Peace Int., Costa Rica | 7375 USB |
| 0800-0900 | | Voice of Hope, Lebanon | 6280 |
| 0800-0900 | | CBN, St. John's, Newfoundland, Can | 6160 |
| 0800-0900 | | CBU, Vancouver, British Columbia | 6160 |
| 0800-0900 | | Radio Havana, Cuba | 11835 |
| 0800-0900 | | Radio Australia (Southwest Pacific) | 6020  6035  6080  9710 |
| 0800-0900 | | WHRI, South Bend, Indiana | 7315  7355 |
| 0800-0900 | | CFCF, Montreal, Quebec, Canada | 6005 |
| 0800-0900 | | CFCN, Calgary, Alberta, Canada | 6030 |
| 0800-0900 | | CHNS, Halifax, Nova Scotia, Canada | 6130 |
| 0800-0900 | | Christian Science World Svc | 9455  9530  9840 13720 15225 15610 |
| 0800-0900 | | CKWX, Vancouver, British Columbia | 6080 |
| 0800-0900 | | CFRB, Toronto, Ontario, Canada | 6070 |
| 0800-0900 | | HCJB,Quito,Ecuador | 6130  9610 11835 |
| 0800-0900 | | HCJB, Quito, Ecuador (alt pro) | 9745 11925 15270 |
| 0800-0900 | | KNLS, Anchor Point, Alaska | 11715 |
| 0800-0900 | | Solomon Islands Broadcasting Co. | 5020 |
| 0800-0900 | | WHRI. South Bend, Indiana | 7355 |
| 0800-0900 | | Voice of Indonesia, Jakarta | 11753 11785 |
| 0800-0900 | | Voice of Nigeria, Lagos | 7255 |
| 0815-0830 | | Radio Korea, Seoul | 9570 13670 |
| 0815-0900 | S | Italian Radio Relay Svc, Milan | 9815 |
| 0815-0900 | A,S | Radio New Zealand, Wellington | 9855 |
| 0830-0855 | M-F | Radio Netherlands Int'l, Hilversum | 15190 |
| 0830-0900 | | Radio Australia, Melbourne | 9580 15240 17630 17750 21525 21775 |
| 0830-0900 | | Radio Netherlands Int'l, Hilversum | 9630 17575 21485 |
| 0830-0900 | | Radio Finland, Helsinki | 21550 17800 |

# BROADCASTING

| | | |
|---|---|---|
| 0830-0900 | Swiss Radio International, Berne | 9560 13685 17670 21695 |
| 0837-0841v | Radio Tikhiy Okean, Vladivostok | 4485 5940 7210 7320 |
| | | 9530 9635 9670 9780 |
| | | 9820 9905 11815 11840 |
| | | 11850 11915 12050 12070 |
| | | 13605 15180 15410 15415 |
| | | 15425 15530 15535 17590 |
| | | 17605 17645 17695 17860 |
| | | 21505 21515 |
| 0840-0850 | Voice of Greece, Athens | 15625 17535 |
| 0845-0900 | KTWR, Agana, Guam | 15210 |

| 0900 UTC | [5:00 AM EDT/2:00 AM PDT] |
|---|---|

| | | |
|---|---|---|
| 0900-0920 | ABC, Perth, Australia | 15425 |
| 0900-0925 | BRT Brussels, Belgium | 9925 |
| 0900-0925 | Radio Netherlands Int'l, Hilversum | 9630 17575 21485 |
| 0900-0930 | Radio Australia (Southwest Pacific) | 6020 6035 6080 9710 |
| 0900-0930 | Radio Australia, Melbourne | 5995 9580 9760 17715 |
| | | 21775 21825 |
| 0900-0930 | KTWR Agana Guam | 15200 |
| 0900-0945 S | Italian Radio Relay Svc, Milan | 9815 |
| 0900-0950 | Deutsche Welle, Koln, West Germany | 6160 9565 11740 15410 |
| | | 17780 17820 21600 21650 |
| | | 21680 |
| 0900-1000 | ABC, Alice Springs, Australia | 2310 (ML) |
| 0900-1000 | Radio Beijing, China | 11755 15440 17710 |
| 0900-1000 | ABC Brisbane, Australia | 9660 |
| 0900-1000 | Solomon Islands Broadcasting Co. | 5020 |
| 0900-1000 | Radio Moscow World Service | 15280 17690 21690 21790 |
| 0900-1000 | ABC, Katherine, Australia | 2485 |
| 0900-1000 | ABC, Tennant Creek, Australia | 2325 (ML) |
| 0900-1000 S | Adventist World Radio, Portugal | 9670 |
| 0900-1000 A | Radio for Peace Int., Costa Rica | 7375 USB |
| 0900-1000 | KTWR, Agana, Guam | 11805 |
| 0900-1000 | Radio New Zealand, Wellington | 9855 |
| 0900-1000 S | Radio Bhutan, Thimpu | 5023v |
| 0900-1000 | Voice of Hope, Lebanon | 6280 |
| 0900-1000 | BBC World Service, London, England | 5975 9740 11750 12095 |
| | | 15070 15190 15360 15400 |
| | | 17640 17705 17790 17885 |
| | | 21470 21660 21715 |
| 0900-1000 | CFCF, Montreal, Quebec, Canada | 6005 |
| 0900-1000 | CFCN, Calgary, Alberta, Canada | 6030 |
| 0900-1000 | CHNS, Halifax, Nova Scotia, Canada | 6130 |
| 0900-1000 | Christian Science World Svc, Boston | 9455 9530 9840 11980 |
| | | 13720 15610 |
| 0900-1000 | CKWX, Vancouver, British Columbia | 6080 |
| 0900-1000 | CFRB, Toronto, Ontario, Canada | 6070 |
| 0900-1000 | FEBC Radio Int'l, Philippines | 11845 |
| 0900-1000 | HCJB, Quito, Ecuador | 6130 |
| 0900-1000 | HCJB, Quito, Ecuador(alt pro) | 9745 11925 |
| 0900-1000 | Radio Japan Australian Svc., Tokyo | 15270 17890 |
| 0900-1000 | Radio Japan General Service, Tokyo | 11840 21610 |
| 0900-1000 | Voice of Nigeria, Lagos | 7255 |
| 0900-1000 | WHRI, Noblesville, Indiana | 7315 7355 |
| 0910-0940 M,W,H,A,S | Radio Ulan Bator, Mongolia | 11850 12015 |

| | | |
|---|---|---|
| 0920-1000 | ABC, Perth, Australia | 6140 |
| 0930-1000 | Radio Afghanistan, Kabul | 4940  9635  17655  21600 |
| 0930-0955 | RRI Surabaya, Jawa Timur, Indonesia | 2377 |
| 0930-1000 | Radio Australia, Melbourne | 5995  9580  9655  9760 |
| | | 17715  21775  21825 |
| 0930-1000 | British Forces Broadcasting Svc, UK | 15205  17695  21735 |
| 0930-1000 | CBN, St. John's, New Foundland | 6160 |
| 0930-1000 | KTWR, Agana, Guam | 11805 |

## 1000 UTC    [6:00 AM EDT/3:00 AM PDT]

| | | |
|---|---|---|
| 1000-1015 | Radio Budapest, Hungary | 15160  15220  11925  9835 |
| | | 9585  6110 |
| 1000-1025 | BRT Brussels, Belgium | 21810  26050 |
| 1000-1030 | Radio Afghanistan, Kabul | 4940  9635  17655  21600 |
| 1000-1030  A | Radio for Peace Int., Costa Rica | 7375 USB |
| 1000-1030 | Voice of Vietnam, Hanoi | 9840  15010 |
| 1000-1030 | Swiss Radio International, Berne | 9560  13685  17670  21695 |
| 1000-1030 | Radio Australia, Melbourne | 5995  9580  9655  17715 |
| | | 21775 |
| 1000-1100 | Radio Korea, Seoul | 15575 |
| 1000-1100 | KHBN Guam | 9830 ML |
| 1000-1100 | WHRI, South Bend, Indiana | 7315  7355 |
| 1000-1100 | Radio Beijing, China | 11755  15440  17710 |
| 1000-1100 | ABC, Alice Springs, Australia | 2310 (ML) |
| 1000-1100 | ABC, Katherine, Australia | 2485 |
| 1000-1100 | Radio Baghdad, Iraq | 11860 |
| 1000-1100 | Solomon Islands Broadcasting Co. | 5020 |
| 1000-1100 | ABC, Perth, Australia | 9610 |
| 1000-1100 | ABC, Tennant Creek, Australia | 2325 (ML) |
| 1000-1100 | KSDA, Guam | 13720 |
| 1000-1100 | Radio Moscow World Service | 11840  17690  21690  21790 |
| 1000-1100 | All India Radio, New Delhi | 15010  15335  17387  17865 |
| | | 21735 |
| 1000-1100 | BBC World Service, London, England | 9410  9740  9750  12095 |
| | | 15070  15190  15360  15420 |
| | | 17705  17790  17885  21660 |
| 1000-1100 | CBN, St. John's, Nfld, Canada | 6160 |
| 1000-1100 | CFCF, Montreal, Quebec, Canada | 6005 |
| 1000-1100 | CFCN, Calgary, Alberta, Canada | 6030 |
| 1000-1100 | CHNS, Halifax, Nova Scotia, Canada | 6130 |
| 1000-1100 | Christian Science World Svc, Boston | 9455  9495  9530  11980 |
| | | 13625  13720 (+11705 A,S) |
| 1000-1100 | CKWX, Vancouver, British Columbia | 6080 |
| 1000-1100 | CFRB, Toronto, Ontario, Canada | 6070 |
| 1000-1100 | FEBC Radio Int'l, Philippines | 11845 |
| 1000-1100 | ABC Brisbane, Australia | 9660 |
| 1000-1100 | WYFR, Okeechobee, Florida | 5950 |
| 1000-1100 | HCJB, Quito, Ecuador | 9745  11925 |
| 1000-1100 | KTWR, Agana, Guam | 11805 |
| 1000-1100 | Voice of America-Caribbean Service | 9590  11915  15120 |
| 1000-1100 | Voice of America-Pacific Service | 5985  11720  15425 |
| 1015-1030 | Radio Korea, Seoul | 7275  11740 |
| 1015-1100  S | Italian Radio Relay Svc, Milan | 9815 |
| 1030-1100 | Radio Austria Int'l, Vienna | 15450  21490 |
| 1030-1100  M-A | Vatican Radio | 6248  9645  11740 |
| 1030-1100 | UAE Radio Dubai | 15320  15435  17865  21605 |

# BROADCASTING

| | | |
|---|---|---|
| 1030-1100 | Radio Tanzania | 5985   6105   7165 |
| 1030-1100 | Radio Korea, Seoul | 11715 |
| 1030-1100 | Radio Netherlands Int'l, Hilversum | 6020 11890 |
| 1030-1100 | Radio Australia, Melbourne | 5995   9580   9655 21775 |
| 1030-1100 | Adventist World Radio, Forli, Italy | 7230 |
| 1030-1045 | Radio Budapest, Hungary | 15160 15220 11925   9835 |
| | | 9585   6110 |
| 1040-1050 | Voice of Greece, Athens | 15625 17535 |
| 1045-1100 | Radio Budapest, Hungary | 7220   9585   9835 11910 |
| 1050-1100 | Radio Finland, Helsinki | 15400 21550 |

| 1100 UTC | [7:00 AM EDT/4:00 AM PDT] |
|---|---|

| | | |
|---|---|---|
| 1100-1115 | Azad Kashmir Radio, Pakistan | 7268   4980   3665 |
| 1100-1115 | Radio Finland, Helsinki | 15400 21550 |
| 1100-1120 | Radio Pakistan | 17565 21520 |
| 1100-1125 | HCJB Quito, Ecuador | 9745 11925 |
| 1100-1125 | Radio Netherlands Int'l, Hilversum | 6020 11890 |
| 1100-1130 | Kol Israel, Jerusalem | 17590 21660 21790 |
| 1100-1130 | Solomon Islands Broadcasting Co. | 5020 |
| 1100-1130 | Radio Mozambique, Maputo | 11835 11818   9525 |
| 1100-1130 | Voice of Vietnam, Hanoi | 9840 15010 |
| 1100-1130 | Radio Australia, Melbourne | 5995   6020   6035   6080 |
| | | 9580   9655   9710 11910 |
| | | 15465 21825 |
| 1100-1130 | Adventist World Radio, Forli, Italy | 7230 |
| 1100-1130 | Swiss Radio International, Berne | 13635 15570 17830 21770 |
| 1100-1150 | Radio Pyongyang, North Korea | 9977 11735 |
| 1100-1150 | Deutsche Welle, Koln, West Germany | 15410 17765 17800 21600 |
| 1100-1200 | ABC, Alice Springs, Australia | 2310 (ML) |
| 1100-1200 | BBC World Service, London, England | 9410   9515   9740   9750 |
| | | 11775 12095 15070 15360 |
| | | 15420 17640 17705 17790 |
| | | 17705 17790 17885 21470 |
| | | 21660 |
| 1100-1200 | Radio Baghdad, Iraq | 11860 |
| 1100-1200 | WHRI, Noblesville, Indiana | 9465 11790 |
| 1100-1200 | KHBN Guam | 9830 ML |
| 1100-1200 A,S | Radio Tanzania | 5985   6105   7165 |
| 1100-1200 | All India Radio, Northeast Svc | 7190 |
| 1100-1200 | WYFR, Okeechobee, Florida | 5950 11580 |
| 1100-1200 | Adventist World Radio, Costa Rica | 9725 11870 |
| 1100-1200 | Radio Moscow World Service | 11840 17690 21690 21790 |
| 1100-1200 | CBC, Montreal | 6160 |
| 1100-1200 | SBC Singapore | 11940 |
| 1100-1200 | ABC, Brisbane, Australia | 9660 |
| 1100-1200 | ABC, Katherine, Australia | 2485 |
| 1100-1200 | ABC, Perth, Australia | 9610 |
| 1100-1200 | ABC, Tennant Creek, Australia | 2325 (ML) |
| 1100-1200 | Trans World Radio, Bonaire | 11815 15345 |
| 1100-1200 | CBN, St. John's, Newfoundland, Can | 6160 |
| 1100-1200 | CFCF, Montreal, Quebec, Canada | 6005 |
| 1100-1200 | CFCN, Calgary, Alberta, Canada | 6030 |
| 1100-1200 | CHNS, Halifax, Nova Scotia, Canada | 6130 |
| 1100-1200 | Christian Science World Svc, Boston | 9455   9495   9530 11980 |
| | 13625 13720 (+11705 A,S) | |
| 1100-1200 | CKWX, Vancouver, British Columbia | 6080 |

| | | |
|---|---|---|
| 1100-1200 | CFRB, Toronto, Ontario, Canada | 6070 |
| 1100-1200 | Radio Japan, Tokyo | 6120 11815 11840 |
| 1100-1200 | Radio Jordan, Amman | 13655 |
| 1100-1200 | Radio RSA, Johannesburg | 9555 11805 11900 17835 |
| 1100-1200 | Voice of America-Caribbean Service | 9590 11915 |
| 1100-1200 | Voice of America-East Asia Service | 5985  6110  9760 11720 15155 15425 |
| 1110-1120vM-F | Radio Botswana | 4830  5995  7255 |
| 1115-1145 | Radio Nepal,Katmandu(External Svc.) | 5005 |
| 1115-1200 | Radio Korea, Seoul | 9750 |
| 1130-1140 | Radio Lesotho | 4800 |
| 1130-1145 | RRI Yogyakarta,Yogyakarta,Indonesia | 5046 |
| 1130-1145 | Radio Budapest, Hungary | 15190  6110  9835 15160 15220 |
| 1130-1200 | HCJB, Quito, Ecuador | 11740 17890 |
| 1130-1200 | Radio Australia, Melbourne | 5995  6020  6035  6080 9580  9710 11720 11910 15465 21825 |
| 1130-1200 | Radio Thailand | 11905 9655  4830 |
| 1130-1200 | Radio Austria International, Vienna | 6155 13730 15430 21490 |
| 1130-1200 | Radio Netherlands Int'l, Hilversum | 5995  6020  9715 11660 17575 21480 21520 |
| 1130-1200 | Voice of Islamic Republic of Iran | 9575  9705 11715 11790 11825 |

NOTE: Non-governmental shortwave broadcasts from ITALY (AWR Fori, Italy; IRRS Milan) are subject to unannounced cessation due to pending legislation in the Italian parliament.  (IRRS= Italian Radio Relay Service.)

## 1200 UTC     [8:00 AM EDT/5:00 AM PDT]

| | | |
|---|---|---|
| 1200-1215 | Radio Korea, Seoul | 9750 |
| 1200-1225 | All India Radio Northeast Svc | 7190 ML |
| 1200-1225 | Radio Netherlands Int'l, Hilversum | 5955  6020  9715 11660 17575 21480 21520 |
| 1200-1225 | Voice of Islamic Republic of Iran | 9575  9705 11715 11790 11825 |
| 1200-1225 M-F | Radio Finland, Helsinki | 15400 21550 |
| 1200-1230 | Vatican Radio, Vatican City | 17865 21515 |
| 1200-1230 | Radio Romania Int'l, Bucharest | 15380 17720 |
| 1200-1230 | Radio Thailand | 11905 9655  4830 |
| 1200-1230 M,W,H,A,S     Radio Ulan Bator, Mongolia | | 11850 12025 |
| 1200-1230 A,S  Radio Norway International, Oslo | | 21735 25730 |
| 1200-1230 | Radio Tashkent, Uzbekistan | 7325  9715 11785 15460 17740 |
| 1200-1230 | Radio Australia, Melbourne | 5995  6020  6035  6080 9580  9710 11720 11910 15465 21825 |
| 1200-1300 | ABC, Alice Springs, Australia | 2310 (ML) |
| 1200-1300 | WWCR Nashville, Tennessee | 15690 |
| 1200-1300 | ABC, Brisbane, Australia | 9660 |
| 1200-1300 | SBC Singapore | 11940 |
| 1200-1300 | ABC, Katherine, Australia | 2485 |
| 1200-1300 | ABC, Perth, Australia | 9610 |
| 1200-1300 | Trans World Radio, Bonaire | 11815 15345 |
| 1200-1300 | ABC, Tennant Creek,  Australia | 2325 (ML) |
| 1200-1300 | Adventist World Radio, Costa Rica | 9725 11870 |

# BROADCASTING

| | | |
|---|---|---|
| 1200-1300 | BBC World Service, London, England | 5965  6195  9515  9740<br>11775 12095 15070 17640<br>17705 17790 17885 21470<br>21660 21710 |
| 1200-1300 | Radio Bras, Brasilia | 11745 |
| 1200-1300 | CBU, Vancouver, British Columbia | 6160 |
| 1200-1300 | CFCF, Montreal, Quebec, Canada | 6005 |
| 1200-1300 | CFCN, Calgary, Alberta, Canada | 6030 |
| 1200-1300 | CHNS, Halifax, Nova Scotia, Canada | 6130 |
| 1200-1300 A,S | Radio Tanzania | 5985  6105  7165 |
| 1200-1300 | KHBN Guam | 9830 ML |
| 1200-1300 | Christian Science World Service | 9495  9895 11930 11980<br>13625 13720 (+21780 A,S) |
| 1200-1300 | CKWX, Vancouver, British Columbia | 6080 |
| 1200-1300 | Radio Moscow World Service | 11840 17690 21690 21790 |
| 1200-1300 | CFRB, Toronto, Ontario | 6070 |
| 1200-1300 | HCJB, Quito, Ecuador | 11740 17890 25950 USB |
| 1200-1300 | Radio Beijing, China | 9530 11600 11660 15450<br>17855 |
| 1200-1300 | Radio Jordan, Amman | 13655 |
| 1200-1300 | Voice of America-East Asia Service | 6110  9760 11715 15155<br>15425  9530 |
| 1200-1300 | WHRI, Noblesville, Indiana | 9465 11790 |
| 1200-1300 | WYFR, Okeechobee, Florida | 5950  6015 11580 17750 |
| 1215-1225 | Radio Bayrak, Northern Cyprus | 6150 |
| 1225-1300 | All India Radio Northeast Svc | 3255 ML |
| 1230-1245 | Radio Korea, Seoul | 7275 11740 |
| 1230-1300 S | Italian Radio Relay Svc, Milan | 9815 (ML) |
| 1230-1300 | Radio Bangladesh, Dacca | 15195 17817 |
| 1230-1300 | Radio France International, Paris | 9805 11670 15155 15195<br>17650 21635 21645 |
| 1230-1300 | Radio Australia, Melbourne | 5995  6020  6035  6080<br>9580 11720 11910 15465 |
| 1230-1300 | Radio Sweden, Stockholm | 15190 21570 17740 |
| 1235-1245 | Voice of Greece, Athens | 15625 15650 17535 |

## 1300 UTC     [9:00 AM EDT/6:00 AM PDT]

| | | |
|---|---|---|
| 1300-1315 | Radio Jordan, Amman | 13655 |
| 1300-1325 | Radio Finland, Helsinki | 15400 21550 |
| 1300-1330 | Radio Yugoslavia, Belgrade | 17740 21555 25795 |
| 1300-1330 | Radio Tirana, Albania | 11855  9500 |
| 1300-1330 A,S | Radio Norway International, Oslo | 9585 9590 |
| 1300-1330 | Radio Australia, Melbourne | 5995  6020  6035  6080<br>9580 11720 11910 15465<br>21825 |
| 1300-1330 S | Trans World Radio, Bonaire | 15345 11815 |
| 1300-1330 | Radio Canada Int'l ( China relay) | 11955 15210 |
| 1300-1330 | Swiss Radio Int'l European Service | 3985  6165  9535 |
| 1300-1350 | Radio Pyongyang, North Korea | 9325  9345 |
| 1300-1400 S | Italian Radio Relay Svc, Milan | 9815 |
| 1300-1400 M-F | Radio Canada Int'l, Montreal | 9635 11855 17820 |
| 1300-1400 | BBC World Service, London, England | 5965  9410  9515  9740<br>11775 12095 15070 17640<br>17705 17790 17885 21470<br>21660 21710 |
| 1300-1400 | ABC, Alice Springs, Australia | 2310 |

| | | |
|---|---|---|
| 1300-1400 | ABC, Brisbane, Australia | 9660 |
| 1300-1400 A,S | Radio Tanzania | 5985  6105  7165 |
| 1300-1400 | KHBN Guam | 9830 ML |
| 1300-1400 | All India Radio Northeast Svc | 3255 ML |
| 1300-1400 | Radio Korea, Seoul | 9570 |
| 1300-1400 | Radio Australia Middle East Svc | 17630 21775 |
| 1300-1400 | ABC, Katherine, Australia | 2485 |
| 1300-1400 | ABC, Perth, Australia | 9610 |
| 1300-1400 | ABC, Tennant Creek, Australia | 2325 (ML) |
| 1300-1400 | Adventist World Radio, Costa Rica | 9725 11870 |
| 1300-1400 | CBC Northern Quebec Service, Canada | 9625 |
| 1300-1400 | CBN, St. John's, Newfoundland | 6160 |
| 1300-1400 | CBU, Vancouver, British Columbia | 6160 |
| 1300-1400 | CFCF, Montreal, Quebec, Canada | 6005 |
| 1300-1400 | CFCN, Calgary, Alberta, Canada | 6030 |
| 1300-1400 | CHNS, Halifax, Nova Scotia, Canada | 6130 |
| 1300-1400 | Christian Science World Service | 9495  9650  9895 11930 11980 13625 (+21780 A,S) |
| 1300-1400 | CKWX, Vancouver, British Columbia | 6080 |
| 1300-1400 | CFRB, Toronto, Ontario, Canada | 6070 |
| 1300-1400 | Radio Moscow World Service | 11840 17690 21690 21790 |
| 1300-1400 | FEBC Radio Int'l, Philippines | 11850 |
| 1300-1400 | HCJB, Quito, Ecuador | 11740 17890 25950 USB |
| 1300-1400 | Radio Beijing, China | 9530 11600 11660 11850 |
| 1300-1400 | Radio Romania Int'l, Bucharest | 11940 15365 17850 21665 |
| 1300-1400 | Voice of America-East Asia Service | 6110  9760 11715 15155 15425 |
| 1300-1400 | WHRI, Noblesville, Indiana | 9465 11790 |
| 1300-1400 | WWCR, Nashville, Tennessee | 15690 |
| 1300-1400 | WYFR, Okeechobee, Florida | 5950  6015 11550 11580 13695 17750 |
| 1315-1400 | Radio Jordan, Amman | 9560 |
| 1315-1400 | Radio Tikhiy Okean, Vladivostok | 5015 |
| 1330-1400 | All India Radio, New Delhi | 11760  9565 |
| 1330-1400 | Radio Austria International, Vienna | 15430 |
| 1330-1345 A,S | Radio Finland, Helsinki | 21550 15400 |
| 1330-1400 | Radio Australia, Melbourne | 5995  6020  6035  6080 9580 |
| 1330-1400 | Laotian National Radio | 7116v |
| 1330-1400 A | Trans World Radio, Bonaire | 11815 15345 |
| 1330-1400 | Voice of Turkey, Ankara | 17785 |
| 1330-1400 M-S | BRT Brussels, Belgium | 21820 |
| 1330-1400 M-F | BRT Brussels, Belgium | 21815 |
| 1330-1400 | All India Radio, New Delhi | 9565 11760 15335 |
| 1330-1400 | British Forces Broadcasting Svc,UK | 15195 17695 21735 |
| 1330-1400 | Radio Tashkent, Uzbekistan | 7325  9715 11785 15460 17740 |
| 1330-1400 | Swiss Radio International, Berne | 9620 11695 15570 17830 21695 25680 |
| 1330-1400 | UAE Radio, Dubai | 15435 17865 21605 |
| 1330-1400 | Voice of Vietnam, Hanoi | 9840 12020 15010 |
| 1345-1400 | Vatican Radio | 7250  9645 11740 |

## 1400 UTC    [10:00 AM EDT/7:00 AM PDT]

| | | |
|---|---|---|
| 1400-1415 | Azad Kashmir Radio, Pakistan | 7268  4980  3665 |
| 1400-1430 | ABC, Alice Springs, Australia | 2310 (ML) |

# BROADCASTING

| Time | Station | Frequencies |
|---|---|---|
| 1400-1430 | Radio Australia, Melbourne | 5995 6020 6035 6060 6080 7215 9580 |
| 1400-1430 | ABC, Tennant Creek, Australia | 2325 (ML) |
| 1400-1430 | Swiss Radio Int'l, Berne | 6165 9535 12030 |
| 1400-1430 | Radio Juba, Sudan | 9540/9550 |
| 1400-1430 | Radio France International, Paris | 11925 21780 |
| 1400-1430 | Radio Polonia, Warsaw, Poland | 6095 7285 |
| 1400-1430 | Radio Sweden, Stockholm | 11905 17740 |
| 1400-1430 | Radio Tirana, Albania | 9500 11895 |
| 1400-1500 | ABC, Brisbane, Australia | 9660 |
| 1400-1500 S | Radio Canada Int'l, Montreal | 11955 17820 |
| 1400-1500 | Radio Sta.Peace & Progress,Moscow (from 1330 add: | 11870 15180 17635 17805 15435 15480 15560 17835) |
| 1400-1500 | Voice of the Mediterranean, Malta | 11925 |
| 1400-1500 | Radio Beijing, China | 11815 11850 15165 |
| 1400-1400 | Radio Jordan, Amman | 9560 |
| 1400-1500 | ABC, Katherine, Australia | 2485 |
| 1400-1500 | ABC, Perth, Australia | 9610 |
| 1400-1500 | All India Radio, New Delhi | 9565 11760 15335 |
| 1400-1500 | BBC World Service, London, England | 9410 11750 12095 15070 17640 17705 17790 17880 |
| 1400-1500 | CBC Northern Quebec Service, Can | 9625 |
| 1400-1500 | CBN, St. John's, Newfoundland | 6160 |
| 1400-1500 | KHBN Guam | 9830 ML |
| 1400-1500 A,S | Radio Tanzania | 5985 6105 7165 |
| 1400-1500 | All India Radio Northeast Svc | 3255 ML |
| 1400-1500 | Radio Australia Middle East Svc | 17630 21775 |
| 1400-1500 M-A | CBU, Vancouver, British Columbia | 6160 |
| 1400-1500 | CFCF, Montreal, Quebec, Canada | 6005 |
| 1400-1500 | CFCN, Calgary, Alberta, Canada | 6030 |
| 1400-1500 | CHNS, Halifax, Nova Scotia, Canada | 6130 |
| 1400-1500 | Christian Science World Service | 9530 11980 13625 13720 21780 (+ 15610 A,S) |
| 1400-1500 | CKWX, Vancouver, British Columbia | 6080 |
| 1400-1500 | CFRB, Toronto, Ontario | 6070 |
| 1400-1500 | FEBC Radio Int'l, Philippines | 11850 |
| 1400-1500 | HCJB, Quito, Ecuador | 11740 17890 25950 USB |
| 1400-1500 | Radio Japan General Service, Tokyo | 11815 11865 |
| 1400-1500 | Radio Moscow World Service | 11840 17690 21690 21790 |
| 1400-1500 | Radio RSA, Johannesburg | 9555 11925 17835 |
| 1400-1500 | Voice of America-East Asia Service | 6110 9760 15155 15425 |
| 1400-1500 | Voice of America-South Asia Service | 7125 9645 9760 15205 15395 |
| 1400-1500 | Voice of Nigeria, Lagos | 7255 |
| 1400-1500 | WHRI, Noblesville, Indiana | 9465 15105 |
| 1400-1500 | WWCR, Nashville, Tennessee | 15690 |
| 1400-1500 | WYFR, Okeechobee, Florida | 5950 6015 11580 13695 17750 |
| 1405-1500 | WYFR, Taiwan | 11550 |
| 1405-1430 | Radio Finland, Helsinki | 15185 21550 11820 |
| 1415-1500 M-A | Radio Bhutan | 5023v |
| 1415-1425 | Radio Nepal, Katmandu | 5005 7165 (alt. 3230) |
| 1430-1500 | Voice of Hope, Lebanon | 6280 |
| 1430-1500 | Voice of Myanmar (Burma) | 5990v |
| 1430-1500 | Radio Australia, Melbourne | 5995 6020 6036 6060 6080 7215 9580 9710 9770 11800 13745 |
| 1430-1500 F | ABC, Alice Springs, Australia | 2310 (ML) |

| | | |
|---|---|---|
| 1430-1500  F | ABC, Tennant Creek, Australia | 2325 (ML) |
| 1430-1500 | Radio Austria International, Vienna | 6155 11780 13730 21490 |
| 1430-1500 | Radio Netherlands Int'l, Hilversum | 5995 13770 15150 17575 |
| | | 17605 |
| 1445-1500 | Radio Korea, Seoul | 7275 |
| 1445-1500 M,W,H,A,S | Radio Ulan Bator, Mongolia | 9795 13780 |

## 1500 UTC    [11:00 AM EDT/8:00 AM PDT]

| | | |
|---|---|---|
| 1530-1600 | Radio Sofia, Bulgaria | 11680 15310 17825 |
| 1500-1515 M,W,H,A,S | Radio Ulan Bator, Mongolia | 9795 13780 |
| 1500-1515 | WYFR, Taiwan | 11550 |
| 1500-1525 | Radio Netherlands Int'l, Hilversum | 5955 13770 15150 17575 |
| | | 17605 |
| 1500-1530 M-A | Vatican Radio, Vatican City | 6248  7250  9645 11740 ML |
| 1500-1530 A,S | Radio Tanzania | 5985  6105  7165 |
| 1500-1530 | Radio Sweden, Stockholm | 17740 11905 |
| 1500-1530 | Radio Romania Inter'l, Bucharest | 11775 11940 15250 15335 |
| | | 17720 17745 |
| 1500-1540 | FEBA, Seychelles | 11865 |
| 1500-1550 | Radio Pyongyang, North Korea | 9325  9640  9977 11760 |
| 1500-1550 | Deutsche Welle, Koln, W. Germany | 9735 11965 17765 21600 |
| 1500-1555 | Radio Beijing, China | 11815 15165 |
| 1500-1600 | Radio Jordan, Amman | 9560 |
| 1500-1600  S | Radio Canada Int'l, Montreal | 11955 17820 |
| 1500-1600 | FEBA, Seychelles | 9590 15330 |
| 1500-1600 | Voice of the Mediterranean, Malta | 11925 |
| 1500-1600 | Voice of Hope, Lebanon | 6280 |
| 1500-1600  F | ABC, Alice Springs, Australia | 2310 (ML) |
| 1500-1600 | ABC, Perth, Australia | 9610 |
| 1500-1600  F | ABC, Tennant Creek, Australia | 2325 (ML) |
| 1500-1600 | BBC World Service, London, England | 9410 11750 11775 12095 |
| | | 15070 15260 17640 17705 |
| | | 17780 21470 21660 21710 |
| 1500-1600 | Voice of Myanmar (Burma) | 5990v |
| 1500-1600  S | Radio Canada Int'l, Montreal | 11955 17820 |
| 1500-1600 | CBC Northern Quebec Service, Can | 9625 (ML) |
| 1500-1600 | CBN, St. John's, Newfoundland | 6160 |
| 1500-1600 | CBU, Vancouver, British Columbia | 6160 |
| 1500-1600 | CFCF, Montreal, Quebec, Canada | 6005 |
| 1500-1600 | CFCN, Calgary, Alberta, Canada | 6030 |
| 1500-1600 | CHNS, Halifax, Nova Scotia, Canada | 6130 |
| 1500-1600 | Christian Science World Service | 9530 11980 13625 13720 |
| | | 21780 (+15610 A,S) |
| 1500-1600 | CKWX, Vancouver, British Columbia | 6080 |
| 1500-1600 | CFRB, Toronto, Ontario | 6070 |
| 1500-1600 | FEBC Radio Int'l, Philippines | 11850 |
| 1500-1600 | HCJB, Quito, Ecuador | 11740 17890 25950 USB |
| 1500-1600 T-S | KNLS, Anchor Point, Alaska | 11715 (or 9750) |
| 1500-1600 | KTWR, Agana, Guam | 11650 |
| 1500-1600 | KUSW, Salt Lake City, Utah | 15590 |
| 1500-1600 | Radio Australia, Melbourne | 5995  6020  6035  6060 |
| | | 6080  7215  9580  9710 |
| | | 9770 11800 13745 |
| 1500-1600 | Radio Japan General Service, Tokyo | 11865 21700 |
| 1500-1600 | Radio Moscow World Service | 11840 17670 21690 21790 |
| 1500-1600 | Radio RSA, Johannesburg S. Africa | 7230 15270 |

# BROADCASTING

| | | |
|---|---|---|
| 1500-1600 | Voice of America-Middle East Service | 9700 15205 15260 21530 |
| 1500-1600 | Voice of America-South Asia Service | 6110 7125 9645 9700 |
| | | 9760 15205 15260 9350 |
| 1500-1600 | Voice of Nigeria, Lagos | 7255 |
| 1500-1600 | All India Radio Northeast Svcs | 3255 ML |
| 1500-1600 | KHBN Guam | 9830 ML |
| 1500-1600 | Radio Korea, Seoul | 5975 9870 |
| 1500-1600 | WHRI, Noblesville, Indiana | 15105 (+ 9465 M-F, |
| | | + 21840 A,S) |
| 1500-1600 | WWCR, Nashville, Tennessee | 15690 |
| 1500-1600 | WYFR, Okeechobee, Florida | 5950 11830 13695 11580 |
| | | 17750 |
| 1515--1530 | RCI European News Svc, Montreal | 9555 11915 11935 15325 |
| | | 21545 (M-A add: 13650 |
| | | 15315 17820) |
| 1530-1540 M-A | Voice of Greece, Athens | 11645 15625 17535 |
| 1530-1600 | Radio Tirana, Albania | 11835 9500 |
| 1530-1600 | Radio Tanzania | 5985 6105 7165 9684 |
| 1530-1600 | Radio Omdurman, Sudan | 11635 9550/9540 |
| 1530-1600 | Radio Sweden, Stockholm | 17880 21500 21655 |
| 1530-1600 | Swiss Radio International, Berne | 13685 15430 17830 21630 |
| 1540-1555 M-A | FEBA, Seychelles | 11865 |
| 1545-1600 | Radio Pakistan | 21740 21480 17895 17580 |
| | | 15605 13665 |
| 1545-1600 | Vatican Radio, Vatican City | 11715 15090 17870 |
| 1555-1600 M,A | FEBA, Seychilles | 11865 |

## 1600 UTC    [12:00 PM EDT/9:00 AM PDT]

| | | |
|---|---|---|
| 1600-1610 | Radio Lesotho | 4800 |
| 1600-1610 M,A | FEBA, Mahe, Seychelles | 11865 |
| 1600-1610 | Vatican Radio, Vatican City | 6248 7250 9645 11740 |
| 1600-1615 | Radio Tanzania | 5985 6105 7165 9684 |
| 1600-1615 | Azad Kashmir Radio, Pakistan | 7268 4980 3665 |
| 1600-1630 | Radio Sofia, Bulgaria | 11680 15310 17825 |
| 1600-1630 | All India Radio Northeastern Svcs | 3255 ML |
| 1600-1630 | Radio Pakistan | 7287 13665 15605 17554 |
| | | 21670 |
| 1600-1630 A,S | Radio Norway International, Oslo | 15220 25730 |
| 1600-1630 | Radio Polonia, Warsaw, Poland | 6135 9540 |
| 1600-1630 M-F | Radio Portugal, Lisbon | 21530 |
| 1600-1630 | Radio Jordan, Amman | 9560 |
| 1600-1630 | Voice of Vietnam, Hanoi | 9840 15010 12020 |
| 1600-1640 | UAE Radio, Dubai | 11795 15320 15435 21605 |
| 1600-1650 | Deutsche Welle, Koln, W. Germany | 6170 7225 15105 15595 |
| | | 17825 21680 |
| 1600-1700 | KSDA, Guam | 11980 |
| 1600-1700 | Radio Baghdad, Iraq | 11860 |
| 1600-1700 F | ABC, Alice Springs, Australia | 2310 (ML) |
| 1600-1700 | BBC World Service, London, England | 9410 11775 12095 15070 |
| | | 15260 17640 17705 21660 |
| 1600-1700 | Radio Australia, Melbourne | 5995 6020 6035 6080 |
| | | 7215 9580 9710 9770 |
| | | 11800 13745 |
| | | (+ 6060 until 1630) |
| 1600-1700 | ABC, Perth, Australia | 9610 |
| 1600-1700 F | ABC, Tennant Creek, Australia | 2325 (ML) |

| | | | |
|---|---|---|---|
| 1600-1700 | | CBC Northern Quebec Service, Can | 9625 (ML) |
| 1600-1700 | | CBN, St. John's, Newfoundland | 6160 |
| 1600-1700 | S | Radio Canada Int'l, Montreal | 11955 17820 |
| 1600-1700 | | Radio Moscow World Service | 7110 9655 9840 11630 |
| | | | 11890 12005 12010 12015 |
| | | | 15375 15540 17600 17670 |
| | | | 17710 21585 21630 21740 |
| | | | (+11840 via Cuba) |
| 1600-1700 | | CBU, Vancouver, British Columbia | 6160 |
| 1600-1700 | | CFCF, Montreal, Quebec, Canada | 6005 |
| 1600-1700 | | CFCN, Calgary, Alberta, Canada | 6030 |
| 1600-1700 | | CHNS, Halifax, Nova Scotia, Canada | 6130 |
| 1600-1700 | | Christian Science World Service | 9530 13625 13745 21640 |
| 1600-1700 | | CKWX, Vancouver, British Columbia | 6080 |
| 1600-1700 | | CFRB, Toronto, Ontario | 6070 |
| 1600-1700 | | KTWR, Agana, Guam | 11650 11910 13720 |
| 1600-1700 | | KUSW, Salt Lake City, Utah | 15590 |
| 1600-1700 | | Radio Beijing, China | 9570 15110 15130 |
| 1600-1700 | | Radio France International, Paris | 6175 11705 12015 15360 |
| | | | 17620 17795 17845 17850 |
| 1600-1700 | | Trans World Radio-Swaziland | 15135 |
| 1600-1700 | | Voice of America-Africa Service | 7195 9575 11920 15410 |
| | | | 15445 15580 15600 17785 |
| | | | 17800 17870 |
| 1600-1700 | | Voice of America-Middle East Service | 3980 9700 15205 15260 |
| 1600-1700 | | Voice of America-Asia Service | 7125 9645 9700 9760 |
| | | | 15205 15260 15395 |
| 1600-1700 | | Radio RSA, Johannesburg | 7230 15270 |
| 1600-1700 | | Voice of Nigeria, Lagos | 7255 |
| 1600-1700 | | WHRI, Noblesville, Indiana | 9465(M-F) 13760(M-A) |
| | | | 15105(S) 21840(A,S) |
| 1600-1700 | | WINB, Red Lion, Pennsylvania | 15295 |
| 1600-1700 | | WRNO New Orleans, Louisiana | 15420 |
| 1600-1700 | | WWCR, Nashville, Tennessee | 15690 |
| 1600-1700 | | WYFR, Okeechobee, Florida | 11830 13695 17750 15566 |
| | | | 11580 17612 21525 21615 |
| 1610-1620 | M-F | Radio Botswana | 3356 4830 7255 |
| 1610-1625 | M | FEBA, Mahe, Seychelles | 11865 |
| 1615-1620 | | Vatican Radio, Vatican City | 9645 11740 |
| 1615-1630 | | Radio Korea, Seoul, South Korea | 9870 |
| 1615-1630 | | Radio Budapest, Hungary | 15160 15220 11910 9835 |
| | | | 9585 7220 |
| 1630-1655 | M-A | BRT Brussels, Belgium | 17580 21810 |
| 1630-1700 | | Radio Netherlands, Hilversum | 15570 6020 |
| 1630-1700 | | Radio Austria Int'l, Vienna | 11780 13730 21490 |
| 1645-1700 | M-F | Radio Botswana | 3356 4830 7255 |
| 1650-1700 | | Radio New Zealand, Wellington | 15485 |

## 1700 UTC    [1:00 PM EDT/10:00 AM PDT]

| | | | |
|---|---|---|---|
| 1700-1725 | | Radio Netherlands, Hilversum | 15570 6020 |
| 1700-1730 | | Radio Prague Int'l, Czechoslovakia | 5930 6055 7345 11990 |
| 1700-1730 | | Radio Sweden, Stockholm | 6065 9615 |
| 1700-1730 | A,S | Radio Norway | 9655 |
| 1700-1750 | | Radio Pyongyang, North Korea | 9325 9640 9977 11760 |
| 1700-1750 | | Radio Bras, Brazil | 15265 |
| 1700-1800 | | ELWA, Monrovia, Liberia | 11800 |

# BROADCASTING

| | | |
|---|---|---|
| 1700-1800 | Radio Beijing, China | 9570 11575 15225 |
| 1700-1800 | Radio Australia, Melbourne | 5995  6020  6035  6080<br>7215  7240  9580  9710<br>9770 11855 |
| 1700-1800 | BBC World Service, London | 9410 11775 12095 15070<br>15260 15310 15400 17640<br>17695 21470 21660 |
| 1700-1800 | Radio Korea, Seoul | 15575 |
| 1700-1800 | Voice of America-Africa Service | 7195  9575 11920 15410<br>15445 15580 15600 17785<br>17800 17870 |
| 1700-1800 | Radio Moscow World Service | 11840 12010 12015 15150<br>15265 17585 17600 17670<br>17695 21585 25375<br>(+11840 via Cuba) |
| 1700-1800 S-F | WMLK Bethel, PA | 9465 |
| 1700-1800 | Voice of America-Middle East Service | 3980  6040  9700  9760<br>11760 15205 15260 |
| 1700-1800 | Voice of America-South Asia Service | 7125  9645  9700 15395 |
| 1700-1800 | WHRI, Noblesville, Indiana | 13760 15105 |
| 1700-1800 | Radio RSA, Johannesburg | 7230 15270 17790 |
| 1700-1800 | Christian Science World Service | 13625 21640<br>(+17555 & 15610 A,S |
| 1700-1800 | Radio New Zealand, Wellington | 15485 |
| 1700-1800 | Radio Baghdad, Iraq | 11860 |
| 1700-1800 | Radio Moscow Africa Service | 11690 11745 11775 11850<br>11960 15230 15330 15415<br>15535 15585 17565 17570<br>17595 17615 17655 21565<br>21630 21715 |
| 1700-1800 | CBC, Montreal | 9625 (ML) |
| 1700-1800 | Radio Surinam Int'l (via Brazil) | 17750 (ML) |
| 1700-1800 | Radio Japan, Tokyo | 9695 11815 11865 |
| 1700-1800 | Radio Pyongyang, North Korea | 9325  9640  9977 11760 |
| 1700-1800 | KUSW Salt Lake City, Utah | 15590 |
| 1700-1800 | WINB, Red Lion, Pennsylvania | 15295 |
| 1700-1800 | WRNO, New Orleans, Louisiana | 15420 |
| 1700-1800 | WWCR, Nashville, Tennessee | 15690 |
| 1700-1800 | WYFR, Okeechobee, Florida | 11830 13695 15440 17750<br>17885 21615 |
| 1715-1730 | Radio Canada Int'l, Montreal | 5995  7235 13650 15325<br>17820 21545 |
| 1715-1800 | Radio Pakistan | 11570 15605 |
| 1730-1740 | Radio Bayrak, Northern Cyprus | 6150 |
| 1730-1755 | BRT Brussells, Belgium | 11695  5910 |
| 1730-1800 | Radio Sta. Peace & Progress, USSR | 6110  9705 11695 11745<br>11775 11850 11910 11980<br>12055 12065 15330 15480<br>15585 17565 17615 17635<br>17655 21715 |
| 1730-1800 | Radio Sofia, Bulgaria | 11680 15310 17825 |
| 1730-1800 | Swiss Radio Int'l, Berne | 9535 |
| 1730-1800 | Vatican Radio African Service | 17710 17730 21650 |
| 1730-1800 | Radio Truth (Clandestine intended<br>for Zimbabwe) | 5015 |
| 1730-1800 | Radio Tirana, Albania | 7155  9480 |
| 1730-1800 | Radio Romania Int'l, Bucharest | 15340 15365 17805 17860 |

| 1800 UTC | [2:00 PM EDT/11:00 AM PDT] |
|---|---|

| | | |
|---|---|---|
| 1800-1815 | Kol Israel | 11585 11655 |
| 1800-1830 | Radio Canada Int'l, Montreal | 13670 15260 17820 |
| 1800-1830 A,S | Radio Norway International, Oslo | 17755 |
| 1800-1830 | Voice of Ethiopia, Addis Ababa | 9660 |
| 1800-1830 | Radio Sweden, Stockholm | 6065 7265 |
| 1800-1830 | Voice of Vietnam, Hanoi | 15010 12010 9840 |
| 1800-1845 | Trans World Radio, Swaziland | 15210 |
| 1800-1845 | All India Radio, New Delhi | 11935 15360 |
| 1800-1855 | Radio Mozambique, Maputo | 9618 4855 3265 |
| 1800-1900 F | ABC, Alice Springs, Australia | 2310 (ML) |
| 1800-1900 F | ABC, Tennant Creek, Australia | 2325 (ML) |
| 1800-1900v | Radio Tanzania | 5985 6105 7165 9684 |
| 1088v-1900 | SLBC World Service, Sri Lanka | 9720 15120 |
| 1800-1900 | KVOH, Rancho Simi, California | 17775 |
| 1800-1900 | BBC World Service, London | 9410 12095 15070 17640 |
| 1800-1900 | Radio Australia, Melbourne | 5995 6020 6035 6080 |
| | | 7205 7215 7240 9580 |
| | | 11855 |
| 1800-1900 | Radio Moscow World Service | 11765 11840 11890 13605 |
| | | 15185 15375 15540 17585 |
| | | 17670 17695 21740 |
| 1800-1900 | Radio New Zealand, Wellington | 15485 |
| 1800-1900 | CBN, St. John's, Newfoundland | 6160 |
| 1800-1900 | CBU, Vancouver, British Columbia | 6160 |
| 1800-1900 | CFCF, Montreal, Quebec, Canada | 6005 |
| 1800-1900 | CFCN, Calgary, Alberta, Canada | 6030 |
| 1800-1900 | CHNS, Halifax, Nova Scotia, Canada | 6130 |
| 1800-1900 | Christian Science World Service | 11650 13625 21640 |
| | | (+21780 M-F) |
| | | (+15610 A,S) |
| 1800-1900 | CKWX, Vancouver, British Columbia | 6080 |
| 1800-1900 | CFRB, Toronto, Ontario | 6070 |
| 1800-1900 | KUSW, Salt Lake City, Utah | 15590 |
| 1800-1900 | CBC Montreal | 9625 |
| 1800-1900 S-F | WMLK Bethel, Pennsylvania | 9465 |
| 1800-1900 | Radio RSA, Johannesburg, S. Africa | 17765 15270 7230 |
| 1800-1900 A,S | Radio for Peace Int'l, Costa Rica | 13630 21566 |
| 1800-1900 | Voice of America-Africa Service | 7195 9575 11920 15410 |
| | | 15445 15580 15600 17785 |
| | | 17800 17870 21485 |
| 1800-1900 | Voice of America-Middle East Service | 6040 9700 9760 11760 |
| | | 15205 |
| 1800-1900 | WHRI, Noblesville, Indiana | 13760 17830 |
| 1800-1900 | Radio Havana Cuba | 15345 |
| 1800-1900 | WINB, Red Lion, Pennsylvania | 15295 |
| 1800-1900 | WRNO, New Orleans, Louisiana | 15420 |
| 1800-1900 | WWCR, Nashville, Tennessee | 15690 |
| 1800-1900 | WYFR, Okeechobee, Florida | 11830 13695 15440 17885 |
| | | 21500 |
| 1815-1900 | Radio Bangladesh, Dacca | 12032 15255 |
| 1830-1845 | Radio Prague Int'l, Czechoslovakia | 6055 7345 |
| 1830-1845 | Radio Finland, Helsinki | 11755 9550 6120 |
| 1830-1855 | BRT Brussels, Belgium | 5910 11695 13675 |
| 1830-1855 | Radio Polonia, Warsaw, Poland | 5995 6135 7125 7285 |
| | | 9525 11840 |
| 1830-1900 | Radio Riyadh, Saudi Arabia | 9705 9720 |

# BROADCASTING

| | | |
|---|---|---|
| 1830-1900 A,S | Radio Canada Int'l, Monreal | 13670 15260 17820 |
| 1830-1900 | Radio Afghanistan, Kabul | 9635 15510 17745 |
| 1830-1900 | Radio Tirana, Albania | 7120 9480 |
| 1830-1900 | Radio Netherlands Int'l, Hilversum | 6020 15560 17605 21685 |
| 1830-1900 | Swiss Radio International, Berne | 9885 11955 |
| 1830-1900 | Swiss Radio Int'l European Service | 3985 6165 9535 |
| 1840-1850 M-A | Voice of Greece, Athens | 11645 12105 15625 |
| 1845-1900 | All India Radio, New Delhi | 7412 9665 9910 11620 |
| | | 11860 11935 |
| 1845-1855vlRR | Africa No. 1, Gabon | 15475 |

| 1900 UTC | [3:00 PM EDT/12:00 PM PDT] |
|---|---|

| | | |
|---|---|---|
| 1900-1910 | Radio Tanzania | 5985 6105 7165 9684 |
| 1900-1910 M-A | Vatican Radio | 6190 6248 7250 9645 |
| | | 17710 17730 21650 |
| 1900-1915 | Sierra Leone Brdcstng.Co.,Freetown | 3316 |
| 1900-1920 | Radio Botswana | 3356 4830 |
| 1900-1920v | Radio Omdurman, Sudan | 11635 |
| 1900-1925 | Radio Netherlands Int'l, Hilversum | 6020 15560 17605 21685 |
| 1900-1930 M-F | Radio Budapest, Hungary | 15160 11910 9835 9585 |
| | | 7220 6110 |
| 1900-1930 | Radio Canada Int'l, Montreal | 13670 15260 17820 |
| 1900-1930 | Radio Sofia, Bulgaria | 11680 15310 17825 |
| 1900-1930 | Radio Afghanistan, Kabul | 9635 15510 17745 |
| 1900-1930 M-F | Radio Canada Int'l, Montreal | 13670 15260 17820 |
| 1900-1930 | Radio Japan General Service, Tokyo | 11850 11865 15270 |
| 1900-1930 A,S | Radio Norway International, Oslo | 15220 15235 21705 25730 |
| 1900-1930 M-F | Radio Portugal, Lisbon | 11740 15250 21530 |
| 1900-1930 | Voice of Vietnam, Hanoi | 9840 12020 15010 |
| 1900-1945 | All India Radio, New Delhi | 7412 9665 9910 11620 |
| | | 11860 11935 |
| 1900-1950 | Deutsche Welle, Koln, W. Germany | 11785 11810 13790 15390 |
| | | 17810 |
| 1900-2000 | ELWA, Monrovia, Liberia | 11800 |
| 1900-2000 | CBC, Montreal | 9625 |
| 1900-2000 | Radio New Zealand, Wellington | 15485 |
| 1900-2000 | Radio Moscow World Service | 11765 11840 12010 12060 |
| | | 13605 15405 15540 15580 |
| | | 17570 17670 21630 21740 |
| | | 21630 |
| 1900-2000 | Radio Moscow African Svc | 11960 12035 15230 15520 |
| | | 17655 |
| | | (in English & Zulu) |
| 1900-2000 M-F | RAE, Buenos Aires, Argentina | 15345 |
| 1900-2000 | Radio Beijing, China | 9440 11515 |
| 1900-2000 | Solomon Islands Broadcasting Co. | 5020 |
| 1900-2000 | KVOH, Rancho Simi, California | 17775 |
| 1900-2000 | BBC World Service, London, England | 9410 12095 15070 15400 |
| | | 17880 |
| 1900-2000 | CBN, St. John's, Newfoundland | 6160 |
| 1900-2000 | CBU, Vancouver, British Columbia | 6160 |
| 1900-2000 | CFCF, Montreal, Quebec, Canada | 6005 |
| 1900-2000 | CFCN, Calgary, Alberta, Canada | 6030 |
| 1900-2000 | CHNS, Halifax, Nova Scotia, Canada | 6130 |
| 1900-2000 | Christian Science World Service | 11650 13625 21640 |
| | | (+17555 & 15610 A,S) |

|  |  | (+21780 M-F) |
|---|---|---|
| 1900-2000 | CKWX, Vancouver, British Columbia | 6080 |
| 1900-2000 | CFRB, Toronto, Ontario | 6070 |
| 1900-2000 | GBC Radio, Accra, Ghana | 6130 |
| 1900-2000 | HJCB European Service, Ecuador | 17790 21480 25950ssb |
| 1900-2000 | KUSW, Salt Lake City, Utah | 15590 |
| 1900-2000 | Radio Algiers, Alger | 9510  9685 15215 |
| 1900-2000 | Radio Australia, Melbourne | 5995  6020  6035  6080 |
|  |  | 7205  7215  7240  9580 |
|  |  | 11855 |
|  |  | (+13745 from 1930) |
| 1900-2000 A,S | Radio for Peace Int'l, Costa Rica | 13630 21566 |
| 1900-2000 | Spanish National Radio, Madrid | 11790 15280 15375 15395 |
| 1900-2000 | Voice of America-Africa Service | 7195 15410 15445 15580 |
|  |  | 15600 17785 17800 17870 |
|  |  | 21485 |
| 1900-2000 | Voice of America-Middle East Service | 6040  9700  9760 11760 |
|  |  | 15205 |
| 1900-2000 | Voice of America-Pacific Service | 9525 11870 15180 |
| 1900-2000 | WHRI, Noblesville, Indiana | 13760 17830 |
| 1900-2000 | WINB, Red Lion, Pennsylvania | 15295 |
| 1900-2000 S-F | WMLK, Bethel, Pennsylvania | 9465 |
| 1900-2000 | WRNO, New Orleans, Louisiana | 15420 |
| 1900-2000 | WWCR, Nashville, Tennessee | 15690 |
| 1900-2000 | WYFR, Okeechobee, Florida | 11830 13695 15440 15566 |
|  |  | 17612 17885 21615 |
| 1920-1930 M-A | Voice of Greece, Athens | 9395 11645 |
| 1930-2000 M | Radio Tallin, Estonia | 5925 |
| 1930-2000 | Radio Austria International, Vienna | 5945  6155 12010 13730 |
| 1930-2000 | Radio Romania Int'l, Bucharest | 5955  9690  9750 11810 |
| 1930-2000 A,S | Radio Budapest, Hungary | 6110  7220  9585  9835 |
|  |  | 11910 15160 |
| 1930-2000 | Radio Sofia, Bulgaria | 11660 11765 15330 |
| 1930-2000 | Radio Yugoslavia, Belgrade | 7215  9660 11735 |
| 1930-2000 M-F | Radio Canada Int'l, Montreal | 5995  7235 11945 15325 |
|  |  | 17875 |
| 1930-2000 | Radio Korea, Seoul | 6480  7550 15575 |
| 1930-2000 | Radio Tikhiy Okean, Vladivostok | 5015  7335  9885 11870 |
|  |  | 11995 15180 15435 15535 |
|  |  | 15560 17645 17850 |
| 1930-2000 | Voice of the Islamic Republic Iran | 6080 n 9022 15084 |
| 1935-1955 | RAI, Rome, Italy | 7275  9710 11800 |
| 1940-2000 M,W,H,A,S | Radio Ulan Bator, Mongolia | 11850 12050 |
| 1945-2000 | All India Radio, New Delhi | 11935 |
| 1945-2000 | Radio Korea, Seoul | 5975  9870 |
| 1950-2000 | Vatican Radio | 6190  7250  9645 |

## 2000 UTC    [4:00 PM EDT/1:00 PM PDT]

| 2000-2010 | Vatican Radio, Vatican City | 6190  7250  9645 |
|---|---|---|
| 2000-2010 M,W,H,A,S | Radio Ulan Bator, Mongolia | 11850 12050 |
| 2000-2010 | Sierra Leone Brdcstng.Co.,Freetown | 3316 |
| 2000-2030 | Kol Israel, Jerusalem | 11605 11745 12077 15090 |
|  |  | 15485 17575 |
| 2000-2030 M-F | Radio Portugal | 15250 |
| 2000-2030 | Radio Prague Int'l, Czechoslovakia | 5930  6055  7345 11990 |
| 2000-2030 | Radio Romania Int'l, Bucharest | 5955  9690  9750 11810 |

# BROADCASTING

| 2000-2030 | Radio Korea, Seoul | 6480 7550 15575 |
| 2000-2030 | Voice of the Islamic Republic Iran | 6080 9022 15084 |
| 2000-2050 | Radio Pyongyang, North Korea | 6576 9345 9640 9977 |
| 2000-2100 | Radio Moscow British Service | 7330 11630 11930 15185 17695 |
| 2000-2100 M-F | Radio for Peace Int'l, Costa Rica | 13630 21566 |
| 2000-2100 | KHBN Guam | 9820 ML |
| 2000-2100 | Voice of Indonesia, Jakarta | 11753 11785 |
| 2000-2100 | Voice of Hope, Lebanon | 6280 |
| 2000-2100 | BBC World Service, London, England | 5975 9410 12095 15070 15260 15400 17755 17760 17880 |
| 2000-2100 | Radio Australia, Melbourne | 6020 6035 7205 7215 7240 9580 11855 13745 (+6080 & 5995 until 2030) |
| 2000-2100 M-A | ABC, Alice Springs, Australia | 2310 (ML) |
| 2000-2100 | ABC, Katherine, Australia | 2485 |
| 2000-2100 M-A | ABC, Tennant Creek, Australia | 2325 (ML) |
| 2000-2100 | CBN, St. John's, Newfoundland | 6160 |
| 2000-2100 | CBU, Vancouver, British Columbia | 6160 |
| 2000-2100 | CFCF, Montreal, Quebec, Canada | 6005 |
| 2000-2100 | Radio Moscow World Service | 7315 11630 11670 11805 11890 12060 13605 15185 15315 15355 15560 17695 |
| 2000-2100 | Radio Moscow Africa Service | 11715 11775 11960 12035 15520 15535 21630 21740 |
| 2000-2100 | CBC, Montreal | 9625 (ML) |
| 2000-2100 | CFCN, Calgary, Alberta, Canada | 6030 |
| 2000-2100 | CHNS, Halifax, Nova Scotia, Canada | 6130 |
| 2000-2100 | Radio Baghdad, Iraq | 13660 |
| 2000-2100 | Christian Science World Service | 9455 9495 11980 13625 13770 15610 17555 |
| 2000-2100 | CKWX, Vancouver, British Columbia | 6080 |
| 2000-2100 | Radio Sta. Peace & Progress, USSR | 9470 9820 11830 11880 11980 15260 |
| 2000-2100 | CFRB, Toronto, Ontario | 6070 |
| 2000-2100 | KUSW, Salt Lake City, Utah | 15590 |
| 2000-2100 | Radio Beijing, China | 9440 9920 11500 11715 15110 |
| 2000-2100 | ELWA, Monrovia, Liberia | 11800 |
| 2000-2100 | Radio Baghdad, Iraq | 11860 |
| 2000-2100 | Radio Havana Cuba | 11800 |
| 2000-2100 | Voice of America-Africa Service | 7195 15410 15445 15580 15600 17785 17800 17870 19480lsb 21485 |
| 2000-2100 | Voice of America-Middle East Service | 6040 9700 9760 11760 15205 |
| 2000-2100 | WHRI, Noblesville, Indiana | 13760 17830 |
| 2000-2100 | WINB, Red Lion, Pennsylvania | 15185 |
| 2000-2100 | WRNO, New Orleans, Louisiana | 15420 |
| 2000-2100 | KVOH, Rancho Simi, California | 17775 |
| 2000-2100 | Solomon Islands Broadcasting Co. | 5020 |
| 2000-2100 | WWCR, Nashville, Tennessee | 15690 |
| 2000-2100 | WYFR, Okeechobee, Florida | 11830 13695 15440 15566 17612 17885 21525 21615 |
| 2000-2100 | Radio New Zealand, Wellington | 15485 |
| 2005-2100 | Radio Damascus, Syria | 12085 15095 |
| 2025-2045 | RAI, Rome, Italy | 7235 9575 11800 |

| | | |
|---|---|---|
| 2030-2100 | Radio Netherlands Int'l, Hilversum | 9860 13700 15560 |
| 2030-2100 | Voice of Vietnam, Hanoi | 9840 12020 15010 |
| 2045-2100 | All India Radio, New Delhi | 7412  9665  9910 11620 |
| | | 11715 15265 |

| **2100 UTC** | **[5:00 PM EDT/2:00 PM PDT]** |
|---|---|

| | | |
|---|---|---|
| 2100-2105 | Radio New Zealand, Wellington | 15485 |
| 2100-2105 | Radio Damascus, Syria | 12085 15095 |
| 2100-2115 | Radio Prague Int'l, Czechoslovakia | 5930  6055  7345 11990 |
| 2100-2125 | Radio Netherlands Int'l, Hilversum | 9860 13700 15560 |
| 2100-2130  M | Radio Ljublijana, Yugoslavia | 5980  7240  9620 |
| 2100-2130 | Vatican Radio | 17710 17730 21650 |
| 2100-2130 A,S | Radio Norway, Oslo | 15165 |
| 2100-2130 | Radio Budapest, Hungary | 11910 15160  9835  9585 |
| | | 7220  6110 |
| 2100-2130 | Sierra Leone Brdcstng.Co.,Freetown | 3316 |
| 2100-2130 | Radio Korea, Seoul | 15575  7550  6480 |
| 2100-2130 | Radio Romania Int'l, Bucharest | 9690  9750 11810 11940 |
| 2100-2130 | Radio Beijing, China | 3985 11715 15110 |
| 2100-2130 | Radio Japan General Service, Tokyo | 11815 11835 15270 17765 |
| | | 17810 17890 |
| 2100-2130 | Radio Sweden, Stockholm | 9655 11705 |
| 2100-2130 | Swiss Radio International, Berne | 9885 13635 15525 12035 |
| 2100-2130 | Radio Finland, Helsinki | 6120 11755 15400 |
| 2100-2150 | Deutsche Welle, Koln, West Germany | 9670  9765 11785 13780 |
| | | 15435 |
| 2100-2200 | ELWA, Monrovia, Liberia | 11800 |
| 2100-2200 | Radio Angola Int'l Svc, Luanda | 3355  9535 |
| 2100-2200v | All India Radio, New Delhi | 7412  9665  9910 11620 |
| | | 11715 15265 |
| 2100-2200 | CBC Montreal | 9625 |
| 2100-2200 | Radio Moscow World Service | 7115  7150  7315  9685 |
| | | 11670 11745 11775 11805 |
| | | 11840 11890 11985 12040 |
| | | 12060 13605 15185 15230 |
| | | 15315 15355 15425 15535 |
| | | 15580 21740 |
| 2100-2200 | Voice of Turkey, Ankara | 9795 |
| 2100-2200 | Radio Kiev, Ukraine | 9865 |
| 2100-2200 | CBN, St. John's, Newfoundland | 6160 |
| 2100-2200 | KHBN Guam | 9820 ML |
| 2100-2200 | CBU, Vancouver, British Columbia | 6160 |
| 2100-2200 | Voice of Hope, Lebanon | 6280 |
| 2100-2200 | CFCF, Montreal, Quebec, Canada | 6005 |
| 2100-2200 | CFCN, Calgary, Alberta, Canada | 6030 |
| 2100-2200 | CHNS, Halifax, Nova Scotia, Canada | 6130 |
| 2100-2200 | Christian Science World Service | 9455  9495 13625 13770 |
| | | 15310 15610 17555 |
| 2100-2200 | Solomon Islands Broadcasting Co. | 5020  9545 |
| 2100-2200 | CKWX, Vancouver, British Columbia | 6080 |
| 2100-2200 | CFRB, Toronto, Ontario | 6070 |
| 2100-2200 T-A | KUSW, Salt Lake City, Utah | 15590 |
| 2100-2200 | Radio Australia, Melbourne | 11880 15465 17795 |
| | | (until 2130: 7215 13745) |
| | | (from  2130: 15240) |
| 2100-2200 | Radio Cairo, Egypt | 9900 |

# BROADCASTING

| | | |
|---|---|---|
| 2100-2200 | KVOH, Rancho Simi, California | 17775 |
| 2100-2200 | Radio Baghdad, Iraq (to Europe) | 13660 |
| 2100-2200 | Radio Baghdad, Iraq | 11860 |
| 2100-2200 | Radio Beijing, China | 9920 11500 |
| 2100-2200 | Radio for Peace, Costa Rica | 13630 21566 |
| 2100-2200 | Radio Havana Cuba | 11800 17860 |
| 2100-2200 | Voice of America-Africa Service | 7195 15410 15445 15580 15600 17785 17800 17870 21485 |
| 2100-2200 | Voice of America-Middle East Service | 6040 9700 9760 11760 15205 11710 |
| 2100-2200 | Voice of America-Pacific Service | 11870 15185 17735 |
| 2100-2200 | WHRI, Noblesville, Indiana | 13760 17830 |
| 2100-2200 | WINB, Red Lion, Pennsylvania | 15185 |
| 2100-2200 | BBC World Service, London, England | 5975 9410 12095 15070 15260 15400 17755 17760 17880 |
| 2100-2200 | WRNO Worldwide, Louisiana | 15420 |
| 2100-2200 | WWCR, Nashville, Tennessee | 15690 |
| 2100-2200 | WYFR, Okeechobee, Florida | 11830 13695 15566 17612 17885 21525 21615 |
| 2105-2200 | Radio New Zealand, Wellington | 17675 |
| 2110-2200 | Radio Damascus, Syria | 12085 15095 |
| 2130-2200 | Radio Sofia, Bulgaria | 11660 11765 15330 |
| 2130-2200 | Radio Canada Int'l, Montreal | 11880 13670 15150 17820 |
| 2130-2200 | Radio Japan, Tokyo | 11815 11835 15270 17765 17810 21610 |
| 2130-2200 | HCJB, Quito, Ecuador | 15270 17790 25950ssb |

## 2200 UTC  [6:00 PM EDT/3:00 PM PDT]

| | | |
|---|---|---|
| 2200-2205 | Radio Damascus, Syria | 12085 15095 |
| 2200-2215 | Sierra Leone Brdcstng.Co.,Freetown | 3316 |
| 2200-2215 M-A | ABC, Alice Springs, Australia | 2310 (ML) |
| 2200-2215 | ABC, Tennant Creek, Australia | 2325 (ML) |
| 2200-2215 M-F | Voice of America-Caribbean Service | 9640 11880 15225 |
| 2200-2225 | RAI, Rome, Italy | 5990 7235 9710 |
| 2200-2230 | BRT Brussels, Belgium | 5910 9925 |
| 2200-2230 | Radio Canada Int'l, Japan relay | 11705 |
| 2200-2230 | ABC, Katherine, Australia | 2485 |
| 2200-2230 S | KGEI, San Francisco, California | 15280 |
| 2200-2230 | All India Radio, New Delhi | 7412 9665 9910 11620 11715 15265 |
| 2200-2230 A,S | Radio Norway International, Oslo | 15195 |
| 2200-2245 | WINB Red Lion , PA | 15295 |
| 2200-2245 | Radio Yugoslavia, Belgrade | 7215 9620 11735 15105 |
| 2200-2300 | KHBN Guam | 9820 ML |
| 2200-2300 | Radio Canada Int'l, Montreal | 9760 11945 |
| 2200-2300 | Radio New Zealand, Wellington | 17675 |
| 2200-2300v | Radio Cairo, Egypt | 9900 |
| 2200-2300 | BBC World Service, London, England | 5975 6005 6175 6195 7325 9410 9590 9915 11750 12095 15070 15260 15400 17750 17830 |
| 2200-2300 | CBC Northern Quebec Svc, Canada | 9625 |
| 2200-2300 | CBN, St. John's, Newfoundland | 6160 |
| 2200-2300 | Radio Korea, Seoul | 15575 |

216

# BROADCASTING

| | | |
|---|---|---|
| 2200-2300 | Radio Moscow North American Svc | 11670 11690 11710 11780 |
| | | 11800 12040 12050 13605 |
| | | 15315 15355 15425 15580 |
| | | 15595 17735 |
| 2200-2300 | Radio Sta. Peace & Progress,USSR | 9470  9820 11830 11880 |
| | | 11980 15260 |
| 2200-2300 | Radio Moscow World Service | 11615 11745 11775 11985 |
| | | 15140 15560 17570 21690 |
| | (from 2230 add: 7315 15480 17655 17850 17890) | |
| 2200-2300 | CBU, Vancouver, British Columbia | 6160 |
| 2200-2300 | CFCF, Montreal, Quebec, Canada | 6005 |
| 2200-2300 | CFCN, Calgary, Alberta, Canada | 6030 |
| 2200-2300 | CHNS, Halifax, Nova Scotia, Canada | 6130 |
| 2200-2300 | Christian Science World Service | 9465 15225 15275 15300 |
| | | 15405 15610 17555 |
| 2200-2300 | CKWX, Vancouver, British Columbia | 6080 |
| 2200-2300 | CFRB, Toronto, Ontario | 6070 |
| 2200-2300 T-A | KUSW, Salt Lake City, Utah | 15590 |
| 2200-2300 | Voice of Hope, Lebanon | 6280 |
| 2200-2300 | Radio Australia, Melbourne | 11880 13605 15240 15465 |
| | | 17715 17795 21740 |
| 2200-2300 | Radio for Peace Int'l, Costa Rica | 13630 21566 |
| 2200-2300 | Radio Tonga, Kingdom of Tonga | 5030v |
| 2200-2300 | Voice of America-East Asia Service | 7120  9770 11760 15185 |
| | | 15290 15305 17735 17820 |
| 2200-2300 | Voice of America-Eur/Pac. Service | 9852 11805 15345 15370 |
| | | 17610 |
| 2200-2300 | Voice of Free China, Taiwan | 17750 21720 |
| 2200-2300 | United Arab Emirates R., Abu Dhabi | 9600 11985 13605 |
| 2200-2300 | WHRI, Noblesville, Indiana | 13760 17830 |
| 2200-2300 | WRNO Worldwide, Louisiana | 15420 |
| 2200-2300 | WWCR, Nashville, Tennessee | 15690 |
| 2200-2300 | WYFR, Okeechobee, Florida | 11580 11830 13695 17612 |
| | | 17885 21525 |
| 2205-2230 | Vatican Radio, Vatican City | 7125  9615 11830 15105 |
| 2230-2300 | Voice of Vietnam, Hanoi | 9840 12020 15010 |
| 2230-2300 | Radio Polonia, Warsaw, Poland | 5995  6135  7125  7270 |
| 2230-2300 | Radio Tirana, Albania | 7215  9480 |
| 2230-2300 | Kol Israel, Jerusalem | 9435 11605 11655 11745 |
| | | 12077 17575 |
| 2230-2300 | Radio Sofia, Bulgaria | 11660 15330 |
| 2230-2300 | Radio Vilnius, Lithuania | 6100  9675 |
| 2230-2300 | Swiss Radio Int'l, European Service | 6190 |
| 2245-2300 | WINB Red Lion, PA | 15145 |

## 2300 UTC  [7:00 PM EDT/4:00 PM PDT]

| | | |
|---|---|---|
| 2300-2310 | Sierra Leone Brdcstng.Co.,Freetown | 3316 |
| 2300-2325 | Radio Finland, Helsinki | 11755 15185 |
| 2300-2330 | Radio Vilnius, Lithuania | 6100  7400  9865 11790 |
| | | 13645 15455 |
| 2300-2330 | Radio Canada Int'l, Montreal | 9755 11730 |
| 2300-2330 | Radio Sofia, Bulgaria | 11660 11720 |
| 2300-2345 | WYFR, Okeechobee, Florida | 5985 11580 15170 |
| 2300-2350 | Radio Pyongyang, North Korea | 11735 13650 |
| 2300-0000 | Radio Havana Cuba | 11930 |
| 2300-0000 | Radio Thailand, Bangkok | 4830  9655 11905 |

# BROADCASTING

| | | | |
|---|---|---|---|
| 2300-0000 | Radio Korea, Seoul | 15575 | |
| 2300-0000 | KHBN Guam | 9820 ML | |
| 2300-0000 | Adventist World Radio, Costa Rica | 9725 11870 | |
| 2300-0000 | Radio Moscow North American Svc. | 7150  7315 11710 11780 | |
| | | 11800 12040 12050 13605 | |
| | | 15315 15355 15425 15580 | |
| | | 15595 17735 | |
| 2300-0000 | Voice of Turkey, Ankara | 9445  9665  9685 17880 | |
| 2300-0000 | Radio Moscow World Service | 12005 15140 15480 15550 | |
| | | 15590 17570 17600 17620 | |
| | | 17655 17730 17850 21585 | |
| | | 21690 21790 | |
| 2300-0000 | CBN, St. John's, Newfoundland | 6160 | |
| 2300-0000 | CBU, Vancouver, British Columbia | 6160 | |
| 2300-0000 | CFCF, Montreal, Quebec, Canada | 6005 | |
| 2300-0000 | CFCN, Calgary, Alberta, Canada | 6030 | |
| 2300-0000 | CHNS, Halifax, Nova Scotia, Canada | 6130 15405 | |
| 2300-0000 | BBC World Service, London, England | 5975  6175  6195  7325 | |
| | | 9410  9590  9915 11750 | |
| | | 15260 | |
| 2300-0000 | Christian Science World Service | 9465 15225 15275 15300 | |
| | | 15405 15610 17555 | |
| 2300-0000 | Radio for Peace Int'l, Costa Rica | 13630 21566 | |
| 2300-0000 | CKWX, Vancouver, British Columbia | 6080 | |
| 2300-0000 | CBC Montreal | 9625 | |
| 2300-0000 | CFRB, Toronto, Ontario | 6070 | |
| 2300-0000 | KSDA, Guam | 15125 | |
| 2300-0000 T-A | KUSW, Salt Lake City, Utah | 15590 | |
| 2300-0000 | Radio Australia, Melbourne | 11880 13605 15240 15465 | |
| | | 17630 17715 17750 17795 | |
| | | 21740 | |
| 2300-0000 | Radio Japan General Service, Tokyo | 11835 15195 17765 17810 | |
| | | 21610 | |
| 2300-0000 | Radio Luxembourg | 6090 | |
| 2300-0000 | Radio Tonga, Kingdom of Tonga | 5030v | |
| 2300-0000 | Voice of America-East Asia Service | 7120  9770 11760 15185 | |
| | | 15290 15305 17735 17820 | |
| 2300-0000 | United Arab Emirates R.,Abu Dhabi | 9600 11985 13605 | |
| 2300-0000 | WHRI, Noblesville, Indiana | 9495 13760 | |
| 2300-0000 | WINB, Red Lion, Pennsylvania | 15145 | |
| 2300-0000 | WRNO, New Orleans, Louisiana | 15420 | |
| 2300-0000 | WWCR, Nashville, Tennessee | 15690 | |
| 2305-2355 | Radio Polonia, Warsaw, Poland | 5995  6135  7125  7145 | |
| | | 7270 | |
| 2315-0000 | All India Radio, New Delhi | 9535  9910 11715 11745 | |
| | | 15110 | |
| 2330-0000 | Voice of Vietnam, Hanoi | 9840 12020 15010 | |
| 2330-0000 | Radio Tirana, Albania | 6120  9760 11825 | |
| 2335-2345 M-A | Voice of Greece, Athens | 9395 11645 | |
| 2345-0000 | Radio Korea, Seoul | 7275 | |

# Pirates and Clandestines

Pirates, or "free radio" broadcasters, share two common denominators: they have no license to operate a broadcasting transmitter, and they are on the air for short periods in the evening, usually over weekends and especially holidays, to avoid apprehension. Best times to listen are 2200-0500 UTC in winter, 0000-0500 UTC summer.

They differ in their choices of format which may be comedy, rock, social commentary, satire or even religious or racial dialogue. Some productions are rather professional, while others are distinctly amateurish.

The most recently-reported pirates at this writing are the following (frequencies may vary to avoid interference):

| Station ID | Freq kHz |
|---|---|
| Hope Radio International | 7385-7410 |
| KPLU | 7415 |
| Radio Clandestine | 7390-7410 |
| Radio USA | 7410-7415 |
| Radio Garbonzo | 7415 |
| Rock-a-Billy Radio | 7415 |
| Voice of Tomorrow (neo-Nazi) | 7410, 6240 |
| WYMN (all female) | 7400 |
| Radio Free Massachis (RFM) | 7400 |
| Tube Radio | 7355, 7400, 7415 |

Clandestine broadcasters, on the other hand, are usually revolutionary, or at least counter to the established government of their broadcast target. Typically, they originate in emerging third world nations, communist block countries and Latin American hot spots. They are apt to be on the air for longer periods of time than the pirates, at least until they get caught.

## Longwave Monitoring

The VLF spectrum, loosely used to refer to all frequencies below the 530 kHz AM broadcast band, is neglected by most listeners even though general coverage receivers are equipped to listen down to at least 100 kHz.

We are indebted to Ken Stryker and the Longwave Club of America for their contributions to monitoring the low frequency spectrum and for their support of this section of the *Shortwave Directory*.

Experimenters and listeners who wish to learn more about the low frequencies may wish to acquire Ken Stryker and Joe Woodlock's *Aero/Marine Beacon Guide* ($15 postpaid from Ken Stryker, 2856 G West Touhy Avenue, Chicago, IL 60645) or join LWCA ($12 per year includes monthly copies of *The Lowdown*, LWCA's magazine; 45 Wildflower Road, Levittown, PA 19057).

## Longwave Frequency Allocations in U.S.

| Band (kHz) | Service* |
|---|---|
| Below 10 | Not allocated |
| 10-14 | Radionavigation (Omega) |
| 14-19.94 | Fixed |
| 19.95-20.05 | Standard frequency (WWVL 20 kHz no longer in service) |
| 59-61 | Standard frequency (WWVB 60 kHz) |
| 70-90 | Fixed; radiolocation |
| 90-110 | Radionavigation (Loran C 100 kHz) |
| 110-130 | Fixed-maritime mobile; radiolocation |
| 130-160 | Fixed; maritime mobile |
| 160-190 | Fixed (1750 meter band) |
| 190-200 | Aeronautical radionavigation |
| 200-275 | Aeronautical radionavigation; aeronautical mobile |
| 275-285 | Aeronautical & maritime radionavigation; aeronautical mobile |
| 285-325 | Maritime radionavigation |
| 325-335 | Aeronautical & maritime radionavigation; aeronautical mobile |
| 335-405 | Aeronautical radionavigation; aeronautical mobile |
| 405-515 | Maritime radionavigation (410 kHz radio direction finding); aeronautical radionavigation; aeronautical mobile |
| 415-490 | Maritime mobile (coast & ship telegraphy) |
| 490-510 | Mobile (500 kHz distress and calling) |
| 510-525 | Aeronautical & maritime radionavigation; (also 512 kHz ship calling) |
| 525-535 | Travelers information (530 kHz) |

\* "Fixed": land-based US military transmitters.
  "Mobile": marine, air or land vehicles
  "Radionavigation": radio beacons

## Longwave Loggings

The following list comprises actual loggings of stations on the air as heard by LWCA members. Corrections and additions from listeners are welcome.

| Freq/kHz | ID | Location | Remarks |
|---|---|---|---|
| .076 | | Clam Lake, Wisconsin | Navy ELF System |
| 10.2 | OMEGA | | Common frequency |
| 11.05 | OMEGA | | Common frequency |

# LONG WAVE

| | | | |
|---|---|---|---|
| 11.33 | OMEGA | | Common frequency |
| 11.8 | OMEGA | Haiku, Hawaii | |
| 11.905 | ALPHA | USSR | |
| 12.0 | OMEGA | Monrovia, Liberia | |
| 12.1 | OMEGA | Aldra, Norway | |
| 12.3 | OMEGA | La Reunion Is | |
| 12.649 | ALPHA | USSR | |
| 12.8 | OMEGA | Tsusshima, Japan | |
| 12.9 | OMEGA | Gulfo Nuevo, Argentina | |
| 13.0 | OMEGA | Australia | |
| 13.1 | OMEGA | LaMoure, North Dakota, USA | |
| 13.6 | OMEGA | | Common frequency |
| 14.881 | ALPHA | Krasnodar, USSR | |
| 15.1 | FUB | Paris, France | RTTY |
| 15.1 | HWU | Leblanc, France | FSK |
| 16.0 | GBR | Rugby, England | |
| 16.4 | JXN | Noviken, Norway | |
| 16.7 | JXZ | Oslo, Norway | Kolsaas Naval Base |
| 16.8 | FUB | St. Arisse, France | Multiplex RTTY |
| 17.1 | UMS | Moscow, USSR | Time signal |
| 17.4 | NDT | Yosami, Japan | |
| 18.5 | DHO38 | Flensburg, West Germany | CW |
| 19.0 | GQD | Anthorn, England | NATO identifier G33DG |
| 19.6 | GBZ | Criggion, Wales FSK | |
| 20.0 | ??? | ??? | FSK |
| 20.27 | ICV | Tavolara, Sardinia | RTTY |
| 20.5 | UTR3 | Gorki, USSR | |
| | UQC3 | Khabarovsk, USSR | |
| | UNW3 | Estonia, USSR | All time signal stations |
| | UPD8 | Murmansk, USSR | |
| | USB2 | Bielorussia | |
| 21.1 | 3SB | China, P.Republic | |
| 21.4 | NSS | Annapolis, MD | U.S. Navy MSK |
| 22.3 | NWC | Exmouth, Australia | U.S. Navy MSK 1000 kW |
| 23.0 | Same as 20.5 | | |
| 23.1 | | | TACAMO aircraft |
| 23.4 | NPM | Lualualue, Hawaii | U.S. Navy MSK 630 kW |
| 24.0 | NAA | Cutler, Maine | U.S. Navy MSK |
| 24.8 | NLK | Jim Creek, Washington | U.S. Navy MSK 130 kW |
| 25.0 | Same as 20.5 | | |
| 25.1 | Same as 20.5 | | |
| 25.5 | Same as 20.5 | | |
| 25.8 | ??? | ??? | FSK/TACAMO aircraft |
| 26.1 | | | TACAMO aircraft |
| 26.5 | ??? | ??? | FSK/TACAMO aircraft |
| 27.1 | | | TACAMO aircraft |
| 27.9 | 3SV | China, P.Republic | |
| 28.5 | NAU | Aguada, Puerto Rico | U.S. Navy MSK 100 kW |
| 37.2 | | Hawes, CA | U.S. Navy MSK |
| 38.0 | OJG | Kemi, Finland | |
| 40.0 | JJF2 | Tokyo, Japan | CW |
| 40.0 | JG2AS | Chiba, Japan | Time signal 10 kw |
| 42.25 | NRK | Keflavik, Iceland | U.S. Navy (JASON) |
| 44.0 | VHB/VIH/VIX | Belconnen, Canberra, AUS | RTTY |
| 44.0 | | | USAF Airborne Command Post |
| 46.25 | DCF46 | Mainflingen, W. Germany | |
| 48.5 | FXL | Silver Creek, Nebraska | Strategic Air Command MSK 110 kW |
| 50.0 | OMA50 | Liblice, Czechoslovakia | Time signal |

| | | | |
|---|---|---|---|
| 50.0 | RTZ | Irkutsk, USSR | Time signal |
| 51.6 | NSS | Annapolis, Maryland | |
| 51.95 | GYA | London, England | |
| 53.0 | NPL | Chollis Heights, CA | U.S. Navy MSK |
| 53.6 | RTO | Moscow, USSR | Fax Weather |
| 54.05 | NAM | Driver, Virginia | U.S. Navy (JASON) |
| 55.25 | DCF55 | Frankfurt, W. Germany | RTTY 75 Bd 0600-1100 UTC News |
| 55.5 | ??? | CA? | MSK |
| 55.5 | GXH | Thurso, Scotland | Royal Navy FSK |
| 56.5 | NPG | Dixon, CA | MSK |
| 57.8/.9 | ??? | | FSK |
| 58.7 | | Ft. Ritchie, MD | Alternate Natl. Mil. Command Center |
| 59.0 | NGR | Greece | U.S. Navy MSK |
| 60.0 | WWVB | Fort Collins, Colorado | Time signal 13 kW |
| 60.0 | MSF | Rugby, England | Time signal |
| 61.55 | ??? | NATO North Atlantic | |
| 61.8 | GIZ | London, England | Royal Navy |
| 62.6 | FTA | Paris, France | French Navy |
| 65.8 | FUE | Brest, France | |
| 66.66 | RBU | Moscow, USSR | Time signal |
| 68.0 | GBY20 | Rugby, England | |
| 68.9 | XPH | Thule, Greenland | |
| 73.6 | CFH | Halifax, Nova Scotia | FSK |
| 75.0 | HBG | Prangins, Switzerland | Time signal 20 kW |
| 76.2 | CKN | Vancouver, British Col. | FSK |
| 77.15 | NAM | Driver, Virginia | U.S. Navy (JASON) |
| 77.5 | DCF77 | Mainflingen, W. Germany | Time signal |
| 80.05 | GXH | Thurso, Scotland | Royal Navy backup for 55.5 |
| 81.0 | MKL | Pitreavie, Scotland | RAF |
| 82.75 | MKL | Pitreavie, Scotland | |
| 83.1 | OFA83 | Helsinki, Finland | FAX weather |
| 83.8 | FTA83 | St.Andre de Corcy, France | CW 45 kW |
| 88.0 | NSS | Annapolis, Maryland | U.S. Navy MSK |
| 91.15 | FTA91 | Saint-Andre-de-corcy, France | |
| 100.0 | LORAN C | Various | Navigational System |
| 106.7 | OLT22 | Prague, Czechoslovakia | Fax weather |
| 110.0 | DPA | Germany | RTTY |
| 110.6 | DCF30 | Hamburg, W. Germany | |
| 110.75 | CKN | Vancouver, B.C. | |
| 111.3 | SOA211 | Warsaw, Poland | RTTY Weather 0618+1827 UTC |
| 111.8 | OLT21 | Prague, Czechoslovakia | FAX Weather |
| 112.15 | CII | Shilo, Manitoba | Can Dept Nat Def FSK 3 kW |
| 113.2 | VER | Ottawa, Ontario | Can Dept Nat Def FSK 3 kW |
| 115.3 | CFH | Halifax, Nova Scotia | |
| 117.4 | DCF37 | Offenbach, W. Germany | Fax weather |
| 119.0 | IDR | Rome, Italy | CW |
| 119.15 | CII | Shilo, Manitoba | Can Dept Nat Def FSK 3 kW |
| 119.15 | ZRH | Cape Naval, S.Africa | RTTY weather |
| 119.85 | NPG | Dixon, CA | U.S. Navy |
| 119.85 | SAY | Norrkoping, Sweden | Fax weather |
| 120.0 | SXA | Spata Attikis, Greece | CW |
| 122.2 | OUA23 | Fredrikshaven, Denmark | 'V' Marker |
| 122.3 | CIF | Borden, Ontario | Can Dept Nat Def FSK 3 kW |
| 122.5 | CFH | Halifax, Nova Scotia | CW weather |
| 122.65 | JMC | Tokyo, Japan | CW weather |
| 123.0 | VEN | Nanaimo, B.C. | Can Dept Nat Def FSK 3 kW |
| 124.0 | CKN | Vancouver, B.C. | CW weather |
| 125.2 | SLZ1 | Sweden | Fax weather |

# LONG WAVE

| | | | |
|---|---|---|---|
| 125.8 | CII | Shilo, Manitoba | Can Dept Nat Def FSK 10 kW |
| 128.25 | NPL | San Diego, CA | U.S. Navy |
| 128.3 | VDD | Debert, Nova Scotia | Can Dept Nat Def FSK 3 kW |
| 128.3 | VEX | Penhold, Alberta | Can Dept Nat Def FSK 3 kW |
| 129.5 | SOA212 | Warsaw, Poland | Meteo 1825-0923 UTC weather |
| 131.05 | FUF | Port de France, Martinique | French Navy |
| 131.4 | VEV | Valcartier, Quebec | Can Dept Nat Def FSK 3 kW |
| 131.8 | FYA31 | Paris, France | Fax weather |
| 133.15 | CFH | Halifax, Nova Scotia | FSK |
| 133.15 | CKN | Vancouver, B.C. | FSK RTTY 25 kW |
| 133.85 | CII | Shilo, Manitoba | Can Dept Nat Def FSK 10 kW |
| 134.2 | DCF54 | Offenbach, W. Germany | Fax weather |
| 134.9 | NSS | Annapolis, Maryland | U.S. Navy (JASON) |
| 135.95 | NPG | Dixon, CA | U.S. Navy |
| 136.0 | NPG | Dixon, CA | 8 channel FDM RTTY |
| 136.0 | TBA | Izmir, Turkey | CW |
| 136.3 | VEV | Valcartier, Quebec | Can Dept Nat Def FSK 3 kW |
| 136.8 | ??? | ??? | RTTY |
| 137.0 | CFH | Halifax, Nova Scotia | RTTY |
| 139.0 | DCF39 | Frankfurt, W. Germany | Fax weather |
| 139.1 | VEV | Valcartier, Quebec | Can Dept Nat Def FSK 3 kW |
| 139.5 | CIF | Borden, Ontario | Can Dept Nat Def FSK 3 kW |
| 140.3 | DCF60 | Frankfurt, W. Germany | |
| 140.5 | VER | Ottawa, Ontario | Can Dept Nat Def FSK 3 kW |
| 142.25 | TFK | Keflavik, Iceland | U.S. Navy (JASON) |
| 142.86 | UIK | Vladivostok, USSR | CW weather |
| 143.5 | VDD | Debert, Nova Scotia | Can Dept Nat Def FSK 3 kW |
| 146.0 | Y7A20 | E. Berlin, E. Germany | 100 Bd RTTY |
| 146.1 | NPM | Lualualei, HI | U.S. Navy |
| 147.2 | DDH47 | Pinneberg, W. Germany | CW weather |
| 148.2 | NPL | San Diego, CA | U.S. Navy |
| 148.55 | VER | Ottawa, Ontario | Can Dept Nat Def FSK 3 kW |
| 162.0 | VEX | Penhold, Alberta | Can Dept Nat Def FSK 3 kW |
| 168.0 | CIF | Borden, Ontario | Can Dept Nat Def FSK 3 kW |
| 168.0 | VEX | Penhold, Alberta | Can Dept Nat Def FSK 3 kW |
| 174.0 | VDD | Debert, Nova Scotia | Can Dept Nat Def FSK 3 kW |
| 174.0 | VER | Ottawa, Ontario | Can Dept Nat Def encrypted FSK RTTY 10 kW |
| 183.0 | CIF | Borden, Ontario | Can Dept Nat Def encrypted FSK RTTY 10 kW |
| 187.0 | VEV | Valcartier, Quebec | Can Dept Nat Def FSK 3 kW |
| 193.0 | CII | Shilo, Manitoba | Can Dept Nat Def FSK 3 kW |
| 197.5 | VER | Ottawa, Ontario | Can Dept Nat Def FSK 3 kW |

Between approximately 160 to 190 kHz--the domain of wireless intercoms and license-free, 1750 meter experimenters--are heard rapidly-pulsating, line-load telemetry signals, a carrier-current system used by electric power generating companies to adjust to consumer demands.

# GLOSSARY

## Abbreviations, Acronyms and Code Names

| | |
|---|---|
| AABNCP | Advanced Airborne National Command Post |
| AACOMS | Army Area Communications System |
| AAF | Army Air Field |
| AB | Air Base |
| ABCCC | Airborne Battlefield Command and Control Center |
| A/C | Aircraft |
| ACD | Automatic Call Distribution System |
| Ace High | Tropospheric Scatter Communications Network (NATO) |
| ACLICS | Airborne Communications Location Identification and Collection System |
| ADSAF | Automatic Data System for the Army in the Field |
| ADSCOM | Advanced Ship Communications |
| ADT | Airborne Data Terminal |
| AF | Air force |
| AFB | Air Force base |
| AFCC | Air Force Communications Center |
| AFRES | Air Force Reserve |
| AFSATCOM | Air Force Satellite Communications |
| ALCC | Airlift Control Center |
| Alligator | Slang for US Navy NTDS ("link 11") data link -- HF, VHF and satellite |
| Alligator Playground | Network of NTDS participating stations |
| ALLOC | Allocation |
| AM | Amplitude modulation (full carrier). |
| AMTOR | Amateur radio TOR |
| AMVER | Automated Merchant Vessel Rescue system. |
| ANDVT | Advanced Narrow-band Digital Voice Terminal |
| ANG | Air National Guard |
| ANGB | Air National Guard Base |
| ANGS | Air National Guard Station |
| ANT | Antenna |
| APACHE | Analysis of Pacific Area Communications for Hardening to EMP |
| ARPANET | Packet Switched Data Communications Network |
| ARPT | Airport |
| ARQ | Automatic request |
| ARR | Arrive, arrival |
| ARRS | Aerospace Rescue and Recovery Service |
| ARS | Air Refueling Squadron |
| ARTADS | Army Tactical Data System |
| ARTCC | Air Route Traffic Control Center |
| AS | Air Station |
| ASCL | Advanced Sonarbuoy Communications Link |
| ASCON | Automatic Switched Communications Network |
| ASW | Anti-Submarine Warfare |
| ATACS | Army Tactical Communications System |
| ATC | Air Traffic Control or Air Training Command |
| ATDS | Airborne Tactical Data System |
| ATOS | Air Terminal Operation Service |
| AUTODIN | Automatic Digital Network |
| AUTOSEVCOM | Automatic Voice Network |
| AWACS | Airborne Warning and Control System |
| Baud (Bd) | Bits of data per second |
| BC | Broadcast |
| BCB | Broadcast band |
| BCI | Broadcast interference |
| BFO | Beat frequency oscillator |
| BUG | Semi-automatic key |
| CALL TAPE | Recorded message sent repeatedly to alert recipient |
| CAMPS | Computer Assisted Message Processing System (NATO) |
| CBW | Chemical & Biological Warfare |
| CCC ($C^3$) | Command, Control and Communications |
| CE | Communications equipment |
| CH | Channel |
| CIC | Commander in Chief |
| Cifax | Enciphered facsimile |
| CINCLANT | Commander in Chief, Atlantic |
| CINCPAC | Commander in Chief, Pacific |
| CIPHER | Encrypted plain test |
| Clansman | British Army Tactical Communications Equipment (UK) |
| Clarinet | Communications Security |
| CNI | Communications, Navigation and Identification |
| COE | Corps of Engineers |
| COMINT | Communications Intelligence |
| COMMS | Communications |
| COMMSTA | U.S. Coast Guard communications stations |
| Compac | Man-portable Satcom Terminal (UK) |
| COMSEC | Communication Security |
| COMSTA | Communications Station |

# GLOSSARY

| | |
|---|---|
| CONUS | Continental United States |
| CQ | General call to anyone monitoring, inviting reply |
| Crazy Cat | Optical Communications Jamming System |
| Crazy Dog | Battlefield Communications Jamming System |
| CRITICOM | Critical Intelligence Communications |
| CS | Communications squadron |
| CSCG | Communications Security Control Group |
| CW | Continuous wave (Morse code) |
| DCA | Defense Communications Agency |
| DCS | Defense Communications System |
| DE | (French "from") This is (id or callsign) |
| DEP | Depart, departure |
| DES | Digital Encryption Standard |
| DOT | Department of Transportation, Department of Transport (in Canada) |
| DSB | Double sideband |
| DSCS | Defense Satellite Communications System |
| DTOI | Date and time of intercept. |
| DX | Distance |
| EAM | Emergency action messages |
| ECSS | Exalted carrier single sideband |
| EMMPS | Enhanced MEECN Message Processing System |
| EMP | Electromagnetic Pulse (from nuclear detonation) |
| EOF | Emergency Operating Facility |
| ERCS | Emergency Rocket Communications System |
| ETA | Estimated Time of Arrival |
| EUCOM | European Command (NATO) |
| FAA | Federal Aviation Administration |
| FACSFAC | U.S. Navy Fleet Area Control and Surveillance Facility |
| FATT | Forward Area Tactical Teletype |
| FAX | Facsimile |
| FCC | Federal Communications Commission |
| FDM | Frequency division multiplex |
| FEC | Forward error correcting (radioteletype TOR) |
| FLTSATCOM | Fleet Satellite Communications |
| FM | Frequency modulation |
| FOCAS | Fibre Optical Communications for Aerospace Systems |
| FREQ | Frequency |
| FSK | Frequency shift keying |
| FSS | Flight Service Station |

| | |
|---|---|
| GCCS | Global Command and Control System |
| Giant Talk | Early Warning Airborne HF Communications System |
| GMT | Greenwich Mean Time (obsolete); see UTC |
| Green | U.S. Navy reference to frequency-hopping voice scrambling |
| Green Pine | Early Warning Airborne UHF Communications System |
| Gryphon | Secure Shore-to-Ship Communications System |
| GWEN | Ground Wave Emergency Network |
| H | On the hour. |
| H+ | Minutes past the hour |
| HEMP | High Altitude Electromagnetic Pulse |
| HF | High frequency |
| HFIC | HF Intra-Task Force Communications |
| HICOM | U.S. Navy High Command Worldwide Voice Network |
| HQ | Headquarters |
| HVIT | High Volume Information Transfer |
| Hydrus | Secure Ship-to-Shore Communications System |
| Hz | Hertz; unit of frequency (formerly cycles per second). |
| IAP | International Airport |
| ICAO | International Civil Aviation Organization |
| ICS 3 | Integrated Communications System (UK) |
| ID | Identification |
| IEMATS | Improved Emergency Message Automation Transmission System |
| IFSS | International Flight Service Station |
| INCA | Integrated Nuclear Communications Assessment |
| INTACS | Integrated Tactical Communications Study |
| IRC | International Reply Coupon |
| IRR | Integrated Radio Room |
| ISB | Independent sideband |
| ITU | International Telecommunications Union (Geneva, Switzerland) |
| IVCS | Integrated Vehicle Communications System |
| Jaguar | Frequency Hopping Radio System (UK) |
| JATCCCS | Joint Advanced Tactical C$^3$ System |

| | |
|---|---|
| JTIDS | Joint Tactical Information Distribution System |
| JUMP | Joint UHF Modernization Project |
| kHz | Kilohertz |
| KW | Kilowatt (1000 watts) |
| LF | Low frequency |
| LL | Landline (telephone call) |
| LORAN-C | Long Range Aid to Navigation |
| LSB | Lower sideband |
| M | Meter(s) |
| MAC | Military Airlift Command |
| MAG | Military Assistance Group |
| MAP | Municipal Airport |
| MARKER | Any continued or repetitive signal sent to keep frequency occupied |
| MASST | Major Ship Satellite Terminal |
| MATS | Mobile Automatic Telephone System (Various) |
| MAW | Military Air Wing |
| MCAS | Marine Corps Air Station |
| MCW | Modulated continuous wave |
| MEDAX | Message Data Exchange Terminal |
| MEECN | Minimum Essential Emergency Communications Network |
| METEO | Meteorological (weather) |
| MF | Medium frequency |
| MHz | Megahertz |
| MKR | Channel markers |
| MM | Maritime mobile |
| MOMCOMS | Man-on-the-Move Communications System |
| MPDS | Message Processing and Distribution System |
| MUF | Maximum usable frequency |
| MULCAST | Multichannel Broadcast |
| MULTOTS | Navy Advanced Tactical $C^3$ Program |
| MW | Medium wave |
| Mystic Link | Digital Information Distribution System |
| Mystic Star | USAF SSB network controlled by Andrews AFB for Air Force 1 and Air Force 2 communications |
| NAS | Naval Air Station |
| NASCOM | NASA Communications Network |
| NAVCOMPARS | Naval Communications Processing and Routing System |
| NAVCOMSTA | Naval Communications Station |
| NAVMACS | Naval Modular Automated Communications System |
| NAVRADSTA | Naval Radio Station |
| NAVTEX | Navigational information teleprinter transmission |
| NBSV | Narrow-Band Secure Voice |
| NCS | Net Control Station |
| NDB | Non-directional beacon |
| NEACP | National Emergency Airborne Command Post (pronounced "kneecap") |
| NFHQ | National Field Headquarters |
| NG | National Guard |
| NICS | NATO Integrated Communications System (NATO) |
| NMCS | National Military Command System |
| NORAD | North American Aerospace Defense Command |
| NORATS | Navy Operated Radio and Telephone System |
| NTDS | Naval Tactical Data System |
| NTM | Notice to mariners |
| NUCO | (NATO Uniform Code for Operations) Voice advisement that encrypted portion of message is to follow. |
| NX | News broadcast or transmission |
| OCCULT | Optical Covert Communications Using Laser Transceivers |
| OM | "Old man"; male operator |
| Ops | Operations |
| PACCS | Post Attack Command and Control System |
| Piccole | Transmission system whereby signal (Morse code or Telex) is converted into 5 or 6 musical tones (UK) |
| PCM | Pulse code modulation |
| PFC | Prepared form card |
| Pilgrim | Secure Communications Development Program |
| PMFR | Pacific Missile Firing Range |
| Pri | Primary Frequency |
| PT | Plain text |
| RAF | Royal Air Force (UK) |
| RAO | Regional Area Office |
| RCC | Rescue Coordination Center or radio common carrier. |
| RCVR | Receiver |
| Red | U.S. Navy reference to unencrypted voice transmission |
| RELOC | Relocation |
| RF | Radio frequency |
| RO | Regional Office |
| RS | Receiver Station |
| RST | Readability, signal strength, tone quality (CW report). |
| RTTY | Radioteletype |
| RX | Receiver |
| SAC | Strategic Air Command |
| SACCS | Strategic Air Command Automated Command Control System |
| SAGE | Semi-Automatic Ground Environment |

# GLOSSARY

| | |
|---|---|
| SAM | Special Air Mission |
| SAMSON | Strategic Automatic Message Switching Operational Network Submarine/Aircraft Optical Communications System |
| SAR | Search and Rescue |
| SARBE | Search and Rescue Beacon Equipment (UK) |
| SASE | Self addressed stamped envelope |
| SATIN | Strategic Air Command Automated Total Information Network |
| SCOT | Shipborne Satcom Terminal (UK) |
| SDS | Satellite Data System |
| SEC | Secondary Frequency |
| SECOM | Secure Communications |
| Seek Bus | Jam-resistant Digital Tactical Communications System |
| Seek Talk | Secure Communications System |
| SELCAL | Selective Calling System |
| SFO | Sector Field Office |
| SHELF | Super Hard ELF System |
| SICOFAA | System of Cooperation Among the American Air Forces (US & Lat.Am) |
| SIGINT | Signals Intelligence |
| SINCGARS V | Single Channel Ground and Airborne Radio System VHF |
| SINPO | (See glossary) |
| SITFA | Interamerican (Air Forces) HF Voice & RTTY network of SICOFAA (US & Lat.Am.) |
| SITOR | Commercial version of TOR |
| SITREP | Situation report |
| SKED | Schedule |
| Skynet | British Military Communications Satellite (UK) |
| SLFCS | Survivable Low Frequency Communications System |
| SOW | Special Operations Wing |
| SPANNER | Special Analysis of Net Radio (US Study) |
| SSB | Single sideband |
| STARCOM | Strategic Army Communications |
| STDL | Submarine Tactical Data Link |
| STFS | Standard Time and Frequency Station |
| STRATCOM | Strategic Communications |
| SURVSATCOM | Survivable Satellite Communications System |
| SW | Shortwave |
| SWBC | Shortwave broadcasting. |
| SWL | Shortwave listener |
| TAC | Tactical Air Command |
| TACAMO | VLF Airborne Radio Relay System |
| TACOM | Tactical Area Communications |

| | |
|---|---|
| | System (FDR) |
| TACSATCOM | Tactical Satellite Communications |
| TACTEC | Totally Advanced Communications Technology |
| TACW | Tactical AirCommand Wing |
| TAS | Tactical Airlift Squadron |
| TAW | Tactical Airlift Wing |
| TCCF | Tactical Communications Control Facility |
| TDM | Time division multiplexing |
| TELEX | A teleprinter activated by telephone lines |
| TES | Transportable Earth Station (UK) |
| TFC | Traffic communications |
| TFW | Tactical Fighter Wing |
| TIS | Travelers Information Station |
| TOR | "Telex over radio" |
| TOS | Tactical Operations System |
| TX | Transmitter |
| UMIB | Urgent marine information broadcast |
| UN-NUCO | Voice advisement that encrypted portion of message is completed (See NUCO). |
| USA | US Army |
| USAF | US Air Force |
| USAR | US Army Reserve |
| USB | Upper sideband |
| USCG | US Coast Guard |
| USMC | US Marine Corps |
| USN | US Navy |
| USNAVFAC | US Naval facility |
| USP | United States and possessions |
| UTC | Coordinated Universal Time |
| Ute | Slang for utilities |
| Veri | Slang for verification form card |
| VLF | Very low frequency |
| VOLMET | French acronym for "flying weather" |
| WARC | World Administrative Radio Conference |
| WHCA | White House Communications Agency |
| WPM | Words per minute (usually referred to in Morse or RTTY) |
| WRS | Weather Reconnaissance Squadron |
| WWMCCS | World-wide Military Command and Control System |
| WX | Weather broadcast |
| XMSN | Transmission |
| XMTR | Transmitter |
| YL | "Young lady"; female operator |
| Z | "Zulu" time - same as UTC and GMT |
| # | number(s) |
| 5F (5L) | Five figure (or letter) group |

# SWL Glossary

| | |
|---|---|
| addressee | The component or individual to whom a message is directed by the originator. |
| aeronautical radio-beacon station | A radiobeacon station intended for the benefit of aircraft. |
| aeronautical station | A land station in the aeronautical mobile service which in certain instances, may be located on board ship or on a platform at sea. |
| agency | A group such as a government bureau empowered to conduct business activity. |
| alpha-numeric | Set of symbols consisting of characters and numbers. |
| amateur service | A radio communication service for the purpose of self-training, intercommunication and technical investigation. (see ham) |
| amplitude modulation (AM) | Mode of transmission where signal is transmitted by using the audio portion to vary the strength of the carrier wave portion. |
| antenna | A device for radiating and receiving electromagnetic waves. (see beam, dipole, longwire, loop, whip) |
| authentication | A security procedure designed to prevent fraudulent transmissions. |
| band | A portion of the radio frequency spectrum. |
| band-slip marker | A call-tape transmission, usually in Morse or RTTY ("QRA DE WLO", etc) |
| bandwidth | The high and low limits of a portion of radio spectrum over which the performance of a device is being measured. |
| base station | Fixed land station communicating with one or more mobile stations |
| baud (bd) | Bits of data per second. (45bd=60wpm; 50bd=67wpm; 57bd=75wpm; 75bd=100wpm |
| Baudot alphabet | A five-unit code applied to teleprinter systems developed by Jean Maurice Emile Baudot. It employs a 32-element alphabet designed particularly for telecommunications wherein each symbol intended for tranmission is represented by a unique arrangement of five mark or space impulses. |
| beam antenna | Antenna with directional properties, usually driven by a rotor so it can be pointed in any desired direction. |
| beacon | Station which transmits signal that is used for navigational purposes by ships and aircraft. |
| beat frequency oscillator (BFO) | Used on radio receivers to permit reception of Morse code, RTTY, FAX, and single sideband voice. |
| blind | Sending traffic without making contact with the recipient. Messages can be acknowledged at a later time and/or via other means. |
| break-in procedure | The receiving station will interrupt the tranmsitting station by sending one or more long dashes. When the transmitting station stops, the receiving station sends the first letter of the group he wants repeated. |
| broadcasting service | A radio communication service in which the transmissions are intended for direct reception by the general public. |
| burst | A short signal transmission often containing information. Message is at a very fast rate. Often used to avoid interception and site location by direction finding. |

# GLOSSARY

bury — To conceal or hide certain words, phrases or messages in the text of a communication.

call sign — Any combination of letters and/or numbers, identifying a communications facility, command, authority, activity, or unit.

carrier — The electromagnetic emanation from a transmitter which may be modulated with information.

CB operator — A person operating a Citizens Band radio station

channel — One of a sequenced set of frequencies arranged for convenient selection.

chatter — Casual exchange between operators of a two-way circuit.

cipher — A code.

Citizens Band — A range of 40 frequencies from 26.965-27.405 MHz allocated to private citizen, short-range communication.

clandestine broadcast — A counter-regime political transmission from an organization that is considered illegal in the target country.

coast station — A land station in the maritime mobile service.

code — A technique to conceal the contents of a transmitted message by rearranging or substituting contents according to a prearranged plan.

code book — A book or document used in a code system, arranged in systematic form, containing units of plain text of varying length (letters, syllables, words, phrases, or sentences) each accompanied by one or more arbitrary groups of symbols used as equivalents in messages.

code test — A formal assessment of a person's skill in receiving and/or transmitting the Morse code at a designated speed.

communication — The transfer of meaningful information from one location to another.

compromise — The availability of classified material to unauthorized persons through loss, theft, capture, recovery by salvage, defections of individuals, unauthorized viewing, or any other physical means.

contact — A successful communication between two stations.

continuous wave (CW) — On-off keying of the carrier wave (used in Morse code)

control — The station located at a headquarters or designated as being in charge of a particular radio net.

Coordinated Universal Time (UTC) — Equivalent to mean solar time at the prime meridian (0 degrees longitude); formerly GMT.

counterpoise — A conductor or system of conductors used as a substitute for the ground in an antenna system.

covert — A concealed or secret relationship.

cryptanalysis — The analysis of encrypted messages and subsequent conversion into plain text without initial knowledge of the key employed in the encryption. (Also cryptoanalysis)

cryptogram — A communication in visible writing which conveys no intelligible meaning in any known language, or which conveys some meaning other than the real meaning.

cryptography — That branch of cryptology which treats of the means, methods, and apparatus for converting or transforming plain text messages into

230

cryptograms, and for reconverting the cryptograms into their original plain text.

**cryptology** — That branch of knowledge which treats of hidden, disguised or encrypted communications.

**cut numbers** — Digits sent in abbreviated fashion. Many different cut number systems are in use throughout the world, ranging from a simple system of only one number being cut (as with zero sent as the letter T), to those with a letter substituted for each of the digits 1-0.

**cycle** — The time interval for the completion of a periodic function such as the electromagnetic property of a radio wave. One cycle equals one hertz.

**decipher** — To convert an enciphered message into its equivalent plain text by a reversal of the cryptographic process used in the encipherment, not by cryptanalysis.

**decode** — To convert an encoded message into its plain text by means of a code book. (This does not include solution by cryptanalysis.)

**decrypt** — See decipher.

**demodulation** — Recovery of original modulator from a modulated radio signal.

**deviation** — The broadening of an FM signal by modulation.

**digraph** — A pair of letters.

**dinome** — A pair of digits.

**dipole antenna** — A center or off center fed antenna cut to a specific length for a particular band (or bands) of frequencies.

**Doppler effect** — An apparent change in frequency caused by motion of either the transmitter or the receiver.

**double sideband (DSB)** — Radio transmission in which both sidebands produced by modulation are transmitted equally.

**doublet** — A digraph or dinome in which a letter or a digit is repeated (e.g., LL, 22, etc.).

**down at ____** — Used by intercept operator to indicate time station ceased transmission.

**dropped at ____** — Used by intercept operator to indicate time monitoring of target was discontinued.

**dummy messages** — False messages transmitted as part of a deception program to mislead enemy traffic and crypto analysis efforts.

**duplex operation** — Simultaneous transmission in both directions of a telecommunication channel by using different frequencies.

**DX'er** — Individual who listens to distant or difficult to receive radio signals; from CW abbreviation "DX" meaning distance.

**emission** — The waves radiated into space by a transmitter.

**encipher** — To convert a plain text message into unintelligible language by means of a code book.

**end product** — Finished intelligence

**exalted carrier single sideband (ECSS)** — A distortion-reducing technique of tuning in a full-carrier AM signal using the receiver's SSB mode.

**facsimile** — A process by which pictures or images are scanned and converted into electrical signal waves, then reproduced.

**fading** — The characteristic of changes in radio wave propagation causing the received amplitude of radio signals to fluctuate in strength.

231

# GLOSSARY

feeder
A point to point transmission of a radio program to a transmitter site intended for subsequent broadcast.

fixed
A permanent installation.

frequency
Symbolized by f, the number of vibrations of electromagnetic energy per second; the unit is the hertz (Hz), formerly cycles per second (cps). Radio frequencies are normally expressed in kilohertz (kHz) below 30,000 kHz, megahertz (MHz) above this frequency, and gigahertz (GHz) above 1,000 MHz.

frequency division multiplex (FDM)
A wideband process for transmitting two or more signals over one frequency band.

frequency modulation (FM)
A carrier wave whose frequency is varied by an amount proportionate to the amplitude of the modulating signal.

garble
An error in transmission reception or encryption which causes words or phrases to be unreadable.

group
A number of digits, letters or characters forming a unit for transmission.

half duplex (semi-duplex)
Transmission on two channels or frequencies, but not simultaneously.

harmonic emission
The unintentional emanation of whole-number multiples of the fundamental (intended) frequency of an oscillator or transmitter, often causing interference to users of those higher frequency services.

hertz
The basic unit of frequency, formerly cycles per second.

ham
A radio amateur. Slang for a licensed radio operator who operates a station as a hobby rather than a business.

high frequency (HF)
Shortwave band 3-30 MHz

identification
Determination of the specific unit (aircraft, ship, land transmitter, etc.), but not necessarily its location.

independent sideband (ISB)
Simultaneous upper and lower sideband, each with different content.

intelligence
The product resulting from collecting and processing information concerning actual or potential situations and conditions relating to criminal activities, enemy-held areas or foreign governments.

intelligence community
The members of the US Intelligence Community are: Central Intelligence Agency, National Security Agency, Defense Intelligence Agency, Bureau of Intelligence and Research of the Department of State, Air Force, Army, Navy and Marine Corps Intelligence elements, Federal Bureau of Investigation, Department of the Treasury, Department of Energy, Department of Defense offices that collect specialized national foreign intelligence through reconnaissance programs, and members of the intelligence community staff.

interception
The process of gaining possession of a communication intended for another without obtaining the consent of the addressee or originator.

interference
The degrading effect of unwanted energy imposed upon reception in a radio communications system.

intermodulation
The generation of spurious signals in a receiving system caused by strong signal overload of the receiver.

international broadcasting
A broadcasting service, employing frequencies

# GLOSSARY

between 5950 kHz and 26100 kHz, whose transmissions are intended to be received directly by the general public in foreign countries. In the United States, sometimes called "world band".

**international Morse code** — A code widely-used in radio telegraphy in which letters and numbers are represented by specific groupings of dots and dashes.

**International Reply Coupon (IRC)** — Purchased at post office and used to reimburse a foreign recipient for his postal cost.

**intrusion** — The intentional insertion of electromagnetic energy into transmission paths, in any manner, with the object of deceiving operators or causing confusion.

**ionosphere** — Electrically-charged (ionized) outer part of the earth's atmosphere which reflects and refracts (bends) radio waves. Multiple reflections of the radio waves alternately by the ionosphere and by the surface of the earth (hops) permit radio waves in the 3 to 30 MHz shortwave or HF band to travel all the way around the world. At times the ionosphere may absorb radio waves and thus destroy or distort them causing reception to be poor or unusable.

**jamming** — Transmission set up to intentionally interfere with or disrupt programming on the same or a nearby frequency. Jamming methods include varying tones, buzzing sounds, warbling, etc.

**key** — A switching device used to send Morse code characters.

**link** — A communication facility located between two points and used to relay a signal, often automatically.

**log** — Record kept by radio operators and listeners. The elements of the record are usually date, time, frequency, callsign and any other notations which may be helpful for future reference.

**long wire antenna** — An end-fed wire antenna longer than one wavelength at the operating frequency.

**loop antenna** — Closed circuit of one or more turns of wire in the same plane.

**low frequency** — That portion of the spectrum from 30 kHz to 300 kHz. Sometimes referred to as longwave. (1000 to 10,000 meter wavelengths)

**lower sideband (LSB)** — The lower of the two frequencies or groups of frequencies produced by the amplitude-modulation process.

**marine broadcast station** — A coast station which makes scheduled broadcasts of time, meteorological and hydrographic information.

**marine radiobeacon station** — A radiobeacon station intended to assist ship navigation.

**maritime mobile service** — A radio service between coast and ship stations, or between ship stations.

**mark impulse** — An impulse used in teleprinter transmission during which current flows through the teleprinter receiving magnet. The other type of impulse is the space impulse.

**marker** — A repeated message in Morse code or voice, usually recorded, identifying a station and sometimes listing frequencies in use by the station.

**marker beacon** — A transmitter in the aeronautical radio navigation service which radiates

233

vertically a distinctive pattern for providing position information to aircraft.

**maximum usable frequency (MUF)** The highest frequency that can be used for radio transmissions without having the signal escape through the ionosphere.

**meaconing** The deliberate transmission of actual or simlulated radio navigation signals for the purpose of confusing navigation.

**medium frequency (MF)** That portion of the spectrum from 300 kHz to 3 MHz, including the standard AM broadcast band (540-1600 kHz).

**medium wave (MW)** Wavelengths of 100 to 1000 meters. This is the same band as Medium Frequency (see preceding entry).

**message** Any thought or idea expressed in plain or secret language, prepared in a form suitable for transmission.

**mike** Popular term for microphone.

**mobile station** A radio station installed on a vehicle, ship, plane, automobile, etc.

**modulation** Process by which essential characteristics of signal wave are impressed on the carrier wave.

**monograph** A single letter

**monome** A single digit

**Morse code** A code developed by Samuel F.B. Morse which uses patterns of long and short pulses of energy to represent letters, numbers, punctuation and symbols.

**multipath** Propagation of radio waves by more than one path from transmitter to receiver; usually causes fading or distortion.

**multiplexing** Use of one common channel to make two or more channels, either by dividing the frequency band of common channel into narrower bands (FDM) or by allotting it alternately to different channels (TDM).

**net (network)** An organization of radio stations which generally reflects the command structures. The headquarter station is usually in charge of the subordinate stations; this station is called the net control station (NCS) and the others are called outstations. The control station is responsible for the supervision of tranmissions, procedures and circuit discipline.

**noise** Any unwanted electrical disturbance or spurious signal which modifies the desired data.

**non-directional beacon (NDB)** Low-powered (usually less than 50 watts) sending an identifier in slow Morse code, used for navigational purposes.

**OMEGA Navigation System** A low frequency (10-20 kHz) navigation system.

**one-time pad** A printed pad with a one-time use key on each page designed to permit the destruction of each page as soon as it is used. An agent in the field may have a pad the size of a postage stamp. Sometimes the pads are called "one-time gammas."

**originator** The command by whose authority a message is sent.

**other end** The station being called.

**overt** Open, without attempt to conceal.

**padding** Extraneous text added to a coded message for the purpose of concealing its length, beginning or ending

**parasitic emission**
or both.
A spuriously-generated signal radiated by a transmitter, usually not a harmonic of the fundamental frequency.

**permutation table**
A table designed for the systematic construction of code groups. It may also be used to correct garbles in groups of code text.

**pirates**
Unlicensed broadcasting stations often run by electronic hobbyists.

**plain text (PT)**
Intelligible meaning in the language in which the message was written with no hidden meaning.

**prepared form card (PFC)**
Sent by a DX'er for a station to fill out to confirm his reception report.

**probable-word method**
The method of solution involving the trial of plain text assumed to be present in a cryptogram.

**propagation**
Manner in which an electromagnetic emission travels outward from its source.

**Q-signals**
An abbreviated form used to send frequently repeated questions and answers.

**quieting**
The suppression of background noise by an FM signal, an indicator of the signal strength.

**QSL card**
Verification of reception sent by a radio station to acknowledge a reception report from a listener. A verification exchanged by licensed amateurs following a contact.

**radiation patterns**
The distribution in space of the electromagnetic energy produced by an antenna.

**radiobeacon station**
A navigation station intended to enable a mobile station to determine its bearing or direction in relation to the radiobeacon station.

**radio frequency (RF)**
Any frequency at which electromagnetic radiation of energy is possible.

**radio wave**
Electromagnetic enery moving from its point of radiation at the speed of light and oscillating at radio frequency.

**raw traffic**
Intercepted traffic before extensive processing for communication(s) intelligence purpose.

**read**
To decrypt, especially as the result of successful cryptanalysis.

**receiver**
The person or device to which information is sent over a communication link.

**reception report**
A report containing information elements of frequency, time, signal strength, call sign and other data describing the transmission so that it can be identified and acknowledged.

**relay**
A transmission forwarded by way of an intermediate station.

**repeater**
A combination receiver/-transmitter which retransmits the amplified received signal in real time.

**RST system**
A numeric rating system for the quality of a received CW signal.

**selectivity**
The ability of a receiver to discriminate between closely related signals.

**sensitivity**
Characteristic of receiver which determine minimum usable input, generally expressed in terms of signal-to-noise ratio, microvolts or decibels.

**ship station**
A mobile station located on board a vessel which is not permanently moored.

# GLOSSARY

shortwave listening
: Abbreviated SWL. The monitoring of tranmissions in the 3-30 MHz HF portion of the spectrum.

signal
: Purposely-transmitted radio energy intended for reception to convey information or perform a task.

signal-to-noise-ratio
: The ratio of the strength of a radio signal to that of the background noise.

simplex operation
: Operating method in which two-way transmission is made alternately in each direction on a telecommunication channel or frequency by manual control.

single sideband (SSB)
: Radio transmission in which either upper or lower sideband is transmitted, but not both.

sliding strip
: A strip of cardboard or similar material which bears a sequence and which can be slid against other such strips to various juxtapositions to encode or decode a crypto message. Also known as the "St. Cyr Slide" after the French military school at which it was developed.

space impulse
: An impulse used in teleprinter transmission during which no current flows through the teleprinter receiving magnet. The other type of impulse is the mark impulse.

spread spectrum
: Frequency hopping technique which decreases interference to other users while achieving privacy. Normally narrowband information is spread over a relatively wide band of frequencies. The receiver correlates the signal to retrieve the original information signal.

spurious emission
: Undesirable and incidental emission on frequencies outside the necessary bandwidth to convey information. Spurious emissions include harmonics, parasitics, intermodulation products, frequency conversion products, but exclude out-of-band emissions.

squirt transmission
: Generally a twenty-second burst of apparent noise which, when decoded, will contain twenty minutes of information.

standard time and frequency station (STFS)
: Transmits precise time signals on precise frequencies.

substitution system
: A system in which the elements of the plain or code text are replaced by other elements for the purpose of encryption.

sunspots
: Cooler spots on the surface of the sun which indirectly produce magnetic and radio propagation disturbances on the earth.

telecommunication
: Any transmission or reception of intelligence by wire, radio, optical, or other electromagnetic system.

telegram
: Written matter transmitted by telegraphy for delivery to the addressee.

telegraphy
: Any transmission of a signal which results in a printed symbol; normally confined to Morse code communications.

telemetry
: The transmission of remote-measuring data signals.

telephony
: The transmission of speech, or other sounds.

teleprinter
: An electrically-operated instrument used in the transmission and reception/printing of messages. Also called teletypewriter, radioprinter or teletype.

telex over radio (TOR)
: Error correcting radioteletype (variously FEC, ARQ, SITOR, AMTOR).

# GLOSSARY

| | |
|---|---|
| tempest | A declassified code name referring to investigations and studies of compromising emanations--incidental radiations from electronic equipment. |
| text | The part of a message containing the basic information which the originator desires to be communicated. |
| time division multiplexing (TDM) | A process by which two or more channels of data are transmitted over the same frequency band by allocating a different time interval for the transmission of each channel. |
| traffic (tfc) | Communications volume during a given period. |
| transcription | A written copy of a previously recorded radio transmission. |
| transmission | The purposeful radiation of a signal from a transmitter. |
| transmitter | Equipment used to generate and radiate a radio signal into space via an antenna. |
| Travellers Information Station (TIS) | A low powered, voice-tape transmission broadcasting to motorists noncommercial road information and directions. Operates on 530 or 1610 kHz. |
| trigraph | A set of three letters |
| trinome | A set of three digits. |
| triplet | A group of three like symbols. |
| tropical bands | The three lowest-frequency SW broadcast bands (2300-2498, 3200-3400, 4750-5060 kHz). |
| upper sideband (USB) | The higher frequency or group of frequencies produced by an amplitude-modulation process. |
| utility communications | Non-broadcast communications, usually |

| | |
|---|---|
| | two-way, and including: aircraft, beacons, experimental, point to point, governmental, land mobile, maritime/coastal, military, radio telephone, amateur, and space satellites. |
| very low frequency (VLF) | The spectrum from 3-30 kHz, but often extended by common usage to approximately 500 kHz. |
| voice mirror | Continuously repeated (recorded) voice identification of a radio circuit. |
| VOLMET | HF station sending regional weather information for large international airports. |
| World Administrative Radio Conference (WARC) | Held periodically by the International Telecommunications Union (ITU) in Geneva, Switzerland, as a means of cooperatively updating radio rules and regulations among member nations. |
| watt | The unit of power. |
| wave angle | The angle above the surface of the earth at which a signal is transmitted from or received by an antenna. |
| wavelength | The distance a radio wave, moving at the speed of light (186,000 miles per second) travels to complete one cycle; usually expressed in meters. |
| whip antenna | Flexible rod supported at one end. |
| working | Stations communicating with each other. In simplex working, stations operate on a common frequency; in complex working, more than one frequency is used. |
| zero beat | The condition between two signals indicating identical frequency; no beat note is heard or seen between the signals. |

237

# GLOSSARY

## Phonetic Alphabets

| | English* | French | German | Italian | Portuguese | Spanish |
|---|---|---|---|---|---|---|
| A | Alpha | Alfa | Anton | Alfa | Antena | Alfa |
| B | Bravo | Bravo | Berta | Bravo | Bateria | Brasil |
| C | Charlie | Charlie | Cäsar | Canada | Condensador | Canada |
| D | Delta | Delta | Dora | Delta | Detector | Delta |
| E | Echo | Echo | Emil | Europa | Estatico | España |
| F | Foxtrot | Foxtrot | Friedrich | Firenze | Filamento | Francia |
| G | Golf | Golf | Gustav | Guatemala | Grade | Guatemala |
| H | Hotel | Hotel | Heinrich | Hotel | Hotel | Hotel |
| I | India | India | Ida | Italia | Itensidade | Italia |
| J | Juliet | Juliett | Julius | Juventus | Juliete | Japón |
| K | Kilo | Kilo | Konrad | Kilometro | Kilo | Kilo |
| L | Lima | Lima | Ludwig | Lima | Lâmpada | Lima |
| M | Mike | Mike | Martha | Messico | Manipulador | Méjico |
| N | November | November | Nordpol | Novembre | Negativo | Noviembre |
| O | Oscar | Oscar | Otto | Otranto | Onda | Oscar |
| P | Papa | Papa | Paula | Palermo | Placa | Papa |
| Q | Quebec | Quebec | Quelle | Quebec | Quadro | Quito |
| R | Romeo | Romeo | Richard | Romeo | Rádio | Radio |
| S | Sierra | Sierra | Siegfried | Santiago | Sintonia | Santiago |
| T | Tango | Tango | Theodor | Tango | Terra | Tango |
| U | Uniform | Uniform | Ulrich | Università | Unidade | Universidad |
| V | Victor | Victor | Viktor | Venezia | Valvula | Victor |
| W | Whisky | Whiskey | Wilhelm | Whisky | Watt | Whisky |
| X | X-ray | X-ray | Xanthippe | Xilofono | Xilofono | Xilófono |
| Y | Yankee | Yankee | Ypsilon | Yokohama | Yucatan | Yucatán |
| Z | Zulu | Zulu | Zeppelin | Zelanda | Zulu | Zulu |

*This ICAO Phonetic Alphabet is also often used by people who do not speak English. It is intended to provide worldwide standardization similar to use of the Q-code. English is used as a worldwide common communications language or lingua franca.

# GLOSSARY

## Voice Procedure Signals

*Note: * indicates primarily military/government use*

| | |
|---|---|
| Acknowledge | I have received and understand your message. |
| Affirmative | Yes or correct. |
| All after* | Repeat portion of message that follows:"_____". |
| All before* | Repeat portion of message that precedes:"___". |
| Break | Pause, interruption. |
| Cambio | The Spanish equivalent of "over". |
| Clear | End of contact. |
| Come back | Used in place of OVER, primarily in CB and pleasure boat communications. |
| Copy | See Info |
| Correction* | Portion in error; correct version is:_____. |
| DNA | Did not answer (or does not answer). Used on Navy MARS phone patch. |
| Disregard this transmission | This transmission is in error. Not to be used to cancel a previous message that has been completed. |
| Do not answer* | Stations called are not to reply or acknowledge transmission, which is followed by OUT. |
| End of message | Message completed. |
| Execute* | Carry out a prescribed action. |
| Execute to follow* | Action on message to be carried out upon receipt of EXECUTE. |
| Figures* | Numbers or numerals follow. |
| Flash* | Highest level of message precedence. |
| From | Originator of message. |
| How copy? | Requesting acknowledgement or signal report. |
| How do you read me? | Signal strength/readability request. |
| Immediate* | High precedence traffic which must not be delayed. |
| Info* | Following addressee designations are addressed for information. |
| I read back* | Following is in response to your instructions to read back. |
| I say again* | I am repeating transmission. |
| I spell* | Next word will be spelled phonetically. |
| I verify* | Following message (or portion) has been verified at your request and is repeated. |
| Mayday | Urgent distress call indicating immediate threat to life. |
| Message follows* | Message upcoming. |
| More to follow | I have additional material which I will communicate to you momentarily. |
| Negative | No. |
| Number | Station serial number |
| Out | My transmission is finished; no reply required. |
| Over | My transmission is finished; reply requested. |
| Pan | In air intercept; calling station has urgent message concerning the safety of a ship, aircraft, vehicle, or some person on board or within sight. |
| Precedence | Prioritized levels of message handling rated by urgency. |
| Priority* | Message handling precedence which requires prompt delivery. |
| Read back* | Repeat my transmission back to me exactly as received. |
| Relay to* | Transmit message to addressee(s) as follows |
| Roger | I received your last transmission. |
| Routine* | Message handling precedence which does not require accelerated treatment. |
| Say again* | Repeat your transmission. |
| Silence* | Stop transmitting immediately. |
| Silence lifted* | Resume transmission. |
| Speak slower* | Reduce speed of transmission. |
| That is correct* | You are correct or what you have transmitted is correct. |

239

# GLOSSARY

| | |
|---|---|
| This is* | Transmission is from station whose callsign or ID follows. |
| Time* | Date/time group of message. |
| To* | Addressee(s) follow(s). |
| Unknown station* | Identity of station with whom I am attempting to contact is unknown. |
| Verify* | Verify message with originator and send correct version. |
| Wait* | Short pause. |
| Wait out* | I have to pause for more than a few seconds. |
| Wilco* | I will comply. |
| Word after* | What is the word following _____. |
| Word before* | What is the word that precedes_____. |
| Words twice* | Phrases or code groups will be repeated twice for clarification (due to poor reception conditions usually). |
| Wrong* | Your last transmission is incorrect. The correct version is_____. |

## CW Prosigns and Abbreviations

While Q signals expedite handling on CW networks, certain dot-dash groupings called prosigns are handy for signalling information quickly to operators. A bar over the letters indicates that there is no space between the letters; they are sent as a single dot-dash group.

| | |
|---|---|
| AA | Unknown Station |
| AA | All After |
| AA | Arabic language |
| AB | All Before |
| ADR | Address (or addressee) |
| AGN | Again |
| AR | End of Transmission |
| AS | Short Wait |
| B | More to Follow |
| BK | Break |
| BN | Between |
| BT | Long Break |
| C | Correct; Yes |
| CC | Chinese language |
| CFM | Confirm |
| CHARAC | Character(s) |
| CK | Sometimes used for group count (GR) |
| CL | Closing Station |
| CLG | Calling |
| COL | Collate |
| CQ | General Call to All Stations |
| CS | Callsigns |
| CT | Net Control Station (NCS) |
| CUD | Could |
| CUL | See you later (later schedule) |
| CUT NBRS | Abbreviated Morse character for numbers; letters are sent for numbers |
| DE | From |
| DIP | Diplomatic |
| dissem | Dissemination |
| DLD | Delivered |
| DR | Dear |
| DTOI | Date/Time of intercept |
| DTG | Date/time group |
| DX | Long distance communications (foreign countries) |
| E | East |
| EE | English language |
| EEEE... | Error |
| EMB | Embassy |
| ES | And |

| | |
|---|---|
| ETA | Estimated Time of Arrival |
| ETD | Estimated Time of Departure |
| F | Do Not Answer |
| FB | Fine business, excellent |
| FF | French language |
| FH | First heard |
| F, FIG | Figure(s) |
| FM | From |
| FOLL | Followed, Following |
| FREQ | Frequency |
| FSK | Frequency shift keying |
| G | Repeat Back |
| GA | Go Ahead |
| GB | Good-by |
| GBA | Give better address |
| GE | Good evening |
| GG | Going |
| GG | German language |
| GM | Good morning |
| GND | Ground |
| GR | Group Count |
| GRP | Group(s) |
| GUD | Good |
| HDNG | Heading |
| HI | The telegraphic laugh, high |
| HQ | Headquarters |
| HR | Here |
| HRD | Heard |
| HV | Have |
| HW | How |
| ID,IDENT | Identification |
| II | Separative Sign |
| IMI | Repeat or question |
| INFO | Information |
| INT | Interogatory |
| INTEL | Intelligence |
| J | Verify with Originator and Repeat |
| K | Invitation to Transmit |
| KA | Starting Signal |
| L, LTRS | Letter(s) |
| LA | Latin America(n) |
| LANG | Language |
| LID | A poor operator |
| M | Deferred |
| MA,MILS | Milliamperes |
| MFA | Ministry of Foreign Affairs |
| MIN | Minute(s) |
| MSG | Message |
| N | Negative; No |
| N | North |
| NCS | Net control station |
| ND | Nothing doing |
| NIL | I have nothing to send |
| NM | No more |

# GLOSSARY

| | | | | |
|---|---|---|---|---|
| NOTAM | Notice to all mariners; Notices to Airmen | | TOI | Time of intercept |
| NR | Number | | TT | That |
| NW | Now | | TU | Thank You |
| NX | News | | TVI | Television interference |
| O | Operational Immediate | | TX | Transmitter |
| OB | Old boy | | TXT | Text |
| OL | Ocean Letter | | UNID | Unidentified |
| OM | Old man | | UR | Your |
| OPR | Operator | | UTE | Utility transmission |
| OPS | Operations | | VE | Understood (sometimes used in place of BK) |
| OS | Out station | | VFO | Variable-frequency oscillator |
| OT | Old-timer, old top | | VY | Very |
| P | Private radiotelegram | | W | West |
| P | Priority | | W | Number of words in message (used in place of CK or GR) |
| PBL | Preamble | | WA | Word After |
| POSS | Possible, possibly | | WB | Word Before |
| PP | Portuguese language | | WD-WDS | Word; words |
| P/P | Phone patch | | WKD-WKG | Worked; working |
| PREV | Previous(ly) | | WL | Well, will |
| PROB | Probably | | WRKG | Working (or WKG) |
| PSE | Please | | WPM | Words per minute (Morse code speed) |
| PT | Plain text | | WUD | Would |
| PTP | Point to point | | WX | Weather Report |
| PWR | Power | | XCVR | Transceiver |
| R | Received as transmitted | | XMTR(TX) | Transmitter |
| RCD | Received | | XMSN | Transmission |
| RCVR | Receiver | | XTAL | Crystal |
| RDO | Radio | | XYL (YF) | Wife |
| REF | Reference | | Y | Emergency |
| RPT | Repeat; report | | YL | Young lady |
| RQ | Request | | Z | Flash |
| RR | Russian language | | 5F, 5L | 5-Figure, 5-Letter |
| RX | Receive (er, ing) | | 73 | Best regards |
| RY's | RYRYRY ....(Radioteletype test tape) | | 88 | Love and kisses |
| S | South | | | |
| SS | Spanish language | | | |
| SED | Said | | | |
| SIG | Signature | | | |
| SINE | Operator's nickname or initials | | | |
| SK | End | | | |
| SKED | Schedule | | | |
| SLT | Radiomarine Letter | | | |
| SPEC | Special | | | |
| SRI | Sorry | | | |
| STN | Station | | | |
| SVC | Service Telegram | | | |
| SWL | Short wave listener | | | |
| T | Station Called; Transmit to all Addressees; zero | | | |
| TFC | Traffic | | | |
| TKS/TNX | Thanks | | | |
| TMW | Tomorrow | | | |
| TO | Action Address | | | |

# The International Q-Code

To expedite traffic handling via Morse code, network operators have devised a system of "Q codes" which have specific meanings. The following abbreviated list will provide the listener with insight as to the meaning of signals copied during CW net operations.

| | |
|---|---|
| QNA | Answer in prearranged order. |
| QNB | All net stations copy. |
| QND | Net is directed. |
| QNE | Entire net stand by. |
| QNF | Net is not controlled. |
| QNG | Take over as net control station. |
| QNH | Your net frequency is high. |
| QNI | Net stations report in. |
| QNJ | Can you copy? |
| QNK | Transmit messages. |
| QNL | Your net frequency is low. |
| QNM | You are interfering; stand by. |
| QNN | Net control station is... |
| QNO | Station is leaving the net. |
| QNP | Unable to copy. |
| QNQ | Move frequency to finish traffic. |
| QNS | Following stations are in the net:... |
| QNT | Request permission to leave the net. |
| QNU | The net has traffic for you. |
| QNV | Establish contact on this frequency. |
| QNW | How do I route messages? |
| QNX | You are excused from the net. |
| QNY | Shift to another frequency to clear traffic. |
| QNZ | Zero beat your signal with mine. |
| QRA | What is the name of your station? |
| QRE | What is your estimated time of arrival? |
| QRG | Will you tell me the exact frequency? |
| QRH | Does my frequency vary? |
| QRI | How is the tone of my transmission? (1)Good, (3)Bad |
| QRK | What is the intelligibility of my signals? (1)Bad, (5)Excellent. |
| QRL | Are you busy? |
| QRM | Is my transmission being interfered with? (1)No, (5)Extreme. |
| QRN | Are you troubled by static? (1)No (5)Extreme. |
| QRO | Shall I increase transmitter power? |
| QRP | Shall I decrease transmitter power? |
| QRQ | Shall I send faster? |
| QRR | Are you ready for automatic operation? |

| | |
|---|---|
| QRRR | EMERGENCY (Amateur only) |
| QRS | Shall I send more slowly? |
| QRT | Shall I stop sending? |
| QRU | Have you anything for me? |
| QRV | Are you ready? |
| QRW | Shall I inform...that you are calling him? |
| QRX | When will you call me again? |
| QRY | When is my turn? |
| QRZ | Who is calling me? |
| QSA | What is my signal strength? |
| QSB | Are my signals fading? |
| QSK | Can you hear me if I break in on your transmission? |
| QSL | Can you acknowledge receipt? |
| QSM | Shall I repeat the last telegram? |
| QSN | Did you hear me? |
| QSO | Can you communicate with...? |
| QSP | Will you relay? |
| QSR | Shall I repeat the call? |
| QSS | What frequency will you use? |
| QSU | Shall I send on this frequency? |
| QSV | Shall I send a series of Vs for adjustment? |
| QSW | Will you send on this frequency? |
| QSX | Will you listen? |
| QSY | Shall I change frequency? |
| QSZ | Shall I send each word or group more than once? |
| QTA | Shall I cancel telegram number...? |
| QTB | Do you agree with my counting of words? |
| QTC | I have...telegrams for you (or for...) |
| QTH | My position is... |
| QST | General call to all amateurs. |

## Jamming

At this writing, several types of intentional interference are used to discourage listening. In most instances the effort is political, designed to irritate the listener so that he will not monitor a broadcast. Jamming transmitters may be classified as:

| | |
|---|---|
| SPARK | Broadband electrical pulse noise capable of wiping out a wide spectrum (sounds like interference from a brush-type motor) |
| SWEEPER | A strong carrier, swept back and forth across the signal. When swept at an audio rate, sounds like a diesel engine. |
| STEP TONE | A constant series of audio tones, usually 3-5, randomly repeated on the signal (sounds like bagpipes) |
| NOISE | White noise transmitted over the jammed signal (sounds like a loud background hiss on a receiver) |

### Broadcast Jammers

For many years Communist bloc countries deliberately jammed transmissions from the West to prevent or discourage reception by their citizens. Although now discontinued, a 1988 multinational effort coordinated by the International Telecommunications Union (ITU) successfully located those jammers which transmitted two-character identifiers every few seconds (ostensibly to determine their effectiveness as reported by monitors).

| Jammer CW ID | Jammer Location |
|---|---|
| A5 | Bulgaria |
| AL | USSR |
| AR | USSR |
| B1 | Czechoslovakia |
| BF | USSR |
| BN | USSR |
| BS | USSR |
| CB | USSR |
| DK | USSR |
| DP | USSR |
| DW | USSR |
| FG | USSR |
| FI | USSR |
| FL | USSR |
| G3 | Bulgaria |
| GA | USSR |
| GD | USSR |
| GF | USSR |
| GL | USSR |
| GS | USSR |
| GU | USSR |
| GV | USSR |
| HD | USSR |
| HP | USSR |
| IL | USSR |
| IR | USSR |
| KM | USSR |
| KU | USSR |
| KV | USSR |
| L4 | Bulgaria |
| LK | USSR |
| LR | USSR |
| LU | USSR |
| M3 | USSR |
| MB | USSR |
| MG | USSR |
| ML | USSR |
| MU | USSR |
| MX | USSR |
| NA | USSR |
| ND | USSR |
| NI | USSR |
| NK | USSR |
| NU | USSR |
| PF | USSR |
| R9 | Czechoslovakia |
| RA | USSR |
| RD | USSR |
| RP | USSR |
| RQ | USSR |
| RT | USSR |
| S5 | Czechoslovakia |
| SF | USSR |
| SU | USSR |
| TF | USSR |
| TK | USSR |
| TU | USSR |
| U7 | Czechoslovakia |
| UA | USSR |
| UD | USSR |
| UR | USSR |
| VL | USSR |
| VN | USSR |
| WA | USSR |
| WQ | USSR |
| WU | USSR |
| WV | USSR |
| XD | USSR |
| Z3 | Czechoslovakia |
| ZA | USSR |
| ZK | USSR |
| ZN | USSR |
| ZT | USSR |

## ...and Other Irritating Interference

The "Russian Woodpeckers" have been around for several years now. Generating incredibly strong pulses, these signals emanate from the USSR and are capable of detecting the intrusion of enemy aircraft into Russian air space. They are used as a form of Over the Horizon Backscatter Radar (OTHB).

The United States Air Force will soon begin transmitting a similar OTHB from Hanscom AFB, Massachusetts.

Occasional industrial, military and scientific frequency sweepers and ionosondes will be heard from time to time due to the long distance propagation qualities of the high frequency spectrum.

# GLOSSARY

## International Callsign Prefixes

| Callsign Prefixes | Country |
|---|---|
| AAA-ALZ | UNITED STATES |
| AMA-AOZ | SPAIN |
| APA-ASZ | PAKISTAN |
| ATA-AWZ | INDIA |
| AXA-AXZ | AUSTRALIA |
| AYA-AZZ | ARGENTINA |
| A2A-A2Z | BOTSWANA |
| A3A-A3Z | TONGA |
| A4A-A4Z | OMAN |
| A5A-A5Z | BHUTAN |
| A6A-A6Z | UNION OF ARAB EMIRATES |
| A7A-A7Z | QATAR |
| A9C-A9Z | BAHRAIN |
| BAA-BZZ | CHINA |
| BVA-BVZ | TAIWAN |
| CAA-CEZ | CHILE |
| CFA-CKZ | CANADA |
| CLA-CMZ | CUBA |
| CNA-CNZ | MOROCCO |
| COA-COZ | CUBA |
| CPA-CPZ | BOLIVIA |
| CQA-CUZ | PORTUGAL |
| CU | CUBA |
| CSA-CSZ | PORTUGAL |
| CVA-CXZ | URUGUAY |
| CYA-CZZ | CANADA |
| C2A-C2Z | NAURU |
| C3A-C3Z | ANDORRA |
| C4A-C4Z | CYPRUS |
| C5A-C5Z | GAMBIA |
| C6A-C6Z | BAHAMAS |
| C7A-C7Z | WORLD METEOROLOGICAL ORGANIZATION |
| C8A-C9Z | MOZAMBIQUE |
| DAA-DRZ | WEST GERMANY |
| DSA-DTZ | SOUTH KOREA |
| DUA-DZZ | PHILIPPINES |
| D2A-D3Z | ANGOLA |
| D4A-D4Z | CAPE VERDE IS. |
| D5A-D5Z | LIBERIA |
| D6A-D6Z | COMOROS |
| D7A-D9Z | SOUTH KOREA (REP) |
| EAA-EHZ | SPAIN |
| EIA-EJZ | IRELAND |
| EKA-EKZ | USSR |
| ELA-ELZ | LIBERIA |
| EMA-EOZ | USSR |
| EPA-EQZ | IRAN |
| ERA-ERZ | USSR |
| ESA-ESZ | ESTONIA |
| ETA-ETZ | ETHIOPIA |
| EUA-EWZ | BIELORUSSIA SSR |
| EXA-EZZ | USSR |
| FAA-FZZ | FRANCE |
| GAA-GZZ | UNITED KINGDOM |
| HAA-HAZ | HUNGARY |
| HBA-HBZ | SWITZERLAND & LIECHTEN-STEIN |
| HCA-HDZ | ECUADOR |
| HEA-HEZ | SWITZERLAND |
| HFA-HFZ | POLAND |
| HGA-HGZ | HUNGARY |
| HHA-HHZ | HAITI |
| HIA-HIZ | DOMINICAN REPUBLIC |
| HJA-HKZ | COLOMBIA |
| HLA-HLZ | SOUTH KOREA |
| HMA-HMZ | NORTH KOREA (DEM.P.R.) |
| HNA-HNZ | IRAQ |
| HOA-HPZ | PANAMA |
| HQA-HRZ | HONDURAS |
| HSA-HSZ | THAILAND |
| HTA-HTZ | NICARAGUA |
| HUA-HUZ | EL SALVADOR |
| HVA-HVZ | VATICAN CITY |
| HWA-HYZ | FRANCE |
| HZA-HZZ | SAUDI ARABIA |
| H2A-H2Z | CYPRUS |
| H3A-H3Z | PANAMA |
| H4A-H4Z | SOLOMON ISLANDS |
| H6A-H7Z | NICARAGUA |
| H8A-H9Z | PANAMA |
| IAA-IZZ | ITALY |
| JAA-JSZ | JAPAN |
| JTA-JVZ | MONGOLIA |
| JWA-JXZ | NORWAY |
| JYA-JYZ | JORDAN |
| JZA-JZZ | WESTERN NEW GUINEA |
| J2A-J2Z | DJIBOUTI |
| J3A-J3Z | GRENADA |
| J4A-J4Z | GREECE |
| J5A-J5Z | GUINEA-BISSAU |
| J6A-J6Z | SAINT LUCIA |
| J7A-J7Z | DOMINICA |
| J8A-K8Z | ST. VINCENT & GRENADIE |
| KAA-KZZ | UNITED STATES |
| LAA-LNZ | NORWAY |
| LOA-LWZ | ARGENTINA |
| LXA-LXZ | LUXEMBOURG |
| LYA-LYZ | LITHUANIA |
| LZA-LZZ | BULGARIA |
| L2A-L9Z | ARGENTINA |

# GLOSSARY

| | |
|---|---|
| MAA-MZZ | UNITED KINGDOM |
| NAA-NZZ | UNITED STATES |
| OAA-OCZ | PERU |
| ODA-ODZ | LEBANON |
| OEA-OEZ | AUSTRIA |
| OFA-OJZ | FINLAND |
| OKA-OMZ | CZECHOSLOVAKIA |
| ONA-OTZ | BELGIUM |
| OUA-OZZ | DENMARK |
| PAA-PIZ | NETHERLANDS |
| PJA-PJZ | NETHERLANDS ANTILLES |
| PKA-POZ | INDONESIA |
| PPA-PYZ | BRAZIL |
| PZA-PZZ | SURINAME |
| P2A-P2Z | PAPUA NEW GUINEA |
| P3A-P3Z | CYPRUS |
| P4A-P4Z | NETHERLANDS ANTILLES |
| P5A-P9Z | NORTH KOREA (DEM.P.R.) |
| QAA-QZZ | INT'L MORSE ABBREVIA-TIONS |
| RAA-RZZ | USSR |
| SAA-SMZ | SWEDEN |
| SNA-SRZ | POLAND |
| SSA-SSM | EGYPT |
| SSN-STZ | SUDAN |
| SUA-SUZ | EGYPT |
| SVA-SZZ | GREECE & ISLANDS |
| S2A-S3Z | BANGLADESH |
| S6A-S6Z | SINGAPORE |
| S7A-S7Z | SEYCHELLES |
| S8A-S9Z | SAO TOME & PRINCIPE |
| TAA-TCZ | TURKEY |
| TDA-TDZ | GUATEMALA |
| TEA-TEZ | COSTA RICA |
| TFA-TFZ | ICELAND |
| TGA-TGZ | GUATEMALA |
| THA-THZ | FRANCE |
| TIA-TIZ | COSTA RICA |
| TJA-TJZ | CAMEROON |
| TKA-TKZ | FRANCE |
| TLA-TLZ | CENTRAL AFRICAN REP. |
| TMA-TMZ | FRANCE |
| TNA-TNZ | CONGO |
| TOA-TQZ | FRANCE |
| TRA-TRZ | GABON REPUBLIC |
| TSA-TSZ | TUNISIA |
| TTA-TTZ | CHAD REPUBLIC |
| TUA-TUZ | IVORY COAST |
| TVA-TXZ | FRANCE |
| TYA-TYZ | BENIN |
| TZA-TZZ | MALI REPUBLIC |
| T2A-T2Z | TUVALU |
| T3A-T3Z | KIRIBATI |
| T4A-T4Z | CUBA |
| T5A-T5Z | SOMALI DEM. REP. |
| T6A-T6Z | AFGHANISTAN |
| T7A-T7Z | SAN MARINO |
| UAA-UQZ | USSR |
| URA-UTZ | UKRAINE |
| VAA-VGZ | CANADA |
| VHA-VNZ | AUSTRALIA |
| VOA-VOZ | CANADA |
| VPA-VSZ | UNITED KINGDOM |
| VTA-VWZ | INDIA |
| VXA-VYZ | CANADA |
| VZA-VZZ | AUSTRALIA |
| V2A-V2Z | ANTIGUA |
| V3A-V3Z | BELIZE |
| V4A-V4Z | ST. CHRISTOPHER & NEVIS |
| V6 | FEDERATED STATES OF MICRONESIA |
| V7 | REPUBLIC OF THE MARSHALL ISLANDS |
| V8A-V8Z | BRUNEI |
| WAA-WZZ | UNITED STATES |
| XAA-XIZ | MEXICO |
| XJA-XOZ | CANADA |
| XPA-XPZ | DENMARK |
| XQA-XRZ | CHILE |
| XSA-XSZ | CHINA |
| XTA-XTZ | BOURKINA FASO |
| XUA-XUZ | CAMBODIA |
| XVA-XVZ | VIET NAM |
| XWA-XWZ | LAOS |
| XXA-XXZ | PORTUGAL |
| XYA-XZZ | BURMA |
| YAA-YAZ | AFGHANISTAN |
| YBA-YHZ | INDONESIA |
| YIA-YIZ | IRAQ |
| YJA-YJZ | VANUATU (NEW HEBRIDES) |
| YKA-YKZ | SYRIA |
| YLA-YLZ | LATVIA |
| YMA-YMZ | TURKEY |
| YNA-YNZ | NICARAGUA |
| YOA-YRZ | RUMANIA |
| YSA-YSZ | EL SALVADOR |
| YTA-YUZ | YUGOSLAVIA |
| YVA-YYZ | VENEZUELA |
| YZA-YZZ | YUGOSLAVIA |
| Y2A-Y9Z | EAST GERMANY (GDR) |
| ZAA-ZAZ | ALBANIA |
| ZBA-ZJZ | UNITED KINGDOM |
| ZKA-ZMZ | NEW ZEALAND |
| ZNA-ZOZ | UNITED KINGDOM |
| ZPA-ZPZ | PARAGUAY |
| ZQA-ZQZ | UNITED KINGDOM |
| ZRA-ZUZ | SOUTH AFRICA |
| ZVA-ZZZ | BRAZIL |

# GLOSSARY

| | | | |
|---|---|---|---|
| Z2A-Z2Z | ZIMBABWE | 6ZA-6ZZ | LIBERIA |
| 2AA-2ZZ | UNITED KINGDOM | 7AA-7IZ | INDONESIA |
| 3AA-3AZ | MONACO | 7JA-7NZ | JAPAN |
| 3BA-3BZ | MAURITIUS | 7OZ-7OZ | YEMEN (DEM.P.R.) |
| 3CA-3CZ | EQUATORIAL GUINEA | 7PA-7PZ | LESOTHO |
| 3DA-3DM | SWAZILAND | 7QA-7QZ | MALAWI |
| 3DN-3DZ | FIJI | 7RA-7RZ | ALGERIA |
| 3EA-3FZ | PANAMA | 7SA-7SZ | SWEDEN |
| 3GA-3GZ | CHILE | 7TA-TYZ | ALGERIA |
| 3HA-EUZ | CHINA | 7ZA-7ZZ | SAUDI ARABIA |
| 3VA-3VZ | TUNISIA | 8AA-8IZ | INDONESIA |
| 3WA-3WZ | VIET NAM | 8JA-8NZ | JAPAN |
| 3XA-3XZ | GUINEA | 8OA-8OZ | BOTSWANA |
| 3YA-3YZ | NORWAY | 8PA-8PZ | BARBADOS |
| 3ZA-3ZZ | POLAND | 8QA-8QZ | MALDIVE ISLANDS |
| 4AA-4CZ | MEXICO | 8RA-8RZ | GUYANA |
| 4DA-4IZ | PHILIPPINES | 8SA-8SZ | SWEDEN |
| 4JA-4LZ | USSR | 8TA-8YZ | INDIA |
| 4MA-4MZ | VENEZUELA | 8ZA-8ZZ | SAUDI ARABIA |
| 4NA-4OZ | YUGOSLAVIA | 9AA-9AZ | SAN MARINO |
| 4PA-4SZ | SRI LANKA | 9BA-8DZ | IRAN |
| 4TA-4TZ | PERU | 9EA-9FZ | ETHIOPIA |
| 4UA-4UZ | UNITED NATIONS | 9GA-9GZ | GHANA |
| 4VA-4VZ | HAITI | 9HA-9HZ | MALTA |
| 4WA-4WZ | YEMEN | 9IA-9JZ | ZAMBIA |
| 4XA-4XZ | ISRAEL | 9KA-9KZ | KUWAIT |
| 4YA-4YZ | INT'L CIVIL AVIATION ORGANIZATION | 9LA-9LZ | SIERRA LEONE |
| | | 9MA-9MZ | MALAYSIA |
| 4ZAQ-4ZZ | ISRAEL | 9NA-9NZ | NEPAL |
| 5AA-5AZ | LIBYA | 9OA-9TZ | ZAIRE |
| 5BA-5BZ | CYPRUS | 9UA-9UZ | BURUNDI |
| 5CA-5GZ | MOROCCO | 9VA-9VZ | SINGAPORE |
| 5HA-5IZ | TANZANIA | 9WA-9WZ | MALAYSIA |
| 5JA-5KZ | COLOMBIA | 9XA-9XZ | RWANDA |
| 5LA-5MZ | LIBERIA | 9YA-9ZZ | TRINIDAD & TOBAGO |
| 5NA-5OZ | NIGERIA | | |
| 5PA-5QZ | DENMARK | | |
| 5RA-5SZ | MADAGASCAR | | |
| 5TA-5TZ | MAURITANIA | | |
| 5UA-5UZ | NIGER | | |
| 5VA-5VZ | TOGO | | |
| 5WA-5WZ | WESTERN SAMOA | | |
| 5XA-5XZ | UGANDA | | |
| 5YA-5ZZ | KENYA | | |
| 6AA-6BZ | EGYPT | | |
| 6CA-6CZ | SYRIA | | |
| 6DA-6JZ | MEXICO | | |
| 6KA-6NZ | KOREA (REPUBLIC) | | |
| 6OA-6OZ | SOMALIA | | |
| 6PA-6SZ | PAKISTAN | | |
| 6TA-6UZ | SUDAN | | |
| 6VA-6WZ | SENEGAL | | |
| 6XA-YXZ | MALAGASY | | |
| 6YA-6YZ | JAMAICA | | |

## U.S. Call Sign Identification

| Call Sign | Service |
|---|---|
| KAA-KZZ, WAA-WZZ | Public coastal station |
| KAA20-KZZ99, WAA20-WZZ99 | Fixed station |
| KAA200-KZZ999, WAA200-WZZ999 | Limited coastal station; land mobile base |
| KAAA-KZZZZ, WAAA-WZZZ | Broadcasting; ship telegraph & telephone |
| KAAA2-KZZZ9, WAAA2-WZZZ9 | Land mobile telegraph disaster station |
| WA2000-WZ9999 | Ship telephone only |
| KA2XAA-KZ9XZZ, WA2XAA-WZ9XZZ | Experimental |
| KAAA20-KZZZ99, WAAA20-WZZZ99 | Ship survival craft |
| KA2000-KZ9999, WA2000-WZ999 | Marine utility; land mobile telephone |
| KAAAA-KZZZZ, WAAAA-WZZZZ | Aircraft telegraph & telephone |
| KAA2-WZZ9, WAA2-WZZ9 | Aeronautical |
| A1AA-A0ZZ, K1AA-K0ZZ | Amateur |
| N1AA-N0ZZ, W1AA-W0ZZ | Amateur |
| KA1AA-KZ0ZZ | Amateur |
| WA1AA-KZ0ZZ | Amateur |

## U.S. Band Plan Primary Allocations

| Frequency | Service |
|---|---|
| 1600-1800 | Radionavigation |
| 1800-2000 | Amateur |
| 2000-2065 | Fixed/mobile |
| 2065-2107 | Maritime mobile |
| 2107-2170 | Maritime mobile |
| 2173-2190 | Mobile |
| 2190-2194 | Maritime mobile |
| 2194-2500 | Fixed/mobile |
| 2500-2850 | Fixed/mobile |
| 2850-3155 | Aeronautical mobile |
| 3155-3400 | Fixed/mobile |
| 3400-3500 | Aeronautical mobile |
| 3500-4000 | Amateur |
| 4000-4063 | Fixed |
| 4063-4438 | Maritime/mobile |
| 4438-4650 | Fixed/aero/mobile |
| 4650-4750 | Aero mobile |
| 4750-4850 | Fixed |
| 4850-5000 | Fixed/mobile |
| 5000-5250 | Fixed |
| 5250-5450 | Fixed/mobile |
| 5450-5730 | Aeronautical mobile |
| 5730-5950 | Fixed |
| 5950-6200 | Broadcasting |
| 6200-6525 | Maritime mobile |
| 6525-6765 | Aeronautical mobile |
| 6765-7000 | Fixed |
| 7000-7300 | Amateur |
| 7300-8195 | Fixed |
| 8195-8815 | Maritime mobile |
| 8815-9040 | Aero mobile |
| 9040-9500 | Fixed |
| 9500-9775 | Broadcasting |
| 9775-10000 | Fixed |
| 10000-10100 | Aero mobile |
| 10100-11175 | Fixed |
| 11175-11400 | Aero mobile |
| 11400-11700 | Fixed |
| 11700-11975 | Broadcasting |
| 11975-12330 | Fixed |
| 12330-13200 | Maritime mobile |
| 13200-13360 | Aero mobile |
| 13360-14000 | Fixed |
| 14000-14350 | Amateur |
| 14350-15000 | Fixed |
| 15010-15100 | Aero mobile |
| 15100-15450 | Broadcasting |
| 15450-16460 | Fixed |
| 16460-17360 | Maritime mobile |
| 17360-17700 | Fixed |
| 17700-17900 | Broadcasting |
| 17900-18030 | Aero mobile |

| | | |
|---|---|---|
| | 18030-21000 | Fixed |
| | 21000-21450 | Amateur |
| | 21450-21650 | Broadcasting |
| | 21750-21850 | Fixed |
| | 21850-21870 | Radio astronomy |
| | 21870-22000 | Aero fixed/mobile |
| | 22000-22720 | Maritime mobile |
| | 22720-23200 | Fixed |
| | 23200-23350 | Aero fixed/mobile |
| | 23350-25000 | Fixed/mobile |
| | 25000-25070 | Land mobile |
| | 25070-25110 | Maritime mobile |
| | 25110-25330 | Land mobile |
| | 25330-25600 | Fixed/mobile |
| | 25600-26100 | Broadcasting |
| | 26100-26480 | Land mobile |
| | 26480-26950 | Fixed/mobile |
| | 26965-27405 | CB |
| | 27405-28000 | Fixed/mobile |
| | 28000-29700 | Amateur |
| | 29700-30000 | Fixed/mobile |

## Emission Classifications

A shorthand method of indicating the mode of transmission has been adopted universally. The three characters indicate basic modulation type, nature of the modulation, and nature of the information.

| | |
|---|---|
| N0N | Unmodulated carrier |
| A1A | Telegraphy by on-off keying (OOK), for aural reception |
| A1B | Telegraphy by on-off keying for automatic reception |
| A2A | Telegraphy by tone-modulation of continuous carrier (MCW) |
| J2A | Telegraphy by tone-modulation of SSB system |
| A3E | Conventional, double-sideband, amplitude modulation |
| H3E | Single sideband, full carrier |
| R3E | Single sideband, reduced carrier (usually 6 dB) |
| J3E | Single sideband, suppressed carrier (USB or LSB) |
| B8E | Independent sideband, analog content (SWBC feeders) |
| B9W | Independent sideband, data or telegraphy in one channel |
| F1B | Radioteletype using frequency-shift keying |
| F1D | Data communication using FSK |
| F3E | Frequency modulated radiotelephone |
| G3E | Phase modulated radiotelephone (sounds the same) |
| J2B | RTTY using audio FSK to SSB (sounds the same) |
| J2C | Quantized (digital) slow-scan television or FAX |
| J3C | Standard, analog, slow-scan television or FAX |
| PON | Pulse |

## Land/Mobile AM/FM Allocations

20 kHz spacing:
    25110-25600   26100-26950   27290-28000

10 kHz spacing:
    29700-29890   29910-30000

PETROLEUM INDUSTRY (FM)
    25020-25320 (20 kHz channel spacing)

RADIO/TV REMOTE BROADCAST
    25870-26470 (40 kHz channel spacing)

FORESTRY NETWORKS (FM)
    29710-29790 (20 kHz channel spacing)

BUSINESS (FM)
    27410-27530 (20 kHz channel spacing)

FEDERAL GOVERNMENT
    27575 27585 (4 watt mobile/walkie-talkie)

# FREQUENCY CROSS REFERENCE

The following exhaustive list of frequencies provides immediate identification of service. For additional information, consult the referenced service or agency as found in the Table of Contents. The cross reference is representative but does not comprehensively list all frequencies found in the book.

Since smuggling, spy numbers, CW beacons, and CB outbander frequencies are irregular and unassigned, they are not included in the frequency list. For the same reason, amateur bands and international radio broadcasters are also not represented in the cross-reference, although you will find they receive thorough treatment in the body of the book.

## Cross Reference Key

### Frequency Code:

*The first letter group indicates the service - the second is the country.* *See the table of contents for specific identifications of the frequency code.*

### Service:

| | | | | | | | |
|---|---|---|---|---|---|---|---|
| A | Aircraft | BC | Broadcasting | G | Government | N | Navy |
| AF | Air Force | CC | Common Carrier | LW | Longwave | PR | Private, Bus., Scient. |
| AR | Army | CG | Coast Guard | M | Maritime | PS | Public Safety |
| | | | | | | SP | Space |

### Country:

| | | | | | | |
|---|---|---|---|---|---|---|
| AFS | REPUBLIC OF SOUTH AFRICA | G | UNITED KINGDOM | NOR | NORWAY |
| | | GHA | GHANA | NZL | NEW ZEALAND |
| AGL | ANGOLA | GRC | GREECE | P | PORTUGAL |
| ALG | ALGERIA | GTM | GUATEMALA | PAK | PAKISTAN |
| ARG | ARGENTINA | GUM | GUAM | PHL | PHILIPPINES |
| ARS | SAUDI ARABIA | HNB | BELIZE | PNR | PANAMA |
| ATN | NETHERLANDS ANTILLES | HND | HONDURAS | POL | POLAND |
| AUS | AUSTRALIA | HNG | HUNGARY | PTR | PUERTO RICO (ALSO PR) |
| AUT | AUSTRIA | HOL | NETHERLANDS | ROU | ROUMANIA |
| B | BRAZIL | HTI | HAITI | S | SWEDEN |
| BAH | BAHAMAS | I | ITALY | SAU | SAUDI ARABIA |
| BEL | BELGIUM | IND | REPUBLIC OF INDIA | SEN | SENEGAL |
| BUL | BULGARIA | INS | INDONESIA | SLV | EL SALVADOR |
| CAN | CANADA | IRL | IRELAND | SMA | AMERICAN SAMOA |
| CHL | CHILE | IRN | IRAN | SUI | SWITZERLAND |
| CHN | CHINA | IRQ | IRAQ | TAI | TAIWAN |
| CLM | COLOMBIA | ISR | ISRAEL | TCH | CZECHOSLOVAKIA |
| COG | CONGO | J | JAPAN | THA | THAILAND |
| CUB | CUBA | JOR | JORDAN | TUN | TUNISIA |
| D | WEST GERMANY (FED REP) | KOR | REPUBLIC OF KOREA | TUR | TURKEY |
| | | KRE | DEM PEOPLE'S REP OF KOREA | UAE | UNITED ARAB EMIRATES |
| DDR | EAST GERMANY (DEM REP) | | | URS | USSR |
| DJ | DJIBOUTI | KWT | KUWAIT | US | UNITED STATES |
| DNK | DENMARK | LBY | LIBYA | VEN | VENEZUELA |
| DOM | DOMINICAN REPUBLIC | M | MEXICO | VTN | VIET NAM |
| E | SPAIN | MAR | MARTINIQUE | YUG | YUGOSLAVIA |
| EGY | EGYPT | MLA | MALASIA | | |
| EQA | ECUADOR | MNG | MONGOLIA | | **Areas:** |
| ETH | ETHIOPIA | MRC | MOROCCO | SAM | SOUTH AMERICA |
| F | FRANCE | NCG | NICARAGUA | NAM | NORTH AMERICA |
| FIN | FINLAND | NGR | NIGER | WW | WORLDWIDE |
| | | NIG | NIGERIA | CRIB | CARIBBEAN |

| Freq | Desig | Freq | Desig | Freq | Desig | Freq | Desig | Freq | Desig |
|---|---|---|---|---|---|---|---|---|---|
| 0.0760 | LW-WW-1 | 73.6000 | LW-WW-1 | 170.6250 | AF-US-5 | 1640.3000 | M-US-1 | 1798.5000 | M-US-1 |
| 10.0000 | G-WW-3 | 75.0000 | G-WW-3 | 171.8750 | AF-US-5 | 1643.0000 | M-US-1 | 1798.5000 | M-US-1 |
| 10.2000 | LW-WW-1 | 75.0000 | LW-WW-1 | 173.1250 | AF-US-5 | 1643.0000 | CC-US-2 | 1799.6000 | M-US-1 |
| 11.0500 | LW-WW-1 | 76.2000 | LW-WW-1 | 174.0000 | LW-WW-1 | 1643.7000 | M-US-1 | 1827.0000 | M-WW-2 |
| 11.3300 | LW-WW-1 | 77.1500 | LW-WW-1 | 177.0000 | G-WW-3 | 1644.0000 | M-CAN-1 | 1841.0000 | M-WW-2 |
| 11.8000 | LW-WW-1 | 77.5000 | G-WW-3 | 183.0000 | LW-WW-1 | 1645.0000 | M-CAN-1 | 1919.5000 | M-US-3 |
| 11.9050 | LW-WW-1 | 77.5000 | LW-WW-1 | 185.0000 | G-WW-3 | 1645.0000 | M-US-2 | 1972.5000 | M-US-3 |
| 12.0000 | LW-WW-1 | 80.0500 | LW-WW-1 | 187.0000 | LW-WW-1 | 1646.0000 | BC-US-1 | 2001.0000 | AR-US-3 |
| 12.1000 | LW-WW-1 | 81.0000 | LW-WW-1 | 193.0000 | LW-WW-1 | 1646.0000 | CC-US-2 | 2003.0000 | CG-US-8 |
| 12.3000 | LW-WW-1 | 82.7500 | LW-WW-1 | 197.5000 | LW-WW-1 | 1646.7000 | M-US-1 | 2003.0000 | CG-CAN |
| 12.6490 | LW-WW-1 | 83.1000 | LW-WW-1 | 201.0000 | G-CAN-2 | 1646.7000 | M-CAN-1 | 2003.0000 | M-US-3 |
| 12.8000 | LW-WW-1 | 83.8000 | LW-WW-1 | 202.0000 | G-CAN-2 | 1647.7000 | M-US-1 | 2003.0000 | PR-US-6 |
| 12.9000 | LW-WW-1 | 88.0000 | G-WW-3 | 315.0000 | G-CAN-2 | 1648.0000 | M-US-1 | 2003.0000 | CC-US-2 |
| 13.0000 | LW-WW-1 | 88.0000 | LW-WW-1 | 316.0000 | G-CAN-2 | 1648.0000 | M-CAN-1 | 2005.0000 | PR-US-6 |
| 13.1000 | LW-WW-1 | 91.1500 | LW-WW-1 | 317.0000 | G-CAN-2 | 1649.0000 | M-US-1 | 2006.0000 | N-AFS |
| 13.6000 | LW-WW-1 | 100.0000 | LW-WW-1 | 318.0000 | G-CAN-2 | 1649.0000 | CC-US-2 | 2006.0000 | PR-US-6 |
| 14.8810 | LW-WW-1 | 106.7000 | LW-WW-1 | 319.0000 | G-CAN-2 | 1649.7000 | M-US-1 | 2006.0000 | CC-US-2 |
| 15.1000 | LW-WW-1 | 110.0000 | LW-WW-1 | 334.0000 | G-CAN-2 | 1650.0000 | M-US-2 | 2009.0000 | M-US-3 |
| 16.0000 | G-WW-3 | 110.6000 | LW-WW-1 | 416.0000 | CG-CAN | 1652.0000 | CC-US-2 | 2010.0000 | AF-US-3 |
| 16.0000 | LW-WW-1 | 111.3000 | LW-WW-1 | 420.0000 | CG-CAN | 1652.7000 | M-US-1 | 2020.0000 | N-AFS |
| 16.4000 | LW-WW-1 | 111.8000 | LW-WW-1 | 427.0000 | CG-CAN | 1653.0000 | M-US-1 | 2020.0000 | PS-AUS |
| 16.7000 | LW-WW-1 | 112.1500 | LW-WW-1 | 430.0000 | CG-CAN | 1655.0000 | M-US-2 | 2025.0000 | N-US-3 |
| 16.8000 | LW-WW-1 | 113.2000 | LW-WW-1 | 432.0000 | CG-CAN | 1657.0000 | CC-US-2 | 2030.0000 | CC-CAN |
| 17.1000 | LW-WW-1 | 117.4000 | LW-WW-1 | 434.0000 | CG-CAN | 1658.0000 | M-US-1 | 2031.0000 | PR-US-6 |
| 17.4000 | LW-WW-1 | 119.0000 | LW-WW-1 | 435.0000 | G-WW-3 | 1658.4250 | M-US-1 | 2031.5000 | M-US-3 |
| 18.5000 | LW-WW-1 | 119.1500 | LW-WW-1 | 440.0000 | CG-CAN | 1660.0000 | M-US-1 | 2037.5000 | M-US-3 |
| 19.0000 | LW-WW-1 | 119.8500 | LW-WW-1 | 444.0000 | CG-CAN | 1660.0000 | CC-US-2 | 2040.0000 | CG-CAN |
| 19.6000 | LW-WW-1 | 120.0000 | LW-WW-1 | 444.3000 | M-ARG | 1660.0150 | M-US-1 | 2049.0000 | AF-US-8 |
| 20.2700 | LW-WW-1 | 122.2000 | LW-WW-1 | 446.0000 | CG-CAN | 1662.0000 | M-US-2 | 2050.0000 | N-J |
| 20.5000 | LW-WW-1 | 122.3000 | LW-WW-1 | 448.0000 | CG-CAN | 1665.0000 | M-US-2 | 2054.0000 | AF-CAN |
| 21.1000 | LW-WW-1 | 122.5000 | AF-CAN | 450.0000 | CG-CAN | 1670.0000 | M-US-2 | 2060.0000 | CC-CAN |
| 21.4000 | LW-WW-1 | 122.5000 | LW-WW-1 | 460.0000 | CG-CAN | 1673.0000 | M-CAN-1 | 2061.5000 | M-US-3 |
| 22.3000 | LW-WW-1 | 122.6500 | LW-WW-1 | 464.0000 | CG-CAN | 1674.0000 | M-CAN-1 | 2064.0000 | AR-US-2 |
| 23.0000 | LW-WW-1 | 123.0000 | LW-WW-1 | 466.0000 | CG-CAN | 1675.0000 | M-US-2 | 2065.0000 | PR-US-6 |
| 23.1000 | LW-WW-1 | 124.0000 | LW-WW-1 | 470.0000 | CG-CAN | 1682.0000 | M-US-2 | 2066.5000 | CG-CAN |
| 23.4000 | LW-WW-1 | 125.2000 | LW-WW-1 | 472.0000 | CG-CAN | 1689.0000 | M-US-2 | 2079.0000 | PR-US-6 |
| 24.0000 | G-WW-3 | 125.8000 | LW-WW-1 | 474.0000 | CG-CAN | 1690.0000 | M-US-2 | 2082.5000 | M-US-3 |
| 24.0000 | LW-WW-1 | 128.2500 | LW-WW-1 | 478.0000 | CG-CAN | 1692.0000 | M-US-2 | 2083.9000 | CG-CAN |
| 24.8000 | LW-WW-1 | 128.3000 | LW-WW-1 | 484.0000 | CG-CAN | 1700.0000 | PR-US-6 | 2086.0000 | M-US-4 |
| 25.0000 | LW-WW-1 | 129.5000 | LW-WW-1 | 489.0000 | CG-CAN | 1705.0000 | M-CAN-1 | 2086.0000 | M-ARG |
| 25.1000 | LW-WW-1 | 131.0500 | LW-WW-1 | 500.0000 | CG-CAN | 1705.0000 | CC-US-2 | 2093.0000 | CG-US-8 |
| 25.5000 | LW-WW-1 | 131.4000 | LW-WW-1 | 510.0000 | CG-US-7 | 1709.0000 | CC-US-2 | 2094.4000 | CG-CAN |
| 25.8000 | LW-WW-1 | 131.8000 | LW-WW-1 | 518.0000 | CG-CAN | 1712.0000 | CC-US-2 | 2094.4000 | G-US-11 |
| 26.1000 | LW-WW-1 | 133.1500 | LW-WW-1 | 524.0000 | M-ARG | 1714.0000 | M-CAN-1 | 2101.4000 | CG-CAN |
| 26.5000 | LW-WW-1 | 133.8500 | LW-WW-1 | 530.0000 | BC-US-2 | 1715.0000 | M-US-2 | 2103.5000 | CG-US-1 |
| 27.1000 | LW-WW-1 | 134.2000 | LW-WW-1 | 1600.0000 | PR-US-6 | 1715.5000 | M-CAN-1 | 2104.0000 | M-WW-2 |
| 27.9000 | LW-WW-1 | 134.9000 | LW-WW-1 | 1602.0000 | M-US-2 | 1716.0000 | M-CAN-1 | 2104.9000 | CG-CAN |
| 28.5000 | LW-WW-1 | 135.9500 | LW-WW-1 | 1610.0000 | M-CAN-1 | 1718.5900 | M-US-1 | 2104.9000 | G-CAN-4 |
| 37.2000 | LW-WW-1 | 136.0000 | LW-WW-1 | 1610.0000 | BC-US-2 | 1722.0000 | M-WW-2 | 2111.0000 | M-WW-2 |
| 38.0000 | LW-WW-1 | 136.3000 | LW-WW-1 | 1613.0000 | M-US-2 | 1729.0000 | M-WW-2 | 2115.0000 | CC-US-2 |
| 40.0000 | G-WW-3 | 136.8000 | LW-WW-1 | 1614.0000 | PR-US-6 | 1736.0000 | M-WW-2 | 2118.0000 | AF-CAN |
| 40.0000 | LW-WW-1 | 137.0000 | LW-WW-1 | 1615.0000 | M-US-2 | 1743.0000 | M-WW-2 | 2118.0000 | M-US-3 |
| 42.2500 | LW-WW-1 | 139.0000 | LW-WW-1 | 1616.0000 | M-CAN-1 | 1745.0000 | M-US-2 | 2118.0000 | CC-US-2 |
| 44.0000 | LW-WW-1 | 139.1000 | LW-WW-1 | 1618.5000 | M-US-1 | 1746.8000 | M-CAN-1 | 2118.0000 | PR-US-6 |
| 46.2500 | LW-WW-1 | 139.5000 | LW-WW-1 | 1618.6000 | M-CAN-1 | 1750.0000 | M-CAN-1 | 2119.4000 | G-CAN-4 |
| 48.5000 | LW-WW-1 | 140.3000 | LW-WW-1 | 1619.6400 | M-US-1 | 1750.0000 | M-WW-2 | 2126.0000 | M-US-1 |
| 50.0000 | G-WW-3 | 140.5000 | LW-WW-1 | 1620.0000 | M-CAN-1 | 1753.0000 | M-CAN-1 | 2126.0000 | M-US-3 |
| 50.0000 | LW-WW-1 | 142.2500 | LW-WW-1 | 1620.0000 | M-US-2 | 1757.0000 | M-CAN-1 | 2130.0000 | N-US-1 |
| 51.6000 | LW-WW-1 | 142.8600 | LW-WW-1 | 1622.0000 | BC-US-1 | 1759.0000 | M-CAN-1 | 2131.0000 | M-US-2 |
| 51.9500 | LW-WW-1 | 143.5000 | LW-WW-1 | 1622.9000 | M-CAN-1 | 1761.5000 | M-CAN-1 | 2131.0000 | M-US-3 |
| 53.0000 | LW-WW-1 | 146.0000 | LW-WW-1 | 1623.0000 | M-US-2 | 1762.5000 | M-CAN-1 | 2132.0000 | PR-US-6 |
| 53.6000 | LW-WW-1 | 146.1000 | LW-WW-1 | 1624.0000 | M-CAN-1 | 1763.0000 | M-CAN-1 | 2134.0000 | AF-CAN |
| 54.0500 | LW-WW-1 | 147.2000 | LW-WW-1 | 1624.0000 | M-AUS | 1764.0000 | M-WW-2 | 2134.0000 | M-US-2 |
| 55.2500 | LW-WW-1 | 148.2000 | LW-WW-1 | 1625.0000 | M-US-2 | 1764.5000 | M-CAN-1 | 2134.0000 | M-US-3 |
| 55.5000 | LW-WW-1 | 148.5500 | LW-WW-1 | 1626.9000 | M-CAN-1 | 1764.8000 | M-CAN-1 | 2141.0000 | CG-US-2 |
| 56.5000 | LW-WW-1 | 150.6250 | AF-US-5 | 1627.0000 | M-CAN-1 | 1765.0000 | M-CAN-1 | 2142.0000 | AF-CAN |
| 57.8000 | LW-WW-1 | 153.1250 | AF-US-5 | 1627.0000 | M-US-2 | 1765.6000 | M-CAN-1 | 2142.0000 | M-US-3 |
| 57.9000 | LW-WW-1 | 154.3750 | AF-US-5 | 1628.5000 | M-CAN-1 | 1766.0000 | M-CAN-1 | 2142.0000 | PR-US-6 |
| 58.7000 | LW-WW-1 | 156.8750 | AF-US-5 | 1630.0000 | M-CAN-1 | 1766.5000 | M-CAN-1 | 2150.0000 | N-US-1 |
| 59.0000 | LW-WW-1 | 158.1250 | AF-US-5 | 1630.0000 | M-US-2 | 1767.5000 | M-CAN-1 | 2153.0000 | M-WW-2 |
| 60.0000 | G-WW-3 | 159.3750 | AF-US-5 | 1632.0000 | M-CAN-1 | 1769.0000 | M-CAN-1 | 2158.0000 | M-US-3 |
| 60.0000 | LW-WW-1 | 160.6250 | AF-US-5 | 1632.0000 | M-US-2 | 1770.0000 | M-CAN-1 | 2158.0000 | M-ATN |
| 61.5500 | LW-WW-1 | 161.8750 | AF-US-5 | 1632.5000 | M-CAN-1 | 1771.0000 | M-CAN-1 | 2158.0000 | PR-US-6 |
| 61.8000 | LW-WW-1 | 162.0000 | LW-WW-1 | 1635.0000 | M-US-2 | 1772.0000 | M-CAN-1 | 2160.0000 | CC-CAN |
| 62.6000 | LW-WW-1 | 163.1250 | AF-US-5 | 1638.0000 | M-US-2 | 1782.0000 | M-AUS | 2166.0000 | M-US-3 |
| 66.6600 | LW-WW-1 | 165.6250 | AF-US-5 | 1638.9000 | M-CAN-1 | 1785.0000 | M-CAN-1 | 2166.0000 | M-US-3 |
| 66.6660 | G-WW-3 | 166.8750 | AF-US-5 | 1639.0000 | M-CAN-1 | 1788.0000 | M-CAN-1 | 2173.5000 | CG-US-1 |
| 68.0000 | LW-WW-1 | 168.0000 | LW-WW-1 | 1640.0000 | M-CAN-1 | 1790.0000 | G-US-10 | 2182.0000 | CG-US-1 |
| 68.9000 | LW-WW-1 | 169.3750 | AF-US-5 | 1640.0000 | M-US-2 | 1792.0000 | G-CAN-4 | 2182.0000 | CG-US-8 |

| Freq | Code | Freq | Code | Freq | Code | Freq | Code | Freq | Code |
|---|---|---|---|---|---|---|---|---|---|
| 2182.0000 | N-I | 2377.0000 | G-US-1 | 2510.0000 | PR-US-6 | 2620.0000 | M-WW-2 | 2716.0000 | N-US-3 |
| 2182.0000 | PS-CRIB | 2381.5000 | PR-WW-2 | 2510.0000 | PR-US-6 | 2621.5000 | CC-CAN | 2716.0000 | N-CAN |
| 2187.5000 | M-WW-3 | 2381.5000 | CC-CAN | 2511.0000 | G-US-7 | 2622.0000 | SP-US-1 | 2716.0000 | SP-US-1 |
| 2194.0000 | PR-US-4 | 2382.0000 | M-US-3 | 2512.0000 | CG-US-8 | 2624.0000 | AF-US-8 | 2720.0000 | N-AUS |
| 2196.0000 | G-US-10 | 2390.0000 | AR-US-1 | 2512.0000 | G-US-4 | 2624.0000 | G-US-2 | 2720.0000 | G-AUS |
| 2197.0000 | PR-US-6 | 2390.0000 | M-US-3 | 2512.0000 | G-AUS | 2629.0000 | CC-US-2 | 2724.0000 | G-US-5 |
| 2198.0000 | M-US-3 | 2395.0000 | PR-US-4 | 2512.0000 | CC-US-2 | 2630.0000 | N-US-1 | 2726.0000 | AR-US-1 |
| 2200.0000 | PR-US-6 | 2396.5000 | PR-US-6 | 2514.0000 | AF-CAN | 2632.0000 | CC-US-2 | 2727.5000 | G-CAN-2 |
| 2201.5000 | G-CAN-4 | 2397.0000 | M-US-2 | 2514.0000 | M-US-3 | 2632.0000 | PS-AUS | 2727.5000 | G-CAN-4 |
| 2203.0000 | M-US-3 | 2397.0000 | M-US-3 | 2514.0000 | PR-US-6 | 2633.5000 | CC-CAN | 2728.0000 | G-AUS |
| 2204.0000 | PR-US-6 | 2398.0000 | G-US-9 | 2522.0000 | M-US-3 | 2634.0000 | CG-US-1 | 2730.5000 | PR-WW-2 |
| 2206.0000 | AF-CAN | 2398.0000 | M-US-1 | 2524.0000 | M-AUS | 2635.0000 | M-WW-2 | 2732.0000 | N-US-1 |
| 2206.0000 | G-CAN-4 | 2398.0000 | PR-US-4 | 2527.0000 | PS-CRIB | 2638.0000 | M-US-3 | 2738.0000 | CG-US-8 |
| 2206.0000 | M-US-3 | 2398.0000 | PR-US-6 | 2530.0000 | AF-CAN | 2638.0000 | CG-US-8 | 2738.0000 | CG-CAN |
| 2206.0000 | PR-US-6 | 2398.0000 | CC-US-1 | 2530.0000 | M-US-3 | 2638.0000 | CG-CAN | 2738.0000 | M-US-3 |
| 2207.0000 | M-WW-2 | 2399.8000 | PR-US-6 | 2530.0000 | M-US-3 | 2638.0000 | G-US-9 | 2740.0000 | G-AUS |
| 2207.4000 | G-CAN-4 | 2400.0000 | M-US-2 | 2530.0000 | PR-US-6 | 2638.0000 | PR-US-6 | 2744.0000 | SP-US-1 |
| 2220.0000 | AR-US-1 | 2400.0000 | M-US-3 | 2535.0000 | G-US-7 | 2648.0000 | CG-US-3 | 2745.0000 | N-US-1 |
| 2220.0000 | AR-US-3 | 2400.0000 | CC-US-2 | 2535.0000 | CC-US-2 | 2649.0000 | M-WW-2 | 2748.0000 | CG-US-2 |
| 2220.0000 | G-US-8 | 2405.0000 | SP-US-1 | 2538.0000 | AF-CAN | 2650.0000 | N-US-3 | 2748.0000 | CG-US-6 |
| 2220.0000 | M-CAN | 2406.0000 | M-US-3 | 2538.0000 | M-US-3 | 2656.0000 | M-WW-2 | 2748.0000 | CG-US-8 |
| 2221.5000 | G-CAN-2 | 2411.0000 | G-US-7 | 2538.0000 | M-US-3 | 2656.0000 | PS-AUS | 2752.0000 | G-CAN-2 |
| 2221.5000 | G-CAN-4 | 2413.0000 | CG-US-8 | 2538.0000 | PR-US-6 | 2658.0000 | G-US-1 | 2752.0000 | G-AUS |
| 2225.0000 | PR-US-6 | 2414.0000 | G-US-7 | 2538.0000 | CC-US-2 | 2659.0000 | CG-US-2 | 2758.0000 | G-US-10 |
| 2234.0000 | CG-US-8 | 2419.0000 | G-US-7 | 2544.0000 | G-US-10 | 2659.0000 | CG-US-8 | 2760.0000 | PS-AUS |
| 2236.0000 | CG-US-3 | 2419.0000 | CC-US-2 | 2550.0000 | N-US-1 | 2660.0000 | CG-US-6 | 2764.0000 | SP-US-1 |
| 2236.5000 | CG-US-3 | 2422.0000 | G-US-7 | 2550.0000 | M-US-3 | 2662.0000 | CG-US-2 | 2773.0000 | CC-US-2 |
| 2237.0000 | AF-CAN | 2422.0000 | N-ARG | 2550.0000 | M-ATN | 2662.0000 | CG-US-8 | 2773.0000 | N-I |
| 2237.0000 | G-CAN-4 | 2422.0000 | PR-US-6 | 2550.0000 | PR-US-6 | 2664.0000 | SP-US-1 | 2776.0000 | G-US-9 |
| 2237.0000 | M-US-2 | 2422.0000 | CC-US-2 | 2558.0000 | AF-CAN | 2667.0000 | CG-US-1 | 2776.0000 | CC-US-2 |
| 2237.0000 | M-US-3 | 2423.0000 | M-WW-2 | 2558.0000 | M-US-3 | 2668.0000 | G-AUS | 2780.0000 | G-AUS |
| 2240.0000 | G-CAN-3 | 2427.0000 | CC-US-2 | 2563.0000 | CC-US-2 | 2670.0000 | CG-US-1 | 2781.0000 | CC-US-2 |
| 2240.0000 | M-US-2 | 2428.0000 | AF/N-G | 2563.0000 | PS-AUS | 2670.0000 | CG-US-2 | 2782.0000 | M-US-4 |
| 2240.0000 | M-US-3 | 2430.0000 | M-US-3 | 2566.0000 | AR-US-1 | 2670.0000 | CG-US-7 | 2782.0000 | PR-US-6 |
| 2242.0000 | PR-US-6 | 2430.0000 | CC-US-2 | 2566.0000 | M-US-3 | 2670.0000 | CG-US-8 | 2784.0000 | CC-US-2 |
| 2243.5000 | CG-US-8 | 2434.0000 | N-US-1 | 2566.0000 | PS-AUS | 2670.0000 | G-CAN-4 | 2788.0000 | G-CAN-1 |
| 2247.5000 | G-CAN-3 | 2436.0000 | N-F | 2566.0000 | CC-US-2 | 2670.0000 | M-US-3 | 2792.0000 | G-AUS |
| 2252.0000 | N-US-1 | 2439.0000 | G-US-7 | 2567.5000 | CC-CAN | 2670.0000 | M-WW-2 | 2792.0000 | PS-AUS |
| 2253.0000 | CC-US-2 | 2442.0000 | CG-US-8 | 2569.0000 | G-US-7 | 2672.0000 | CG-US-8 | 2800.0000 | N-ISR |
| 2256.0000 | CC-US-2 | 2442.0000 | M-US-3 | 2569.0000 | PS-AUS | 2674.0000 | CG-US-4 | 2800.0000 | SP-US-1 |
| 2258.0000 | AR-US-1 | 2444.0000 | CG-US-3 | 2570.0000 | N-J | 2674.0000 | CG-US-3 | 2801.0000 | G-US-7 |
| 2259.5000 | AR-US-3 | 2445.0000 | G-US-1 | 2572.0000 | M-US-3 | 2675.0000 | CG-US-2 | 2804.0000 | G-US-7 |
| 2260.0000 | PS-AUS | 2446.5000 | G-US-2 | 2572.0000 | PS-AUS | 2675.0000 | CG-US-8 | 2805.0000 | PS-AUS |
| 2261.0000 | CG-US-2 | 2447.0000 | CC-US-2 | 2574.0000 | PS-AUS | 2678.0000 | CG-US-2 | 2808.0000 | G-AUS |
| 2269.5000 | G-US-9 | 2450.0000 | G-US-3 | 2575.0000 | PS-AUS | 2678.0000 | CG-US-8 | 2808.5000 | G-US-12 |
| 2270.0000 | G-AUS | 2450.0000 | M-US-3 | 2580.0000 | G-AUS | 2678.0000 | SP-US-1 | 2809.0000 | CG-US-6 |
| 2280.0000 | PS-AUS | 2450.0000 | CC-US-2 | 2582.0000 | AF-CAN | 2680.0000 | CG-US-8 | 2810.0000 | PR-US-6 |
| 2287.5000 | G-US-2 | 2456.0000 | M-US-1 | 2582.0000 | M-US-3 | 2680.0000 | G-AUS | 2812.0000 | G-US-7 |
| 2289.0000 | PR-US-4 | 2458.0000 | AF-CAN | 2582.0000 | M-WW-2 | 2682.0000 | CG-US-4 | 2813.5000 | AR-US-3 |
| 2292.0000 | G-CAN-4 | 2458.0000 | G-CAN-4 | 2583.4000 | G-CAN-4 | 2682.0000 | CG-US-3 | 2815.0000 | AF-CAN |
| 2292.0000 | PR-US-4 | 2458.0000 | M-US-3 | 2586.0000 | N-US1 | 2683.0000 | CG-US-2 | 2824.0000 | M-WW-2 |
| 2295.0000 | G-US-8 | 2458.0000 | PR-US-6 | 2587.0000 | G-US-7 | 2683.0000 | CG-US-8 | 2826.0000 | CG-US-1 |
| 2300.0000 | AR-US-1 | 2463.0000 | G-US-7 | 2590.0000 | CC-CAN | 2686.0000 | CG-US-2 | 2830.0000 | N-US-3 |
| 2300.0000 | AR-US-2 | 2463.0000 | CC-US-2 | 2590.0000 | M-US-3 | 2686.0000 | CG-US-8 | 2830.0000 | CG-US-8 |
| 2308.0000 | AR-US-3 | 2463.5000 | G-US-9 | 2591.0000 | AF/N-G | 2689.5000 | CC-CAN | 2830.0000 | M-US-3 |
| 2309.0000 | M-US-2 | 2466.0000 | G-US-7 | 2593.0000 | G-WW-3 | 2690.0000 | CG-US-8 | 2836.0000 | N-US-1 |
| 2309.0000 | M-US-3 | 2466.0000 | M-US-3 | 2598.0000 | CG-CAN | 2691.0000 | CG-US-2 | 2836.0000 | G-AUS |
| 2312.0000 | M-US-2 | 2466.0000 | CC-US-2 | 2598.0000 | M-US-3 | 2691.0000 | CG-US-8 | 2836.0000 | SP-US-1 |
| 2312.0000 | M-US-3 | 2470.0000 | G-AUS | 2598.0000 | PR-US-6 | 2691.0000 | CC-US-2 | 2840.0000 | G-WW-3 |
| 2312.0000 | CC-US-2 | 2471.0000 | G-US-7 | 2598.5000 | G-CAN-2 | 2692.0000 | G-AUS | 2840.0000 | G-AUS |
| 2316.0000 | PS-AUS | 2471.0000 | CC-US-2 | 2599.0000 | PR-US-6 | 2694.0000 | CG-US-2 | 2845.0000 | CG-US-4 |
| 2318.0000 | CG-CAN | 2474.0000 | G-US-7 | 2600.0000 | G-CAN-4 | 2694.0000 | CG-US-8 | 2845.0000 | G-US-10 |
| 2319.4000 | G-CAN-2 | 2474.0000 | CC-US-2 | 2600.0000 | G-AUS | 2694.0000 | CC-US-2 | 2846.5000 | G-CAN-2 |
| 2320.0000 | G-US-1 | 2479.0000 | CC-US-2 | 2600.0000 | PR-US-6 | 2698.0000 | CG-US-4 | 2848.0000 | M-US-1 |
| 2326.0000 | AR-US-2 | 2482.0000 | M-US-3 | 2601.0000 | CC-US-2 | 2698.0000 | CG-US-8 | 2850.0000 | G-AUS |
| 2326.0000 | G-US-7 | 2482.0000 | CC-US-2 | 2602.0000 | AR-US-2 | 2699.0000 | CG-US-2 | 2851.0000 | A-US-1 |
| 2328.0000 | CG-US-8 | 2484.0000 | M-CAN | 2604.0000 | CC-US-2 | 2699.0000 | CG-US-8 | 2851.0000 | A-WW-1 |
| 2330.0000 | CG-US-3 | 2487.0000 | G-US-7 | 2605.0000 | AR-US-2 | 2702.0000 | CG-US-2 | 2851.0000 | A-WW-2 |
| 2332.0000 | G-US-12 | 2488.0000 | G-AUS | 2606.0000 | CG-US-8 | 2702.0000 | CG-US-8 | 2851.0000 | A-US-2 |
| 2340.0000 | AF-CAN | 2490.0000 | M-US-3 | 2608.0000 | CG-US-4 | 2705.0000 | CG-US-3 | 2851.0000 | PR-US-6 |
| 2340.0000 | G-CAN-4 | 2500.0000 | G-WW-3 | 2610.5000 | G-US-2 | 2707.0000 | CG-US-2 | 2860.0000 | A-WW-5 |
| 2348.5000 | AR-US-2 | 2505.0000 | SP-US-1 | 2612.0000 | CG-US-4 | 2707.0000 | CG-US-8 | 2861.0000 | G-US-10 |
| 2350.0000 | AR-US-2 | 2505.0000 | PR-US-1 | 2612.0000 | G-CAN-4 | 2708.0000 | G-AUS | 2861.0000 | A-WW-2 |
| 2360.0000 | AR-US-1 | 2506.0000 | M-US-3 | 2612.0000 | G-AUS | 2710.0000 | AR-US-1 | 2863.0000 | A-WW-5 |
| 2360.0000 | G-US-1 | 2506.0000 | CC-US-2 | 2613.0000 | G-US-9 | 2710.0000 | CG-US-2 | 2872.0000 | A-WW-2 |
| 2366.0000 | M-US-3 | 2508.0000 | G-CAN-2 | 2614.0000 | G-US-9 | 2710.0000 | CG-US-8 | 2878.0000 | A-US-2 |
| 2368.0000 | N-US-1 | 2508.0000 | G-CAN-3 | 2616.0000 | CC-US-2 | 2713.0000 | AR-US-1 | 2878.0000 | A-WW-2 |
| 2371.0000 | AF-US-7 | 2509.0000 | CC-US-2 | 2617.5000 | G-CAN-2 | 2714.0000 | N-US-1 | 2881.0000 | A-WW-5 |
| 2374.0000 | AF-US-7 | 2510.0000 | M-US-1 | 2620.0000 | G-AUS | 2716.0000 | N-US-1 | 2887.0000 | A-WW-2 |

| Freq | ID | Freq | ID | Freq | ID | Freq | ID | Freq | ID |
|---|---|---|---|---|---|---|---|---|---|
| 2932.0000 | A-WW-2 | 3167.5000 | A-WW-5 | 3289.0000 | AR-US-3 | 3397.5000 | CC-CAN | 4005.5000 | N-URS |
| 2941.0000 | A-WW-5 | 3167.5000 | CC-US-2 | 3290.0000 | AR-US-2 | 3404.0000 | A-WW-5 | 4007.0000 | N-US-4 |
| 2944.0000 | A-WW-2 | 3170.0000 | N-J | 3290.0000 | G-AUS | 3407.0000 | G-US-9 | 4008.5000 | N-US-4 |
| 2950.0000 | A-WW-5 | 3170.0000 | G-WW-3 | 3290.5000 | M-US-1 | 3407.0000 | A-WW-5 | 4009.0000 | CC-CAN |
| 2962.0000 | A-WW-2 | 3170.0000 | PR-US-4 | 3292.0000 | AF-US-1 | 3411.0000 | G-US-4 | 4010.0000 | N-URS |
| 2965.0000 | A-WW-5 | 3171.0000 | CC-CAN | 3292.0000 | AF-US-8 | 3413.0000 | A-WW-2 | 4010.0000 | PS-AUS |
| 2971.0000 | A-WW-2 | 3174.5000 | G-CAN-3 | 3294.5000 | M-US-1 | 3413.0000 | 1-WW-5 | 4011.0000 | N-US-4 |
| 2980.0000 | A-WW-5 | 3175.0000 | AR-US-1 | 3295.0000 | AF-US-1 | 3419.0000 | A-WW-2 | 4012.0000 | AR-US-3 |
| 2982.5000 | CG-US-8 | 3176.0000 | G-AUS | 3296.0000 | AF-US-8 | 3420.0000 | N-ARG | 4012.5000 | N-US-4 |
| 2992.0000 | A-WW-2 | 3180.0000 | CC-US-2 | 3296.0000 | G-US-4 | 3425.0000 | A-WW-1 | 4013.5000 | N-US-4 |
| 2995.0000 | AF-US-2 | 3183.0000 | CC-US-2 | 3296.5000 | M-US-1 | 3428.0000 | G-US-10 | 4014.0000 | N-US-1 |
| 2998.0000 | A-WW-2 | 3186.0000 | N-G | 3299.0000 | AF-US-8 | 3432.0000 | A-WW-5 | 4015.0000 | N-US-4 |
| 3001.0000 | A-WW-5 | 3186.0000 | G-US-3 | 3299.5000 | CC-CAN | 3434.0000 | A-US-2 | 4016.5000 | N-US-4 |
| 3004.0000 | A-WW-2 | 3187.0000 | N-CAN | 3300.0000 | CC-CAN | 3443.0000 | A-US-2 | 4020.0000 | AR-US-3 |
| 3004.0000 | A-US-2 | 3187.0000 | SP-US-1 | 3300.5000 | M-US-1 | 3443.0000 | PR-US-6 | 4020.0000 | G-US-3 |
| 3004.0000 | PR-US-6 | 3191.0000 | PR-US-6 | 3302.0000 | AR-US-2 | 3452.0000 | A-WW-2 | 4025.0000 | AR-US-3 |
| 3007.0000 | A-US-1 | 3191.5000 | G-CAN-2 | 3303.0000 | CC-US-2 | 3452.0000 | A-WW-3 | 4027.0000 | G-US-10 |
| 3007.0000 | A-WW-1 | 3192.5000 | PR-US-1 | 3303.0000 | G-US-1 | 3454.0000 | AF-US-8 | 4030.0000 | AR-US-1 |
| 3008.0000 | A-WW-2 | 3194.0000 | G-US-10 | 3304.5000 | G-US-1 | 3455.0000 | A-WW-2 | 4030.0000 | AR-US-3 |
| 3010.0000 | A-US-1 | 3194.5000 | G-AUS | 3305.0000 | AR-US-2 | 3458.0000 | A-WW-1 | 4030.0000 | G-US-12 |
| 3010.0000 | A-WW-1 | 3196.0000 | G-AUS | 3306.5000 | M-US-1 | 3458.0000 | A-WW-3 | 4030.0000 | PS-AUS |
| 3013.0000 | A-US-1 | 3198.0000 | CC-US-2 | 3308.0000 | AF-US-8 | 3460.0000 | A-WW-2 | 4035.0000 | AR-US-1 |
| 3016.0000 | A-WW-2 | 3199.5000 | G-US-1 | 3310.0000 | G-US-3 | 3461.0000 | A-WW-5 | 4035.0000 | AR-US-3 |
| 3019.0000 | A-US-2 | 3201.0000 | AR-US-1 | 3310.0000 | CC-CAN | 3467.0000 | A-WW-2 | 4035.0000 | CC-US-2 |
| 3019.0000 | A-WW-2 | 3201.0000 | G-US-4 | 3311.0000 | AF-US-8 | 3470.0000 | A-WW-2 | 4036.5000 | CC-CAN |
| 3023.0000 | A-US-3 | 3201.0000 | CC-US-2 | 3311.5000 | CC-CAN | 3476.0000 | A-WW-2 | 4040.0000 | N-US-5 |
| 3023.0000 | N-US-1 | 3205.0000 | AR-US-1 | 3312.0000 | AF-US-8 | 3479.0000 | A-WW-2 | 4040.0000 | N-AUS |
| 3023.0000 | CG-US-1 | 3206.5000 | CC-CAN | 3315.0000 | AF-US-8 | 3479.0000 | A-WW-3 | 4040.0000 | N-AFS |
| 3023.0000 | CG-US-8 | 3209.0000 | CG-US-3 | 3315.0000 | G-AUS | 3482.0000 | A-WW-3 | 4040.0000 | N-US-1 |
| 3023.5000 | N-AUS | 3213.5000 | CC-CAN | 3319.0000 | N-US-3 | 3485.0000 | A-WW-2 | 4040.0000 | PR-US-6 |
| 3024.0000 | N-J | 3214.0000 | G-CAN-3 | 3320.4000 | M-US-1 | 3485.0000 | A-WW-5 | 4041.0000 | N-US-4 |
| 3024.0000 | SP-US-1 | 3214.0000 | PR-WW-2 | 3329.0000 | G-US-1 | 3488.0000 | A-US-3 | 4042.5000 | N-US-4 |
| 3024.5000 | N-J | 3214.0000 | CC-CAN | 3330.0000 | G-WW-3 | 3494.0000 | A-US-1 | 4044.0000 | PR-US-6 |
| 3025.0000 | N-J | 3215.0000 | G-US-4 | 3331.0000 | G-US-10 | 3496.0000 | AF-US-8 | 4045.0000 | N-US-1 |
| 3030.0000 | AF-US-2 | 3223.5000 | G-US-4 | 3333.0000 | G-US-9 | 3496.5000 | PR-WW-2 | 4045.0000 | N-US-4 |
| 3032.0000 | AF-US-2 | 3224.0000 | CC-CAN | 3335.0000 | G-US-2 | 3497.0000 | A-US-1 | 4045.0000 | PS-AUS |
| 3032.0000 | AF-US-2 | 3228.0000 | AF-US-8 | 3336.0000 | G-US-10 | 3497.0000 | A-WW-1 | 4048.0000 | G-US-3 |
| 3032.0000 | AF-AUS/NZL | 3230.0000 | G-US-4 | 3340.0000 | G-AUS | 3503.0000 | AF-US-3 | 4048.5000 | CG-US-8 |
| 3035.0000 | N-J | 3231.0000 | AF-US-8 | 3340.0000 | PR-US-6 | 3545.0000 | AF-US-8 | 4050.0000 | CG-US-6 |
| 3046.0000 | AF-CAN | 3235.5000 | AR-US-3 | 3341.0000 | G-US-1 | 3561.0000 | N-URS | 4050.0000 | G-US-3 |
| 3046.0000 | AF/N-G | 3235.5000 | G-US-10 | 3343.0000 | G-US-4 | 3564.5000 | N-URS | 4053.0000 | PS-S |
| 3050.0000 | N-US-1 | 3237.0000 | N-US-1 | 3347.0000 | AF-US-8 | 3593.0000 | G-WW-3 | 4054.0000 | AF-US-8 |
| 3053.0000 | N-US-1 | 3237.0000 | N-US-3 | 3347.0000 | AR-US-3 | 3616.0000 | PS-CRIB | 4055.0000 | G-US-10 |
| 3060.0000 | AF-US-2 | 3237.0000 | AR-US-3 | 3353.0000 | G-US-10 | 3705.0000 | G-WW-3 | 4055.0000 | PS-AUS |
| 3067.0000 | AF-US-3 | 3238.0000 | CC-US-2 | 3354.0000 | CC-US-2 | 3714.0000 | G-WW-3 | 4059.0000 | N-AFS |
| 3081.0000 | AF-US-3 | 3238.5000 | G-US-6 | 3354.5000 | PR-WW-2 | 3720.0000 | G-AUS | 4060.0000 | G-US-3 |
| 3088.0000 | N-US-1 | 3240.0000 | PR-US-6 | 3357.0000 | CC-US-2 | 3724.0000 | G-AUS | 4060.0000 | G-US-10 |
| 3092.0000 | AF-CAN | 3241.0000 | CG-US-2 | 3359.0000 | CC-CAN | 3729.0000 | PS-AUS | 4061.0000 | AF-US-8 |
| 3095.0000 | AF/N-G | 3241.0000 | CG-US-8 | 3362.0000 | CC-US-2 | 3732.0000 | PS-AUS | 4063.0000 | M-WW-1 |
| 3095.0000 | AF-SAU | 3241.0000 | CC-US-2 | 3363.0000 | G-US-9 | 3735.0000 | PS-AUS | 4063.0000 | M-US-4 |
| 3095.0000 | N-US-1 | 3241.5000 | CC-CAN | 3365.0000 | SP-US-1 | 3737.0000 | N-AFS | 4063.0000 | PR-US-6 |
| 3102.0000 | N-US-3 | 3245.0000 | AR-US-3 | 3365.0000 | PR-US-6 | 3743.0000 | PS-AUS | 4066.1000 | N-US-5 |
| 3102.0000 | PR-US-6 | 3250.0000 | N-AFS | 3365.0000 | CC-US-2 | 3746.0000 | PS-AUS | 4066.1000 | M-WW-1 |
| 3104.0000 | PR-US-6 | 3252.0000 | PS-AUS | 3367.4000 | G-US-1 | 3752.0000 | PS-AUS | 4067.0000 | N-G |
| 3109.0000 | AF/N-G | 3253.0000 | CG-US-2 | 3368.5000 | G-CAN-2 | 3801.5000 | PS-WW-1 | 4069.2000 | CG-CAN |
| 3109.0000 | N-US-1 | 3253.0000 | CG-US-8 | 3369.0000 | AF-US-1 | 3802.5000 | AF-US-8 | 4069.2000 | M-WW-1 |
| 3113.0000 | AF-US-1 | 3253.0000 | CG-CAN | 3370.5000 | AF-US-8 | 3810.0000 | G-WW-3 | 4072.0000 | PR-US-6 |
| 3113.0000 | G-US-1 | 3253.0000 | G-US-10 | 3372.0000 | G-AUS | 3815.0000 | PS-CRIB | 4072.3000 | M-WW-1 |
| 3116.0000 | AF/N-G | 3253.5000 | G-CAN-3 | 3375.0000 | G-AUS | 3822.0000 | G-US-3 | 4072.4000 | M-US-3 |
| 3116.0000 | AF-AFS | 3256.0000 | CG-US-8 | 3379.0000 | G-US-1 | 3830.0000 | G-US-3 | 4075.4000 | M-US-3 |
| 3116.0000 | AF-US-10 | 3257.0000 | AR-US-3 | 3380.0000 | SP-US-1 | 3836.0000 | G-AUS | 4075.4000 | M-WW-1 |
| 3116.0000 | A-WW-5 | 3258.0000 | PR-US-6 | 3382.0000 | CG-US-2 | 3841.0000 | AF-US-8 | 4078.5000 | M-WW-1 |
| 3120.0000 | AF/N-G | 3258.0000 | CC-US-2 | 3382.0000 | CG-US-8 | 3842.0000 | G-WW-3 | 4081.5000 | G-CAN-3 |
| 3120.0000 | CG-US-2 | 3260.0000 | G-AUS | 3385.0000 | CG-US-1 | 3848.0000 | G-AUS | 4081.6000 | CG-US-8 |
| 3120.0000 | SP-US-1 | 3261.0000 | N-US-1 | 3385.0000 | G-US-5 | 3875.0000 | AF-US-8 | 4081.6000 | M-WW-1 |
| 3123.0000 | CG-US-2 | 3261.0000 | AR-US-1 | 3385.0000 | G-US-4 | 3885.0000 | AF/N-G | 4082.0000 | N-US-1 |
| 3130.0000 | AF-AUS/NZL | 3261.0000 | CC-US-2 | 3385.0000 | SP-US-1 | 3888.0000 | AF-US-8 | 4084.0000 | G-CAN-3 |
| 3130.0000 | N-US-1 | 3265.0000 | N-US-1 | 3387.0000 | CG-US-8 | 3932.0000 | AF-AFS | 4084.7000 | G-US-9 |
| 3144.0000 | AF-US-3 | 3270.0000 | CC-CAN | 3388.0000 | G-US-1 | 3935.0000 | AF/N-G | 4084.7000 | M-WW-1 |
| 3152.0000 | G-AUS | 3273.0000 | G-CAN-4 | 3388.8000 | G-US-1 | 3939.0000 | A-US-3 | 4087.8000 | M-US-4 |
| 3155.0000 | G-AUS | 3273.5000 | AR-US-3 | 3389.0000 | CG-US-4 | 3939.0000 | AF/N-G | 4087.8000 | M-WW-1 |
| 3155.0000 | PR-US-4 | 3276.5000 | AR-US-3 | 3389.0000 | CG-US-3 | 3966.0000 | AF-US-8 | 4090.9000 | M-WW-1 |
| 3156.0000 | PR-US-1 | 3277.0000 | G-US-4 | 3390.0000 | N-AFS | 3990.0000 | G-US-5 | 4094.0000 | M-WW-1 |
| 3158.0000 | G-US-10 | 3278.0000 | G-CAN-3 | 3390.0000 | G-US-10 | 4000.0000 | AR-US-1 | 4097.1000 | M-WW-1 |
| 3158.0000 | G-AUS | 3281.0000 | M-US-1 | 3393.0000 | G-AUS | 4001.0000 | N-US-4 | 4100.0000 | PR-US-6 |
| 3161.5000 | G-US-10 | 3281.0000 | PR-US-6 | 3394.0000 | G-US-10 | 4001.5000 | AR-US-1 | 4100.2000 | M-WW-1 |
| 3162.0000 | SP-US-1 | 3287.0000 | AF-CAN | 3395.0000 | G-US-1 | 4001.5000 | AR-US-3 | 4103.3000 | M-WW-1 |
| 3164.5000 | CC-US-2 | 3287.0000 | AR-US-2 | 3395.0000 | G-US-10 | 4001.5000 | N-US-4 | 4106.4000 | CG-US-8 |
| 3166.0000 | G-CAN-3 | 3288.5000 | M-US-1 | 3395.0000 | SP-US-1 | 4004.0000 | AR-US-3 | 4106.4000 | M-WW-1 |

| Freq | ID | Freq | ID | Freq | ID | Freq | ID | Freq | ID |
|---|---|---|---|---|---|---|---|---|---|
| 4109.5000 | M-WW-1 | 4233.0000 | AR-US-1 | 4371.0000 | M-US-3 | 4464.0000 | AF/N-G | 4575.0000 | N-J |
| 4110.8000 | M-ATN | 4235.0000 | CG-US-3 | 4372.9000 | M-WW-1 | 4464.5000 | AF-US-7 | 4575.0000 | CG-US-2 |
| 4112.6000 | M-WW-1 | 4235.0000 | CG-CAN | 4373.0000 | N-US-1 | 4465.0000 | CC-CAN | 4575.0000 | CG-US-4 |
| 4115.7000 | M-US-4 | 4236.5000 | CG-CAN | 4376.0000 | CG-US-4 | 4465.0000 | BC-WW-1 | 4575.0000 | CG-US-6 |
| 4115.7000 | M-WW-1 | 4238.0000 | M-US-3 | 4376.0000 | CG-US-8 | 4466.0000 | AF-US-8 | 4575.0000 | PR-US-4 |
| 4117.2000 | M-US-3 | 4241.4000 | N-ISR | 4376.0000 | CG-CAN | 4467.5000 | AF-US-7 | 4576.0000 | PS-AUS |
| 4118.8000 | M-WW-1 | 4245.0000 | CG-US-3 | 4376.0000 | M-WW-1 | 4468.0000 | G-US-3 | 4577.0000 | AF-US-8 |
| 4120.4000 | M-US-3 | 4247.0000 | M-US-3 | 4376.0000 | SP-US-1 | 4470.5000 | N-US-4 | 4580.0000 | G-CAN-2 |
| 4121.9000 | M-WW-1 | 4250.0000 | AR-US-1 | 4377.0000 | N-US-1 | 4472.0000 | N-US-4 | 4582.0000 | AF-US-7 |
| 4125.0000 | CG-US-1 | 4250.0000 | M-US-3 | 4379.1000 | CG-CAN | 4474.0000 | G-CAN-2 | 4585.0000 | AF-US-7 |
| 4125.0000 | CG-US-8 | 4250.0000 | M-WW-2 | 4379.1000 | G-US-9 | 4475.0000 | G-AUS | 4588.5000 | AF-US-8 |
| 4125.0000 | CG-CAN | 4253.0000 | N-US-1 | 4379.1000 | M-WW-1 | 4475.0000 | G-US-10 | 4589.0000 | G-US-3 |
| 4125.0000 | G-US-4 | 4255.0000 | AF-CAN | 4380.0000 | G-US-9 | 4475.0000 | BC-WW-1 | 4590.0000 | AF-US-8 |
| 4125.0000 | M-WW-1 | 4256.0000 | M-US-3 | 4382.2000 | M-WW-1 | 4476.0000 | G-US-10 | 4593.5000 | AF-US-8 |
| 4125.0000 | N-AFS | 4260.0000 | CG-CAN | 4385.3000 | CG-CAN | 4477.0000 | AF/N-G | 4596.0000 | G-US-3 |
| 4125.0000 | PR-WW-2 | 4262.0000 | CG-US-3 | 4385.3000 | M-WW-1 | 4480.0000 | G-US-3 | 4599.5000 | AF-US-7 |
| 4125.0000 | PR-US-6 | 4262.0000 | M-WW-2 | 4388.4000 | M-WW-1 | 4480.5000 | G-US-2 | 4600.0000 | G-US-2 |
| 4128.1000 | M-WW-1 | 4263.5000 | AR-US-1 | 4391.5000 | M-WW-1 | 4483.0000 | G-US-8 | 4602.5000 | AF-US-7 |
| 4130.0000 | M-US-3 | 4268.0000 | AF-CAN | 4394.6000 | M-WW-1 | 4486.0000 | SP-US-1 | 4602.6000 | AF-US-8 |
| 4131.2000 | M-WW-1 | 4268.0000 | M-US-3 | 4397.7000 | M-WW-1 | 4487.0000 | AF-US-10 | 4603.0000 | PR-WW-2 |
| 4133.2000 | M-CAN | 4268.0000 | M-WW-2 | 4400.8000 | CG-US-1 | 4490.0000 | N-ARG | 4603.0000 | CC-CAN |
| 4134.3000 | CG-US-8 | 4269.0000 | M-WW-2 | 4400.8000 | CG-US-8 | 4491.0000 | N-US-1 | 4610.0000 | AR-US-1 |
| 4134.3000 | M-WW-1 | 4271.0000 | AF-CAN | 4400.8000 | M-WW-1 | 4492.0000 | AF-US-1 | 4610.0000 | G-US-10 |
| 4137.4000 | G-US-9 | 4274.0000 | M-US-3 | 4403.9000 | M-WW-1 | 4495.0000 | AF-US-1 | 4610.0000 | CC-CAN |
| 4137.4000 | M-WW-1 | 4275.5000 | M-WW-2 | 4404.4000 | G-US-9 | 4496.5000 | G-CAN-2 | 4610.5000 | PR-US-4 |
| 4139.5000 | G-US-10 | 4276.0000 | M-J | 4407.0000 | M-WW-1 | 4500.0000 | AF-US-8 | 4613.5000 | PR-US-4 |
| 4140.0000 | AF/N-G | 4277.0000 | AF-US-3 | 4409.4000 | M-ATN | 4500.0000 | CG-US-4 | 4616.0000 | G-US-12 |
| 4140.5000 | G-US-9 | 4277.0000 | M-WW-2 | 4410.0000 | G-AUS | 4500.0000 | G-US-3 | 4617.5000 | G-US-6 |
| 4140.5000 | M-WW-1 | 4280.0000 | AR-US-1 | 4410.0000 | M-US-4 | 4500.0000 | G-WW-3 | 4621.5000 | PR-WW-2 |
| 4140.9000 | G-US-9 | 4280.0000 | N-I | 4410.1000 | CC-CAN | 4500.0000 | SP-US-1 | 4622.5000 | N-US-1 |
| 4143.6000 | M-WW-1 | 4283.0000 | M-WW-2 | 4410.1000 | M-US-4 | 4504.5000 | AF-US-7 | 4625.0000 | N-US-4 |
| 4143.6000 | M-WW-2 | 4283.0000 | N-AFS | 4410.1000 | M-WW-1 | 4505.0000 | AF-US-10 | 4625.5000 | G-CAN-2 |
| 4143.6000 | N-B | 4285.0000 | CG-CAN | 4413.2000 | M-WW-1 | 4505.0000 | BC-WW-1 | 4626.0000 | G-US-3 |
| 4145.0000 | M-WW-2 | 4286.0000 | G-WW-3 | 4415.0000 | AR-US-1 | 4507.5000 | AF-US-7 | 4627.0000 | AF-US-7 |
| 4145.0000 | PR-US-6 | 4289.0000 | N-ISR | 4415.8000 | M-US-3 | 4510.0000 | G-CAN-2 | 4630.0000 | AF-US-7 |
| 4152.0000 | N-US-1 | 4290.0000 | AR-US-1 | 4416.0000 | N-US-1 | 4510.0000 | SP-US-1 | 4630.0000 | M-J |
| 4155.0000 | N-AFS | 4291.0000 | G-WW-3 | 4416.3000 | M-US-3 | 4515.0000 | N-US-1 | 4630.0000 | CC-CAN |
| 4164.1000 | CG-US-5 | 4292.0000 | M-I | 4419.0000 | M-US-3 | 4517.0000 | AF-US-8 | 4632.5000 | G-WW-3 |
| 4164.7000 | CG-US-5 | 4294.0000 | M-US-3 | 4419.0000 | PR-US-6 | 4520.0000 | AR-US-1 | 4633.0000 | AF-US-8 |
| 4165.0000 | CC-CAN | 4298.0000 | CG-US-4 | 4419.4000 | CG-US-8 | 4520.0000 | G-CAN-2 | 4634.5000 | PR-WW-2 |
| 4165.3000 | CG-US-5 | 4298.0000 | CG-US-3 | 4419.4000 | M-WW-1 | 4524.0000 | G-AUS | 4634.5000 | PR-US-4 |
| 4169.0000 | PR-US-6 | 4298.0000 | G-WW-3 | 4419.4000 | M-WW-2 | 4525.0000 | G-WW-3 | 4635.0000 | PS-AUS |
| 4170.0000 | CG-US-8 | 4310.0000 | M-US-3 | 4419.4000 | PR-WW-2 | 4529.0000 | M-US-3 | 4637.5000 | PR-WW-2 |
| 4171.0000 | G-CAN-4 | 4316.0000 | CG-US-3 | 4422.5000 | M-WW-1 | 4530.0000 | G-US-3 | 4637.5000 | PR-US-4 |
| 4172.0000 | CG-US-1 | 4320.0000 | M-I | 4425.6000 | M-WW-1 | 4531.0000 | CG-US-6 | 4638.0000 | AF-US-8 |
| 4172.0000 | CG-US-4 | 4325.0000 | M-WW-2 | 4428.6000 | M-US-3 | 4532.0000 | G-CAN-2 | 4640.0000 | AR-US-1 |
| 4172.0000 | CG-US-3 | 4331.0000 | M-US-3 | 4428.7000 | CG-US-2 | 4532.0000 | M-US-3 | 4647.0000 | PR-US-4 |
| 4172.0000 | M-US-3 | 4334.0000 | CG-CAN | 4428.7000 | M-WW-1 | 4535.0000 | G-CAN-2 | 4654.0000 | A-US-1 |
| 4173.5000 | G-CAN-4 | 4335.0000 | CG-US-2 | 4428.7000 | G-AUS | 4535.0000 | M-US-3 | 4654.0000 | A-WW-1 |
| 4174.5000 | G-CAN-4 | 4340.0000 | N-CAN | 4431.8000 | G-US-9 | 4538.0000 | M-US-3 | 4657.0000 | A-WW-1 |
| 4176.0000 | CG-US-1 | 4342.6500 | M-US-3 | 4431.8000 | M-WW-1 | 4538.5000 | PR-US-4 | 4663.0000 | A-WW-4 |
| 4176.0000 | CG-US-4 | 4346.0000 | CG-US-2 | 4431.8000 | M-CAN | 4540.0000 | AF/N-G | 4665.0000 | A-WW-3 |
| 4176.0000 | CG-US-3 | 4347.0000 | AF-CAN | 4433.2000 | G-US-9 | 4540.0000 | N-J | 4666.0000 | A-WW-2 |
| 4176.0000 | PR-US-6 | 4349.0000 | M-WW-2 | 4434.9000 | G-US-9 | 4543.5000 | CC-CAN | 4666.0000 | G-CAN-3 |
| 4176.5000 | M-US-3 | 4350.0000 | PS-AUS | 4434.9000 | M-WW-1 | 4548.0000 | G-US-3 | 4667.0000 | A-WW-5 |
| 4177.0000 | M-US-3 | 4351.5000 | CG-US-1 | 4435.0000 | G-US-3 | 4548.5000 | PR-US-4 | 4668.0000 | A-WW-5 |
| 4177.5000 | CG-US-3 | 4351.5000 | CG-US-4 | 4438.0000 | PR-US-1 | 4549.5000 | CG-US-6 | 4669.0000 | A-WW-2 |
| 4177.5000 | CG-US-8 | 4351.5000 | CG-US-3 | 4438.0000 | PR-US-4 | 4550.0000 | PR-WW-2 | 4669.5000 | G-US-2 |
| 4178.0000 | CG-US-4 | 4351.5000 | M-US-3 | 4438.0000 | PR-US-6 | 4550.1000 | CG-US-6 | 4672.0000 | A-US-2 |
| 4178.0000 | CG-US-3 | 4353.0000 | CG-CAN | 4440.0000 | AR-US-1 | 4554.0000 | G-CAN-2 | 4675.0000 | G-CAN-2 |
| 4178.0000 | M-US-3 | 4353.0000 | G-CAN-4 | 4441.0000 | G-CAN-2 | 4555.0000 | BC-WW-1 | 4675.0000 | A-WW-3 |
| 4178.0000 | PR-US-6 | 4354.0000 | G-CAN-4 | 4444.0000 | AF-US-10 | 4557.0000 | AF-US-8 | 4675.0000 | A-WW-5 |
| 4178.5000 | CG-US-3 | 4355.5000 | CG-US-1 | 4444.0000 | G-WW-3 | 4557.0000 | PS-AUS | 4675.0000 | G-US-10 |
| 4179.0000 | CG-US-4 | 4355.5000 | CG-US-4 | 4445.0000 | AF-US-10 | 4560.0000 | AF-US-8 | 4678.0000 | A-WW-2 |
| 4179.0000 | CG-US-3 | 4355.5000 | CG-US-3 | 4445.0000 | G-WW-3 | 4560.0000 | N-CAN | 4687.0000 | A-US-1 |
| 4181.8000 | CG-US-1 | 4356.0000 | M-US-3 | 4446.5000 | AR-US-3 | 4560.0000 | PS-AUS | 4687.0000 | A-WW-1 |
| 4182.2000 | CG-US-1 | 4356.5000 | M-US-3 | 4447.0000 | AF-US-8 | 4565.0000 | G-CAN-3 | 4700.0000 | N-US-1 |
| 4189.5000 | CG-US-1 | 4357.4000 | M-WW-1 | 4448.0000 | CC-CAN | 4565.0000 | BC-WW-1 | 4702.0000 | N-US-1 |
| 4190.0000 | PR-US-6 | 4359.0000 | N-US-1 | 4448.8000 | G-AUS | 4567.0000 | PS-AUS | 4703.0000 | AF-AUS/NZL |
| 4195.0000 | G-US-11 | 4360.0000 | N-US-5 | 4449.0000 | AF-US-5 | 4568.4000 | N-US-3 | 4703.0000 | AF-AFS |
| 4198.0000 | CG-US-1 | 4360.5000 | M-WW-1 | 4449.0000 | AF-US-1 | 4571.0000 | G-AUS | 4704.0000 | AF-CAN |
| 4205.0000 | CG-US-1 | 4360.5000 | N-US-5 | 4450.0000 | AF-US-8 | 4572.0000 | G-US-10 | 4704.0000 | N-US-1 |
| 4213.5000 | G-US-9 | 4363.6000 | CG-CAN | 4450.0000 | G-AUS | 4572.5000 | G-US-1 | 4704.0000 | SP-US-1 |
| 4221.2000 | N-G | 4363.6000 | M-WW-1 | 4455.0000 | AF-US-10 | 4573.0000 | CG-US-6 | 4705.0000 | N-ARG |
| 4225.0000 | AF-CAN | 4365.0000 | AR-US-1 | 4455.0000 | PR-WW-2 | 4573.0000 | G-CAN-3 | 4705.5000 | AF-US-7 |
| 4225.0000 | PR-US-6 | 4366.7000 | M-WW-1 | 4457.0000 | AF-US-5 | 4573.5000 | PR-WW-2 | 4707.0000 | AF/N-G |
| 4228.0000 | M-US-3 | 4369.8000 | M-US-3 | 4460.0000 | C-CAN-2 | 4573.5000 | CC-CAN | 4707.0000 | N-US-1 |
| 4230.0000 | CG-US-4 | 4369.8000 | M-WW-1 | 4462.0000 | G-CAN-2 | 4574.0000 | G-AUS | 4710.0000 | A-WW-5 |
| 4232.0000 | CG-US-3 | 4369.8000 | M-WW-2 | 4463.0000 | G-US-3 | 4575.0000 | AF-US-8 | 4710.0000 | AF/N-G |

| Freq | Code | Freq | Code | Freq | Code | Freq | Code | Freq | Code |
|---|---|---|---|---|---|---|---|---|---|
| 4710.5000 | N-I | 4860.0000 | PS-AUS | 5060.0000 | SP-US-1 | 5204.0000 | G-US-3 | 5345.0000 | G-AUS |
| 4710.5000 | N-US-1 | 4861.5000 | G-CAN-2 | 5060.0000 | PR-US-1 | 5204.5000 | CC-US-2 | 5346.0000 | AR-US-2 |
| 4711.0000 | N-US-1 | 4861.5000 | G-CAN-4 | 5060.0000 | PR-US-6 | 5205.0000 | G-US-5 | 5346.5000 | G-CAN-2 |
| 4714.0000 | SP-US-1 | 4863.0000 | G-US-4 | 5061.5000 | CC-CAN | 5207.5000 | CC-US-2 | 5347.5000 | G-CAN-2 |
| 4717.0000 | AF/N-G | 4865.0000 | AF-US-10 | 5061.5000 | PR-US-4 | 5208.0000 | AR-US-3 | 5350.0000 | G-US-1 |
| 4718.0000 | N-US-3 | 4865.0000 | CC-CAN | 5063.0000 | CG-US-6 | 5208.0000 | G-WW-3 | 5350.0000 | SP-US-1 |
| 4719.0000 | AF/N-G | 4870.0000 | AR-US-1 | 5063.6000 | CG-US-6 | 5211.0000 | G-US-1 | 5355.0000 | PS-AUS |
| 4721.0000 | AF-US-3 | 4872.0000 | AF-US-8 | 5066.0000 | CG-US-6 | 5212.5000 | G-US-2 | 5357.5000 | G-US-10 |
| 4721.0000 | AF-US-10 | 4873.0000 | N-US-3 | 5067.5000 | PR-US-4 | 5214.0000 | AR-US-1 | 5360.0000 | PS-AUS |
| 4722.0000 | AF/N-G | 4878.0000 | AF-US-8 | 5074.5000 | PR-US-4 | 5215.0000 | AF-US-1 | 5362.5000 | G-US-10 |
| 4725.0000 | AF-US-1 | 4880.0000 | G-US-3 | 5078.0000 | AF-US-10 | 5215.0000 | G-CAN-3 | 5370.0000 | G-AUS |
| 4730.0000 | AF-CAN | 4880.0000 | G-CAN-2 | 5079.0000 | M-US-3 | 5217.0000 | AR-US-3 | 5370.0000 | PS-AUS |
| 4730.0000 | AF/N-G | 4880.0000 | G-AUS | 5080.0000 | N-G | 5220.0000 | G-US-3 | 5370.0000 | CC-US-2 |
| 4730.0000 | N-US-1 | 4880.0000 | PS-AUS | 5080.0000 | N-US-1 | 5220.0000 | G-US-10 | 5370.5000 | AF-US-6 |
| 4732.0000 | AF/N-G | 4882.0000 | AF-US-8 | 5080.0000 | N-US-3 | 5222.5000 | G-CAN-3 | 5373.0000 | G-US-8 |
| 4735.0000 | N-US-1 | 4883.0000 | CG-US-6 | 5080.0000 | CG-US-1 | 5222.5000 | M-US-2 | 5374.7000 | PR-WW-2 |
| 4739.0000 | AF/N-G | 4883.5000 | G-CAN-2 | 5082.0000 | M-US-3 | 5224.0000 | SG-UD-6 | 5377.5000 | G-US-2 |
| 4742.0000 | AF-US-2 | 4885.0000 | AF-US-8 | 5085.0000 | G-US-10 | 5224.0000 | AF-US-6 | 5379.0000 | G-US-3 |
| 4742.0000 | AF/N-G | 4885.0000 | AR-US-1 | 5085.0000 | M-US-3 | 5226.0000 | CG-US-6 | 5380.0000 | G-CAN-4 |
| 4744.0000 | AF/N-G | 4885.0000 | G-CAN-4 | 5085.0000 | N-AFS | 5227.0000 | PS-AUS | 5384.5000 | CG-US-3 |
| 4746.0000 | AF-US-3 | 4886.0000 | G-US-3 | 5088.0000 | M-US-3 | 5230.0000 | PS-AUS | 5385.0000 | N-US-5 |
| 4747.0000 | AF-US-3 | 4886.0000 | G-US-3 | 5090.0000 | AR-US-1 | 5230.8000 | BC-WW-1 | 5388.5000 | G-US-12 |
| 4749.0000 | AF/N-G | 4886.5000 | G-CAN-4 | 5095.0000 | AF/N-G | 5235.0000 | AR-US-1 | 5390.0000 | G-CAN-3 |
| 4752.0000 | AF-CAN | 4895.0000 | AR-US-1 | 5097.0000 | N-CAN | 5235.0000 | SP-US-1 | 5390.0000 | BC-WW-1 |
| 4755.0000 | SP-US-1 | 4896.0000 | AF-US-1 | 5099.0000 | PR-US-4 | 5240.0000 | G-CAN-2 | 5395.0000 | AF-US-10 |
| 4758.0000 | AF-US-8 | 4898.0000 | AR-US-1 | 5100.0000 | PR-US-6 | 5242.5000 | AF-US-6 | 5396.0000 | AR-US-3 |
| 4760.0000 | AF-US-5 | 4898.0000 | G-CAN-3 | 5102.0000 | PR-US-4 | 5242.5000 | G-US-4 | 5396.0000 | G-CAN-3 |
| 4760.0000 | AF-US-10 | 4900.0000 | G-US-3 | 5104.0000 | G-WW-3 | 5243.0000 | AF-US-1 | 5396.0000 | M-US-2 |
| 4760.0000 | SP-US-1 | 4900.0000 | SP-US-1 | 5107.5000 | G-CAN-3 | 5245.0000 | A-WW-3 | 5397.0000 | AR-US-1 |
| 4764.0000 | N-AFS | 4902.0000 | G-US-1 | 5110.0000 | G-US-3 | 5246.0000 | AF-US-6 | 5398.0000 | AF-US-10 |
| 4765.0000 | AF-US-8 | 4903.0000 | G-US-3 | 5110.0000 | PS-AUS | 5246.0000 | SP-US-1 | 5398.0000 | G-US-3 |
| 4765.0000 | G-CAN-1 | 4910.0000 | G-US-3 | 5110.0000 | AF-US-1 | 5248.5000 | CC-CAN | 5400.0000 | N-AUS |
| 4765.0000 | G-CAN-2 | 4912.0000 | AF-US-8 | 5110.0000 | PS-AUS | 5250.0000 | G-US-3 | 5400.0000 | N-I |
| 4766.5000 | G-CAN-3 | 4925.0000 | AF-US-10 | 5113.0000 | G-US-3 | 5250.0000 | PR-US-6 | 5400.0000 | N-NZL |
| 4768.0000 | N-US-3 | 4926.0000 | PS-AUS | 5115.0000 | AR-US-3 | 5251.0000 | PR-US-6 | 5400.0000 | AR-US-2 |
| 4769.0000 | G-CAN-4 | 4927.5000 | G-CAN-4 | 5115.0000 | N-US-4 | 5255.0000 | G-US-1 | 5401.0000 | AR-US-3 |
| 4770.0000 | N-ARG | 4938.0000 | AF-US-10 | 5115.0000 | G-CAN-4 | 5260.0000 | PS-AUS | 5402.0000 | G-US-1 |
| 4770.0000 | N-AFS | 4940.0000 | PS-AUS | 5116.5000 | PR-WW-2 | 5270.0000 | AF-US-10 | 5402.0000 | G-US-10 |
| 4776.5000 | G-US-2 | 4946.5000 | G-US-2 | 5120.0000 | A-WW-3 | 5270.0000 | G-US-3 | 5405.0000 | CC-CAN |
| 4776.5000 | G-CAN-1 | 4950.0000 | G-CAN-3 | 5123.0000 | CG-US-6 | 5270.0000 | PR-US-6 | 5410.0000 | G-US-3 |
| 4780.0000 | AR-US-1 | 4960.0000 | AR-US-1 | 5124.0000 | G-CAN-2 | 5270.5000 | G-US-3 | 5410.0000 | G-CAN-3 |
| 4780.0000 | G-US-1 | 4964.5000 | CC-CAN | 5125.0000 | G-US-3 | 5276.5000 | G-CAN-3 | 5410.0000 | PS-AUS |
| 4785.0000 | G-CAN-1 | 4965.0000 | G-US-1 | 5125.0000 | BC-WW-1 | 5278.5000 | PR-US-6 | 5411.0000 | CC-CAN |
| 4791.5000 | CC-US-2 | 4967.0000 | G-CAN-3 | 5130.0000 | PS-AUS | 5280.0000 | A-WW-5 | 5412.0000 | AF-US-10 |
| 4792.5000 | M-US-2 | 4971.1000 | G-CAN-3 | 5133.0000 | G-US-8 | 5281.5000 | CC-CAN | 5419.5000 | CG-US-8 |
| 4795.5000 | PR-US-6 | 4980.0000 | PS-AUS | 5134.0000 | CC-CAN | 5283.0000 | PR-WW-2 | 5420.0000 | A-US-3 |
| 4797.0000 | PR-US-6 | 4982.0000 | AF-US-10 | 5134.5000 | CC-US-2 | 5286.0000 | G-US-10 | 5422.0000 | CG-US-6 |
| 4797.0000 | PR-US-7 | 4982.0000 | G-CAN-2 | 5135.0000 | G-US-7 | 5287.5000 | CG-US-4 | 5422.5000 | CG-US-1 |
| 4797.0000 | CC-US-1 | 4983.0000 | M-US-5 | 5135.0000 | CC-US-2 | 5287.5000 | G-US-4 | 5424.0000 | G-US-1 |
| 4798.5000 | G-CAN-1 | 4990.0000 | G-CAN-4 | 5137.5000 | CC-US-2 | 5289.5000 | CC-CAN | 5426.0000 | G-US-3 |
| 4798.5000 | PR-US-6 | 4991.0000 | PR-US-6 | 5140.0000 | G-CAN-3 | 5290.0000 | BC-WW-1 | 5430.0000 | N-US-1 |
| 4800.0000 | N-AFS | 4992.0000 | SP-US-1 | 5140.0000 | G-US-7 | 5295.0000 | BC-WW-1 | 5430.0000 | G-CAN-3 |
| 4803.5000 | N-US-4 | 4992.5000 | PR-US-6 | 5145.0000 | PS-AUS | 5297.0000 | AF-US-4 | 5431.0000 | G-US-3 |
| 4805.5000 | N-US-4 | 4996.0000 | G-WW-3 | 5149.0000 | CC-CAN | 5300.0000 | G-US-3 | 5432.0000 | AF-US-10 |
| 4807.0000 | A-WW-3 | 5000.0000 | G-WW-3 | 5150.0000 | G-US-5 | 5300.0000 | N-AFS | 5434.0000 | AF-US-10 |
| 4807.0000 | A-WW-1 | 5004.0000 | G-WW-3 | 5150.0000 | G-US-4 | 5300.0000 | N-CHL | 5434.0000 | BC-WW-1 |
| 4808.0000 | G-US-3 | 5005.0000 | PR-US-1 | 5152.0000 | AF-US-10 | 5300.0000 | PS-AUS | 5436.0000 | SP-US-1 |
| 4812.0000 | AF-US-5 | 5005.0000 | PR-US-4 | 5158.0000 | N-US-2 | 5300.0000 | PR-US-6 | 5436.5000 | CC-CAN |
| 4812.5000 | G-CAN-1 | 5005.0000 | PR-US-6 | 5159.5000 | N-US-4 | 5302.0000 | N-NOR | 5437.0000 | AF-US-10 |
| 4813.5000 | N-US-1 | 5008.0000 | G-US-1 | 5161.0000 | G-US-10 | 5305.0000 | AF-US-10 | 5437.0000 | AR-US-2 |
| 4815.0000 | AF-US-8 | 5009.0000 | G-CAN-3 | 5162.5000 | G-US-10 | 5305.0000 | G-US-3 | 5441.0000 | AF/N-G |
| 4820.0000 | N-US-4 | 5009.0000 | M-US-2 | 5164.5000 | CC-US-2 | 5305.5000 | N-URS | 5441.5000 | CC-CAN |
| 4820.0000 | CC-CAN | 5010.0000 | G-US-10 | 5167.5000 | G-US-4 | 5305.5000 | G-WW-3 | 5443.0000 | G-US-3 |
| 4821.0000 | G-US-1 | 5010.0000 | PS-AUS | 5167.5000 | CC-US-2 | 5308.0000 | G-US-2 | 5445.0000 | G-CAN-1 |
| 4825.0000 | SP-US-1 | 5011.0000 | AR-US-2 | 5170.0000 | G-CAN-2 | 5313.0000 | CG-US-8 | 5445.0000 | PS-AUS |
| 4826.5000 | N-US-4 | 5013.5000 | G-CAN-2 | 5171.0000 | AF-US-1 | 5313.5000 | CC-CAN | 5446.0000 | N-US-1 |
| 4832.0000 | AF-US-8 | 5015.0000 | AR-US-2 | 5180.0000 | PS-AUS | 5313.0000 | PR-US-4 | 5450.0000 | AF/N-G |
| 4835.0000 | N-US-1 | 5020.0000 | AF-US-1 | 5180.0000 | SP-US-1 | 5315.5000 | CG-US-6 | 5450.0000 | N-F |
| 4837.0000 | CC-CAN | 5024.5000 | G-US-1 | 5181.0000 | G-CAN-3 | 5317.0000 | CG-US-6 | 5450.0000 | G-US-3 |
| 4837.5000 | G-WW-3 | 5026.0000 | AF-US-1 | 5186.6000 | CC-CAN | 5320.0000 | CG-US-2 | 5450.0000 | G-AUS |
| 4840.0000 | AR-US-1 | 5026.0000 | AF-US-10 | 5187.0000 | SP-US-1 | 5320.0000 | CG-US-3 | 5451.0000 | A-US-2 |
| 4840.0000 | G-US-3 | 5031.0000 | G-US-1 | 5190.0000 | SP-US-1 | 5320.0000 | CG-US-6 | 5455.0000 | BC-WW-1 |
| 4840.0000 | G-CAN-3 | 5031.0000 | G-CAN-2 | 5192.0000 | G-US-7 | 5320.0000 | CG-US-8 | 5460.0000 | BC-WW-1 |
| 4842.0000 | AF-US-8 | 5031.0000 | G-CAN-3 | 5194.0000 | G-US-3 | 5326.0000 | G-US-6 | 5462.0000 | AF/N-G |
| 4850.0000 | AR-US-2 | 5046.5000 | PR-US-4 | 5195.0000 | G-US-4 | 5327.0000 | AR-US-2 | 5463.0000 | A-US-2 |
| 4855.5000 | G-WW-3 | 5052.0000 | PR-US-4 | 5195.0000 | G-US-7 | 5328.0000 | AF-US-1 | 5469.0000 | A-US-2 |
| 4856.0000 | SP-US-1 | 5053.0000 | G-CAN-2 | 5195.0000 | BC-WW-1 | 5330.0000 | G-US-1 | 5469.0000 | PR-US-6 |
| 4857.5000 | CG-US-6 | 5055.5000 | PR-US-4 | 5197.0000 | AF-US-3 | 5340.0000 | AF-US-10 | 5470.0000 | AF/N-G |
| 4860.0000 | SP-US-1 | 5058.5000 | G-US-12 | 5200.0000 | PS-AUS | 5341.5000 | G-CAN-2 | 5470.0000 | G-US-3 |

| Freq | ID | Freq | ID | Freq | ID | Freq | ID | Freq | ID |
|---|---|---|---|---|---|---|---|---|---|
| 5470.0000 | G-US-5 | 5680.0000 | A-US-3 | 5803.0000 | AF-CAN | 6049.0000 | G-US-1 | 6273.3000 | CG-US-1 |
| 5470.0000 | BC-WW-1 | 5680.0000 | AF-AFS | 5803.0000 | A-WW-5 | 6060.0000 | PR-US-6 | 6274.0000 | N-AFS |
| 5475.0000 | G-US-3 | 5680.0000 | N-US-1 | 5805.0000 | AF-US-5 | 6060.0000 | CC-US-1 | 6275.0000 | N-IND |
| 5475.0000 | A-WW-5 | 5680.0000 | AF-CAN | 5805.0000 | G-US-10 | 6060.0000 | PR-US-6 | 6284.2500 | CG-US-1 |
| 5475.0000 | M-WW-2 | 5680.0000 | CG-US-1 | 5806.0000 | G-US-3 | 6100.0000 | AF-US-5 | 6285.0000 | PR-US-6 |
| 5481.0000 | CG-US-2 | 5680.0000 | CG-US-2 | 5810.0000 | N-US-3 | 6100.0000 | G-WW-3 | 6292.5000 | CG-CAN |
| 5481.0000 | CG-US-4 | 5680.0000 | CG-US-8 | 5810.0000 | SP-US-1 | 6100.0000 | PR-US-6 | 6297.0000 | CG-US-1 |
| 5486.5000 | CG-CAN | 5680.0000 | SP-US-1 | 5810.0000 | CC-CAN | 6106.0000 | G-US-1 | 6320.2500 | G-US-9 |
| 5490.0000 | A-WW-3 | 5684.0000 | AF-CAN | 5810.0000 | BC-WW-1 | 6108.0000 | G-US-1 | 6326.5000 | M-US-3 |
| 5490.0000 | N-J | 5684.0000 | AF-US-1 | 5811.5000 | G-CAN-2 | 6147.5000 | M-US-3 | 6330.0000 | AF-CAN |
| 5493.0000 | A-WW-2 | 5685.0000 | AF/N-G | 5811.5000 | G-CAN-4 | 6151.0000 | G-US-1 | 6335.5000 | CG-CAN |
| 5498.0000 | A-WW-2 | 5688.0000 | AF-US-3 | 5812.0000 | N-AFS | 6151.5000 | PR-US-6 | 6339.0000 | PR-US-6 |
| 5498.0000 | A-WW-5 | 5688.0000 | AF-AUS/NZL | 5815.0000 | BC-WW-1 | 6153.0000 | PR-US-6 | 6340.0000 | AF-CAN |
| 5499.0000 | A-WW-5 | 5690.0000 | AF-CAN | 5820.0000 | AF-US-10 | 6153.0000 | CC-US-1 | 6340.5000 | CG-US-4 |
| 5508.0000 | A-US-2 | 5691.0000 | AF-D | 5820.0000 | AR-US-1 | 6154.5000 | PR-US-6 | 6340.5000 | CG-US-3 |
| 5520.0000 | A-WW-2 | 5691.0000 | A-WW-5 | 5821.0000 | G-US-1 | 6160.0000 | PR-US-6 | 6341.0000 | CG-US-2 |
| 5526.0000 | AF-SAU | 5692.0000 | CG-US-2 | 5822.0000 | AF-US-10 | 6160.0000 | CC-US-1 | 6344.0000 | CG-US-3 |
| 5526.0000 | A-WW-1 | 5695.0000 | A-US-3 | 5822.0000 | SP-US-1 | 6171.0000 | PR-US-6 | 6348.0000 | M-US-3 |
| 5526.0000 | A-WW-2 | 5695.0000 | AF-AUS/NZL | 5822.5000 | G-US-3 | 6172.5000 | CC-US-1 | 6348.0000 | N-F |
| 5526.0000 | A-WW-3 | 5695.0000 | AF-AFS | 5826.0000 | AF-US-1 | 6172.5000 | PR-US-6 | 6351.5000 | CG-CAN |
| 5529.0000 | A-US-1 | 5695.5000 | AF/N-G | 5827.0000 | AF-US-5 | 6174.0000 | PR-US-6 | 6352.0000 | N-F |
| 5529.0000 | A-WW-1 | 5696.0000 | N-US-3 | 5827.0000 | G-AUS | 6174.5000 | PR-US-6 | 6355.0000 | CG-US-4 |
| 5532.0000 | A-US-1 | 5696.0000 | CG-US-2 | 5830.0000 | G-US-3 | 6176.0000 | G-US-1 | 6361.0000 | M-WW-2 |
| 5532.0000 | A-WW-1 | 5700.0000 | AF-US-1 | 5830.0000 | BC-WW-1 | 6190.0000 | AF/N-G | 6364.5000 | M-WW-2 |
| 5535.0000 | A-US-1 | 5700.0000 | SP-US-1 | 5833.0000 | PS-AUS | 6200.0000 | CG-US-8 | 6365.5000 | M-US-3 |
| 5535.0000 | A-WW-1 | 5703.0000 | AF-US-2 | 5835.0000 | AF-AUS/NZL | 6200.0000 | M-WW-1 | 6369.0000 | M-US-3 |
| 5535.0000 | A-WW-3 | 5703.0000 | AF-US-3 | 5835.0000 | N-AUS | 6200.0000 | M-WW-2 | 6370.0000 | CG-CAN |
| 5536.5000 | A-WW-1 | 5703.0000 | AF-US-1 | 5841.0000 | PR-US-6 | 6200.0000 | PR-US-6 | 6371.2000 | N-G |
| 5538.0000 | A-US-1 | 5703.0000 | AF-US-5 | 5845.0000 | PS-AUS | 6203.1000 | G-US-9 | 6372.5000 | M-WW-2 |
| 5538.0000 | A-WW-1 | 5703.0000 | AF-CAN | 5845.0000 | BC-WW-1 | 6203.1000 | M-WW-1 | 6376.0000 | M-US-3 |
| 5540.0000 | G-CAN-2 | 5707.0000 | AF-AUS/NZL | 5850.0000 | AR-US-1 | 6206.2000 | CG-US-8 | 6381.0000 | CG-US-2 |
| 5541.0000 | A-US-1 | 5710.0000 | AF-US-3 | 5850.0000 | N-CAN | 6206.2000 | M-WW-1 | 6381.0000 | CG-US-8 |
| 5541.0000 | A-WW-1 | 5710.0000 | A-WW-1 | 5850.0000 | PS-AUS | 6209.3000 | M-US-4 | 6382.2200 | M-WW-2 |
| 5544.0000 | A-US-1 | 5716.5000 | N-US-1 | 5856.0000 | G-US-3 | 6209.3000 | M-WW-1 | 6383.0000 | CG-US-1 |
| 5544.0000 | A-WW-1 | 5718.0000 | AF-CAN | 5856.0000 | G-CAN-3 | 6211.8000 | G-US-9 | 6383.0000 | M-WW-2 |
| 5546.0000 | G-US-3 | 5718.0000 | N-US-1 | 5856.0000 | M-US-2 | 6212.4000 | M-US-4 | 6384.5000 | AF-CAN |
| 5547.0000 | A-WW-2 | 5721.0000 | AF/N-G | 5860.0000 | G-US-10 | 6212.4000 | M-WW-1 | 6385.0000 | AR-US-1 |
| 5550.0000 | A-WW-2 | 5724.0000 | N-US-1 | 5865.0000 | N-US-4 | 6215.5000 | CG-US-1 | 6386.5000 | M-WW-2 |
| 5553.0000 | A-WW-1 | 5725.0000 | G-US-3 | 5865.0000 | PS-AUS | 6215.5000 | M-WW-1 | 6386.5000 | N-AFS |
| 5559.0000 | CC-US-1 | 5726.0000 | N-US-3 | 5868.0000 | G-US-3 | 6218.6000 | CG-US-4 | 6388.0000 | N-E |
| 5561.0000 | A-WW-5 | 5729.0000 | AF/N-G | 5875.0000 | BC-WW-1 | 6218.6000 | CG-US-8 | 6389.0000 | N-P |
| 5562.0000 | G-US-9 | 5730.0000 | G-CAN-3 | 5875.5000 | AF-CAN | 6218.6000 | M-WW-1 | 6389.6500 | M-US-3 |
| 5565.0000 | A-WW-2 | 5730.0000 | PR-US-1 | 5880.0000 | BC-WW-1 | 6218.6000 | M-WW-2 | 6390.0000 | BC-WW-1 |
| 5565.0000 | A-WW-3 | 5730.0000 | PR-US-6 | 5885.0000 | G-CAN-4 | 6218.6000 | N-B | 6390.3000 | N-I |
| 5568.0000 | A-WW-2 | 5731.0000 | G-US-3 | 5895.0000 | G-WW-3 | 6220.0000 | PR-US-6 | 6391.5000 | N-BEL |
| 5571.0000 | A-US-2 | 5731.0000 | PS-AUS | 5895.0000 | PS-AUS | 6221.6000 | CG-US-8 | 6393.0000 | G-US-9 |
| 5571.0000 | G-US-10 | 5732.0000 | G-US-3 | 5900.0000 | G-AUS | 6221.6000 | AF-CAN | 6393.5000 | M-WW-2 |
| 5571.0000 | PR-US-6 | 5733.0000 | SP-US-1 | 5906.0000 | G-US-10 | 6221.6000 | M-WW-1 | 6397.0000 | N-US-3 |
| 5574.0000 | A-WW-2 | 5735.0000 | PS-AUS | 5910.0000 | AF-US-10 | 6223.0000 | G-US-11 | 6397.0000 | G-US-9 |
| 5575.0000 | A-WW-5 | 5735.0000 | BC-WW-1 | 5912.0000 | PR-US-6 | 6246.1000 | CG-US-5 | 6400.5000 | M-WW-2 |
| 5580.0000 | A-WW-5 | 5736.5000 | PR-US-1 | 5912.5000 | G-US-6 | 6246.7000 | CG-US-5 | 6402.0000 | N-AFS |
| 5598.0000 | A-WW-2 | 5737.0000 | N-AFS | 5913.0000 | G-US-12 | 6247.0000 | CC-CAN | 6404.0000 | M-US-3 |
| 5601.0000 | A-WW-1 | 5740.0000 | AF-US-8 | 5915.0000 | PS-AUS | 6247.3000 | CG-US-5 | 6406.0000 | M-WW-2 |
| 5601.0000 | A-WW-2 | 5740.0000 | PS-AUS | 5916.5000 | PR-WW-2 | 6252.0000 | PR-US-6 | 6409.5000 | M-I |
| 5601.0000 | A-WW-5 | 5742.5000 | BC-WW-1 | 5920.0000 | A-US-1 | 6258.0000 | CG-US-1 | 6410.0000 | N-AFS |
| 5604.0000 | A-WW-3 | 5745.0000 | A-WW-1 | 5920.0000 | BC-WW-1 | 6258.0000 | CG-US-4 | 6414.5000 | M-US-3 |
| 5610.0000 | A-WW-1 | 5745.0000 | AF-US-5 | 5920.0000 | G-US-3 | 6258.0000 | CG-US-3 | 6416.0000 | M-US-3 |
| 5610.0000 | A-WW-3 | 5745.0000 | BC-WW-1 | 5924.0000 | G-US-3 | 6258.5000 | M-US-3 | 6418.2000 | M-I |
| 5616.0000 | A-WW-2 | 5745.0000 | CG-US-6 | 5925.0000 | G-US-4 | 6259.5000 | G-CAN-4 | 6421.5000 | CG-US-4 |
| 5628.0000 | A-WW-2 | 5747.0000 | AF/N-G | 5925.0000 | G-US-9 | 6260.0000 | CG-US-8 | 6421.5000 | CG-US-3 |
| 5631.0000 | G-US-10 | 5749.0000 | G-US-1 | 5925.0000 | G-US-10 | 6260.5000 | G-CAN-4 | 6432.0000 | M-WW-2 |
| 5634.0000 | A-WW-2 | 5751.0000 | G-US-2 | 5925.0000 | CC-CAN | 6262.0000 | CG-US-1 | 6435.5000 | M-I |
| 5640.0000 | A-WW-5 | 5755.0000 | G-US-3 | 5937.0000 | AF-US-10 | 6262.0000 | CG-US-3 | 6438.0000 | CG-CAN |
| 5643.0000 | G-US-3 | 5755.5000 | G-US-1 | 5940.0000 | AF-CAN | 6262.5000 | M-US-3 | 6441.0000 | M-WW-2 |
| 5643.0000 | A-WW-2 | 5760.0000 | AF-US-10 | 5940.0000 | N-J | 6265.6000 | G-US-9 | 6450.0000 | AF-CAN |
| 5645.0000 | A-WW-1 | 5760.0000 | AR-US-3 | 5945.0000 | G-AUS | 6266.0000 | CG-US-1 | 6452.0000 | AF-CAN |
| 5645.0000 | PR-WW-2 | 5760.0000 | PR-US-6 | 5945.0000 | G-US-10 | 6266.0000 | CG-US-4 | 6455.0000 | M-US-3 |
| 5646.0000 | A-WW-2 | 5760.0000 | CC-US-1 | 5947.0000 | N-J | 6266.0000 | CG-US-3 | 6455.0000 | N-AFS |
| 5649.0000 | A-WW-2 | 5760.0000 | BC-WW-1 | 5947.0000 | G-US-2 | 6266.5000 | M-US-3 | 6456.0000 | AF-CAN |
| 5652.0000 | A-WW-1 | 5765.0000 | CG-US-3 | 5950.0000 | PR-US-6 | 6267.5000 | CG-US-8 | 6457.0000 | CG-US-4 |
| 5652.0000 | A-WW-2 | 5775.0000 | AF-US-10 | 5960.0000 | PR-US-6 | 6268.0000 | CG-US-3 | 6460.0000 | AR-US-1 |
| 5655.0000 | A-WW-2 | 5775.0000 | SP-US-1 | 5960.0000 | CC-US-1 | 6268.0000 | PR-US-6 | 6462.0000 | AF-F |
| 5658.0000 | A-WW-2 | 5785.0000 | N-NZL | 5961.0000 | G-US-1 | 6268.4000 | CG-US-3 | 6467.0000 | CG-US-4 |
| 5661.0000 | A-WW-2 | 5790.0000 | BC-WW-1 | 5985.0000 | PR-US-6 | 6268.5000 | CG-US-4 | 6467.0000 | CG-US-3 |
| 5664.0000 | A-WW-2 | 5793.5000 | G-US-3 | 5987.0000 | AF-US-1 | 6268.5000 | M-US-3 | 6470.0000 | BC-WW-1 |
| 5666.0000 | A-WW-2 | 5795.0000 | G-CAN-4 | 5997.0000 | AR-US-1 | 6269.0000 | CG-US-3 | 6474.0000 | CG-US-3 |
| 5667.0000 | A-WW-2 | 5797.5000 | N-US-4 | 6020.0000 | AR-US-2 | 6269.5000 | CG-US-4 | 6477.5000 | CG-US-3 |
| 5670.0000 | A-US-3 | 5800.0000 | AF-US-1 | 6040.0000 | A-US-1 | 6269.5000 | CG-US-3 | 6477.5000 | M-US-3 |
| 5676.0000 | A-WW-5 | 5800.0000 | AF-US-10 | 6040.0000 | N-ARG | 6272.7000 | CG-US-1 | 6491.5000 | CG-US-3 |

| | | | | | | | | | | | |
|---|---|---|---|---|---|---|---|---|---|---|---|
| 6491.5000 | CG-CAN | 6655.0000 | A-WW-2 | 6760.0000 | AF/N-G | 6855.0000 | PR-US-4 | 6980.0000 | G-US-3 | 6980.0000 | BC-WW-1 |
| 6493.0000 | CG-CAN | 6669.0000 | CC-US-1 | 6761.0000 | AF-US-1 | 6858.0000 | PR-US-4 | 6982.0000 | G-US-3 | 6982.5000 | G-US-2 |
| 6496.0000 | CG-US-1 | 6673.0000 | G-US-9 | 6762.0000 | G-US-3 | 6860.0000 | G-US-3 | 6982.0000 | SP-US-1 | 6985.0000 | PS-AUS |
| 6496.0000 | CG-US-4 | 6673.0000 | A-WW-1 | 6763.0000 | PR-US-4 | 6861.0000 | PR-US-4 | 6987.0000 | G-US-3 | 6987.5000 | BC-WW-1 |
| 6496.0000 | CG-US-3 | 6673.0000 | A-WW-2 | 6765.0000 | AF/N-G | 6863.0000 | AF-US-1 | 6988.0000 | AR-US-1 | 6988.0000 | AR-US-3 |
| 6496.5000 | M-US-3 | 6676.0000 | AF-AUS/NZL | 6765.0000 | G-US-3 | 6866.0000 | G-US-3 | 6989.0000 | AF-US-10 | 6991.0000 | G-US-3 |
| 6497.5000 | CG-CAN | 6676.0000 | A-WW-5 | 6765.0000 | PR-US-1 | 6866.0000 | PS-AUS | 6993.0000 | AF-US-10 | 6993.0000 | BC-WW-1 |
| 6497.5000 | G-CAN-4 | 6679.0000 | A-WW-5 | 6766.0000 | AR-US-1 | 6866.5000 | G-US-3 | 6994.0000 | AF-US-8 | 6994.0000 | AR-US-1 |
| 6498.5000 | M-WW-2 | 6680.0000 | A-WW-5 | 6766.0000 | AR-US-3 | 6870.0000 | AF-US-1 | 6995.0000 | AF-US-8 | 6995.0000 | BC-WW-1 |
| 6500.0000 | CG-US-1 | 6680.0000 | AF-US-1 | 6766.0000 | G-US-3 | 6870.0000 | G-US-10 | 6995.5000 | AR-US-3 | 6996.0000 | AF-US-8 |
| 6500.0000 | CG-US-3 | 6680.0000 | N-US-4 | 6769.0000 | AF-US-5 | 6871.0000 | G-US-3 | 6997.0000 | N-B | 6997.5000 | AR-US-3 |
| 6500.5000 | M-US-3 | 6683.0000 | AF-US-3 | 6770.0000 | BC-WW-1 | 6873.0000 | BC-WW-1 | 6998.5000 | PS-WW-1 | 7047.0000 | N-AFS |
| 6503.5000 | G-US-9 | 6683.0000 | AF-US-10 | 6771.0000 | G-US-3 | 6874.0000 | G-US-10 | 7160.0000 | M-AUS | 7220.0000 | PS-CRIB |
| 6504.0000 | CG-US-1 | 6686.0000 | AF/N-G | 6771.5000 | G-CAN-2 | 6874.0000 | N-AFS | 7300.0000 | PR-US-1 | 7300.0000 | PR-US-4 |
| 6504.0000 | CG-US-4 | 6687.0000 | AF-AFS | 6775.0000 | AF-US-8 | 6874.5000 | CC-CAN | 7300.0000 | PR-US-6 | 7301.5000 | N-US-4 |
| 6504.0000 | CG-US-3 | 6687.0000 | N-US-1 | 6776.0000 | G-US-10 | 6878.5000 | N-US-4 | 7302.0000 | AF-US-8 | 7305.0000 | AF-US-8 |
| 6504.5000 | M-US-3 | 6690.0000 | AF/N-G | 6780.0000 | AF-US-9 | 6880.0000 | SP-US-1 | 7305.0000 | AF-US-10 | 7306.5000 | G-CAN-2 |
| 6505.5000 | M-WW-2 | 6693.0000 | AF-CAN | 6780.0000 | N-AFS | 6880.0000 | PS-AUS | 7307.0000 | PS-AUS | 7309.5000 | AR-US-3 |
| 6506.0000 | M-WW-2 | 6693.0000 | AF/N-G | 6780.0000 | G-US-3 | 6885.0000 | PR-US-4 | 7310.0000 | CC-CAN | 7311.0000 | AR-US-3 |
| 6506.4000 | CG-US-8 | 6693.0000 | N-US-1 | 6780.0000 | PR-US-3 | 6886.0000 | AF-US-1 | 7312.0000 | AF-US-8 | 7313.0000 | SP-US-1 |
| 6506.4000 | M-WW-1 | 6693.0000 | A-WW-5 | 6783.5000 | G-US-6 | 6888.0000 | N-F | 7313.5000 | AF-US-8 | 7313.5000 | AR-US-3 |
| 6506.4000 | M-WW-2 | 6697.0000 | AF/N-G | 6785.0000 | AR-US-2 | 6888.0000 | PR-US-4 | 7315.0000 | AF-US-8 | 7316.0000 | AF-US-10 |
| 6509.5000 | G-US-9 | 6697.0000 | N-US-1 | 6785.0000 | PS-AUS | 6888.5000 | BC-WW-1 | 7321.5000 | G-US-3 | 7324.0000 | AF-US-8 |
| 6509.5000 | M-WW-1 | 6699.5000 | AF-AUS/NZL | 6787.0000 | G-US-3 | 6889.0000 | SP-US-1 | 7326.5000 | G-CAN-3 | 7329.0000 | AF-US-8 |
| 6509.6000 | M-WW-2 | 6700.5000 | AF-PHL | 6788.5000 | G-US-3 | 6890.0000 | AF-US-5 | 7330.0000 | AF-US-1 | 7330.0000 | PS-AUS |
| 6512.6000 | CG-US-8 | 6705.0000 | AF-CAN | 6790.0000 | AF-US-10 | 6890.0000 | PS-AUS | 7331.0000 | AF-US-8 | 7335.0000 | G-WW-3 |
| 6512.6000 | CG-CAN | 6705.0000 | AF/N-G | 6790.0000 | AF/N-G | 6895.0000 | AF-US-5 | 7340.0000 | N-US-3 | 7340.0000 | PS-AUS |
| 6512.6000 | M-WW-1 | 6705.0000 | N-US-1 | 6790.0000 | AR-US-2 | 6895.0000 | G-US-10 | 7346.5000 | N-US-4 | 7347.5000 | G-CAN-2 |
| 6513.0000 | CG-US-4 | 6706.0000 | G-US-3 | 6790.0000 | CC-CAN | 6896.0000 | SP-US-1 | 7348.0000 | G-US-1 | 7349.5000 | CG-US-8 |
| 6515.7000 | CC-CAN | 6707.0000 | AF/N-G | 6792.0000 | G-WW-3 | 6900.0000 | PR-US-6 | 7349.5000 | G-US-2 | 7350.0000 | N-US-4 |
| 6515.7000 | M-US-4 | 6708.0000 | AF-CAN | 6792.0000 | G-CAN-1 | 6902.5000 | BC-WW-1 | 7350.0000 | PR-WW-2 | 7351.5000 | CG-US-4 |
| 6515.7000 | M-WW-1 | 6708.0000 | AF/N-G | 6793.0000 | AF-US-10 | 6905.0000 | AR-CAN | 7354.0000 | PR-US-6 | 7355.0000 | G-US-2 |
| 6516.8000 | G-US-9 | 6708.0000 | N-US-1 | 6795.0000 | G-US-3 | 6905.0000 | G-WW-3 | 7357.0000 | AF-US-8 | 7360.0000 | AF-US-8 |
| 6518.8000 | CG-CAN | 6708.0000 | N-US-3 | 6799.0000 | N-US-1 | 6905.0000 | PS-AUS | 7360.0000 | AR-US-1 | 7360.0000 | N-US-4 |
| 6518.8000 | M-US-4 | 6712.0000 | AF-US-1 | 6799.0000 | G-US-3 | 6910.0000 | AR-US-1 | 7360.0000 | AR-US-1 | 7360.0000 | AR-US-3 |
| 6518.8000 | M-WW-1 | 6712.0000 | AF-US-3 | 6800.0000 | AF-US-5 | 6910.0000 | AR-US-3 | 7365.0000 | N-US-3 | 7365.0000 | N-US-4 |
| 6521.9000 | CG-US-8 | 6714.0000 | AF-US-5 | 6800.0000 | G-US-3 | 6910.0000 | BC-WW-1 | 7367.5000 | N-US-4 | 7368.5000 | CC-US-2 |
| 6521.9000 | M-WW-1 | 6715.0000 | AF/N-G | 6800.0000 | PR-US-4 | 6918.5000 | BC-US-1 | 7370.0000 | N-US-3 | 7370.0000 | N-US-4 |
| 6521.9000 | M-WW-2 | 6715.0000 | AF-US-5 | 6800.0000 | PS-WW-1 | 6919.0000 | SP-US-1 | 7372.5000 | N-US-4 | 7373.0000 | N-US-4 |
| 6521.9000 | PR-US-1 | 6716.0000 | AF-US-10 | 6801.5000 | N-URS | 6920.0000 | PS-AUS | 7373.5000 | G-US-1 | | |
| 6522.0000 | PR-US-6 | 6716.0000 | AF-CAN | 6802.0000 | G-US-3 | 6925.0000 | G-US-3 | | | | |
| 6525.0000 | AF-PHL | 6720.0000 | AF-AUS/NZL | 6802.5000 | N-URS | 6925.0000 | PS-AUS | | | | |
| 6526.0000 | A-WW-1 | 6720.0000 | N-US-1 | 6803.0000 | N-URS | 6925.5000 | G-US-3 | | | | |
| 6532.0000 | A-WW-2 | 6720.0000 | SP-US-1 | 6803.0000 | G-US-3 | 6927.0000 | AF-US-10 | | | | |
| 6535.0000 | A-WW-2 | 6723.0000 | AF-AUS/NZL | 6803.0000 | PR-US-4 | 6928.0000 | G-US-3 | | | | |
| 6538.0000 | A-WW-5 | 6723.0000 | N-US-1 | 6806.0000 | PR-US-4 | 6930.5000 | G-US-2 | | | | |
| 6550.0000 | A-US-2 | 6726.0000 | G-US-3 | 6808.0000 | BC-WW-1 | 6935.0000 | CG-US-6 | | | | |
| 6550.0000 | PR-US-6 | 6728.0000 | AF-US-10 | 6809.0000 | G-US-1 | 6936.0000 | G-US-3 | | | | |
| 6550.0000 | BC-WW-1 | 6730.0000 | AF-US-1 | 6810.0000 | G-US-10 | 6937.0000 | SP-US-1 | | | | |
| 6553.0000 | A-WW-1 | 6730.0000 | AF-US-2 | 6810.0000 | SP-US-1 | 6940.0000 | AF-US-5 | | | | |
| 6556.0000 | A-WW-2 | 6730.0000 | AF-US-3 | 6810.0000 | BC-WW-1 | 6940.0000 | G-AUS | | | | |
| 6559.0000 | A-WW-1 | 6730.0000 | AF-US-10 | 6812.0000 | AF-US-9 | 6942.5000 | BC-WW-1 | | | | |
| 6562.0000 | A-WW-2 | 6730.0000 | AF/N-G | 6812.0000 | AF-US-10 | 6942.8000 | M-WW-2 | | | | |
| 6562.0000 | AF/N-G | 6730.0000 | A-WW-5 | 6812.0000 | CG-US-6 | 6943.5000 | CG-US-6 | | | | |
| 6574.0000 | A-WW-1 | 6731.0000 | AF-US-10 | 6817.0000 | AF-US-10 | 6945.0000 | CG-US-6 | | | | |
| 6575.0000 | A-WW-2 | 6733.0000 | AF/N-G | 6820.0000 | N-AFS | 6945.0000 | PS-AUS | | | | |
| 6577.0000 | A-WW-2 | 6738.0000 | AF-US-3 | 6820.0000 | SP-US-1 | 6948.5000 | CC-US-2 | | | | |
| 6580.0000 | A-WW-5 | 6738.0000 | AF/N-G | 6823.0000 | BC-WW-1 | 6950.0000 | PS-AUS | | | | |
| 6586.0000 | A-WW-2 | 6740.0000 | AF/N-G | 6824.0000 | G-US-3 | 6950.0000 | PR-US-6 | | | | |
| 6592.0000 | A-WW-2 | 6740.0000 | AF-US-2 | 6825.0000 | AF/N-G | 6952.5000 | G-US-12 | | | | |
| 6594.0000 | G-US-12 | 6741.0000 | AF/N-G | 6825.0000 | AR-US-3 | 6955.0000 | G-CAN-3 | | | | |
| 6598.0000 | A-WW-2 | 6742.0000 | N-US-1 | 6825.0000 | PS-AUS | 6960.0000 | PS-AUS | | | | |
| 6600.0000 | N-NOR | 6746.0000 | AF-CAN | 6825.0000 | CC-CAN | 6965.0000 | PS-AUS | | | | |
| 6603.0000 | A-WW-5 | 6746.0000 | N-US-1 | 6825.0000 | BC-WW-1 | 6966.0000 | CG-US-1 | | | | |
| 6604.0000 | A-WW-5 | 6748.0000 | A-URS | 6826.0000 | AF-US-1 | 6966.5000 | CC-CAN | | | | |
| 6610.0000 | N-AUS | 6748.0000 | AF/N-G | 6828.0000 | G-US-3 | 6968.5000 | N-US-4 | | | | |
| 6617.0000 | A-WW-5 | 6750.0000 | AF-US-3 | 6830.0000 | CC-CAN | 6969.0000 | AF-US-1 | | | | |
| 6628.0000 | A-WW-2 | 6750.0000 | SP-US-1 | 6835.0000 | N-US-1 | 6970.0000 | N-US-3 | | | | |
| 6631.0000 | A-WW-2 | 6751.0000 | AF/N-G | 6838.0000 | BC-WW-1 | 6970.0000 | N-US-4 | | | | |
| 6634.0000 | A-WW-3 | 6753.0000 | AF-US-3 | 6839.4000 | G-US-3 | 6970.0000 | G-US-9 | | | | |
| 6637.0000 | A-US-1 | 6753.0000 | AF-US-2 | 6840.0000 | AF-US-1 | 6970.0000 | BC-WW-1 | | | | |
| 6637.0000 | A-WW-1 | 6753.0000 | AF-CAN | 6840.0000 | G-WW-3 | 6972.0000 | G-US-3 | | | | |
| 6640.0000 | A-US-1 | 6753.0000 | SP-US-1 | 6840.0000 | PS-AUS | 6974.0000 | G-US-3 | | | | |
| 6640.0000 | A-WW-1 | 6754.0000 | G-US-3 | 6840.0000 | BC-WW-1 | 6975.0000 | PS-AUS | | | | |
| 6643.0000 | A-US-1 | 6756.0000 | AF-US-3 | 6845.0000 | PS-AUS | 6975.0000 | BC-WW-1 | | | | |
| 6643.0000 | A-WW-1 | 6756.0000 | AF-US-10 | 6845.0000 | PR-US-6 | 6976.0000 | G-US-6 | | | | |
| 6646.0000 | A-US-1 | 6757.0000 | AF-US-1 | 6847.5000 | BC-WW-1 | 6977.5000 | G-US-9 | | | | |
| 6646.5000 | G-US-2 | 6757.0000 | AF-US-3 | 6850.0000 | N-AUS | 6977.5000 | PS-CRIB | | | | |
| 6649.0000 | A-WW-2 | 6760.0000 | A-US-3 | 6850.0000 | G-US-10 | 6978.0000 | G-US-3 | | | | |

| Frequency | Station | Frequency | Station | Frequency | Station | Frequency | Station | Frequency | Station |
|---|---|---|---|---|---|---|---|---|---|
| 7375.0000 | N-US-4 | 7530.0000 | N-ARG | 7714.0000 | G-US-3 | 7878.0000 | G-US-5 | 8067.0000 | CC-US-2 |
| 7377.5000 | CG-US-8 | 7532.0000 | G-WW-3 | 7717.5000 | CG-US-6 | 7880.0000 | G-US-9 | 8070.0000 | PR-WW-2 |
| 7380.0000 | N-US-4 | 7535.0000 | N-US-1 | 7719.0000 | G-US-3 | 7882.0000 | G-US-3 | 8070.0000 | CC-US-2 |
| 7382.5000 | N-US-4 | 7538.0000 | G-US-3 | 7720.0000 | AR-US-3 | 7885.0000 | N-US-1 | 8077.0000 | SP-US-1 |
| 7385.0000 | N-US-4 | 7540.0000 | AF-US-8 | 7723.5000 | G-US-2 | 7887.5000 | G-US-3 | 8080.0000 | AF-US-10 |
| 7387.5000 | N-US-4 | 7540.0000 | N-US-3 | 7724.0000 | G-US-3 | 7893.0000 | N-US-1 | 8084.0000 | AF/N-G |
| 7392.0000 | PS-AUS | 7541.0000 | A-WW-1 | 7724.0000 | BC-WW-1 | 7893.0000 | BC-WW-1 | 8084.0000 | G-US-10 |
| 7392.2000 | M-US-3 | 7541.5000 | BC-WW-1 | 7726.5000 | AR-US-4 | 7894.0000 | G-US-3 | 8086.0000 | AF-US-10 |
| 7393.0000 | N-US-3 | 7545.0000 | AF-US-8 | 7726.5000 | G-US-1 | 7903.5000 | G-US-12 | 8090.0000 | N-US-3 |
| 7394.5000 | N-URS | 7546.0000 | G-CAN-3 | 7726.5000 | G-US-3 | 7906.0000 | G-WW-3 | 8090.0000 | CG-US-3 |
| 7398.0000 | M-US-3 | 7547.5000 | CC-CAN | 7727.0000 | G-US-3 | 7907.0000 | AF-US-10 | 8092.0000 | CC-CAN |
| 7400.0000 | A-WW-3 | 7548.5000 | BC-WW-1 | 7731.0000 | PR-US-6 | 7910.0000 | SP-US-1 | 8093.0000 | G-US-3 |
| 7400.0000 | G-CAN-3 | 7549.0000 | PR-US-4 | 7735.0000 | AF-US-10 | 7911.0000 | G-US-3 | 8097.0000 | G-US-3 |
| 7401.0000 | G-WW-3 | 7550.0000 | PS-AUS | 7736.0000 | G-US-1 | 7913.5000 | AF-US-8 | 8101.0000 | AF-US-1 |
| 7401.5000 | CC-CAN | 7552.0000 | PR-US-4 | 7737.0000 | G-US-3 | 7918.0000 | BC-WW-1 | 8105.5000 | G-US-3 |
| 7405.0000 | AR-US-3 | 7555.0000 | G-US-3 | 7742.0000 | G-CAN-4 | 7918.5000 | AF-US-7 | 8110.0000 | N-AUS |
| 7405.0000 | G-US-10 | 7555.0000 | PR-US-4 | 7742.0000 | SP-US-1 | 7918.5000 | CG-US-6 | 8110.0000 | BC-WW-1 |
| 7407.0000 | AF-US-8 | 7555.5000 | G-CAN-4 | 7743.0000 | G-US-1 | 7919.0000 | SP-US-1 | 8111.0000 | CG-US-6 |
| 7408.5000 | BC-WW-1 | 7555.5000 | G-CAN-3 | 7750.0000 | N-US-3 | 7922.0000 | AF-US-10 | 8115.0000 | G-US-3 |
| 7410.0000 | PS-AUS | 7558.0000 | PR-US-4 | 7750.0000 | N-BEL | 7922.0000 | N-AUS | 8115.0000 | G-US-5 |
| 7411.0000 | PR-US-6 | 7558.5000 | G-CAN-3 | 7765.0000 | N-URS | 7922.5000 | AF-US-6 | 8119.0000 | G-US-10 |
| 7412.0000 | SP-US-1 | 7559.0000 | PR-US-4 | 7765.0000 | SP-US-1 | 7924.0000 | N-AUS | 8120.0000 | G-US-3 |
| 7419.0000 | G-US-1 | 7562.0000 | PR-US-4 | 7767.5000 | G-US-2 | 7925.0000 | BC-WW-1 | 8120.0000 | G-US-10 |
| 7420.0000 | N-F | 7565.0000 | PS-AUS | 7768.5000 | BC-WW-1 | 7929.5000 | N-US-3 | 8122.0000 | G-US-10 |
| 7429.5000 | G-US-2 | 7567.0000 | AF-US-6 | 7770.0000 | G-US-10 | 7930.0000 | AF-US-10 | 8123.0000 | G-US-3 |
| 7430.0000 | AF-US-3 | 7580.0000 | CG-US-6 | 7771.0000 | N-J | 7932.0000 | AR-US-1 | 8125.0000 | G-US-10 |
| 7434.0000 | G-US-3 | 7580.0000 | PR-US-2 | 7775.0000 | N-G | 7932.0000 | G-US-7 | 8125.0000 | BC-WW-1 |
| 7440.0000 | BC-WW-1 | 7582.0000 | G-US-1 | 7775.0000 | PR-US-2 | 7933.0000 | G-US-3 | 8127.0000 | G-US-3 |
| 7441.0000 | CG-US-6 | 7590.0000 | AR-US-3 | 7775.0000 | PS-WW-1 | 7934.0000 | PS-US | 8129.0000 | G-US-3 |
| 7442.0000 | G-US-3 | 7597.1000 | PR-US-3 | 7778.0000 | G-US-3 | 7935.0000 | G-US-5 | 8130.0000 | G-US-3 |
| 7442.5000 | BC-WW-1 | 7600.0000 | G-WW-3 | 7778.5000 | G-US-12 | 7935.0000 | G-US-7 | 8130.0000 | G-US-10 |
| 7445.0000 | G-US-3 | 7604.0000 | G-US-8 | 7779.0000 | N-URS | 7937.0000 | G-US-3 | 8135.0000 | G-US-3 |
| 7452.0000 | G-US-3 | 7605.0000 | G-US-3 | 7780.0000 | G-CAN-1 | 7938.0000 | AF-US-8 | 8135.0000 | G-CAN-4 |
| 7453.5000 | PS-CRIB | 7605.0000 | SP-US-1 | 7780.0000 | PR-US-6 | 7952.0000 | M-US-5 | 8144.0000 | PS-AUS |
| 7455.0000 | N-US-5 | 7611.0000 | G-US-10 | 7790.0000 | G-US-8 | 7953.0000 | CC-CAN | 8147.0000 | PS-AUS |
| 7457.0000 | AF-US-8 | 7620.0000 | G-US-3 | 7791.0000 | G-US-3 | 7955.0000 | AF-US-10 | 8148.0000 | N-DNK |
| 7460.0000 | AF-US-5 | 7620.0000 | BC-WW-1 | 7799.0000 | AF-US-8 | 7955.0000 | BC-WW-1 | 8150.0000 | N-G |
| 7461.0000 | SP-US-1 | 7622.5000 | G-US-3 | 7800.0000 | G-US-3 | 7956.0000 | G-US-10 | 8150.0000 | N-US-4 |
| 7461.5000 | G-CAN-3 | 7623.0000 | N-G | 7800.0000 | BC-WW-1 | 7960.0000 | G-US-3 | 8150.0000 | CG-US-2 |
| 7462.0000 | G-US-3 | 7626.0000 | CG-US-1 | 7800.0000 | PS-AUS | 7965.0000 | N-URS | 8150.0000 | PS-AUS |
| 7465.0000 | G-CAN-3 | 7627.0000 | G-US-3 | 7802.0000 | G-US-7 | 7971.0000 | PR-WW-2 | 8151.5000 | G-CAN-4 |
| 7465.0000 | PS-AUS | 7628.0000 | SP-US-1 | 7804.0000 | SP-US-1 | 7972.0000 | G-US-3 | 8154.0000 | G-US-3 |
| 7465.5000 | G-US-2 | 7630.0000 | G-CAN-4 | 7804.0000 | CC-CAN | 7973.0000 | BC-WW-1 | 8160.0000 | G-US-3 |
| 7465.5000 | A-WW-5 | 7632.0000 | AF-US-8 | 7805.0000 | G-US-7 | 7975.0000 | AF-US-10 | 8161.0000 | N-J |
| 7466.0000 | AF-US-10 | 7633.5000 | AF-US-8 | 7812.0000 | G-US-3 | 7977.0000 | BC-WW-1 | 8162.0000 | AF-US-10 |
| 7469.0000 | AF-US-10 | 7634.0000 | G-US-3 | 7812.5000 | AR-US-4 | 7980.5000 | M-US-3 | 8167.5000 | G-WW-3 |
| 7470.0000 | G-US-3 | 7635.0000 | AF-US-7 | 7813.0000 | AF-US-10 | 7982.0000 | AF-US-10 | 8169.0000 | CG-US-1 |
| 7470.0000 | BC-WW-1 | 7640.0000 | G-US-3 | 7816.0000 | G-US-3 | 7983.0000 | AF-US-1 | 8170.0000 | A-US-1 |
| 7473.0000 | CG-US-6 | 7645.0000 | N-US-1 | 7818.5000 | CC-CAN | 7983.0000 | G-US-3 | 8170.0000 | A-WW-3 |
| 7474.0000 | G-US-3 | 7645.0000 | G-US-3 | 7821.0000 | G-US-1 | 7985.0000 | SP-US-1 | 8170.0000 | AF-US-10 |
| 7475.0000 | AF-US-1 | 7647.0000 | G-US-3 | 7821.0000 | G-US-3 | 7991.0000 | BC-WW-1 | 8170.0000 | G-US-3 |
| 7475.0000 | G-US-10 | 7648.5000 | BC-WW-1 | 7822.0000 | G-US-3 | 7992.5000 | N-US-4 | 8172.0000 | G-US-3 |
| 7475.0000 | PS-AUS | 7651.0000 | BC-WW-1 | 7825.0000 | G-US-3 | 7993.0000 | N-US-3 | 8174.0000 | G-US-3 |
| 7477.0000 | G-US-7 | 7652.0000 | G-US-3 | 7825.0000 | G-CAN-4 | 7997.0000 | AF-US-10 | 8176.0000 | AF-US-10 |
| 7478.0000 | BC-WW-1 | 7653.5000 | BC-WW-1 | 7827.0000 | AF-US-10 | 8003.5000 | BC-WW-1 | 8176.0000 | G-US-3 |
| 7480.0000 | G-US-7 | 7655.0000 | PR-US-6 | 7828.0000 | AF-F | 8008.0000 | G-US-3 | 8177.0000 | AF-US-8 |
| 7480.0000 | PR-US-4 | 7657.0000 | PR-US-6 | 7830.0000 | G-US-3 | 8013.5000 | G-US-2 | 8178.5000 | AF-US-8 |
| 7483.0000 | PR-US-4 | 7657.0000 | PS-AUS | 7831.0000 | AF-US-1 | 8015.0000 | PR-US-6 | 8180.0000 | AR-US-1 |
| 7484.0000 | G-CAN-4 | 7658.5000 | CC-CAN | 7831.5000 | AF-US-8 | 8021.5000 | CG-US-6 | 8181.0000 | N-J |
| 7485.0000 | N-US-1 | 7660.0000 | AF-US-10 | 7832.0000 | G-WW-3 | 8025.0000 | G-US-5 | 8182.0000 | N-AFS |
| 7486.0000 | PR-US-4 | 7660.0000 | PS-AUS | 7833.0000 | SP-US-1 | 8029.0000 | AF-US-10 | 8185.0000 | A-WW-3 |
| 7490.0000 | N-NZL | 7662.0000 | G-US-3 | 7835.0000 | AR-US-1 | 8030.0000 | G-WW-3 | 8185.0000 | G-US-3 |
| 7490.0000 | BC-WW-1 | 7669.5000 | G-US-1 | 7836.0000 | CG-US-6 | 8033.0000 | N-US-4 | 8185.0000 | N-AFS |
| 7497.0000 | AF-US-10 | 7675.0000 | G-CAN-4 | 7836.6000 | CG-US-6 | 8035.0000 | G-US-3 | 8185.0000 | PS-AUS |
| 7495.0000 | N-US-4 | 7676.0000 | SP-US-1 | 7839.0000 | N-US-1 | 8035.0000 | PS-AUS | 8186.0000 | N-J |
| 7500.0000 | G-WW-3 | 7678.0000 | G-US-3 | 7840.0000 | AF-US-9 | 8038.0000 | G-WW-3 | 8190.0000 | AF/N-G |
| 7500.0000 | N-US-4 | 7680.0000 | AF-US-8 | 7840.5000 | G-US-2 | 8039.0000 | AF-US-10 | 8194.5000 | G-US-3 |
| 7507.0000 | N-US-1 | 7684.0000 | N-US-4 | 7846.0000 | BC-WW-1 | 8040.0000 | AR-US-1 | 8195.0000 | CG-US-8 |
| 7508.5000 | G-US-10 | 7685.0000 | N-US-4 | 7849.5000 | AR-US-3 | 8045.0000 | AF-US-10 | 8195.0000 | M-WW-1 |
| 7508.5000 | CC-CAN | 7690.0000 | AF-US-10 | 7853.0000 | AF-US-5 | 8045.0000 | G-US-10 | 8195.0000 | PR-US-6 |
| 7512.5000 | CG-US-6 | 7690.0000 | G-US-3 | 7860.0000 | SP-US-1 | 8045.0000 | G-WW-3 | 8196.4000 | CG-US-1 |
| 7517.0000 | G-US-3 | 7690.5000 | G-US-2 | 7861.0000 | AR-US-1 | 8046.5000 | G-CAN-3 | 8198.1000 | M-WW-1 |
| 7517.0000 | PS-AUS | 7691.0000 | BC-WW-1 | 7862.5000 | AF-US-8 | 8047.0000 | AF-US-10 | 8201.0000 | PR-US-6 |
| 7520.0000 | G-US-3 | 7697.0000 | SP-US-1 | 7862.5000 | PR-WW-2 | 8050.0000 | AF-US-10 | 8201.2000 | M-US-4 |
| 7525.0000 | SP-US-1 | 7697.0000 | PR-US-4 | 7865.0000 | G-US-3 | 8055.0000 | AF-US-1 | 8201.2000 | M-WW-1 |
| 7527.0000 | AF-US-8 | 7700.0000 | G-US-2 | 7870.0000 | AF-US-6 | 8060.0000 | AF-US-10 | 8204.3000 | M-WW-1 |
| 7527.0000 | CG-US-4 | 7700.0000 | G-US-3 | 7873.0000 | AF-US-10 | 8060.0000 | AR-US-1 | 8207.4000 | M-WW-1 |
| 7530.0000 | CG-US-3 | 7705.0000 | SP-US-1 | 7875.0000 | G-US-3 | 8063.0000 | CG-US-6 | 8210.5000 | M-WW-1 |
| 7530.0000 | BC-WW-1 | 7710.0000 | CG-CAN | 7875.0000 | G-US-5 | 8063.6000 | CG-US-6 | 8210.8000 | M-US-3 |

| Freq | Desig | Freq | Desig | Freq | Desig | Freq | Desig | Freq | Desig |
|---|---|---|---|---|---|---|---|---|---|
| 8213.6000 | M-US-4 | 8353.0000 | G-US-9 | 8570.0000 | M-WW-2 | 8737.5000 | M-US-4 | 8873.0000 | A-WW-1 |
| 8213.6000 | M-WW-1 | 8353.5000 | CG-US-1 | 8574.0000 | CG-US-1 | 8737.5000 | M-WW-1 | 8876.0000 | G-US-9 |
| 8216.7000 | M-WW-1 | 8353.5000 | CG-US-4 | 8574.0000 | CG-US-4 | 8740.6000 | M-WW-1 | 8879.0000 | A-WW-2 |
| 8217.0000 | G-US-3 | 8353.5000 | CG-US-3 | 8574.0000 | M-WW-2 | 8743.7000 | M-WW-1 | 8885.0000 | A-WW-3 |
| 8219.8000 | M-WW-1 | 8354.0000 | M-US-3 | 8582.0000 | CG-US-1 | 8746.8000 | M-WW-1 | 8888.0000 | A-WW-1 |
| 8222.9000 | M-WW-1 | 8355.0000 | CG-US-8 | 8582.0000 | CG-US-4 | 8749.9000 | M-WW-1 | 8888.0000 | A-WW-5 |
| 8226.0000 | M-WW-1 | 8355.5000 | G-CAN-4 | 8585.0000 | M-WW-2 | 8753.0000 | CG-CAN | 8891.0000 | A-WW-2 |
| 8229.1000 | G-US-9 | 8356.0000 | G-CAN-4 | 8586.0000 | M-US-3 | 8753.0000 | G-US-4 | 8893.0000 | A-US-3 |
| 8229.1000 | M-WW-1 | 8357.0000 | CG-US-1 | 8598.4000 | N-NZL | 8753.0000 | G-US-9 | 8893.0000 | A-WW-1 |
| 8232.2000 | M-WW-1 | 8357.0000 | CG-US-4 | 8603.5000 | AR-US-1 | 8753.0000 | M-WW-1 | 8894.0000 | A-WW-2 |
| 8233.5000 | N-US-1 | 8357.0000 | CG-US-3 | 8615.5000 | AR-US-1 | 8754.0000 | G-US-9 | 8895.0000 | A-WW-1 |
| 8235.3000 | M-WW-1 | 8363.6000 | CG-US-1 | 8615.5000 | M-US-3 | 8756.1000 | M-WW-1 | 8903.0000 | A-WW-2 |
| 8236.4000 | M-CAN | 8364.0000 | A-US-3 | 8618.0000 | M-US-3 | 8757.0000 | N-US-1 | 8906.0000 | A-WW-2 |
| 8236.4000 | M-ATN | 8364.0000 | N-J | 8618.0000 | M-WW-2 | 8759.2000 | CC-CAN | 8907.0000 | AF-AUS/NZL |
| 8238.4000 | M-WW-1 | 8364.0000 | CG-US-1 | 8622.0000 | M-WW-2 | 8759.2000 | M-WW-1 | 8912.0000 | G-US-10 |
| 8241.5000 | CG-US-8 | 8364.0000 | G-US-9 | 8623.0000 | N-NZL | 8762.3000 | M-WW-1 | 8917.0000 | G-US-3 |
| 8241.5000 | M-WW-1 | 8364.4000 | CG-US-1 | 8624.0000 | AR-US-1 | 8765.4000 | CG-US-2 | 8917.0000 | A-WW-2 |
| 8244.6000 | M-WW-1 | 8379.5000 | CG-US-1 | 8625.0000 | N-F | 8765.4000 | CG-US-8 | 8918.0000 | A-WW-2 |
| 8245.0000 | CG-US-8 | 8380.0000 | PR-US-6 | 8625.2000 | N-G | 8765.4000 | M-WW-1 | 8921.0000 | A-US-1 |
| 8247.7000 | N-US-5 | 8396.5000 | CG-US-1 | 8630.0000 | M-US-3 | 8765.4000 | M-WW-2 | 8921.0000 | A-WW-1 |
| 8247.7000 | M-WW-1 | 8410.4000 | CG-US-1 | 8632.0000 | M-WW-2 | 8768.5000 | CG-US-8 | 8924.0000 | A-US-1 |
| 8250.8000 | M-WW-1 | 8417.0000 | SP-URS | 8634.0000 | G-WW-3 | 8768.5000 | M-WW-1 | 8924.0000 | A-WW-1 |
| 8253.9000 | M-WW-1 | 8418.0000 | SP-URS | 8636.0000 | CG-US-3 | 8769.0000 | CG-US-4 | 8924.0000 | A-WW-2 |
| 8257.0000 | M-WW-1 | 8427.0000 | G-US-9 | 8640.0000 | G-US-3 | 8770.4000 | M-CAN | 8927.0000 | A-US-1 |
| 8257.0000 | N-AFS | 8427.5000 | G-US-9 | 8642.0000 | M-US-3 | 8770.4000 | M-ATN | 8927.0000 | A-WW-1 |
| 8260.1000 | M-WW-1 | 8437.3000 | N-ISR | 8645.5000 | N-URS | 8771.0000 | N-US-1 | 8930.0000 | A-US-1 |
| 8263.2000 | M-WW-1 | 8440.0000 | CG-CAN | 8646.0000 | G-US-9 | 8771.6000 | N-US-5 | 8930.0000 | A-WW-1 |
| 8264.0000 | N-AFS | 8441.0000 | CG-US-3 | 8646.0000 | M-WW-2 | 8771.6000 | M-WW-1 | 8932.0000 | G-US-3 |
| 8265.0000 | N-AFS | 8443.0000 | CG-CAN | 8648.0000 | CG-US-2 | 8774.7000 | M-WW-1 | 8933.0000 | A-US-1 |
| 8266.3000 | M-WW-1 | 8444.5000 | M-US-3 | 8650.0000 | CG-US-1 | 8777.8000 | M-WW-1 | 8933.0000 | A-WW-1 |
| 8269.4000 | M-WW-1 | 8445.0000 | M-US-3 | 8650.0000 | G-WW-3 | 8778.0000 | N-US-1 | 8936.0000 | A-US-1 |
| 8270.0000 | N-AFS | 8450.0000 | PR-US-6 | 8650.0000 | M-WW-2 | 8778.6000 | M-I | 8936.0000 | A-WW-1 |
| 8272.5000 | M-WW-1 | 8450.0000 | BC-WW-1 | 8654.4000 | M-US-3 | 8780.9000 | M-WW-1 | 8938.0000 | N-AUS |
| 8275.6000 | M-WW-1 | 8451.0000 | CG-US-3 | 8654.4000 | M-WW-2 | 8780.9000 | M-WW-2 | 8938.0000 | A-WW-5 |
| 8278.7000 | G-US-9 | 8453.0000 | CG-CAN | 8658.0000 | M-US-3 | 8783.2000 | M-US-3 | 8939.0000 | A-WW-5 |
| 8278.7000 | M-WW-1 | 8455.0000 | CG-US-2 | 8660.0000 | BC-WW-1 | 8784.0000 | M-WW-1 | 8939.5000 | PS-AUS |
| 8281.8000 | M-WW-1 | 8457.0000 | CG-US-3 | 8662.0000 | AF-CAN | 8787.1000 | CG-CAN | 8942.0000 | A-WW-2 |
| 8282.6000 | G-US-9 | 8459.0000 | CG-US-4 | 8666.0000 | M-US-3 | 8787.1000 | M-WW-1 | 8951.0000 | A-WW-2 |
| 8284.4000 | AF-CAN | 8461.0000 | CG-US-4 | 8669.9000 | M-I | 8789.7000 | G-US-9 | 8952.0000 | A-WW-1 |
| 8284.9000 | M-WW-1 | 8461.0000 | G-WW-3 | 8674.0000 | M-WW-2 | 8790.2000 | M-WW-1 | 8952.0000 | A-WW-5 |
| 8288.0000 | M-WW-1 | 8463.0000 | AF-CAN | 8675.4000 | M-WW-2 | 8793.2000 | M-US-3 | 8957.0000 | A-WW-5 |
| 8291.0000 | PR-US-6 | 8465.0000 | CG-US-1 | 8678.0000 | M-WW-2 | 8793.3000 | CG-CAN | 8960.0000 | A-WW-3 |
| 8291.1000 | M-WW-1 | 8465.0000 | CG-US-4 | 8682.0000 | CG-US-4 | 8793.3000 | M-WW-1 | 8960.0000 | N-F |
| 8291.1000 | N-B | 8465.0000 | M-WW-2 | 8682.0000 | CG-US-3 | 8796.4000 | M-US-3 | 8963.0000 | A-WW-1 |
| 8294.2000 | N-US-3 | 8471.0000 | CG-US-1 | 8682.0000 | M-WW-2 | 8796.4000 | M-WW-1 | 8964.0000 | AF-US-2 |
| 8294.2000 | M-WW-1 | 8473.0000 | G-WW-3 | 8690.0000 | M-WW-2 | 8796.4000 | M-WW-2 | 8964.0000 | AF-US-3 |
| 8294.5000 | N-AUS | 8473.0000 | M-WW-2 | 8697.0000 | AF-CAN | 8796.4000 | M-I | 8967.0000 | AF-US-3 |
| 8295.0000 | CG-US-8 | 8475.6000 | N-F | 8697.0000 | CG-CAN | 8799.5000 | M-WW-1 | 8967.0000 | AF/N-G |
| 8295.6000 | G-US-11 | 8476.0000 | M-WW-2 | 8706.5000 | CG-US-4 | 8800.6000 | G-US-9 | 8967.0000 | AF-SAU |
| 8296.0000 | PR-US-6 | 8478.0000 | M-WW-2 | 8706.5000 | CG-US-3 | 8802.6000 | G-US-9 | 8971.0000 | AF-AUS/NZL |
| 8297.1000 | CG-US-8 | 8490.0000 | CG-US-3 | 8708.5000 | CG-US-2 | 8802.6000 | M-WW-1 | 8972.0000 | N-US-1 |
| 8297.6000 | CG-US-3 | 8490.0000 | BC-WW-1 | 8710.0000 | AR-US-1 | 8805.7000 | M-WW-1 | 8974.0000 | N-US-3 |
| 8298.0000 | PR-US-6 | 8492.0000 | M-J | 8710.5000 | CG-US-1 | 8805.7000 | M-WW-2 | 8975.0000 | AF/N-G |
| 8298.1000 | CG-US-4 | 8498.0000 | M-WW-2 | 8710.5000 | CG-US-4 | 8808.8000 | M-WW-1 | 8975.0000 | AF-AUS/NZL |
| 8298.1000 | CG-US-3 | 8500.0000 | AR-US-1 | 8710.5000 | CG-US-3 | 8811.9000 | M-WW-1 | 8975.0000 | AF-AFS |
| 6498.5000 | G-CAN-4 | 8502.0000 | CG-US-3 | 8711.0000 | M-US-3 | 8815.0000 | G-US-3 | 8976.0000 | N-US-1 |
| 8298.1000 | M-US-3 | 8504.0000 | M-WW-2 | 8713.0000 | M-US-3 | 8819.0000 | A-WW-1 | 8977.0000 | AF/N-G |
| 8298.6000 | CG-US-3 | 8508.0000 | M-US-3 | 8714.0000 | G-US-9 | 8819.0000 | A-WW-3 | 8980.0000 | AF-AUS/NZL |
| 8299.1000 | CG-US-4 | 8510.0000 | N-NOR | 8714.5000 | CG-US-1 | 8819.0000 | A-WW-5 | 8980.0000 | N-ARG |
| 8299.1000 | CG-US-3 | 8510.0000 | CG-US-1 | 8714.5000 | CG-US-4 | 8822.0000 | A-US-2 | 8981.5000 | N-US-1 |
| 8300.0000 | N-AFS | 8510.0000 | CG-US-4 | 8714.5000 | CG-US-3 | 8822.0000 | PR-US-6 | 8984.0000 | AF-AUS/NZL |
| 8302.0000 | G-US-3 | 8514.0000 | M-US-3 | 8715.0000 | M-US-3 | 8823.9000 | G-US-3 | 8984.0000 | CG-US-2 |
| 8329.6000 | CG-US-5 | 8518.0000 | N-ISR | 8716.5000 | CG-CAN | 8825.0000 | A-WW-2 | 8987.0000 | AF/N-G |
| 8330.2000 | CG-US-5 | 8518.1000 | N-ISR | 8716.0000 | CG-US-2 | 8828.0000 | A-WW-5 | 8989.0000 | AF-US-3 |
| 8330.5000 | CC-CAN | 8525.0000 | M-US-3 | 8716.5000 | G-CAN-4 | 8828.0000 | N-CHL | 8990.0000 | AF/N-G |
| 8330.8000 | CG-US-5 | 8527.5000 | M-WW-2 | 8717.0000 | G-CAN-4 | 8837.0000 | A-WW-1 | 8990.0000 | AF-SAU |
| 8334.0000 | G-US-3 | 8530.0000 | M-I | 8718.9000 | CG-US-8 | 8842.0000 | A-URS | 8990.0000 | CG-US-2 |
| 8334.0000 | PR-US-6 | 8536.0000 | CG-US-3 | 8718.9000 | M-WW-1 | 8843.0000 | A-WW-2 | 8992.0000 | AF-AUS/NZL |
| 8340.0000 | PS-AUS | 8539.0000 | G-WW-3 | 8720.0000 | M-A | 8843.0000 | A-WW-3 | 8992.0000 | AF-AFS |
| 8340.5000 | G-CAN-4 | 8542.0000 | AF-CAN | 8721.0000 | G-WW-3 | 8846.0000 | A-WW-2 | 8993.0000 | AF/N-G |
| 8343.5000 | CG-US-8 | 8542.0000 | G-WW-3 | 8722.0000 | M-WW-1 | 8852.0000 | A-WW-5 | 8993.0000 | AF-US-3 |
| 8345.0000 | BC-WW-1 | 8554.0000 | CG-US-3 | 8725.1000 | M-US-4 | 8855.0000 | A-WW-2 | 8993.0000 | SP-US-1 |
| 8345.5000 | CG-US-4 | 8558.0000 | G-WW-3 | 8725.1000 | M-WW-1 | 8855.0000 | A-WW-3 | 8997.0000 | N-I |
| 8345.5000 | CG-US-3 | 8558.4000 | M-US-3 | 8726.0000 | M-WW-2 | 8861.0000 | A-WW-2 | 8997.0000 | N-US-3 |
| 8349.5000 | CG-US-1 | 8562.0000 | M-WW-2 | 8728.2000 | M-WW-1 | 8861.0000 | A-WW-3 | 9002.0000 | N-US-1 |
| 8349.5000 | CG-US-4 | 8565.0000 | AR-US-1 | 8731.0000 | AF-US-5 | 8861.0000 | A-WW-5 | 9003.0000 | A-WW-1 |
| 8349.5000 | CG-US-3 | 8566.0000 | M-WW-2 | 8731.3000 | M-WW-1 | 8864.0000 | A-WW-2 | 9005.0000 | AF-AUS/NZL |
| 8350.0000 | M-US-3 | 8570.0000 | CG-US-4 | 8734.4000 | M-WW-1 | 8867.0000 | A-WW-2 | 9006.0000 | AF-CAN |
| 8352.0000 | M-US-3 | 8570.0000 | M-US-3 | 8737.5000 | CG-CAN | 8870.0000 | A-WW-1 | 9006.0000 | N-US-1 |

| Freq | Code | Freq | Code | Freq | Code | Freq | Code | Freq | Code |
|---|---|---|---|---|---|---|---|---|---|
| 9007.0000 | AF-US-10 | 9127.0000 | CG-US-4 | 9414.0000 | AF-US-10 | 9985.0000 | N-AFS | 10183.0000 | G-US-3 |
| 9010.0000 | AF-CAN | 9130.0000 | AR-US-2 | 9419.0000 | AR-US-3 | 9989.0000 | BC-WW-1 | 10185.0000 | AF-US-6 |
| 9011.0000 | AF/N-G | 9130.0000 | CG-US-3 | 9425.0000 | AF-US-10 | 9990.0000 | AR-US-3 | 10190.0000 | PR-US-1 |
| 9011.0000 | AF-US-3 | 9130.0000 | G-US-3 | 9435.0000 | N-NZL | 9991.0000 | AF-US-10 | 10190.0000 | BC-WW-1 |
| 9014.0000 | AF/N-G | 9132.0000 | SP-US-1 | 9435.0000 | G-US-6 | 9996.0000 | G-WW-3 | 10194.0000 | AF-US-4 |
| 9014.0000 | AF-US-2 | 9138.0000 | SP-US-1 | 9440.0000 | G-US-3 | 9997.0000 | G-US-3 | 10194.0000 | G-US-1 |
| 9014.0000 | AF-US-3 | 9140.0000 | G-US-3 | 9445.0000 | G-US-3 | 10000.0000 | G-WW-3 | 10195.0000 | G-US-1 |
| 9015.0000 | G-US-12 | 9140.0000 | BC-WW-1 | 9445.0000 | CC-CAN | 10004.0000 | G-WW-3 | 10200.0000 | PR-US-6 |
| 9017.0000 | AF-US-1 | 9145.0000 | G-US-3 | 9462.0000 | G-US-1 | 10015.0000 | G-US-9 | 10200.0000 | CC-US-1 |
| 9017.0000 | AF-US-10 | 9145.0000 | BC-WW-1 | 9462.0000 | CC-CAN | 10017.0000 | A-WW-2 | 10202.0000 | AF-US-10 |
| 9018.0000 | AF-US-3 | 9148.5000 | CC-CAN | 9473.5000 | CG-US-6 | 10017.0000 | A-WW-5 | 10205.0000 | AF-US-6 |
| 9018.0000 | AF-CUB | 9150.0000 | BC-WW-1 | 9474.0000 | CG-US-6 | 10018.0000 | A-WW-2 | 10205.4000 | AF-US-6 |
| 9018.0000 | SP-US-1 | 9162.0000 | G-US-3 | 9475.0000 | AF-US-10 | 10024.0000 | G-US-3 | 10208.0000 | AF-US-6 |
| 9020.0000 | N-D | 9169.0000 | G-US-1 | 9481.0000 | G-US-3 | 10024.0000 | A-WW-2 | 10210.0000 | G-US-3 |
| 9022.0000 | SP-US-1 | 9170.0000 | SP-US-1 | 9497.0000 | PR-US-6 | 10027.0000 | A-US-1 | 10211.3000 | AF-US-6 |
| 9023.0000 | AF-US-1 | 9170.0000 | BC-WW-1 | 9497.4000 | PR-US-4 | 10027.0000 | A-WW-1 | 10215.0000 | SP-US-1 |
| 9023.0000 | AF-US-4 | 9172.0000 | G-CAN-2 | 9500.0000 | PR-US-6 | 10030.0000 | A-US-1 | 10215.5000 | N-NRS |
| 9024.0000 | AF/N-G | 9172.5000 | AF-US-6 | 9630.0000 | CG-US-6 | 10030.0000 | A-WW-1 | 10225.0000 | G-US-1 |
| 9025.0000 | A-US-3 | 9181.0000 | AR-US-3 | 9650.0000 | AR-US-1 | 10033.0000 | A-US-1 | 10230.0000 | G-US-3 |
| 9025.0000 | AF/N-G | 9183.5000 | G-US-12 | 9655.5000 | PR-US-6 | 10039.0000 | A-WW-2 | 10230.0000 | SP-US-1 |
| 9027.0000 | AF-US-1 | 9185.0000 | G-US-1 | 9657.0000 | PR-US-6 | 10040.0000 | CG-US-3 | 10235.0000 | N-US-3 |
| 9029.0000 | SP-US-1 | 9188.0000 | AF-US-10 | 9657.0000 | CC-US-1 | 10040.0000 | G-US-3 | 10235.0000 | BC-WW-1 |
| 9032.0000 | AF/N-G | 9197.0000 | G-US-1 | 9668.5000 | PR-US-6 | 10042.0000 | G-US-3 | 10237.5000 | BC-WW-1 |
| 9032.0000 | N-US-1 | 9198.0000 | AF-US-6 | 9771.0000 | G-US-1 | 10042.0000 | A-WW-3 | 10248.0000 | G-US-3 |
| 9036.0000 | AF/N-G | 9200.0000 | G-WW-3 | 9775.0000 | PR-US-1 | 10045.0000 | CG-US-3 | 10255.0000 | G-US-3 |
| 9037.0000 | N-US-1 | 9200.0000 | G-CAN-1 | 9778.0000 | N-AFS | 10045.0000 | G-US-3 | 10255.0000 | N-US-4 |
| 9040.0000 | G-US-3 | 9200.0000 | BC-WW-1 | 9784.0000 | G-US-3 | 10045.0000 | A-US-2 | 10259.5000 | N-US-4 |
| 9040.0000 | PR-US-1 | 9210.0000 | BC-WW-1 | 9785.0000 | G-AUS | 10045.0000 | PR-US-6 | 10265.0000 | PR-US-2 |
| 9040.0000 | PR-US-6 | 9210.0000 | N-US-4 | 9785.0000 | N-US-3 | 10048.0000 | A-WW-2 | 10267.0000 | AF-US-8 |
| 9040.5000 | G-US-3 | 9212.0000 | SP-US-1 | 9785.0000 | G-US-10 | 10048.0000 | A-WW-5 | 10270.0000 | AF-US-8 |
| 9040.6000 | PR-WW-2 | 9215.0000 | AF-US-10 | 9786.5000 | N-US-3 | 10051.0000 | A-WW-5 | 10270.0000 | SP-US-1 |
| 9043.0000 | AF/N-G | 9215.0000 | N-US-3 | 9788.0000 | N-F | 10057.0000 | A-WW-5 | 10273.5000 | AF-US-8 |
| 9043.0000 | SP-US-1 | 9220.0000 | AF-US-1 | 9793.0000 | AF-US-4 | 10066.0000 | A-WW-2 | 10285.0000 | G-US-3 |
| 9044.4000 | PR-US-4 | 9223.0000 | CG-US-6 | 9796.5000 | G-US-10 | 10067.0000 | A-WW-5 | 10285.5000 | N-URS |
| 9046.0000 | G-US-3 | 9223.6000 | CG-US-6 | 9802.0000 | CG-US-4 | 10069.0000 | A-US-1 | 10290.0000 | G-CAN-4 |
| 9047.0000 | AF-US-8 | 9224.0000 | AF-US-8 | 9802.0000 | PR-US-6 | 10069.0000 | A-WW-1 | 10291.0000 | A-WW-3 |
| 9047.5000 | G-US-3 | 9224.0000 | G-US-3 | 9810.0000 | AR-US-3 | 10072.0000 | A-US-1 | 10295.0000 | G-WW-3 |
| 9050.0000 | CG-US-7 | 9229.0000 | G-US-3 | 9811.0000 | G-US-3 | 10072.0000 | A-WW-1 | 10295.0000 | PS-AUS |
| 9051.0000 | G-US-3 | 9234.0000 | AF-US-1 | 9813.0000 | G-CAN-2 | 10075.0000 | A-US-1 | 10298.0000 | G-US-3 |
| 9052.0000 | G-US-3 | 9236.0000 | AF-US-6 | 9821.0000 | G-WW-3 | 10075.0000 | A-WW-1 | 10301.0000 | SP-US-1 |
| 9052.0000 | N-US-4 | 9238.5000 | G-US-12 | 9827.0000 | G-US-3 | 10078.0000 | CG-US-3 | 10305.0000 | AF-US-6 |
| 9055.0000 | G-US-3 | 9238.5000 | PR-US-6 | 9830.0000 | N-US-3 | 10078.0000 | A-US-1 | 10305.0000 | SP-US-1 |
| 9057.0000 | AF-US-1 | 9240.0000 | PR-US-6 | 9840.0000 | G-US-3 | 10078.0000 | A-WW-1 | 10310.0000 | G-US-3 |
| 9062.0000 | G-US-3 | 9240.0000 | BC-WW-1 | 9842.0000 | G-US-3 | 10084.0000 | A-WW-2 | 10310.0000 | SP-US-1 |
| 9065.0000 | G-US-3 | 9249.2000 | M-US-3 | 9843.0000 | N-AFS | 10087.0000 | A-WW-5 | 10310.0000 | PR-US-2 |
| 9065.0000 | N-G | 9250.0000 | BC-WW-1 | 9846.0000 | G-US-10 | 10090.0000 | A-WW-5 | 10315.0000 | AF-US-6 |
| 9072.0000 | G-AUS | 9257.0000 | N-US-1 | 9858.0000 | BC-WW-1 | 10093.0000 | A-WW-1 | 10315.0000 | BC-WW-1 |
| 9073.0000 | N-US-3 | 9260.0000 | N-US-1 | 9862.5000 | G-US-3 | 10096.0000 | G-US-3 | 10318.0000 | G-US-3 |
| 9073.0000 | CG-US-6 | 9265.0000 | G-CAN-4 | 9866.0000 | AF-US-10 | 10096.0000 | A-WW-2 | 10320.0000 | BC-WW-1 |
| 9074.5000 | G-US-1 | 9265.0000 | N-B | 9881.0000 | G-US-3 | 10096.0000 | A-WW-3 | 10322.0000 | G-US-3 |
| 9078.0000 | G-US-3 | 9268.0000 | N-F | 9885.0000 | AF-US-10 | 10100.0000 | G-US-3 | 10325.0000 | G-US-3 |
| 9079.0000 | G-US-3 | 9268.0000 | G-US-3 | 9905.0000 | BC-WW-1 | 10100.0000 | PS-CRIB | 10327.0000 | SP-US-1 |
| 9081.5000 | CC-CAN | 9270.0000 | AF-US-9 | 9910.0000 | G-US-3 | 10100.0000 | PR-US-1 | 10333.6000 | CG-US-6 |
| 9086.0000 | CG-US-3 | 9270.0000 | AF-US-10 | 9910.0000 | SP-US-1 | 10100.0000 | PR-US-6 | 10335.4000 | BC-WW-1 |
| 9090.0000 | BC-WW-1 | 9272.0000 | N-BEL | 9914.0000 | G-US-10 | 10101.0000 | G-US-3 | 10337.5000 | CG-US-6 |
| 9092.0000 | PS-S | 9278.0000 | G-US-3 | 9915.0000 | G-US-3 | 10110.0000 | G-US-3 | 10338.0000 | BC-WW-1 |
| 9100.0000 | N-G | 9278.5000 | CG-US-8 | 9918.0000 | G-US-1 | 10112.0000 | AF-US-10 | 10344.0000 | AF-US-1 |
| 9100.0000 | G-US-3 | 9285.0000 | G-WW-3 | 9918.0000 | G-US-2 | 10112.0000 | N-F | 10346.0000 | N-G |
| 9100.0000 | BC-WW-1 | 9290.0000 | BC-WW-1 | 9920.0000 | G-US-3 | 10113.0000 | N-F | 10352.0000 | SP-US-1 |
| 9104.0000 | G-US-3 | 9297.0000 | CG-US-6 | 9921.0000 | AF-US-10 | 10118.8000 | AF-US-6 | 10357.0000 | AF-US-10 |
| 9105.0000 | G-WW-3 | 9297.0000 | G-US-3 | 9930.0000 | G-US-1 | 10120.0000 | PR-US-6 | 10365.0000 | G-US-3 |
| 9105.0000 | G-CAN-1 | 9300.0000 | PS-AUS | 9940.0000 | N-AUS | 10120.0000 | BC-WW-1 | 10368.5000 | CG-US-6 |
| 9106.0000 | N-CHL | 9303.0000 | CG-US-6 | 9941.5000 | N-AUS | 10125.0000 | BC-WW-1 | 10382.5000 | BC-WW-1 |
| 9106.0000 | N-G | 9305.0000 | AR-US-3 | 9941.5000 | PR-US-1 | 10130.0000 | N-US-5 | 10385.0000 | BC-WW-1 |
| 9106.0000 | N-D | 9311.5000 | G-US-12 | 9942.0000 | G-US-3 | 10136.0000 | CG-US-8 | 10390.0000 | G-US-3 |
| 9108.0000 | CG-US-8 | 9315.0000 | CC-CAN | 9946.0000 | G-US-6 | 10138.0000 | CG-US-3 | 10390.0000 | G-WW-3 |
| 9110.0000 | CG-US-3 | 9320.0000 | AF-US-9 | 9947.5000 | G-US-9 | 10140.0000 | AF-US-8 | 10390.0000 | G-CAN-1 |
| 9112.0000 | G-US-3 | 9320.0000 | AF-US-10 | 9950.0000 | G-US-10 | 10145.0000 | G-US-3 | 10393.0000 | G-US-3 |
| 9114.0000 | G-CAN-3 | 9320.0000 | BC-WW-1 | 9950.0000 | N-ARG | 10146.0000 | G-US-3 | 10398.0000 | G-US-3 |
| 9115.0000 | SP-US-1 | 9350.0000 | BC-WW-1 | 9955.0000 | G-US-3 | 10150.0000 | G-US-3 | 10400.0000 | PR-US-6 |
| 9115.0000 | BC-WW-1 | 9359.5000 | G-US-2 | 9958.0000 | AF-US-10 | 10153.0000 | AF-US-1 | 10400.0000 | CC-US-1 |
| 9115.5000 | G-US-2 | 9370.5000 | N-URS | 9960.0000 | N-AFS | 10153.0000 | AF-US-10 | 10402.0000 | G-US-3 |
| 9115.5000 | G-CAN-3 | 9380.0000 | N-US-1 | 9966.0000 | G-US-3 | 10157.0000 | CG-CAN | 10402.0000 | BC-WW-1 |
| 9121.8000 | G-US-1 | 9389.5000 | G-CAN-3 | 9973.0000 | G-US-3 | 10159.0000 | SP-US-1 | 10405.0000 | PR-WW-2 |
| 9122.0000 | G-US-1 | 9391.0000 | G-CAN-3 | 9977.0000 | CG-US-6 | 10163.0000 | G-US-10 | 10415.0000 | G-US-3 |
| 9122.5000 | AR-US-1 | 9392.5000 | G-CAN-3 | 9982.0000 | G-US-3 | 10164.0000 | G-US-3 | 10415.0000 | BC-WW-1 |
| 9122.5000 | AR-US-2 | 9396.0000 | CG-US-6 | 9982.5000 | G-US-10 | 10165.0000 | AR-US-3 | 10418.0000 | BC-WW-1 |
| 9125.0000 | CG-US-8 | 9410.0000 | G-US-3 | 9985.0000 | G-US-3 | 10166.0000 | CG-US-8 | 10423.0000 | G-US-3 |
| 9125.0000 | G-US-3 | 9410.0000 | G-US-10 | 9985.0000 | N-ARG | 10168.0000 | CG-US-4 | 10427.0000 | AF-US-10 |

| Frequency | Code | Frequency | Code | Frequency | Code | Frequency | Code | Frequency | Code |
|---|---|---|---|---|---|---|---|---|---|
| 10430.0000 | G-US-3 | 10689.0000 | BC-WW-1 | 11037.0000 | G-US-3 | 11228.0000 | AF-US-3 | 11400.0000 | G-US-3 |
| 10434.0000 | G-US-3 | 10690.0000 | BC-WW-1 | 11042.0000 | G-US-3 | 11229.0000 | AF-US-10 | 11400.0000 | PR-US-1 |
| 10436.0000 | G-US-10 | 10705.0000 | G-US-3 | 11045.0000 | G-US-1 | 11230.0000 | CG-US-3 | 11401.0000 | A-WW-1 |
| 10437.5000 | G-US-10 | 10707.0000 | G-US-3 | 11046.0000 | G-US-3 | 11230.0000 | G-US-3 | 11401.0000 | N-US-4 |
| 10440.5000 | G-WW-3 | 10707.0000 | PR-US-6 | 11055.0000 | AF-US-10 | 11233.0000 | AF-CAN | 11402.5000 | N-US-4 |
| 10452.0000 | AF-US-1 | 10710.0000 | BC-WW-1 | 11055.0000 | G-CAN-4 | 11234.0000 | AF/N-G | 11404.4000 | PR-US-4 |
| 10452.0000 | AF-US-3 | 10712.0000 | N-CHL | 11055.0000 | N-G | 11234.0000 | N-US-1 | 11406.0000 | G-US-3 |
| 10452.0000 | AF-US-4 | 10714.0000 | G-US-3 | 11056.0000 | AF-US-10 | 11235.0000 | AF-AUS/NZL | 11407.0000 | AF-US-8 |
| 10454.0000 | BC-WW-1 | 10720.0000 | AF-US-10 | 11058.0000 | AF-US-10 | 11236.0000 | AF-US-3 | 11407.0000 | AF-US-10 |
| 10456.5000 | BC-WW-1 | 10720.0000 | G-US-3 | 11070.0000 | G-US-3 | 11237.0000 | AF-AFS | 11407.0000 | SP-US-1 |
| 10464.0000 | G-US-3 | 10725.0000 | G-US-3 | 11070.0000 | BC-WW-1 | 11239.0000 | AF-US-3 | 11408.0000 | AF-US-1 |
| 10475.0000 | G-US-3 | 10730.0000 | N-US-1 | 11073.5000 | PR-US-6 | 11240.0000 | G-US-3 | 11410.0000 | N-US-1 |
| 10475.0000 | SP-US-1 | 10730.0000 | G-US-3 | 11075.0000 | G-US-12 | 11243.0000 | AF-US-1 | 11411.0000 | G-US-3 |
| 10476.0000 | BC-WW-1 | 10732.0000 | G-US-3 | 11076.0000 | CG-US-4 | 11244.0000 | AF/N-G | 11411.0000 | N-AFS |
| 10480.0000 | G-US-3 | 10737.0000 | G-US-3 | 11076.0000 | PR-US-6 | 11246.0000 | AF-US-3 | 11413.0000 | AF-US-10 |
| 10482.0000 | G-US-3 | 10740.0000 | BC-WW-1 | 11077.5000 | G-CAN-4 | 11247.0000 | AF-AUS/NZL | 11414.0000 | SP-US-1 |
| 10484.0000 | G-US-3 | 10748.0000 | G-US-3 | 11086.0000 | G-US-3 | 11249.0000 | AF-CAN | 11418.0000 | G-US-3 |
| 10493.0000 | G-US-1 | 10755.0000 | G-US-3 | 11090.0000 | G-US-3 | 11250.0000 | AF/N-G | 11422.0000 | G-US-3 |
| 10498.0000 | AF-US-1 | 10760.0000 | N-NZL | 11090.0000 | G-US-10 | 11250.0000 | G-US-3 | 11423.0000 | G-US-3 |
| 10502.0000 | PS-AUS | 10760.0000 | G-US-3 | 11090.0000 | BC-WW-1 | 11252.0000 | N-US-1 | 11424.0000 | G-US-3 |
| 10505.0000 | PS-AUS | 10760.0000 | BC-WW-1 | 11092.0000 | BC-WW-1 | 11252.0000 | SP-US-1 | 11425.0000 | AR-US-2 |
| 10507.0000 | CG-US-3 | 10765.0000 | AF-US-10 | 11095.0000 | G-US-3 | 11255.0000 | N-US-1 | 11433.0000 | G-US-3 |
| 10510.0000 | AF-US-1 | 10770.0000 | G-US-3 | 11095.0000 | BC-WW-1 | 11255.0000 | N-US-3 | 11434.0000 | CG-US-8 |
| 10510.0000 | AF-US-3 | 10780.0000 | AF-US-10 | 11098.5000 | AF-US-8 | 11255.0000 | N-G | 11434.0000 | G-US-3 |
| 10510.0000 | SP-US-1 | 10780.0000 | G-US-3 | 11100.0000 | AF-US-1 | 11255.0000 | G-US-3 | 11435.0000 | PS-AUS |
| 10520.0000 | G-US-3 | 10780.0000 | SP-US-1 | 11104.0000 | N-US-3 | 11256.0000 | A-WW-1 | 11436.0000 | CG-US-4 |
| 10520.0000 | CC-CAN | 10790.0000 | G-US-3 | 11104.0000 | SP-US-1 | 11257.0000 | AF/N-G | 11436.0000 | CG-US-3 |
| 10520.0000 | BC-WW-1 | 10792.0000 | G-US-3 | 11105.0000 | G-US-3 | 11261.0000 | N-US-1 | 11437.0000 | CC-US-2 |
| 10522.0000 | G-US-3 | 10800.0000 | AR-UN | 11106.0000 | G-US-3 | 11266.0000 | AF-US-3 | 11440.0000 | AF-US-5 |
| 10523.6000 | CG-US-6 | 10800.0000 | PR-US-6 | 11118.0000 | AF-US-1 | 11266.0000 | AF-D | 11440.0000 | G-WW-3 |
| 10530.0000 | AF-US-10 | 10800.0000 | CC-US-1 | 11118.0000 | AF-US-10 | 11267.0000 | N-US-1 | 11441.0000 | AF-US-4 |
| 10530.0000 | BC-WW-1 | 10804.0000 | G-US-3 | 11121.0000 | AF-US-8 | 11271.0000 | AF-F | 11441.0000 | AF-US-10 |
| 10531.5000 | AF-US-6 | 10804.0000 | SP-US-1 | 11121.5000 | G-US-10 | 11279.0000 | A-WW-2 | 11441.0000 | G-US-3 |
| 10536.0000 | AF-CAN | 10815.0000 | AR-US-3 | 11122.0000 | G-US-3 | 11279.0000 | A-WW-5 | 11445.0000 | BC-WW-1 |
| 10538.0000 | G-US-3 | 10832.0000 | G-US-3 | 11123.0000 | G-US-3 | 11282.0000 | A-WW-2 | 11447.0000 | G-US-3 |
| 10538.0000 | BC-WW-1 | 10850.0000 | SP-US-1 | 11125.0000 | G-US-3 | 11284.0000 | AF-US-4 | 11448.0000 | AF-US-10 |
| 10547.0000 | G-US-3 | 10853.0000 | G-US-3 | 11126.5000 | G-US-2 | 11287.0000 | CC-US-1 | 11448.0000 | G-US-3 |
| 10550.0000 | AF-US-10 | 10855.0000 | BC-WW-1 | 11142.0000 | G-US-3 | 11288.0000 | A-US-2 | 11449.0000 | G-US-3 |
| 10550.0000 | N-NZL | 10860.0000 | BC-WW-1 | 11145.0000 | N-AFS | 11288.0000 | G-US-10 | 11452.4000 | PR-US-4 |
| 10550.0000 | AF-US-10 | 10865.0000 | BC-WW-1 | 11145.0000 | G-US-3 | 11288.0000 | PR-US-6 | 11455.0000 | G-US-3 |
| 10550.0000 | G-US-3 | 10866.5000 | BC-WW-1 | 11148.0000 | G-US-3 | 11291.0000 | A-WW-2 | 11459.0000 | G-US-3 |
| 10550.0000 | G-US-12 | 10869.0000 | BC-WW-1 | 11152.0000 | G-US-3 | 11297.0000 | A-WW-5 | 11459.0000 | BC-WW-1 |
| 10554.0000 | G-US-3 | 10869.0000 | G-US-1 | 11152.5000 | BC-WW-1 | 11300.0000 | N-F | 11460.0000 | N-US-1 |
| 10555.0000 | N-URS | 10870.0000 | N-ARG | 11153.0000 | AF-US-10 | 11300.0000 | A-WW-2 | 11460.0000 | G-US-3 |
| 10555.5000 | G-US-2 | 10870.7000 | G-US-1 | 11155.5000 | N-URS | 11305.0000 | G-US-3 | 11461.0000 | G-US-3 |
| 10560.0000 | G-US-3 | 10871.5000 | G-US-2 | 11156.0000 | AF-US-10 | 11306.0000 | A-US-2 | 11461.5000 | N-US-3 |
| 10563.0000 | G-US-10 | 10872.0000 | BC-WW-1 | 11165.0000 | G-US-6 | 11306.0000 | A-WW-2 | 11462.0000 | AR-G |
| 10567.0000 | G-US-3 | 10877.5000 | BC-WW-1 | 11166.0000 | G-US-3 | 11306.0000 | A-WW-3 | 11462.0000 | GK-US-3 |
| 10576.0000 | AF-US-8 | 10880.0000 | AF-US-10 | 11170.0000 | G-US-3 | 11306.0000 | PR-US-6 | 11463.0000 | N-US-1 |
| 10583.0000 | AF-US-10 | 10880.0000 | SP-US-1 | 11175.0000 | BC-WW-1 | 11309.0000 | A-WW-2 | 11463.0000 | G-US-3 |
| 10586.0000 | AF-US-10 | 10880.0000 | BC-WW-1 | 11176.0000 | AF-US-3 | 11312.0000 | A-URS | 11464.0000 | G-US-3 |
| 10586.0000 | AR-US-1 | 10881.0000 | AF-US-10 | 11179.0000 | AF-US-3 | 11312.0000 | A-WW-1 | 11466.0000 | AF-US-10 |
| 10587.0000 | G-US-3 | 10890.0000 | N-ARG | 11181.0000 | G-US-3 | 11315.0000 | A-WW-5 | 11470.0000 | G-CAN-4 |
| 10588.0000 | G-US-1 | 10891.0000 | G-US-1 | 11185.0000 | AF/N-G | 11318.0000 | A-WW-5 | 11470.0000 | PR-WW-2 |
| 10590.0000 | AF-US-10 | 10900.0000 | G-US-3 | 11187.0000 | AF-D | 11319.0000 | A-WW-2 | 11471.0000 | G-US-10 |
| 10595.0000 | BC-WW-1 | 10902.0000 | G-US-8 | 11190.0000 | N-US-1 | 11330.0000 | A-WW-2 | 11474.0000 | G-US-3 |
| 10600.0000 | PR-US-6 | 10905.0000 | SP-US-1 | 11191.0000 | N-US-1 | 11336.0000 | A-WW-2 | 11475.0000 | G-US-3 |
| 10600.0000 | CC-US-1 | 10915.0000 | G-US-12 | 11193.0000 | A-URS | 11339.0000 | A-WW-1 | 11478.0000 | G-US-3 |
| 10602.0000 | G-US-3 | 10918.0000 | G-US-1 | 11195.0000 | N-US-1 | 11342.0000 | A-US-1 | 11484.0000 | AF-US-10 |
| 10603.0000 | AF-US-10 | 10922.0000 | BC-WW-1 | 11195.0000 | PR-US-6 | 11345.0000 | A-US-1 | 11488.5000 | G-US-12 |
| 10607.0000 | G-US-3 | 10922.5000 | G-US-3 | 11197.0000 | A-WW-1 | 11345.0000 | A-WW-1 | 11494.0000 | AF-US-1 |
| 10618.5000 | G-US-3 | 10935.0000 | AF-US-3 | 11198.0000 | N-US-1 | 11348.0000 | A-US-1 | 11494.0000 | AF-US-10 |
| 10620.0000 | CG-US-3 | 10935.0000 | BC-WW-1 | 11198.0000 | CG-US-2 | 11348.0000 | A-URS | 11495.0000 | BC-WW-1 |
| 10620.0000 | BC-WW-1 | 10938.5000 | CG-US-6 | 11200.0000 | AF/N-G | 11351.0000 | A-US-1 | 11498.0000 | AF-US-10 |
| 10621.0000 | G-US-3 | 10942.0000 | G-US-3 | 11201.0000 | AF-AUS/NZL | 11351.0000 | A-WW-1 | 11500.0000 | G-US-3 |
| 10627.0000 | G-US-3 | 10945.0000 | N-CAN | 11201.0000 | CG-US-2 | 11354.0000 | A-US-1 | 11500.0000 | N-M |
| 10637.0000 | G-US-3 | 10949.0000 | SP-US-1 | 11204.0000 | AF/N-G | 11354.0000 | A-WW-1 | 11505.0000 | G-US-3 |
| 10640.0000 | G-US-3 | 10958.0000 | G-US-3 | 11205.0000 | N-US-1 | 11360.0000 | A-WW-2 | 11505.0000 | G-CAN-4 |
| 10642.5000 | G-US-3 | 10958.5000 | CG-US-6 | 11210.0000 | G-US-12 | 11362.0000 | G-US-3 | 11508.0000 | G-US-3 |
| 10643.5000 | N-URS | 10972.0000 | BC-WW-1 | 11212.0000 | AF/N-G | 11369.0000 | A-WW-5 | 11510.0000 | AR-US-1 |
| 10649.0000 | N-BEL | 10985.0000 | G-US-10 | 11214.0000 | AF-US-2 | 11375.0000 | A-WW-2 | 11510.0000 | SP-US-1 |
| 10655.0000 | G-US-8 | 10990.0000 | BC-WW-1 | 11214.0000 | AF-US-4 | 11384.0000 | A-WW-1 | 11513.5000 | CG-US-8 |
| 10660.0000 | SP-US-1 | 10997.5000 | AR-US-3 | 11214.0000 | AF-AUS/NZL | 11384.0000 | A-WW-2 | 11518.7000 | G-US-1 |
| 10660.0000 | BC-WW-1 | 11000.0000 | N-AFS | 11216.0000 | G-US-1 | 11387.0000 | A-WW-2 | 11520.0000 | G-US-3 |
| 10675.0000 | CG-US-3 | 11000.0000 | N-SAM | 11220.0000 | AF-US-1 | 11387.0000 | A-WW-5 | 11522.5000 | G-US-3 |
| 10675.0000 | G-US-3 | 11000.0000 | G-US-3 | 11220.0000 | AF-US-10 | 11391.0000 | A-WW-5 | 11531.0000 | G-US-3 |
| 10675.0000 | BC-WW-1 | 11005.0000 | G-US-3 | 11220.0000 | G-US-3 | 11396.0000 | A-WW-2 | 11532.5000 | N-J |
| 10680.0000 | G-US-3 | 11028.0000 | CG-US-6 | 11224.0000 | AF/N-G | 11397.5000 | G-US-2 | 11538.0000 | G-WW-3 |
| 10688.0000 | N-NZL | 11028.0000 | G-US-1 | 11226.0000 | AF-US-10 | 11398.0000 | G-US-9 | 11538.4000 | N-US-4 |

| Freq | | Freq | | Freq | | Freq | | Freq | |
|---|---|---|---|---|---|---|---|---|---|
| 11539.0000 | N-US-4 | 12009.0000 | G-US-1 | 12225.0000 | CG-US-3 | 12370.3000 | M-WW-1 | 12569.0000 | CG-US-1 |
| 11539.5000 | N-US-4 | 12020.5000 | G-US-2 | 12225.0000 | G-US-3 | 12373.0000 | PR-US-6 | 12570.0000 | N-URS |
| 11540.0000 | N-US-4 | 12022.5000 | G-US-3 | 12230.0000 | CG-US-3 | 12373.4000 | M-WW-1 | 12570.0000 | PR-US-6 |
| 11541.0000 | G-US-3 | 12029.0000 | AF-US-10 | 12230.0000 | G-US-3 | 12375.5000 | M-CAN | 12594.5000 | CG-US-1 |
| 11543.0000 | G-US-3 | 12029.0000 | N-US-3 | 12230.0000 | N-ARG | 12375.5000 | M-ATN | 12597.0000 | N-F |
| 11545.0000 | AF-US-10 | 12030.0000 | G-US-3 | 12233.5000 | G-US-2 | 12376.5000 | M-WW-1 | 12605.0000 | G-US-3 |
| 11548.0000 | SP-US-1 | 12049.0000 | N-US-4 | 12235.0000 | CG-US-3 | 12379.6000 | CG-US-8 | 12615.5000 | CG-US-1 |
| 11550.0000 | G-US-3 | 12058.0000 | AF-US-10 | 12235.0000 | G-US-3 | 12379.6000 | M-WW-1 | 12640.5000 | G-US-9 |
| 11550.0000 | N-F | 12060.0000 | AF-US-10 | 12238.5000 | G-US-3 | 12382.7000 | M-WW-1 | 12660.0000 | M-US-3 |
| 11553.0000 | N-F | 12060.0000 | AR-US-1 | 12240.0000 | AF-US-10 | 12385.8000 | G-US-9 | 12660.0000 | N-ARG |
| 11553.0000 | N-US-3 | 12060.0000 | G-US-3 | 12240.0000 | AR-US-1 | 12385.8000 | M-WW-1 | 12660.0000 | PR-US-6 |
| 11555.0000 | G-US-2 | 12064.5000 | G-US-1 | 12240.0000 | G-US-10 | 12388.9000 | M-WW-1 | 12664.5000 | CG-US-4 |
| 11565.0000 | N-J | 12067.5000 | G-US-3 | 12240.0000 | BC-WW-1 | 12392.0000 | M-WW-1 | 12664.5000 | CG-US-3 |
| 11570.0000 | N-US-1 | 12070.0000 | AF-US-1 | 12242.0000 | AF-US-10 | 12392.0000 | N-AFS | 12665.0000 | N-F |
| 11570.0000 | N-NZL | 12072.0000 | AR-US-3 | 12246.5000 | BC-WW-1 | 12395.0000 | M-WW-1 | 12671.0000 | CG-CAN |
| 11570.0000 | N-SAM | 12075.0000 | AF-US-1 | 12250.0000 | AF-US-10 | 12398.2000 | M-WW-1 | 12695.5000 | M-US-3 |
| 11572.0000 | G-US-3 | 12078.0000 | G-US-3 | 12250.0000 | BC-WW-1 | 12401.3000 | M-WW-1 | 12700.0000 | CG-US-1 |
| 11573.0000 | G-US-3 | 12080.0000 | CC-CAN | 12253.0000 | BC-WW-1 | 12401.4000 | G-US-9 | 12700.0000 | M-WW-2 |
| 11574.0000 | G-US-3 | 12090.0000 | AF-US-10 | 12254.0000 | AF-US-10 | 12404.4000 | M-WW-1 | 12702.0000 | AF-CAN |
| 11575.0000 | N-F | 12090.0000 | AR-US-1 | 12254.0000 | G-US-3 | 12407.5000 | M-WW-1 | 12711.0000 | M-WW-2 |
| 11575.0000 | BC-WW-1 | 12094.5000 | G-US-1 | 12255.0000 | N-AUS | 12410.6000 | M-WW-1 | 12714.0000 | CG-US-4 |
| 11578.0000 | AR-US-1 | 12101.0000 | G-US-3 | 12255.0000 | AR-US-1 | 12413.7000 | M-WW-1 | 12714.0000 | M-WW-2 |
| 11580.0000 | N-US-3 | 12105.0000 | AF-US-10 | 12256.5000 | N-AUS | 12416.8000 | M-WW-1 | 12718.5000 | CG-US-1 |
| 11585.0000 | BC-WW-1 | 12105.0000 | G-US-3 | 12256.5000 | PR-US-1 | 12419.9000 | M-WW-1 | 12718.5000 | M-WW-2 |
| 11590.0000 | G-US-3 | 12106.0000 | AF-US-10 | 12257.0000 | G-US-3 | 12420.0000 | G-US-3 | 12724.0000 | G-WW-3 |
| 11601.5000 | CC-US-2 | 12107.0000 | AF-US-10 | 12258.0000 | AF-US-10 | 12422.4000 | G-US-9 | 12725.0000 | N-BEL |
| 11602.5000 | N-NOR | 12107.0000 | SP-US-1 | 12260.0000 | G-US-3 | 12423.0000 | CC-AUS | 12726.0000 | AF-CAN |
| 11605.0000 | G-US-1 | 12111.5000 | G-US-3 | 12265.0000 | G-US-3 | 12423.0000 | M-WW-1 | 12726.0000 | CG-CAN |
| 11606.0000 | CG-US-6 | 12115.0000 | CG-US-3 | 12267.0000 | AR-US-2 | 12424.5000 | AF-CAN | 12727.5000 | M-WW-2 |
| 11607.0000 | AF-US-1 | 12115.0000 | G-US-3 | 12268.5000 | G-US-3 | 12425.9000 | G-US-9 | 12730.0000 | CG-US-2 |
| 11615.0000 | AF-US-10 | 12124.0000 | N-US-4 | 12270.0000 | AR-US-1 | 12426.1000 | CG-US-8 | 12732.0000 | CG-US-3 |
| 11619.5000 | AF-US-8 | 12124.0000 | G-US-3 | 12272.0000 | G-US-3 | 12426.1000 | M-WW-1 | 12740.0000 | M-WW-2 |
| 11621.0000 | SP-US-1 | 12126.0000 | G-US-3 | 12277.0000 | AF-US-10 | 12429.0000 | PR-US-6 | 12743.0000 | CG-US-1 |
| 11623.0000 | G-CAN-3 | 12126.5000 | G-CAN-2 | 12277.0000 | SP-US-1 | 12429.2000 | N-US-3 | 12743.0000 | CG-US-4 |
| 11627.0000 | AF-US-10 | 12127.0000 | BC-WW-1 | 12280.0000 | N-US-3 | 12429.2000 | M-WW-1 | 12743.0000 | M-WW-2 |
| 11634.0000 | AF-US-10 | 12127.5000 | N-US-4 | 12282.0000 | G-US-3 | 12432.0000 | PR-US-6 | 12745.0000 | G-WW-3 |
| 11634.0000 | SP-US-1 | 12129.0000 | AF-US-1 | 12287.0000 | SP-US-1 | 12432.3000 | M-WW-1 | 12748.1000 | CG-US-3 |
| 11635.0000 | G-US-3 | 12129.0000 | G-US-1 | 12290.0000 | CG-US-3 | 12433.7000 | G-US-11 | 12750.0000 | CG-US-4 |
| 11636.0000 | G-CAN-3 | 12130.0000 | G-US-3 | 12290.0000 | G-US-3 | 12435.4000 | M-WW-1 | 12750.0000 | CG-US-3 |
| 11637.0000 | G-US-10 | 12131.0000 | G-US-3 | 12290.0000 | BC-WW-1 | 12442.0000 | N-AFS | 12753.0000 | AF-CAN |
| 11640.5000 | G-US-3 | 12133.0000 | A-WW-3 | 12295.0000 | CG-US-3 | 12460.0000 | N-AFS | 12754.0000 | CG-US-3 |
| 11650.0000 | G-US-3 | 12135.0000 | BC-WW-1 | 12295.0000 | G-US-3 | 12481.1000 | CG-US-5 | 12755.5000 | M-WW-2 |
| 11651.5000 | CC-CAN | 12138.5000 | PR-US-6 | 12296.0000 | BC-WW-1 | 12481.7000 | CG-US-5 | 12761.0000 | N-US-1 |
| 11653.0000 | G-US-3 | 12150.0000 | GC-US-6 | 12297.0000 | G-US-3 | 12482.0000 | CC-CAN | 12763.5000 | M-ARG |
| 11655.0000 | N-US-4 | 12152.0000 | N-BEL | 12300.0000 | N-G | 12482.3000 | CG-US-5 | 12768.0000 | M-US-3 |
| 11660.0000 | N-CHL | 12155.0000 | AF-US-10 | 12300.0000 | G-US-3 | 12490.0000 | PR-US-6 | 12768.0000 | M-WW-2 |
| 11660.0000 | N-NOR | 12158.0000 | G-US-1 | 12304.0000 | G-US-3 | 12491.0000 | CG-US-8 | 12786.0000 | CG-US-4 |
| 11667.0000 | AF-US-10 | 12160.0000 | SP-US-1 | 12305.0000 | N-ARG | 12493.0000 | CG-US-4 | 12786.0000 | CG-US-3 |
| 11676.0000 | G-US-3 | 12165.0000 | BC-WW-1 | 12306.0000 | G-US-3 | 12493.0000 | CG-US-3 | 12790.0000 | CG-US-3 |
| 11682.5000 | AF-US-6 | 12168.0000 | AR-US-1 | 12307.0000 | G-WW-3 | 12497.0000 | CG-US-1 | 12799.5000 | M-WW-2 |
| 11688.0000 | G-US-3 | 12168.0000 | G-US-3 | 12310.0000 | G-US-3 | 12497.0000 | CG-US-4 | 12804.0000 | G-WW-3 |
| 11689.0000 | N-US-5 | 12170.0000 | AF-US-10 | 12311.0000 | G-US-3 | 12497.0000 | CG-US-3 | 12808.5000 | M-US-3 |
| 11695.0000 | G-CAN-4 | 12171.0000 | G-US-1 | 12315.0000 | N-US-1 | 12497.0000 | M-US-3 | 12814.0000 | AF-CAN |
| 11698.4000 | PR-US-4 | 12178.7000 | G-US-1 | 12317.0000 | AF-US-10 | 12500.5000 | G-US-9 | 12824.2000 | N-G |
| 11700.0000 | PR-US-6 | 12179.0000 | BC-WW-1 | 12324.0000 | AF-US-10 | 12501.0000 | CG-US-1 | 12826.5000 | M-US-3 |
| 11721.0000 | G-US-1 | 12180.0000 | G-US-10 | 12325.0000 | N-ARG | 12501.0000 | CG-US-4 | 12844.5000 | M-US-3 |
| 11761.5000 | PR-US-6 | 12181.0000 | CC-CAN | 12327.0000 | M-US-5 | 12501.0000 | CG-US-3 | 12849.0000 | M-WW-2 |
| 11763.0000 | PR-US-6 | 12182.0000 | N-SAM | 12329.0000 | N-DNK | 12501.0000 | G-CAN-4 | 12850.0000 | N-NOR |
| 11763.0000 | CC-US-1 | 12183.0000 | N-SAM | 12330.0000 | M-WW-1 | 12501.5000 | M-US-3 | 12853.5000 | M-WW-2 |
| 11764.5000 | AR-US-1 | 12187.0000 | G-US-10 | 12330.0000 | PR-US-6 | 12502.5000 | CG-US-8 | 12870.0000 | N-CHN |
| 11764.5000 | PR-US-6 | 12188.5000 | G-US-10 | 12333.1000 | M-US-4 | 12503.5000 | CG-US-3 | 12871.5000 | M-WW-2 |
| 11801.0000 | G-US-1 | 12196.0000 | G-US-10 | 12333.1000 | M-WW-1 | 12504.5000 | CG-US-1 | 12874.0000 | CG-CAN |
| 11957.0000 | G-US-1 | 12201.0000 | AF-US-6 | 12336.2000 | M-WW-1 | 12504.5000 | CG-US-4 | 12876.0000 | SP-US-1 |
| 11975.0000 | PR-US-1 | 12205.0000 | CG-US-6 | 12339.3000 | M-WW-1 | 12510.5000 | G-CAN-4 | 12876.0000 | M-WW-2 |
| 11975.0000 | PR-US-6 | 12205.0000 | G-US-3 | 12342.4000 | CG-US-8 | 12511.5000 | G-CAN-4 | 12876.0000 | CG-CAN |
| 11980.0000 | G-US-3 | 12205.0000 | BC-WW-1 | 12342.4000 | M-WW-1 | 12518.5000 | CG-US-8 | 12880.5000 | M-WW-2 |
| 11984.0000 | SP-US-1 | 12207.0000 | N-B | 12342.4000 | M-WW-2 | 12519.5000 | CG-US-8 | 12887.5000 | CG-US-2 |
| 11985.0000 | N-F | 12210.0000 | G-US-3 | 12344.0000 | CG-US-3 | 12520.0000 | CG-US-3 | 12887.5000 | M-WW-2 |
| 11985.0000 | G-US-3 | 12210.0000 | BC-WW-1 | 12344.0000 | G-US-3 | 12520.0000 | PR-US-6 | 12889.5000 | CG-US-1 |
| 11987.0000 | G-US-3 | 12211.0000 | G-US-3 | 12345.5000 | M-WW-1 | 12520.5000 | CG-US-4 | 12894.0000 | G-US-9 |
| 11987.0000 | N-AFS | 12211.5000 | G-US-3 | 12348.6000 | M-WW-1 | 12520.5000 | CG-US-3 | 12899.5000 | M-WW-2 |
| 11988.0000 | AF-US-10 | 12213.0000 | G-US-3 | 12351.7000 | M-WW-1 | 12521.0000 | CG-US-3 | 12908.5000 | AR-US-1 |
| 11988.0000 | SP-US-1 | 12215.0000 | N-AUS | 12354.8000 | M-WW-1 | 12521.5000 | CG-US-4 | 12926.5000 | M-US-3 |
| 11990.0000 | AR-US-3 | 12215.0000 | N-US-1 | 12355.0000 | AR-US-1 | 12521.5000 | M-US-3 | 12934.5000 | CG-US-3 |
| 11995.0000 | AF-US-10 | 12215.0000 | G-US-3 | 12357.9000 | M-WW-1 | 12523.0000 | PR-US-6 | 12934.5000 | M-WW-2 |
| 11995.0000 | G-US-3 | 12220.0000 | CG-US-4 | 12361.0000 | M-WW-1 | 12530.0000 | N-G | 12942.0000 | M-J |
| 12000.0000 | AR-US-1 | 12222.0000 | PR-US-6 | 12364.1000 | M-WW-1 | 12545.4000 | CG-US-1 | 12948.0000 | M-WW-2 |
| 12008.0000 | G-US-3 | 12223.0000 | BC-WW-1 | 12367.2000 | M-WW-1 | 12546.6000 | CG-US-1 | 12961.5000 | M-US-3 |
| 12008.0000 | G-WW-3 | 12223.0000 | N-US-4 | 12370.3000 | G-US-9 | 12568.0000 | N-J | 12961.5000 | M-WW-2 |

| Freq | ID | Freq | ID | Freq | ID | Freq | ID | Freq | ID |
|---|---|---|---|---|---|---|---|---|---|
| 12966.0000 | M-WW-2 | 13156.6000 | M-WW-1 | 13285.0000 | A-WW-1 | 13385.0000 | G-US-3 | 13590.0000 | BC-WW-1 |
| 12980.0000 | CG-US-3 | 13159.7000 | M-WW-1 | 13285.0000 | A-WW-3 | 13386.0000 | G-US-3 | 13595.0000 | G-CAN-4 |
| 12984.0000 | N-ISR | 13162.8000 | M-WW-1 | 13288.0000 | A-WW-2 | 13387.0000 | G-US-3 | 13600.0000 | SP-US-1 |
| 12984.2000 | N-ISR | 13162.8000 | M-WW-2 | 13290.0000 | N-ARG | 13388.0000 | G-US-3 | 13604.0000 | AF-US-10 |
| 12985.0000 | AF-CAN | 13165.9000 | M-WW-1 | 13290.0000 | G-US-3 | 13390.0000 | G-US-3 | 13605.0000 | G-US-3 |
| 12988.5000 | M-WW-2 | 13169.0000 | M-WW-1 | 13291.0000 | A-WW-2 | 13390.0000 | N-NZL | 13606.5000 | G-US-3 |
| 12997.5000 | M-US-3 | 13169.5000 | N-US-1 | 13294.0000 | A-WW-2 | 13395.0000 | G-US-3 | 13608.0000 | CG-US-6 |
| 13002.0000 | M-US-3 | 13172.1000 | M-WW-1 | 13295.0000 | G-US-3 | 13395.0000 | BC-WW-1 | 13608.8000 | CG-US-6 |
| 13005.0000 | CG-US-3 | 13175.2000 | M-WW-1 | 13297.0000 | A-WW-2 | 13396.0000 | G-US-3 | 13610.0000 | G-US-3 |
| 13011.0000 | M-US-3 | 13178.3000 | M-WW-1 | 13300.0000 | G-US-3 | 13397.4000 | PR-US-4 | 13614.0000 | AF-US-8 |
| 13015.3000 | M-I | 13180.4000 | G-US-9 | 13300.0000 | A-WW-2 | 13397.5000 | G-US-3 | 13617.0000 | G=US-3 |
| 13020.0000 | CG-US-3 | 13181.0000 | N-US-1 | 13302.0000 | CG-US-3 | 13401.0000 | G-US-3 | 13620.0000 | G-US-3 |
| 13020.0000 | G-WW-3 | 13181.4000 | M-WW-1 | 13303.0000 | A-WW-2 | 13402.0000 | N-AFS | 13626.0000 | G-US-3 |
| 13024.9000 | M-US-3 | 13184.5000 | M-WW-1 | 13306.0000 | A-WW-2 | 13403.0000 | N-AFS | 13626.0000 | G-US-10 |
| 13027.5000 | CG-US-3 | 13187.6000 | M-WW-1 | 13309.0000 | A-WW-2 | 13410.0000 | BC-WW-1 | 13629.5000 | N-US-1 |
| 13031.2000 | N-MAR | 13190.7000 | M-WW-1 | 13310.0000 | G-US-3 | 13412.0000 | AF-US-10 | 13630.0000 | G-US-3 |
| 13033.5000 | M-US-3 | 13191.5000 | AR-US-1 | 13310.0000 | PR-WW-2 | 13412.0000 | G-US-3 | 13643.0000 | N-US-4 |
| 13038.0000 | M-US-3 | 13193.8000 | M-WW-1 | 13311.0000 | G-US-3 | 13415.0000 | N-CAN | 13644.5000 | N-US-4 |
| 13042.5000 | N-D | 13196.9000 | CG-US-8 | 13312.0000 | A-WW-2 | 13416.8000 | G-US-3 | 13647.0000 | G-US-3 |
| 13051.5000 | M-US-3 | 13196.9000 | M-WW-1 | 13312.0000 | A-US-2 | 13420.0000 | A-WW-3 | 13655.0000 | CG-US-2 |
| 13056.0000 | M-WW-2 | 13201.0000 | AF-US-3 | 13312.0000 | PR-US-6 | 13420.0000 | A-WW-1 | 13655.0000 | G-US-3 |
| 13065.0000 | G-US-9 | 13204.0000 | AF-US-2 | 13312.0000 | G-US-10 | 13423.0000 | CG-US-6 | 13657.0000 | N-I |
| 13065.0000 | M-WW-2 | 13204.0000 | AF-US-10 | 13314.0000 | G-US-3 | 13425.0000 | N-AFS | 13658.5000 | PR-US-6 |
| 13073.0000 | CG-US-4 | 13205.0000 | AF-US-1 | 13314.0000 | CG-US-3 | 13432.5000 | G-US-1 | 13660.0000 | G-US-12 |
| 13073.0000 | CG-US-3 | 13205.0000 | AF-AUS/NZL | 13315.0000 | A-WW-2 | 13434.0000 | G-US-1 | 13664.9000 | CG-US-6 |
| 13077.0000 | CG-US-1 | 13205.0000 | A-WW-1 | 13315.9000 | G-US-3 | 13434.0000 | G-US-3 | 13665.5000 | CG-US-6 |
| 13077.0000 | CG-US-4 | 13207.0000 | AF-US-9 | 13316.0000 | G-US-3 | 13436.0000 | G-US-3 | 13676.0000 | SP-US-1 |
| 13077.0000 | CG-US-3 | 13210.0000 | G-US-3 | 13318.0000 | A-WW-2 | 13440.0000 | AF-US-10 | 13690.0000 | G-US-3 |
| 13077.0000 | M-US-3 | 13210.0000 | SP-US-1 | 13322.0000 | G-US-3 | 13455.0000 | AF-US-10 | 13690.0000 | BC-WW-1 |
| 13080.5000 | G-US-9 | 13211.0000 | AF-US-1 | 13324.0000 | A-US-1 | 13457.0000 | AF-US-10 | 13699.0000 | G-US-3 |
| 13081.0000 | CG-US-1 | 13211.0000 | AF-US-10 | 13324.0000 | A-WW-1 | 13457.0000 | G-US-10 | 13700.0000 | N-SAM |
| 13081.0000 | CG-US-4 | 13211.0000 | SP-US-1 | 13327.0000 | A-US-1 | 13462.0000 | BC-WW-1 | 13700.0000 | G-US-3 |
| 13081.0000 | CG-US-3 | 13214.0000 | AF-US-3 | 13327.0000 | A-WW-1 | 13468.0000 | SP-US-1 | 13700.0000 | G-CAN-4 |
| 13081.5000 | M-US-3 | 13215.0000 | AF-US-3 | 13330.0000 | A-US-1 | 13473.0000 | AF-US-6 | 13702.0000 | CG-US-3 |
| 13082.5000 | CG-US-2 | 13216.0000 | AF/N-G | 13330.0000 | A-WW-1 | 13473.0000 | N-G | 13704.7000 | N-SAM |
| 13084.5000 | CG-US-1 | 13217.0000 | AF-US-1 | 13333.0000 | G-US-3 | 13479.0000 | AR-US-3 | 13707.0000 | CG-US-6 |
| 13084.5000 | CG-US-4 | 13217.0000 | AF-US-10 | 13333.0000 | A-US-1 | 13484.0000 | AF-US-10 | 13710.0000 | AF-US-10 |
| 13090.5000 | CG-CAN | 13218.0000 | SP-US-1 | 13333.0000 | A-WW-1 | 13485.0000 | AF-US-10 | 13710.0000 | G-US-3 |
| 13090.5000 | G-CAN-4 | 13220.0000 | A-WW-1 | 13336.0000 | A-US-1 | 13485.0000 | N-US-4 | 13710.0000 | BC-WW-1 |
| 13091.5000 | G-CAN-4 | 13224.0000 | N-US-1 | 13336.0000 | A-WW-1 | 13485.0000 | G-US-3 | 13717.5000 | G-CAN-3 |
| 13092.0000 | M-WW-2 | 13225.0000 | A-WW-1 | 13339.0000 | A-US-1 | 13486.0000 | G-US-3 | 13718.0000 | G-US-3 |
| 13100.8000 | CG-CAN | 13227.0000 | N-US-1 | 13339.0000 | A-WW-1 | 13486.0000 | BC-WW-1 | 13720.0000 | BC-WW-1 |
| 13100.8000 | M-WW-1 | 13227.0000 | SP-US-1 | 13339.5000 | N-URS | 13491.0000 | G-US-3 | 13722.0000 | G-US-3 |
| 13103.9000 | M-US-4 | 13231.0000 | AF-CAN | 13340.0000 | G-US-3 | 13491.5000 | BC-WW-1 | 13727.0000 | PS-AUS |
| 13103.9000 | M-WW-1 | 13234.0000 | AF/N-G | 13342.0000 | G-US-3 | 13493.0000 | G-US-1 | 13728.0000 | AF-US-6 |
| 13105.0000 | G-WW-3 | 13235.0000 | N-ARG | 13342.0000 | A-US-1 | 13493.0000 | G-US-8 | 13730.0000 | PS-AUS |
| 13107.0000 | M-WW-1 | 13237.0000 | AF/N-G | 13342.0000 | A-WW-1 | 13495.0000 | SP-US-1 | 13735.0000 | SP-US-1 |
| 13109.9000 | M-US-4 | 13237.0000 | N-US-1 | 13345.0000 | G-US-3 | 13496.0000 | G-US-3 | 13742.0000 | SP-US-1 |
| 13110.1000 | M-WW-1 | 13237.0000 | SP-US-1 | 13345.0000 | A-US-1 | 13498.0000 | AF-US-8 | 13743.0000 | AR-US-3 |
| 13113.2000 | CG-US-2 | 13239.0000 | AF-F | 13345.0000 | A-WW-1 | 13502.0000 | G-US-3 | 13745.0000 | CC-CAN |
| 13113.2000 | CG-US-8 | 13241.0000 | AF-US-1 | 13347.0000 | G-US-3 | 13505.0000 | AR-US-3 | 13747.0000 | G-WW-3 |
| 13113.2000 | M-WW-1 | 13244.0000 | AF-US-3 | 13348.0000 | A-US-1 | 13507.0000 | G-US-3 | 13752.0000 | AF-US-10 |
| 13113.2000 | M-WW-2 | 13244.0000 | SP-US-1 | 13350.0000 | G-US-3 | 13508.5000 | AR-US-3 | 13756.0000 | SP-US-1 |
| 13115.0000 | G-US-3 | 13245.0000 | AF-AUS/NZL | 13351.0000 | A-US-1 | 13510.0000 | AF-CAN | 13759.0000 | G-US-3 |
| 13116.3000 | M-WW-1 | 13246.0000 | AF-D | 13351.0000 | A-WW-1 | 13511.0000 | AR-US-3 | 13760.0000 | AF-US-10 |
| 13119.4000 | CG-CAN | 13247.0000 | AF-US-1 | 13352.0000 | G-US-3 | 13512.0000 | BC-WW-1 | 13760.0000 | BC-WW-1 |
| 13119.4000 | M-WW-1 | 13247.0000 | AF-US-10 | 13352.0000 | A-WW-5 | 13514.0000 | AR-US-3 | 13761.0000 | G-US-3 |
| 13122.5000 | M-WW-1 | 13247.0000 | AF-US-3 | 13354.0000 | A-WW-2 | 13515.0000 | N-CAN | 13765.0000 | AF-US-10 |
| 13125.6000 | M-WW-1 | 13248.0000 | AF-D | 13357.0000 | A-WW-2 | 13520.0000 | G-WW-3 | 13770.0000 | G-US-3 |
| 13125.6000 | M-I | 13250.0000 | AF/N-G | 13357.0000 | A-WW-3 | 13522.5000 | G-US-3 | 13770.0000 | BC-WW-1 |
| 13128.7000 | M-WW-1 | 13250.0000 | G-US-3 | 13360.0000 | PR-US-1 | 13523.0000 | G-US-3 | 13776.0000 | G-US-3 |
| 13131.8000 | M-WW-1 | 13251.0000 | AF-AUS/NZL | 13360.0000 | PR-US-6 | 13524.5000 | AR-US-1 | 13776.0000 | G-US-2 |
| 13134.9000 | M-WW-1 | 13251.0000 | N-US-1 | 13364.0000 | PR-US-4 | 13530.0000 | N-US-4 | 13781.0000 | G-US-3 |
| 13138.0000 | CG-CAN | 13251.0000 | N-US-3 | 13371.0000 | G-US-3 | 13538.0000 | G-US-3 | 13797.0000 | BC-WW-1 |
| 13138.0000 | M-WW-1 | 13252.0000 | G-US-3 | 13371.5000 | N-SAM | 13540.0000 | N-US-4 | 13803.5000 | G-US-2 |
| 13138.0000 | M-WW-2 | 13254.0000 | AF-CAN | 13373.0000 | N-AFS | 13540.0000 | AR-US-1 | 13804.0000 | G-AUS |
| 13141.1000 | G-US-9 | 13257.0000 | AF-CAN | 13373.4000 | PR-US-4 | 13545.0000 | N-US-3 | 13805.0000 | G-US-3 |
| 13141.1000 | M-WW-1 | 13257.0000 | AF/N-G | 13374.0000 | G-US-3 | 13547.0000 | AF-US-1 | 13810.0000 | BC-WW-1 |
| 13142.0000 | G-US-9 | 13261.5000 | G-US-2 | 13375.0000 | G-US-3 | 13550.0000 | AR-US-4 | 13813.0000 | G-US-3 |
| 13144.2000 | M-WW-1 | 13264.0000 | A-WW-5 | 13377.0000 | G-US-3 | 13550.0000 | N-SAM | 13818.0000 | BC-WW-1 |
| 13147.0000 | N-US-1 | 13267.0000 | G-US-9 | 13377.5000 | G-US-3 | 13550.0000 | G-US-3 | 13820.0000 | G-WW-3 |
| 13147.3000 | M-WW-1 | 13267.0000 | A-WW-5 | 13378.0000 | G-US-3 | 13551.5000 | N-US-3 | 13820.0000 | BC-WW-1 |
| 13150.0000 | CG-US-4 | 13270.0000 | A-WW-5 | 13379.0000 | G-US-3 | 13555.0000 | N-SAM | 13823.0000 | AF-US-10 |
| 13150.4000 | CG-US-8 | 13273.0000 | A-WW-2 | 13380.0000 | N-US-4 | 13555.0000 | AR-US-1 | 13825.0000 | AF-US-10 |
| 13150.4000 | M-WW-1 | 13275.0000 | G-US-3 | 13380.0000 | G-US-3 | 13560.0000 | N-AFS | 13826.0000 | N-US-3 |
| 13153.5000 | M-WW-1 | 13279.0000 | A-WW-5 | 13380.0000 | SP-US-1 | 13560.0000 | PR-US-3 | 13826.0000 | N-US-4 |
| 13154.5000 | M-CAN | 13280.0000 | N-ARG | 13380.0000 | BC-WW-1 | 13565.0000 | AF-US-10 | 13830.0000 | G-US-8 |
| 13154.5000 | M-ATN | 13282.0000 | A-WW-5 | 13382.0000 | G-US-3 | 13582.0000 | AF-US-10 | 13850.0000 | G-US-3 |
| 13156.6000 | G-US-9 | 13284.0000 | G-US-3 | 13385.0000 | N-URS | 13585.0000 | AF-US-10 | 13855.0000 | A-WW-3 |

| Freq | Desig | Freq | Desig | Freq | Desig | Freq | Desig | Freq | Desig |
|---|---|---|---|---|---|---|---|---|---|
| 13855.0000 | BC-WW-1 | 14364.0000 | AF-US-10 | 14470.0000 | N-NOR | 14678.5000 | G-US-3 | 14895.0000 | N-J |
| 13856.0000 | G-US-3 | 14365.0000 | G-US-3 | 14470.0000 | N-US-4 | 14681.0000 | G-US-3 | 14896.0000 | AF-US-10 |
| 13860.0000 | BC-WW-1 | 14365.0000 | BC-WW-1 | 14471.5000 | N-US-4 | 14684.0000 | G-US-3 | 14896.0000 | SP-US-1 |
| 13865.0000 | A-WW-3 | 14365.0000 | N-B | 14472.0000 | G-US-3 | 14686.0000 | CG-US-4 | 14897.0000 | AF-US-10 |
| 13865.0000 | BC-WW-1 | 14366.0000 | N-AFS | 14477.0000 | N-US-4 | 14686.0000 | PR-US-6 | 14899.0000 | G-US-1 |
| 13872.0000 | G-US-3 | 14367.0000 | G-US-3 | 14477.5000 | N-URS | 14687.2000 | N-ARG | 14902.0000 | AF-US-10 |
| 13878.0000 | AF-US-10 | 14370.0000 | G-US-3 | 14478.0000 | G-US-3 | 14690.0000 | PR-US-6 | 14902.0000 | G-US-3 |
| 13878.0000 | G-US-3 | 14371.0000 | CG-US-4 | 14478.5000 | N-US-4 | 14694.0000 | G-US-3 | 14905.0000 | AF-US-7 |
| 13878.0000 | SP-US-1 | 14372.5000 | AF-US-8 | 14480.0000 | N-US-4 | 14695.0000 | G-US-3 | 14908.0000 | G-US-1 |
| 13885.0000 | BC-WW-1 | 14373.0000 | G-US-3 | 14482.0000 | G-US-3 | 14696.0000 | G-US-3 | 14912.0000 | G-US-3 |
| 13894.0000 | G-US-3 | 14374.0000 | G-US-3 | 14483.5000 | N-US-4 | 14697.2000 | N-ARG | 14913.0000 | AF-US-10 |
| 13900.0000 | SP-US-1 | 14377.0000 | G-US-3 | 14485.0000 | G-US-3 | 14700.0000 | N-US-3 | 14913.0000 | G-US-3 |
| 13903.5000 | G-US-10 | 14382.0000 | G-US-3 | 14485.0000 | N-US-4 | 14707.0000 | G-WW-3 | 14915.0000 | N-G |
| 13904.5000 | N-SAM | 14383.5000 | N-US-4 | 14485.5000 | AR-US-3 | 14710.0000 | G-US-3 | 14920.0000 | G-US-3 |
| 13905.0000 | G-US-3 | 14385.0000 | AR-CAN | 14487.0000 | AR-US-3 | 14713.0000 | G-US-3 | 14926.0000 | G-US-3 |
| 13905.0000 | CG-US-3 | 14387.0000 | G-US-3 | 14488.5000 | AR-US-3 | 14713.0000 | BC-WW-1 | 14930.0000 | AR-US-3 |
| 13907.0000 | AF-US-1 | 14389.0000 | AF-US-8 | 14490.0000 | G-US-3 | 14715.0000 | AF-US-10 | 14934.0000 | N-US-4 |
| 13914.0000 | G-US-3 | 14390.0000 | G-US-3 | 14493.5000 | G-US-12 | 14716.0000 | AF-US-1 | 14936.5000 | AR-US-3 |
| 13915.0000 | PS-WW-1 | 14390.5000 | AF-US-8 | 14495.0000 | N-AFS | 14728.0000 | G-US-3 | 14937.0000 | SP-US-1 |
| 13916.0000 | G-US-3 | 14392.5000 | CC-CAN | 14497.0000 | AF-US-10 | 14729.0000 | G-US-9 | 14938.0000 | AR-US-3 |
| 13920.0000 | G-US-3 | 14393.0000 | G-US-3 | 14497.5000 | SP-US-1 | 14730.0000 | G-US-3 | 14939.5000 | AR-US-3 |
| 13925.0000 | G-US-3 | 14393.0000 | N-US-4 | 14498.0000 | G-US-3 | 14731.0000 | G-US-3 | 14940.0000 | N-F |
| 13926.0000 | G-US-3 | 14398.0000 | BC-WW-1 | 14500.0000 | G-US-3 | 14733.5000 | G-CAN-3 | 14945.0000 | G-US-3 |
| 13927.0000 | AF-US-8 | 14400.0000 | G-US-3 | 14505.0000 | AF-US-8 | 14734.0000 | G-US-3 | 14950.0000 | G-US-10 |
| 13927.5000 | N-US-4 | 14400.5000 | G-US-2 | 14510.0000 | AR-US-3 | 14738.0000 | G-US-3 | 14953.0000 | G-US-1 |
| 13930.0000 | G-US-3 | 14402.0000 | AF-US-8 | 14513.0000 | AR-US-3 | 14744.0000 | AF-US-1 | 14955.0000 | AF-US-1 |
| 13936.0000 | G-US-3 | 14402.0000 | G-US-3 | 14513.0000 | G-US-3 | 14745.0000 | G-US-3 | 14955.0000 | G-US-3 |
| 13937.0000 | AF-US-3 | 14402.0000 | AR-US-3 | 14515.0000 | G-US-3 | 14745.0000 | PS-AUS | 14957.0000 | AF-US-10 |
| 13940.0000 | G-US-3 | 14402.5000 | N-ARG | 14516.0000 | A-WW-3 | 14755.0000 | G-US-3 | 14959.0000 | G-US-3 |
| 13945.0000 | G-US-3 | 14403.0000 | G-US-3 | 14520.0000 | G-US-3 | 14757.5000 | G-US-3 | 14960.0000 | G-US-3 |
| 13945.0000 | PR-US-2 | 14405.0000 | AF-US-8 | 14522.0000 | G-US-3 | 14761.5000 | N-US-3 | 14962.0000 | G-US-3 |
| 13946.0000 | G-US-3 | 14405.0000 | G-US-3 | 14526.0000 | BC-WW-1 | 14761.5000 | N-US-4 | 14967.0000 | SP-US-1 |
| 13950.0000 | AF-US-3 | 14406.5000 | AR-US-3 | 14527.0000 | AF-US-8 | 14766.0000 | AF-US-10 | 14968.0000 | N-CUB |
| 13950.0000 | G-US-3 | 14407.0000 | G-US-3 | 14529.5000 | AF-US-8 | 14767.6000 | N-US-4 | 14972.0000 | G-US-3 |
| 13951.0000 | G-US-3 | 14408.0000 | AF-US-8 | 14530.0000 | AF-US-8 | 14768.0000 | N-US-3 | 14983.0000 | N-URS |
| 13955.0000 | G-US-3 | 14410.0000 | G-US-3 | 14534.0000 | G-US-12 | 14772.0000 | G-US-3 | 14985.0000 | BC-WW-1 |
| 13960.0000 | BC-WW-1 | 14411.0000 | AF-US-8 | 14537.0000 | G-US-3 | 14775.0000 | AF-US-1 | 14987.0000 | SP-US-1 |
| 13965.0000 | G-US-3 | 14412.0000 | AF-US-10 | 14542.0000 | G-US-3 | 14775.0000 | AF-US-3 | 14989.0000 | G-US-3 |
| 13965.0000 | PS-WW-1 | 14415.0000 | N-AUS | 14548.0000 | G-US-3 | 14776.0000 | G-US-1 | 14996.0000 | G-WW-3 |
| 13965.0000 | PS-CRIB | 14415.0000 | N-AUS | 14550.0000 | G-US-3 | 14782.0000 | G-US-3 | 15000.0000 | G-WW-3 |
| 13970.0000 | AF-US-1 | 14417.0000 | AF-US-10 | 14558.0000 | N-URS | 14790.0000 | CC-CAN | 15004.0000 | G-WW-3 |
| 13970.0000 | AF-CAN | 14417.0000 | G-US-3 | 14560.0000 | G-US-3 | 14800.0000 | BC-WW-1 | 15010.0000 | PR-US-1 |
| 13970.0000 | G-US-3 | 14417.5000 | AF-US-6 | 14562.0000 | G-US-3 | 14809.0000 | G-US-3 | 15013.0000 | AF/N-G |
| 13971.0000 | AR-CAN | 14426.0000 | G-US-3 | 14569.0000 | G-US-3 | 14812.0000 | G-US-3 | 15014.0000 | AF-US-3 |
| 13973.0000 | PS-WW-1 | 14430.0000 | N-SAM | 14570.0000 | N-J | 14814.0000 | G-US-3 | 15014.0000 | G-US-3 |
| 13974.0000 | N-US-3 | 14433.0000 | G-US-3 | 14572.0000 | AF-US-10 | 14815.0000 | G-US-3 | 15015.0000 | AF-US-3 |
| 13974.0000 | N-US-4 | 14434.0000 | N-F | 14576.0000 | AF-US-10 | 14817.0000 | G-US-3 | 15018.0000 | AF-US-10 |
| 13975.0000 | G-US-3 | 14438.0000 | G-US-3 | 14577.5000 | G-US-6 | 14817.5000 | G-WW-3 | 15021.0000 | N-US-1 |
| 13975.0000 | N-US-4 | 14440.0000 | AR-US-3 | 14580.0000 | N-NZL | 14817.5000 | G-CAN-1 | 15021.0000 | SP-US-1 |
| 13977.0000 | AF-US-8 | 14440.0000 | N-ARG | 14580.5000 | AR-US-3 | 14818.0000 | G-US-3 | 15024.0000 | A-URS |
| 13977.0000 | BC-WW-1 | 14440.0000 | N-F | 14585.0000 | G-US-6 | 14818.5000 | N-US-4 | 15031.0000 | AF-US-3 |
| 13981.0000 | G-US-3 | 14440.5000 | N-ARG | 14585.0000 | SP-US-1 | 14820.0000 | G-US-3 | 15031.0000 | AF-CAN |
| 13985.0000 | AF-US-8 | 14441.5000 | N-US-4 | 14605.0000 | PR-US-2 | 14821.0000 | G-US-3 | 15031.0000 | AF/N-G |
| 13985.0000 | G-US-3 | 14442.0000 | G-US-3 | 14606.0000 | AF-US-8 | 14823.0000 | G-US-3 | 15035.0000 | AF-US-1 |
| 13986.0000 | AR-UN | 14443.0000 | G-US-3 | 14607.5000 | G-US-10 | 14824.0000 | G-US-3 | 15035.0000 | AF-CAN |
| 13989.0000 | G-US-3 | 14443.0000 | N-US-4 | 14607.5000 | G-WW-3 | 14827.0000 | G-WW-3 | 15036.0000 | AF-US-3 |
| 13990.0000 | G-US-8 | 14445.0000 | N-AFS | 14615.0000 | SP-US-1 | 14829.0000 | AF-US-8 | 15036.0000 | AF-US-10 |
| 13991.0000 | G-US-3 | 14445.0000 | PR-WW-2 | 14616.0000 | G-US-3 | 14829.0000 | AF-US-10 | 15039.0000 | AF/N-G |
| 13992.0000 | G-US-8 | 14446.0000 | PR-US-6 | 14619.0000 | G-US-3 | 14829.0000 | G-US-3 | 15040.0000 | G-US-3 |
| 13993.0000 | AF-US-8 | 14448.0000 | AF-US-1 | 14620.0000 | G-CAN-1 | 14832.0000 | AF-US-8 | 15041.0000 | AF-US-1 |
| 13995.0000 | BC-WW-1 | 14448.0000 | AF-US-10 | 14625.0000 | G-US-3 | 14832.0000 | G-US-3 | 15041.0000 | AF-US-10 |
| 13996.0000 | AF-US-8 | 14448.0000 | G-US-3 | 14628.0000 | G-US-3 | 14837.0000 | G-US-1 | 15044.0000 | AF-US-1 |
| 13996.5000 | PS-CRIB | 14450.0000 | N-J | 14630.0000 | N-CHL | 14840.0000 | N-US-4 | 15049.0000 | G-US-3 |
| 13997.0000 | G-US-3 | 14450.0000 | G-US-1 | 14638.0000 | G-US-3 | 14841.0000 | G-US-3 | 15051.0000 | N-US-1 |
| 13997.5000 | AR-US-3 | 14450.0000 | G-US-3 | 14645.0000 | G-CAN-3 | 14847.5000 | AR-US-3 | 15056.0000 | AF-AFS |
| 13997.5000 | BC-WW-1 | 14455.0000 | SP-US-1 | 14646.0000 | AF-US-6 | 14850.0000 | BC-WW-1 | 15056.0000 | PR-US-6 |
| 13998.5000 | PS-WW-1 | 14455.0000 | PR-US-6 | 14649.0000 | G-US-3 | 14855.5000 | AR-US-3 | 15067.0000 | N-US-1 |
| 14192.0000 | G-US-3 | 14456.0000 | BC-WW-1 | 14650.0000 | AF-US-5 | 14856.0000 | G-US-3 | 15077.0000 | N-US-1 |
| 14303.0000 | PS-CRIB | 14458.5000 | AR-CAN | 14650.0000 | G-US-3 | 14863.0000 | AF-US-10 | 15077.8000 | CG-US-4 |
| 14314.0000 | G-US-3 | 14460.0000 | G-US-3 | 14650.0000 | SP-US-1 | 14867.0000 | AF-US-10 | 15081.0000 | CG-US-2 |
| 14350.0000 | G-US-3 | 14460.0000 | G-US-12 | 14652.0000 | G-US-3 | 14875.0000 | G-US-3 | 15083.5000 | AF-US-8 |
| 14350.0000 | PR-US-1 | 14461.0000 | G-US-3 | 14655.0000 | N-NZL | 14877.0000 | AF-US-8 | 15084.0000 | CG-US-2 |
| 14350.0000 | PR-US-6 | 14463.5000 | N-US-4 | 14657.0000 | G-US-2 | 14877.5000 | AR-US-3 | 15087.0000 | N-US-1 |
| 14353.5000 | G-US-3 | 14464.0000 | G-US-10 | 14657.0000 | G-US-3 | 14879.5000 | AF-US-6 | 15087.0000 | CG-US-2 |
| 14355.0000 | G-US-3 | 14465.5000 | N-US-4 | 14662.0000 | G-US-3 | 14880.0000 | G-US-3 | 15091.0000 | AF-US-1 |
| 14357.5000 | CC-CAN | 14466.5000 | N-ARG | 14665.0000 | AR-US-3 | 14885.0000 | G-US-10 | 15091.0000 | AF-US-2 |
| 14360.0000 | G-US-3 | 14467.0000 | N-US-4 | 14670.0000 | AF-US-10 | 14886.0000 | G-US-1 | 15100.0000 | PR-US-6 |
| 14360.5000 | G-US-3 | 14468.0000 | N-US-4 | 14670.0000 | N-F | 14887.0000 | AF-US-10 | 15212.5000 | AR-US-1 |
| 14364.0000 | AF-US-4 | 14470.0000 | N-US-3 | 14670.0000 | G-WW-3 | 14890.0000 | A-WW-3 | 15233.0000 | AR-US-1 |

| Frequency | Station | Frequency | Station | Frequency | Station | Frequency | Station | Frequency | Station |
|---|---|---|---|---|---|---|---|---|---|
| 15341.5000 | PR-US-6 | 15734.0000 | AF-US-1 | 16041.0000 | AF-US-10 | 16300.0000 | N-US-4 | 16472.4000 | M-WW-1 |
| 15343.0000 | PR-US-6 | 15738.0000 | G-WW-3 | 16044.5000 | CG-US-6 | 16300.0000 | N-AFS | 16475.5000 | CG-US-8 |
| 15343.0000 | CC-US-1 | 15742.0000 | AF-US-10 | 16050.0000 | NF | 16300.0000 | G-US-3 | 16475.5000 | M-WW-1 |
| 15344.5000 | PR-US-6 | 15752.0000 | BC-WW-1 | 16057.5000 | N-US-4 | 16301.0000 | G-US-3 | 16478.6000 | M-WW-1 |
| 15400.0000 | G-US-3 | 15757.0000 | AF-US-2 | 16060.0000 | G-US-2 | 16302.0000 | G-US-3 | 16481.7000 | M-WW-1 |
| 15406.0000 | PR-US-6 | 15757.5000 | AF-US-6 | 16060.0000 | PR-US-6 | 16308.0000 | G-US-3 | 16484.8000 | M-WW-1 |
| 15407.5000 | PR-US-6 | 15760.0000 | AF-US-10 | 16060.0000 | CC-US-1 | 16310.0000 | G-US-3 | 16485.9000 | G-US-9 |
| 15407.5000 | CC-US-1 | 15760.0000 | PR-US-6 | 16065.0000 | G-US-3 | 16320.0000 | AF-US-10 | 16487.9000 | M-WW-1 |
| 15409.0000 | PR-US-6 | 15760.0000 | CC-US-1 | 16065.0000 | BC-WW-1 | 16320.0000 | AR-US-1 | 16491.0000 | M-WW-1 |
| 15449.0000 | US-AF-1 | 15760.0000 | N-G | 16065.0000 | N-NZL | 16325.0000 | G-US-3 | 16494.1000 | G-US-9 |
| 15450.0000 | PR-US-1 | 15765.0000 | BC-WW-1 | 16065.6000 | G-US-2 | 16325.0000 | N-AFS | 16494.1000 | M-WW-1 |
| 15450.0000 | PR-US-6 | 15768.0000 | G-US-3 | 16066.0000 | G-US-3 | 16329.0000 | N-AFS | 16497.2000 | CG-US-8 |
| 15454.4000 | PR-US-4 | 15770.0000 | BC-WW-1 | 16070.0000 | AF-US-10 | 16330.0000 | BC-WW-1 | 16497.2000 | M-WW-1 |
| 15454.5000 | G-US-2 | 15775.0000 | BC-WW-1 | 16077.0000 | AF-US-10 | 16330.0000 | G-US-1 | 16500.3000 | M-WW-1 |
| 15458.0000 | N-AFS | 15776.0000 | AF-US-10 | 16077.0000 | AR-US-2 | 16331.0000 | G-US-3 | 16503.4000 | M-WW-1 |
| 15470.0000 | N-CHL | 15780.0000 | G-US-3 | 16078.0000 | G-US-3 | 16335.0000 | G-US-3 | 16505.0000 | M-CAN |
| 15470.0000 | N-NOR | 15780.0000 | PR-US-2 | 16080.0000 | AF-US-10 | 16340.0000 | AR-US-1 | 16505.5000 | M-ATN |
| 15482.0000 | G-US-3 | 15780.0000 | BC-WW-1 | 16082.0000 | G-US-3 | 16343.0000 | PR-US-6 | 16506.5000 | M-WW-1 |
| 15487.0000 | G-US-3 | 15781.0000 | G-US-3 | 16088.0000 | G-US-3 | 16348.0000 | G-US-10 | 16509.6000 | M-WW-1 |
| 15492.0000 | G-US-3 | 15784.0000 | G-US-3 | 16100.0000 | AF-US-10 | 16349.0000 | G-US-3 | 16510.0000 | N-AFS |
| 15497.2000 | G-US-9 | 15790.0000 | BC-WW-1 | 16100.0000 | G-US-3 | 16350.0000 | G-US-3 | 16511.0000 | PR-US-6 |
| 15498.5000 | N-US-4 | 15793.0000 | SP-US-1 | 16107.0000 | G-US-3 | 16351.0000 | G-US-3 | 16512.7000 | M-WW-1 |
| 15500.0000 | AF-US-10 | 15800.0000 | BC-WW-1 | 16108.5000 | G-US-3 | 16352.0000 | G-US-3 | 16515.8000 | M-WW-1 |
| 15500.0000 | G-US-3 | 15804.0000 | G-US-3 | 16117.0000 | AF-US-10 | 16355.0000 | G-US-3 | 16518.9000 | M-US-4 |
| 15502.5000 | G-WW-3 | 15805.5000 | AF-US-8 | 16120.0000 | AF-US-10 | 16356.0000 | G-CAN-4 | 16518.9000 | M-WW-1 |
| 15513.5000 | AF-US-8 | 15806.0000 | G-US-3 | 16121.0000 | AF-US-10 | 16358.0000 | G-US-3 | 16522.0000 | M-WW-1 |
| 15519.0000 | G-US-3 | 15807.0000 | AF-US-8 | 16125.0000 | G-US-3 | 16360.0000 | G-US-3 | 16522.0000 | N-AFS |
| 15520.0000 | N-US-1 | 15811.0000 | G-US-3 | 16127.0000 | G-US-3 | 16363.0000 | G-US-3 | 16525.1000 | M-WW-1 |
| 15522.0000 | N-US-1 | 15821.0000 | AF-US-10 | 16132.0000 | G-US-3 | 16366.0000 | G-US-3 | 16528.2000 | M-WW-1 |
| 15540.0000 | G-US-3 | 15831.0000 | G-US-3 | 16133.0000 | G-US-3 | 16368.0000 | G-US-3 | 16531.3000 | M-WW-1 |
| 15542.0000 | G-US-3 | 15845.0000 | N-AUS | 16135.0000 | G-US-3 | 16370.0000 | A-WW-3 | 16534.4000 | CG-US-8 |
| 15544.0000 | AF-US-1 | 15846.0000 | N-AUS | 16135.0000 | G-US-10 | 16370.0000 | G-US-3 | 16534.4000 | M-WW-1 |
| 15555.0000 | AF-US-10 | 15849.0000 | BC-WW-1 | 16135.6000 | G-US-10 | 16372.0000 | G-US-3 | 16537.5000 | M-WW-1 |
| 15564.0000 | N-US-3 | 15851.0000 | G-US-10 | 16140.0000 | AF-US-6 | 16376.0000 | G-US-12 | 16540.6000 | M-WW-1 |
| 15570.0000 | N-CHN | 15860.0000 | PR-US-6 | 16140.0000 | BC-WW-1 | 16376.0000 | PR-US-6 | 16542.0000 | G-US-3 |
| 15575.5000 | AF-US-6 | 15860.0000 | CC-US-1 | 16145.0000 | G-US-3 | 16377.0000 | G-US-3 | 16543.7000 | M-WW-1 |
| 15576.0000 | BC-WW-1 | 15870.0000 | BC-WW-1 | 16149.5000 | G-US-3 | 16378.0000 | G-US-3 | 16546.8000 | M-WW-1 |
| 15582.0000 | G-US-3 | 15871.0000 | G-US-3 | 16150.0000 | G-US-3 | 16380.0000 | G-US-3 | 16547.0000 | CG-US-2 |
| 15592.0000 | G-WW-3 | 15873.0000 | G-US-3 | 16152.0000 | N-US-3 | 16385.0000 | AF-US-10 | 16549.9000 | M-WW-1 |
| 15595.0000 | AF-US-10 | 15875.0000 | CG-US-6 | 16153.5000 | N-US-3 | 16387.5000 | N-US-4 | 16551.0000 | CG-US-3 |
| 15595.0000 | G-US-3 | 15875.0000 | BC-WW-1 | 16158.5000 | G-US-3 | 16392.0000 | G-US-3 | 16552.0000 | N-AFS |
| 15600.0000 | G-US-3 | 15877.5000 | BC-WW-1 | 16160.0000 | AF-US-10 | 16395.0000 | G-US-3 | 16553.0000 | M-WW-1 |
| 15604.4000 | PR-US-4 | 15895.0000 | G-US-3 | 16160.0000 | PR-US-6 | 16397.0000 | G-US-3 | 16556.1000 | M-WW-1 |
| 15611.5000 | N-SAM | 15908.0000 | AF-US-10 | 16160.0000 | CC-US-1 | 16398.0000 | G-US-3 | 16559.2000 | M-WW-1 |
| 15619.0000 | N-ARG | 15910.0000 | BC-WW-1 | 16162.0000 | G-US-3 | 16399.0000 | G-US-3 | 16562.0000 | G-US-3 |
| 15620.5000 | AF-US-6 | 15910.0000 | G-US-1 | 16165.0000 | G-US-3 | 16400.0000 | G-US-3 | 16562.3000 | M-WW-1 |
| 15626.0000 | G-US-5 | 15911.0000 | G-US-3 | 16170.0000 | G-US-3 | 16401.0000 | G-US-3 | 16565.4000 | M-WW-1 |
| 15632.0000 | AF-US-8 | 15913.0000 | BC-WW-1 | 16174.0000 | N-US-4 | 16405.0000 | G-US-3 | 16566.4000 | G-US-9 |
| 15645.0000 | G-US-3 | 15917.0000 | G-US-3 | 16175.0000 | G-US-3 | 16407.0000 | AF-US-10 | 16568.5000 | M-WW-1 |
| 15650.0000 | BC-WW-1 | 15918.0000 | G-US-3 | 16180.0000 | G-WW-3 | 16412.5000 | G-US-5 | 16571.6000 | M-WW-1 |
| 15654.5000 | CG-US-8 | 15920.0000 | BC-WW-1 | 16186.0000 | AF-US-10 | 16413.0000 | G-US-3 | 16573.4000 | G-US-9 |
| 15656.5000 | CG-US-4 | 15920.0000 | N-CAN | 16190.0000 | BC-WW-1 | 16418.0000 | G-US-3 | 16574.7000 | CG-US-8 |
| 15656.5000 | CG-US-3 | 15922.0000 | CG-US-6 | 16194.0000 | N-SAM | 16420.0000 | G-US-3 | 16574.7000 | M-WW-1 |
| 15664.0000 | AF-US-10 | 15930.0000 | N-B | 16199.0000 | G-US-3 | 16425.0000 | G-US-3 | 16577.8000 | M-WW-1 |
| 15664.0000 | G-US-3 | 15935.0000 | G-US-3 | 16201.0000 | G-US-1 | 16425.0000 | N-AFS | 16580.9000 | M-WW-1 |
| 15670.0000 | BC-WW-1 | 15940.0000 | PR-US-2 | 16201.0000 | G-US-3 | 16430.0000 | G-US-1 | 16584.0000 | M-WW-1 |
| 15672.0000 | AF-US-10 | 15941.5000 | G-US-3 | 16202.0000 | G-US-3 | 16430.0000 | G-US-3 | 16587.0000 | PR-US-6 |
| 15673.0000 | BC-WW-1 | 15955.0000 | G-US-12 | 16211.5000 | G-US-1 | 16432.0000 | G-US-3 | 16587.1000 | M-WW-1 |
| 15674.0000 | G-US-10 | 15955.0000 | PR-US-6 | 16216.0000 | SP-US-1 | 16433.0000 | BC-WW-1 | 16590.2000 | G-US-9 |
| 15680.0000 | AF-US-10 | 15957.0000 | G-US-3 | 16222.0000 | BC-WW-1 | 16441.5000 | AR-US-3 | 16590.2000 | M-US-5 |
| 15684.0000 | G-WW-3 | 15960.0000 | G-US-3 | 16234.0000 | G-US-10 | 16443.0000 | G-US-3 | 16590.2000 | M-WW-1 |
| 15687.0000 | AF-US-10 | 15960.0000 | PR-US-6 | 16240.0000 | G-US-3 | 16448.0000 | AF-US-6 | 16591.6000 | G-US-5 |
| 15698.0000 | SP-US-1 | 15960.0000 | CC-US-1 | 16240.0000 | BC-WW-1 | 16450.0000 | G-US-3 | 16593.3000 | M-WW-1 |
| 15701.0000 | G-US-3 | 15962.0000 | AF-US-1 | 16242.0000 | G-US-3 | 16450.0000 | M-US-5 | 16595.0000 | G-US-3 |
| 15704.0000 | G-US-3 | 15963.0000 | G-US-3 | 16243.0000 | G-US-3 | 16451.0000 | G-US-3 | 16602.0000 | G-US-3 |
| 15705.0000 | G-US-3 | 15966.0000 | G-US-3 | 16246.0000 | AF-US-10 | 16452.0000 | AF-US-8 | 16608.0000 | G-US-3 |
| 15710.0000 | AF-US-10 | 15969.0000 | G-US-1 | 16246.0000 | SP-US-1 | 16452.0000 | G-US-3 | 16623.0000 | G-US-3 |
| 15711.0000 | G-US-3 | 15980.0000 | G-US-3 | 16247.0000 | BC-WW-1 | 16456.4000 | PR-US-6 | 16634.0000 | G-US-3 |
| 15712.0000 | AF-US-8 | 15981.0000 | G-US-1 | 16250.0000 | G-US-3 | 16457.0000 | G-US-3 | 16638.1000 | CG-US-5 |
| 15715.0000 | BC-WW-1 | 15987.0000 | G-US-3 | 16250.0000 | BC-WW-1 | 16458.0000 | G-US-3 | 16638.7000 | CG-US-5 |
| 15720.0000 | G-US-3 | 15988.0000 | G-US-3 | 16251.0000 | G-US-3 | 16459.0000 | G-US-3 | 16639.0000 | CC-CAN |
| 15720.0000 | BC-WW-1 | 15990.0000 | G-US-3 | 16255.0000 | G-US-3 | 16460.0000 | G-US-3 | 16639.3000 | CG-US-5 |
| 15720.0000 | N-ARG | 15996.0000 | G-US-3 | 16256.0000 | G-US-3 | 16460.0000 | M-WW-1 | 16641.0000 | G-US-9 |
| 15722.4000 | CG-US-6 | 16010.0000 | G-US-3 | 16260.0000 | G-US-3 | 16460.0000 | PR-US-6 | 16648.0000 | PR-US-6 |
| 15723.0000 | CG-US-6 | 16011.0000 | AF-US-10 | 16268.0000 | G-US-3 | 16463.1000 | M-WW-1 | 16660.0000 | CG-US-8 |
| 15724.0000 | AF-US-10 | 16024.0000 | G-US-3 | 16269.0000 | G-US-3 | 16466.2000 | M-WW-1 | 16662.0000 | CG-US-4 |
| 15727.0000 | G-US-3 | 16030.0000 | BC-WW-1 | 16275.0000 | G-US-3 | 16469.3000 | M-WW-1 | 16662.0000 | CG-US-4 |
| 15728.0000 | G-US-3 | 16035.0000 | AF-US-6 | 16280.0000 | G-US-10 | 16470.0000 | G-US-3 | 16666.0000 | CG-US-1 |
| 15733.0000 | AF-US-10 | 16040.0000 | G-US-3 | 16298.5000 | G-US-3 | 16471.0000 | N-AFS | 16666.0000 | CG-US-4 |

| Frequency | Code | Frequency | Code | Frequency | Code | Frequency | Code | Frequency | Code |
|---|---|---|---|---|---|---|---|---|---|
| 16666.0000 | CG-US-3 | 17026.0000 | M-US-3 | 17270.1000 | CG-US-8 | 17450.0000 | G-US-3 | 17902.5000 | G-US-2 |
| 16666.5000 | M-US-3 | 17045.6000 | M-WW-2 | 17270.1000 | G-US-9 | 17457.0000 | AF-US-10 | 17904.0000 | A-WW-2 |
| 16669.5000 | G-US-9 | 17050.4000 | N-ISR | 17270.1000 | M-WW-1 | 17458.0000 | G-US-3 | 17907.0000 | A-WW-2 |
| 16670.0000 | CG-US-1 | 17061.0000 | A-WW-2 | 17273.2000 | M-WW-1 | 17459.0000 | AR-US-3 | 17910.0000 | A-WW-1 |
| 16670.0000 | CG-US-4 | 17064.8000 | M-WW-2 | 17276.3000 | M-WW-1 | 17460.0000 | AF-US-10 | 17910.0000 | A-WW-3 |
| 16670.0000 | CG-US-3 | 17074.4000 | M-WW-2 | 17279.4000 | M-WW-1 | 17460.0000 | AR-US-1 | 17916.0000 | A-US-1 |
| 16670.5000 | M-US-3 | 17079.4000 | M-WW-2 | 17280.8000 | M-WW-2 | 17460.0000 | G-US-3 | 17916.0000 | A-WW-1 |
| 16671.5000 | CG-US-8 | 17084.0000 | AF-CAN | 17280.9000 | G-US-9 | 17470.0000 | SP-US-1 | 17919.0000 | A-US-1 |
| 16673.5000 | CG-US-1 | 17084.0000 | G-US-9 | 17282.5000 | M-WW-1 | 17480.0000 | AF-US-10 | 17919.0000 | A-WW-1 |
| 16673.5000 | CG-US-4 | 17096.0000 | G-WW-3 | 17285.6000 | M-WW-1 | 17480.0000 | N-AUS | 17922.0000 | A-US-1 |
| 16673.5000 | CG-US-3 | 17100.0000 | N-AFS | 17285.7000 | M-WW-2 | 17480.0000 | N-F | 17922.0000 | A-WW-1 |
| 16675.5000 | G-CAN-4 | 17104.2000 | M-WW-2 | 17288.7000 | M-WW-1 | 17481.5000 | N-AUS | 17925.0000 | A-US-1 |
| 16676.0000 | G-CAN-4 | 17105.0000 | G-US-9 | 17291.8000 | M-US-4 | 17485.0000 | G-US-3 | 17928.0000 | A-WW-1 |
| 16680.5000 | M-US-3 | 17117.6000 | M-US-3 | 17291.8000 | M-WW-1 | 17487.0000 | AF-US-8 | 17931.0000 | A-US-1 |
| 16693.0000 | CG-US-8 | 17126.0000 | G-US-3 | 17294.9000 | M-WW-1 | 17488.0000 | CG-US-6 | 17931.0000 | A-WW-1 |
| 16694.5000 | CG-US-8 | 17128.0000 | G-US-3 | 17294.9000 | M-WW-2 | 17488.6000 | CG-US-6 | 17934.0000 | A-US-1 |
| 16695.0000 | CG-US-3 | 17132.0000 | M-WW-2 | 17298.0000 | M-WW-1 | 17490.0000 | SP-US-1 | 17934.0000 | A-WW-1 |
| 16695.0000 | PR-US-6 | 17146.4000 | CG-US-1 | 17300.5000 | M-CAN | 17490.0000 | BC-WW-1 | 17936.0000 | A-URS |
| 16695.5000 | CG-US-4 | 17146.4000 | CG-US-4 | 17300.5000 | M-ATN | 17496.0000 | AF-US-10 | 17937.0000 | A-US-1 |
| 16695.5000 | CG-US-3 | 17146.4000 | CG-US-3 | 17301.1000 | M-WW-1 | 17498.5000 | AR-US-3 | 17937.0000 | A-WW-1 |
| 16696.0000 | CG-US-3 | 17146.4000 | M-WW-2 | 17304.2000 | M-WW-1 | 17501.5000 | AR-US-3 | 17937.0000 | A-WW-3 |
| 16696.5000 | CG-US-4 | 17151.3000 | CG-US-2 | 17304.2000 | M-I | 17503.0000 | G-US-3 | 17940.0000 | A-US-1 |
| 16696.5000 | CG-US-3 | 17160.8000 | M-I | 17307.3000 | CG-US-2 | 17507.0000 | G-US-3 | 17940.0000 | A-WW-1 |
| 16697.5000 | SP-URS | 17161.0000 | G-US-3 | 17307.3000 | CG-US-8 | 17517.0000 | G-US-3 | 17946.0000 | A-WW-2 |
| 16698.0000 | M-US-3 | 17165.6000 | M-WW-2 | 17307.3000 | M-WW-1 | 17519.0000 | G-US-1 | 17952.0000 | G-US-10 |
| 16727.2000 | CG-US-1 | 17170.4000 | M-US-3 | 17310.4000 | M-WW-1 | 17520.0000 | AR-US-3 | 17955.0000 | A-WW-2 |
| 16728.7000 | CG-US-1 | 17170.4000 | M-WW-2 | 17313.5000 | M-WW-1 | 17520.0000 | AR-US-4 | 17958.0000 | A-WW-2 |
| 16758.5000 | CG-US-1 | 17175.2000 | CG-CAN | 17316.6000 | M-WW-1 | 17523.0000 | G-US-3 | 17961.0000 | A-WW-2 |
| 16760.0000 | PR-US-6 | 17175.0000 | G-US-9 | 17319.7000 | M-WW-1 | 17525.0000 | G-US-1 | 17964.0000 | A-US-2 |
| 16792.5000 | CG-US-1 | 17175.2000 | M-WW-2 | 17322.8000 | M-WW-1 | 17528.0000 | G-US-3 | 17964.0000 | PR-US-6 |
| 16800.9000 | M-WW-2 | 17176.0000 | M-WW-2 | 17325.9000 | M-WW-1 | 17532.0000 | G-US-3 | 17972.0000 | AF-US-1 |
| 16802.0000 | G-US-3 | 17180.0000 | G-WW-3 | 17329.0000 | M-WW-1 | 17538.0000 | G-US-3 | 17975.0000 | A-WW-1 |
| 16820.5000 | CG-US-1 | 17184.8000 | M-US-3 | 17332.1000 | M-WW-1 | 17541.0000 | G-US-3 | 17975.0000 | AF-US-1 |
| 16835.0000 | SP-URS | 17184.8000 | M-WW-2 | 17335.2000 | CG-CAN | 17542.0000 | G-US-3 | 17975.0000 | AF-US-5 |
| 16836.0000 | SP-URS | 17189.6000 | CG-US-3 | 17335.2000 | M-WW-1 | 17547.0000 | AF-US-10 | 17980.0000 | AF/N-G |
| 16837.0000 | SP-URS | 17192.0000 | CG-US-3 | 17336.0000 | G-US-3 | 17548.0000 | N-F | 17983.0000 | AF-AUS/NZL |
| 16854.5000 | G-US-9 | 17194.4000 | G-WW-3 | 17338.3000 | M-WW-1 | 17551.1000 | G-WW-3 | 17985.0000 | AF-US-10 |
| 16861.7000 | M-US-3 | 17199.0000 | CG-US-1 | 17341.4000 | M-WW-1 | 17552.0000 | G-US-3 | 17985.0000 | N-US-1 |
| 16871.3000 | M-US-3 | 17199.0000 | CG-US-3 | 17341.4000 | M-WW-2 | 17554.0000 | AF-US-10 | 17992.0000 | AF-D |
| 16880.0000 | AF-CAN | 17203.0000 | CG-US-1 | 17344.5000 | M-WW-1 | 17554.0000 | G-US-3 | 17993.0000 | AF-AFS |
| 16880.9000 | CG-US-1 | 17203.0000 | CG-US-4 | 17347.3000 | CG-US-8 | 17554.0000 | SP-US-1 | 17995.0000 | AF-CAN |
| 16880.9000 | CG-US-4 | 17203.0000 | CG-US-3 | 17347.6000 | M-WW-1 | 17556.0000 | G-US-3 | 18000.0000 | SP-URS |
| 16890.0000 | PR-US-6 | 17203.0000 | CG-US-2 | 17348.0000 | N-AFS | 17560.0000 | BC-WW-1 | 18001.5000 | AF-F |
| 16895.3000 | M-I | 17203.5000 | M-US-3 | 17350.7000 | M-WW-1 | 17562.0000 | AF-US-10 | 18002.0000 | AF-US-3 |
| 16902.0000 | M-WW-2 | 17206.5000 | G-US-9 | 17353.8000 | M-WW-1 | 17562.0000 | G-US-3 | 18005.0000 | AF-US-1 |
| 16904.8000 | N-D | 17207.0000 | CG-US-1 | 17356.9000 | M-WW-1 | 17565.0000 | AF-US-10 | 18005.0000 | AF-US-3 |
| 16918.6000 | N-AUS | 17207.0000 | CG-US-4 | 17360.0000 | PR-US-1 | 17570.0000 | G-US-3 | 18006.0000 | AF-D |
| 16926.5000 | AF-CAN | 17207.0000 | CG-US-3 | 17360.0000 | PR-US-6 | 17576.0000 | G-US-3 | 18009.0000 | N-US-1 |
| 16926.5000 | CG-CAN | 17207.5000 | M-US-3 | 17361.5000 | N-US-3 | 17580.0000 | G-US-3 | 18009.0000 | SP-US-1 |
| 16933.2000 | M-US-3 | 17210.5000 | CG-US-1 | 17362.0000 | G-US-3 | 17580.0000 | BC-WW-1 | 18012.0000 | AF-CAN |
| 16945.0000 | M-WW-2 | 17210.5000 | CG-US-4 | 17365.0000 | G-US-3 | 17585.0000 | G-US-3 | 18014.0000 | G-US-3 |
| 16948.0000 | M-WW-2 | 17210.5000 | CG-US-3 | 17367.0000 | G-US-3 | 17590.0000 | N-SAM | 18016.5000 | N-URS |
| 16948.5000 | CG-CAN | 17212.5000 | CG-CAN | 17368.0000 | G-US-3 | 17590.0000 | N-URS | 18018.0000 | AF/N-G |
| 16951.5000 | AF-F | 17212.5000 | G-CAN-4 | 17375.0000 | G-US-3 | 17594.0000 | AR-US-3 | 18019.0000 | AF-US-2 |
| 16952.4000 | M-WW-2 | 17213.0000 | G-CAN-4 | 17380.0000 | N-F | 17601.0000 | PR-US-6 | 18019.0000 | AF-US-3 |
| 16959.2000 | CG-US-3 | 17217.5000 | M-US-3 | 17385.0000 | G-US-3 | 17603.0000 | G-US-3 | 18022.0000 | SP-US-1 |
| 16960.0000 | AF-CAN | 17232.9000 | M-WW-1 | 17386.0000 | AF-US-10 | 17603.0000 | G-US-12 | 18023.0000 | AF-US-10 |
| 16961.5000 | N-F | 17232.9000 | M-WW-2 | 17390.0000 | G-US-3 | 17617.0000 | AF-US-1 | 18027.0000 | AF-CAN |
| 16961.5000 | N-MAR | 17235.0000 | M-A | 17395.0000 | G-US-3 | 17621.0000 | G-US-10 | 18027.0000 | AF-AUS/NZL |
| 16968.0000 | CG-US-4 | 17236.0000 | M-WW-1 | 17397.0000 | G-US-2 | 17623.0000 | G-US-3 | 18027.0000 | AF-US-4 |
| 16968.8000 | CG-US-3 | 17239.1000 | M-WW-1 | 17403.0000 | N-US-3 | 17624.0000 | G-US-3 | 18030.0000 | G-US-3 |
| 16972.0000 | M-US-3 | 17239.1000 | M-I | 17404.0000 | G-US-3 | 17625.0000 | AF-US-10 | 18030.0000 | PR-US-1 |
| 16973.4500 | M-US-3 | 17242.2000 | CG-CAN | 17405.0000 | A-WW-3 | 17625.0000 | G-US-3 | 18030.0000 | PR-US-6 |
| 16976.0000 | CG-US-1 | 17242.2000 | M-WW-1 | 17405.0000 | G-US-12 | 17625.0000 | G-US-10 | 18034.4000 | PR-US-4 |
| 16976.0000 | M-WW-2 | 17245.3000 | M-WW-1 | 17413.5000 | G-US-3 | 17649.0000 | G-US-1 | 18035.0000 | G-US-3 |
| 16983.0000 | CG-US-1 | 17248.4000 | CG-US-8 | 17419.0000 | G-US-3 | 17665.0000 | G-US-3 | 18037.0000 | G-US-3 |
| 16983.2000 | CG-US-4 | 17248.4000 | M-WW-1 | 17421.0000 | G-US-1 | 17665.0000 | M-WW-2 | 18040.0000 | G-US-3 |
| 16987.3000 | N-G | 17248.4000 | M-I | 17426.0000 | G-US-3 | 17668.0000 | SP-US-1 | 18042.0000 | G-US-3 |
| 16997.6000 | M-US-3 | 17251.5000 | M-WW-1 | 17427.0000 | AF-US-10 | 17669.0000 | G-US-10 | 18043.0000 | G-US-3 |
| 17000.0000 | N-AFS | 17254.6000 | CG-CAN | 17428.0000 | G-US-3 | 17670.0000 | AF-US-8 | 18045.0000 | G-US-3 |
| 17002.4000 | CG-US-3 | 17254.6000 | M-WW-1 | 17428.0000 | SP-US-1 | 17670.0000 | G-US-3 | 18046.0000 | AF-US-1 |
| 17005.0000 | M-I | 17255.0000 | G-US-3 | 17430.0000 | G-US-3 | 17675.0000 | PS-AUS | 18046.0000 | N-AFS |
| 17007.2000 | M-US-3 | 17257.7000 | M-WW-1 | 17433.0000 | AF-US-10 | 17679.0000 | AF-US-10 | 18049.0000 | AF-US-1 |
| 17007.2000 | M-WW-2 | 17260.8000 | M-WW-1 | 17435.0000 | G-US-3 | 17695.0000 | N-F | 18050.0000 | G-US-3 |
| 17015.5000 | N-URS | 17263.9000 | M-WW-1 | 17437.5000 | G-US-3 | 17700.0000 | PR-US-6 | 18050.0000 | G-US-8 |
| 17016.5000 | M-US-3 | 17267.0000 | G-US-9 | 17444.0000 | G-US-3 | 17720.0000 | G-US-3 | 18051.0000 | G-US-3 |
| 17016.8000 | M-US-3 | 17267.0000 | M-WW-1 | 17445.0000 | BC-WW-1 | 17850.0000 | G-US-3 | 18051.0000 | SP-US-1 |
| 17018.0000 | G-WW-3 | 17268.0000 | G-US-9 | 17448.0000 | G-US-3 | 17894.0000 | G-US-3 | 18052.0000 | G-US-3 |
| 17021.6000 | M-US-3 | 17270.0000 | G-US-4 | 17450.0000 | N-US-3 | 17902.0000 | G-US-9 | | |

| Freq | Code | Freq | Code | Freq | Code | Freq | Code | Freq | Code |
|---|---|---|---|---|---|---|---|---|---|
| 18054.0000 | CG-US-3 | 18337.0000 | G-US-3 | 18715.0000 | G-US-3 | 19131.0000 | PR-US-6 | 19514.0000 | G-US-3 |
| 18054.0000 | G-US-3 | 18347.5000 | N-URS | 18716.0000 | G-US-1 | 19133.0000 | G-US-3 | 19518.0000 | G-US-3 |
| 18055.0000 | G-US-3 | 18348.0000 | G-US-3 | 18723.0000 | G-US-3 | 19143.0000 | SP-US-1 | 19521.0000 | G-US-3 |
| 18060.0000 | AF-US-10 | 18351.0000 | G-US-3 | 18743.0000 | G-US-3 | 19146.0000 | G-US-3 | 19522.0000 | BC-WW-1 |
| 18060.0000 | G-US-3 | 18354.0000 | SP-US-1 | 18744.0000 | G-US-1 | 19149.5000 | SP-US-1 | 19531.0000 | AR-US-3 |
| 18060.0000 | SP-URS | 18355.0000 | G-US-3 | 18750.0000 | G-US-3 | 19155.0000 | BC-WW-1 | 19532.5000 | BC-WW-1 |
| 18062.0000 | G-US-3 | 18362.0000 | G-US-3 | 18755.8000 | G-WW-3 | 19160.0000 | AF-US-10 | 19532.9000 | N-F |
| 18063.0000 | G-US-3 | 18365.0000 | G-US-3 | 18758.0000 | G-US-3 | 19160.0000 | G-US-3 | 19542.0000 | CG-US-3 |
| 18064.4000 | PR-US-4 | 18380.0000 | G-WW-3 | 18758.2000 | G-US-3 | 19180.0000 | AF-US-10 | 19542.0000 | SP-URS |
| 18065.0000 | G-US-3 | 18390.0000 | AF-US-10 | 18760.0000 | G-US-3 | 19187.5000 | N-US-4 | 19571.0000 | G-US-3 |
| 18068.0000 | G-US-3 | 18392.0000 | G-US-3 | 18762.0000 | G-US-3 | 19200.0000 | AF-US-8 | 19572.5000 | G-CAN-3 |
| 18081.5000 | M-WW-2 | 18400.0000 | AF-US-10 | 18764.0000 | G-US-10 | 19200.0000 | N-AFS | 19585.0000 | G-US-3 |
| 18092.0000 | G-US-3 | 18400.0000 | G-US-3 | 18769.0000 | SP-US-1 | 19210.0000 | G-US-3 | 19612.0000 | AF-US-8 |
| 18095.0000 | AF-US-10 | 18401.0000 | G-US-3 | 18782.5000 | BC-WW-1 | 19210.0000 | BC-WW-1 | 19615.0000 | AF-US-8 |
| 18100.0000 | AF-US-10 | 18402.0000 | G-US-3 | 18785.0000 | G-US-3 | 19223.0000 | G-US-1 | 19616.0000 | N-SAM |
| 18110.0000 | G-US-3 | 18403.0000 | G-US-1 | 18790.0000 | AF-US-10 | 19226.0000 | AF-US-8 | 19617.0000 | G-US-3 |
| 18112.0000 | G-US-3 | 18403.0000 | G-US-3 | 18795.0000 | AF-US-10 | 19230.0000 | G-US-3 | 19630.0000 | G-US-3 |
| 18114.0000 | G-US-3 | 18417.4000 | G-US-2 | 18801.0000 | SP-US-1 | 19230.0000 | G-US-8 | 19632.0000 | G-US-3 |
| 18115.0000 | G-US-3 | 18420.0000 | G-US-3 | 18802.0000 | G-US-3 | 19231.0000 | G-US-3 | 19635.0000 | G-US-3 |
| 18120.0000 | G-US-3 | 18421.0000 | G-US-3 | 18803.0000 | G-US-3 | 19237.0000 | AF-US-10 | 19636.0000 | G-US-3 |
| 18124.0000 | N-P | 18430.0000 | G-US-3 | 18805.0000 | G-US-3 | 19241.0000 | G-US-3 | 19640.0000 | SP-US-1 |
| 18129.5000 | AR-US-3 | 18434.0000 | SP-US-1 | 18808.0000 | G-US-3 | 19242.0000 | G-US-3 | 19653.0000 | AF-US-10 |
| 18137.5000 | BC-WW-1 | 18435.0000 | G-US-3 | 18810.0000 | G-US-3 | 19250.0000 | N-NZL | 19662.0000 | G-US-3 |
| 18143.0000 | G-US-3 | 18448.0000 | G-US-3 | 18830.0000 | BC-WW-1 | 19250.0000 | G-US-3 | 19662.0000 | CG-US-3 |
| 18145.0000 | G-US-3 | 18450.0000 | G-US-3 | 18842.0000 | G-US-3 | 19256.6000 | N-AUS | 19668.0000 | AF-US-10 |
| 18146.0000 | AF-US-10 | 18454.0000 | G-US-3 | 18890.0000 | G-US-3 | 19261.5000 | BC-WW-1 | 19668.0000 | G-US-3 |
| 18148.0000 | G-US-3 | 18460.0000 | BC-WW-1 | 18902.0000 | G-US-3 | 19270.0000 | AF-US-10 | 19671.5000 | N-US-4 |
| 18150.0000 | G-US-3 | 18460.5000 | G-US-3 | 18909.0000 | G-US-3 | 19270.0000 | G-US-3 | 19676.0000 | G-US-3 |
| 18150.0000 | BC-WW-1 | 18462.0000 | G-US-3 | 18933.0000 | G-US-3 | 19285.0000 | AF-US-6 | 19700.0000 | G-US-3 |
| 18155.0000 | AF-US-8 | 18468.0000 | G-US-3 | 18941.0000 | G-US-3 | 19291.0000 | BC-WW-1 | 19704.0000 | G-US-3 |
| 18157.0000 | BC-WW-1 | 18478.0000 | AF-US-10 | 18962.2000 | CG-US-6 | 19295.0000 | G-US-3 | 19707.0000 | G-US-3 |
| 18160.0000 | G-US-3 | 18480.0000 | G-US-3 | 18963.0000 | CG-US-6 | 19295.0000 | CC-CAN | 19715.0000 | G-US-3 |
| 18161.2000 | G-US-3 | 18488.0000 | G-US-3 | 18964.0000 | G-US-3 | 19296.0000 | G-US-3 | 19721.5000 | BC-WW-1 |
| 18162.5000 | N-US-1 | 18490.0000 | AF-US-10 | 18968.0000 | G-US-3 | 19296.5000 | CG-US-6 | 19725.0000 | BC-WW-1 |
| 18169.0000 | G-US-3 | 18499.0000 | G-US-3 | 18969.0000 | G-US-3 | 19297.1000 | CG-US-6 | 19728.0000 | G-US-3 |
| 18171.0000 | PR-US-6 | 18500.0000 | G-US-3 | 18972.0000 | G-US-3 | 19303.0000 | G-US-3 | 19730.0000 | G-US-3 |
| 18172.0000 | BC-WW-1 | 18505.0000 | N-J | 18973.0000 | G-US-3 | 19303.0000 | SP-US-1 | 19747.0000 | G-US-3 |
| 18173.0000 | G-US-12 | 18505.0000 | G-US-3 | 18980.0000 | AF-US-10 | 19317.0000 | G-US-3 | 19755.0000 | BC-WW-1 |
| 18175.0000 | AF-US-10 | 18515.0000 | BC-WW-1 | 18985.0000 | N-AFS | 19320.0000 | G-US-3 | 19757.0000 | G-US-1 |
| 18176.0000 | G-US-3 | 18515.0000 | N-AFS | 18988.0000 | N-SAM | 19320.0000 | BC-WW-1 | 19758.0000 | AF-US-10 |
| 18180.0000 | CC-CAN | 18525.0000 | G-US-3 | 18990.0000 | N-SAM | 19321.0000 | G-US-3 | 19765.0000 | G-US-3 |
| 18185.0000 | G-US-3 | 18530.0000 | PR-US-2 | 18990.0000 | SP-US-1 | 19330.0000 | G-US-3 | 19780.0000 | PR-US-2 |
| 18187.0000 | G-US-3 | 18531.0000 | G-US-3 | 19002.0000 | G-US-3 | 19335.0000 | PR-US-2 | 19785.0000 | G-US-3 |
| 18190.0000 | G-WW-3 | 18532.0000 | AF-US-10 | 19004.5000 | AR-US-3 | 19340.0000 | AF-US-10 | 19791.5000 | G-US-2 |
| 18195.0000 | BC-WW-1 | 18545.0000 | G-US-3 | 19005.0000 | AF-US-10 | 19360.0000 | G-WW-3 | 19800.0000 | AF-US-10 |
| 18195.5000 | CG-US-8 | 18572.0000 | G-US-3 | 19006.0000 | G-US-3 | 19360.0000 | G-CAN-1 | 19800.0000 | G-US-3 |
| 18196.0000 | SP-US-1 | 18585.0000 | G-US-3 | 19007.0000 | G-US-3 | 19371.0000 | SP-US-1 | 19803.0000 | G-US-3' |
| 18197.5000 | CG-US-3 | 18586.0000 | G-US-3 | 19008.0000 | AF-US-10 | 19373.0000 | AF-US-10 | 19805.0000 | G-US-3 |
| 18205.0000 | G-US-3 | 18590.0000 | AF-US-10 | 19010.0000 | N-AFS | 19385.0000 | G-US-3 | 19808.0000 | CG-US-3 |
| 18210.0000 | A-WW-3 | 18590.0000 | AR-UN | 19010.5000 | AR-US-3 | 19390.0000 | G-US-3 | 19809.0000 | G-US-3 |
| 18210.0000 | BC-WW-1 | 18592.0000 | G-US-3 | 19012.0000 | G-US-3 | 19390.0000 | SP-US-1 | 19810.0000 | G-US-3 |
| 18212.5000 | AR-US-3 | 18594.0000 | AF-US-1 | 19013.5000 | AR-US-3 | 19395.0000 | G-US-3 | 19815.0000 | G-US-3 |
| 18215.0000 | BC-WW-1 | 18607.5000 | BC-WW-1 | 19014.5000 | G-US-3 | 19395.0000 | N-ARG | 19816.0000 | G-US-3 |
| 18218.0000 | AF-US-10 | 18612.0000 | G-US-3 | 19015.0000 | G-US-3 | 19401.0000 | G-US-3 | 19821.0000 | G-US-3 |
| 18226.0000 | AF-US-6 | 18615.0000 | G-US-10 | 19016.0000 | G-US-3 | 19405.0000 | G-WW-3 | 19840.0000 | AR-US-3 |
| 18235.0000 | G-US-3 | 18628.0000 | G-US-3 | 19020.0000 | G-US-3 | 19410.0000 | G-US-10 | 19845.0000 | BC-WW-1 |
| 18236.0000 | G-US-3 | 18629.0000 | AF-US-10 | 19022.0000 | G-US-3 | 19420.0000 | G-US-3 | 19850.0000 | G-US-3 |
| 18247.0000 | G-US-3 | 18630.0000 | G-US-3 | 19024.0000 | AR-US-3 | 19425.0000 | AR-UN | 19878.0000 | G-US-3 |
| 18248.5000 | G-US-3 | 18638.0000 | G-US-3 | 19029.0000 | AR-US-3 | 19425.0000 | G-US-3 | 19882.0000 | G-US-3 |
| 18250.0000 | G-US-3 | 18640.0000 | G-US-3 | 19042.0000 | G-US-3 | 19428.0000 | AF-US-10 | 19882.0000 | N-US-4 |
| 18265.0000 | BC-WW-1 | 18643.0000 | G-US-3 | 19047.0000 | AF-US-10 | 19429.0000 | G-US-3 | 19885.0000 | G-US-3 |
| 18270.0000 | AF-US-10 | 18645.0000 | G-US-3 | 19049.0000 | PR-US-6 | 19430.0000 | N-US-4 | 19895.0000 | AF-US-10 |
| 18271.0000 | A-US-3 | 18650.0000 | G-US-3 | 19050.0000 | AF-US-10 | 19435.0000 | G-US-3 | 19900.0000 | G-US-3 |
| 18275.0000 | BC-WW-1 | 18653.0000 | BC-WW-1 | 19050.0000 | G-US-3 | 19440.0000 | G-US-3 | 19904.0000 | A-WW-1 |
| 18285.0000 | BC-WW-1 | 18653.5000 | BC-WW-1 | 19055.0000 | G-US-3 | 19445.0000 | G-US-3 | 19905.0000 | G-US-3 |
| 18297.0000 | G-US-3 | 18662.0000 | G-US-3 | 19060.0000 | AF-US-10 | 19447.0000 | AF-US-10 | 19912.5000 | BC-WW-1 |
| 18300.0000 | G-US-3 | 18666.0000 | CG-US-4 | 19060.0000 | G-US-3 | 19452.0000 | G-US-3 | 19915.0000 | BC-WW-1 |
| 18300.0000 | BC-WW-1 | 18666.0000 | PR-US-6 | 19065.0000 | G-US-3 | 19455.0000 | BC-WW-1 | 19918.0000 | BC-WW-1 |
| 18306.0000 | G-US-3 | 18667.0000 | G-US-3 | 19074.0000 | G-US-3 | 19458.0000 | G-US-3 | 19918.5000 | N-US-4 |
| 18306.5000 | G-US-3 | 18668.0000 | G-US-3 | 19075.0000 | BC-WW-1 | 19463.0000 | G-US-3 | 19920.0000 | G-US-3 |
| 18310.0000 | BC-WW-1 | 18675.0000 | AF-US-10 | 19083.0000 | N-AFS | 19465.0000 | G-US-3 | 19922.0000 | BC-WW-1 |
| 18310.0000 | SP-US-1 | 18675.0000 | N-J | 19090.0000 | AR-US-1 | 19480.0000 | BC-WW-1 | 19928.0000 | SP-US-1 |
| 18317.0000 | G-US-3 | 18682.0000 | G-US-3 | 19092.0000 | AF-US-10 | 19495.0000 | G-US-3 | 19934.0000 | G-US-1 |
| 18320.0000 | AF-US-10 | 18685.0000 | G-US-3 | 19094.0000 | G-US-3 | 19500.0000 | G-US-3 | 19937.0000 | AF-US-8 |
| 18320.0000 | G-US-3 | 18689.0000 | G-US-3 | 19101.0000 | G-US-3 | 19501.0000 | G-US-3 | 19940.0000 | PR-US-2 |
| 18321.0000 | G-US-3 | 18700.0000 | G-US-3 | 19124.0000 | G-US-3 | 19505.0000 | BC-WW-1 | 19945.0000 | G-US-3 |
| 18323.0000 | G-US-3 | 18700.0000 | SP-US-1 | 19126.0000 | SP-US-1 | 19510.0000 | A-WW-3 | 19950.0000 | G-US-3 |
| 18331.0000 | AF-US-10 | 18701.0000 | G-US-3 | 19130.0000 | G-WW-3 | 19512.0000 | G-US-3 | 19954.0000 | SP-URS |
| 18331.0000 | SP-US-1 | 18707.0000 | G-US-3 | 19130.0000 | G-CAN-1 | 19513.0000 | AF-US-10 | 19955.0000 | AF-US-6 |

| Freq | | Freq | | Freq | | Freq | | Freq | |
|---|---|---|---|---|---|---|---|---|---|
| 19956.5000 | N-US-4 | 20229.0000 | G-US-3 | 20625.0000 | N-US-4 | 20938.0000 | G-US-3 | 21842.0000 | G-US-3 |
| 19963.0000 | SP-US-1 | 20240.0000 | N-F | 20631.0000 | AF-US-1 | 20939.0000 | G-US-3 | 21843.0000 | AF-US-10 |
| 19966.0000 | SP-US-1 | 20245.0000 | G-US-3 | 20632.0000 | G-US-3 | 20940.0000 | G-US-3 | 21843.0000 | G-US-3 |
| 19969.0000 | G-US-1 | 20247.0000 | CG-US-3 | 20640.0000 | AF-US-10 | 20942.0000 | G-US-3 | 21845.0000 | N-F |
| 19973.0000 | G-US-3 | 20259.0000 | G-US-3 | 20650.0000 | BC-WW-1 | 20942.0000 | PS-WW-1 | 21846.0000 | G-US-3 |
| 19977.0000 | G-US-3 | 20261.0000 | SP-US-1 | 20650.0000 | AF-US-10 | 20945.0000 | AR-UN | 21847.0000 | G-US-3 |
| 19985.0000 | G-US-3 | 20262.5000 | CG-US-6 | 20650.5000 | AR-US-3 | 20945.0000 | G-US-3 | 21854.5000 | G-US-3 |
| 19989.0000 | SP-URS | 20264.0000 | G-US-3 | 20655.0000 | AR-US-3 | 20948.0000 | G-US-3 | 21856.0000 | G-US-3 |
| 19990.0000 | G-US-3 | 20265.0000 | N-J | 20664.0000 | N-P | 20950.0000 | G-US-3 | 21860.0000 | AF-AFS |
| 19994.0000 | SP-URS | 20266.0000 | SP-US-1 | 20665.0000 | AF-US-10 | 20953.0000 | AF-US-10 | 21862.5000 | G-US-3 |
| 20000.0000 | G-WW-3 | 20272.0000 | SP-US-1 | 20665.0000 | AR-US-3 | 20954.0000 | G-US-3 | 21870.0000 | PR-US-1 |
| 20001.0000 | G-US-3 | 20280.0000 | BC-WW-1 | 20680.0000 | N-US-4 | 20955.0000 | G-US-3 | 21870.0000 | PR-US-6 |
| 20002.0000 | G-US-3 | 20313.0000 | AF-US-10 | 20682.0000 | G-US-3 | 20957.0000 | AR-CAN | 21876.0000 | G-US-3 |
| 20004.0000 | G-US-3 | 20319.0000 | G-US-3 | 20690.0000 | SP-US-1 | 20957.0000 | G-US-3 | 21882.0000 | G-US-3 |
| 20007.0000 | G-US-3 | 20321.0000 | G-US-3 | 20695.0000 | G-US-3 | 20958.0000 | G-US-3 | 21886.0000 | AF/N-G |
| 20008.0000 | SP-URS | 20322.0000 | G-US-3 | 20700.0000 | G-US-3 | 20960.0000 | AF-US-8 | 21890.0000 | G-US-3 |
| 20010.0000 | G-US-3 | 20325.0000 | BC-WW-1 | 20707.0000 | G-US-3 | 20960.0000 | G-US-3 | 21919.0000 | G-US-1 |
| 20010.0000 | PR-US-1 | 20327.0000 | G-US-3 | 20712.0000 | G-US-3 | 20962.0000 | AR-CAN | 21931.0000 | A-US-2 |
| 20010.0000 | PR-US-6 | 20330.0000 | G-US-3 | 20720.0000 | G-US-3 | 20964.0000 | G-US-3 | 21931.0000 | PR-US-6 |
| 20011.0000 | G-US-3 | 20340.0000 | AF-US-10 | 20722.0000 | G-US-3 | 20970.0000 | G-US-3 | 21931.0000 | CC-US-1 |
| 20012.0000 | G-US-3 | 20345.0000 | BC-WW-1 | 20728.0000 | G-US-3 | 20970.0000 | AR-CAN | 21937.0000 | G-US-9 |
| 20014.0000 | G-US-3 | 20348.0000 | G-US-3 | 20732.0000 | G-US-3 | 20972.0000 | AF-US-10 | 21940.0000 | A-US-1 |
| 20016.0000 | AF-US-10 | 20348.5000 | PR-US-6 | 20735.0000 | G-US-3 | 20973.0000 | G-US-3 | 21940.0000 | A-WW-1 |
| 20020.0000 | G-US-3 | 20350.0000 | N-SAM | 20735.0000 | PS-WW-1 | 20975.0000 | N-BEL | 21943.0000 | A-US-1 |
| 20021.0000 | G-US-3 | 20353.0000 | G-US-3 | 20737.0000 | AF-US-1 | 20975.0000 | AR-US-3 | 21943.0000 | A-WW-1 |
| 20022.0000 | G-US-3 | 20365.0000 | G-US-3 | 20737.0000 | AF-US-3 | 20977.0000 | G-US-3 | 21946.0000 | A-US-1 |
| 20023.0000 | G-US-3 | 20369.0000 | G-US-3 | 20737.0000 | AF-US-8 | 20980.0000 | G-US-3 | 21946.0000 | A-WW-1 |
| 20025.0000 | BC-WW-1 | 20373.0000 | G-US-3 | 20740.0000 | AF-US-3 | 20986.0000 | G-US-3 | 21949.0000 | A-US-1 |
| 20027.0000 | G-US-1 | 20375.0000 | G-US-3 | 20740.0000 | G-US-3 | 20988.5000 | N-US-4 | 21949.0000 | A-WW-1 |
| 20028.0000 | G-US-3 | 20375.0000 | N-US-4 | 20750.0000 | G-US-3 | 20990.0000 | G-US-3 | 21952.0000 | A-US-1 |
| 20030.0000 | G-US-3 | 20378.0000 | G-US-3 | 20752.0000 | G-US-3 | 20991.0000 | AF-US-8 | 21952.0000 | A-WW-1 |
| 20035.0000 | G-US-3 | 20381.0000 | G-US-3 | 20753.0000 | PS-WW-1 | 20991.5000 | N-URS | 21953.0000 | G-US-3 |
| 20038.0000 | G-US-3 | 20390.0000 | SP-US-1 | 20763.0000 | AF-US-8 | 20992.5000 | AF-US-8 | 21955.0000 | A-US-1 |
| 20043.7000 | G-US-3 | 20395.0000 | AF-US-10 | 20765.0000 | G-US-3 | 20992.5000 | AR-US-3 | 21955.0000 | A-WW-1 |
| 20044.0000 | G-US-3 | 20398.0000 | G-US-3 | 20770.0000 | G-US-3 | 20993.0000 | N-URS | 21958.0000 | A-US-1 |
| 20050.0000 | G-US-3 | 20400.0000 | AF-US-10 | 20775.0000 | G-US-3 | 20993.5000 | N-URS | 21958.0000 | A-WW-1 |
| 20053.0000 | AF-US-10 | 20402.0000 | G-US-3 | 20778.0000 | G-US-3 | 20993.5000 | PS-WW-1 | 21961.0000 | A-US-1 |
| 20054.0000 | G-US-3 | 20405.0000 | G-US-3 | 20781.0000 | G-US-3 | 20994.0000 | AF-US-8 | 21965.0000 | G-US-3 |
| 20060.0000 | G-US-3 | 20405.5000 | G-US-2 | 20788.0000 | G-US-3 | 20995.5000 | AR-US-3 | 21967.0000 | A-US-1 |
| 20060.0000 | BC-WW-1 | 20413.0000 | G-US-3 | 20792.0000 | G-US-3 | 20996.0000 | G-US-3 | 21967.0000 | A-WW-1 |
| 20063.0000 | G-US-1 | 20415.0000 | AF-US-10 | 20797.0000 | G-US-3 | 20998.5000 | N-US-4 | 21970.0000 | A-US-1 |
| 20065.0000 | A-WW-3 | 20417.0000 | G-US-3 | 20802.0000 | G-US-3 | 21003.0000 | G-US-3 | 21970.0000 | A-WW-1 |
| 20073.0000 | G-US-3 | 20419.0000 | G-US-3 | 20805.0000 | G-US-3 | 21008.0000 | G-US-3 | 21973.0000 | A-US-1 |
| 20095.0000 | G-US-1 | 20420.0000 | G-US-3 | 20807.0000 | AF-US-8 | 21010.0000 | G-US-3 | 21973.0000 | A-WW-1 |
| 20099.0000 | G-US-3 | 20422.0000 | G-US-3 | 20808.0000 | G-US-3 | 21300.0000 | G-US-3 | 21976.0000 | A-US-1 |
| 20102.0000 | G-US-3 | 20450.0000 | G-US-3 | 20810.0000 | G-US-3 | 21450.0000 | PR-US-6 | 21979.0000 | A-US-1 |
| 20105.5000 | AR-US-3 | 20452.0000 | AF-US-10 | 20812.0000 | AR-US-3 | 21455.0000 | BC-WW-1 | 21979.0000 | A-WW-1 |
| 20112.5000 | N-F | 20455.0000 | G-US-3 | 20815.0000 | PS-WW-1 | 21725.5000 | PR-US-6 | 21980.0000 | G-US-3 |
| 20123.0000 | G-US-3 | 20455.0000 | BC-WW-1 | 20825.0000 | G-US-3 | 21727.0000 | PR-US-6 | 21982.0000 | A-US-1 |
| 20124.0000 | AF-US-1 | 20458.0000 | G-US-3 | 20830.0000 | G-US-3 | 21727.0000 | CC-US-1 | 21985.0000 | A-US-1 |
| 20124.0000 | AF-US-10 | 20473.0000 | G-US-3 | 20840.0000 | G-US-3 | 21728.5000 | PR-US-6 | 21985.0000 | A-WW-1 |
| 20125.0000 | G-US-3 | 20475.0000 | AF-US-10 | 20843.0000 | G-US-1 | 21750.0000 | PR-US-6 | 21988.0000 | A-US-1 |
| 20125.0000 | BC-WW-1 | 20475.0000 | N-J | 20846.0000 | AF-US-1 | 21751.0000 | G-US-3 | 21988.0000 | A-WW-1 |
| 20129.0000 | G-US-3 | 20475.0000 | SP-US-1 | 20850.0000 | G-US-3 | 21755.0000 | G-US-3 | 21991.0000 | A-US-1 |
| 20130.0000 | G-US-3 | 20480.0000 | G-US-3 | 20852.0000 | G-US-10 | 21760.0000 | CG-US-2 | 21991.0000 | A-WW-1 |
| 20130.0000 | G-US-10 | 20480.0000 | G-US-10 | 20855.0000 | AF-US-4 | 21760.0000 | BC-WW-1 | 21994.0000 | A-US-1 |
| 20148.0000 | G-US-3 | 20492.0000 | G-US-3 | 20860.0000 | G-US-3 | 21763.0000 | G-US-3 | 21994.0000 | A-WW-1 |
| 20149.0000 | G-US-3 | 20496.0000 | G-US-3 | 20868.0000 | G-US-3 | 21764.0000 | G-US-3 | 21997.0000 | A-US-1 |
| 20150.0000 | G-US-3 | 20500.0000 | BC-WW-1 | 20870.0000 | AF-US-8 | 21765.0000 | A-WW-3 | 22000.0000 | M-WW-1 |
| 20153.0000 | AF-US-9 | 20506.0000 | G-US-3 | 20873.0000 | AF-US-7 | 21767.0000 | AF-US-6 | 22000.0000 | PR-US-6 |
| 20155.0000 | G-US-3 | 20511.0000 | AF/N-G | 20873.0000 | AF-US-8 | 21770.0000 | AF-US-10 | 22003.1000 | M-WW-1 |
| 20161.0000 | G-US-3 | 20520.0000 | AR-US-3 | 20874.0000 | G-US-3 | 21773.0000 | G-US-3 | 22006.2000 | M-WW-1 |
| 20164.0000 | G-US-3 | 20520.0000 | M-WW-2 | 20875.0000 | PR-US-1 | 21785.0000 | G-WW-3 | 22009.3000 | M-WW-1 |
| 20165.0000 | AF-US-10 | 20525.0000 | G-US-3 | 20876.3000 | PR-US-1 | 21785.0000 | G-CAN-1 | 22012.4000 | M-WW-1 |
| 20165.0000 | G-US-3 | 20252.0000 | N-US-4 | 20880.0000 | G-US-3 | 21792.0000 | N-SAM | 22015.5000 | CG-US-8 |
| 20167.0000 | AF-US-1 | 20527.5000 | G-US-3 | 20885.0000 | AF-US-3 | 21800.0000 | BC-WW-1 | 22015.5000 | M-WW-1 |
| 20170.0000 | G-US-3 | 20529.5000 | G-CAN-3 | 20888.0000 | G-US-3 | 21804.0000 | N-US-3 | 22018.6000 | N-US-3 |
| 20177.0000 | G-US-3 | 20535.0000 | AF-US-10 | 20890.0000 | AF-US-1 | 21805.5000 | N-US-3 | 22021.7000 | M-WW-1 |
| 20186.0000 | SP-US-1 | 20545.0000 | BC-WW-1 | 20895.0000 | G-US-3 | 21807.5000 | G-WW-3 | 22024.8000 | M-WW-1 |
| 20187.0000 | G-US-3 | 20560.0000 | AR-US-3 | 20900.0000 | G-US-3 | 21807.5000 | G-CAN-1 | 22025.9000 | G-US-9 |
| 20188.5000 | AF-US-8 | 20568.0000 | G-US-3 | 20908.0000 | G-US-3 | 21810.0000 | G-US-3 | 22027.9000 | M-WW-1 |
| 20190.0000 | G-US-3 | 20569.0000 | G-US-3 | 20915.0000 | G-US-3 | 21810.0000 | SP-US-1 | 22030.0000 | G-US-3 |
| 20192.0000 | BC-WW-1 | 20571.0000 | G-US-3 | 20917.0000 | G-US-3 | 21811.5000 | G-US-3 | 22031.0000 | M-WW-1 |
| 20192.0000 | SP-US-1 | 20580.0000 | N-US-4 | 20920.0000 | G-US-3 | 21815.0000 | AF-US-1 | 22034.1000 | M-WW-1 |
| 20195.0000 | AF-US-10 | 20605.0000 | BC-WW-1 | 20921.5000 | AR-US-3 | 21817.0000 | G-US-3 | 22037.2000 | M-WW-1 |
| 20198.0000 | SP-US-1 | 20610.0000 | G-US-3 | 20929.0000 | G-US-3 | 21825.5000 | AR-US-3 | 22040.3000 | M-WW-1 |
| 20215.0000 | BC-WW-1 | 20620.0000 | G-US-3 | 20929.5000 | G-US-3 | 21828.5000 | G-US-2 | 22043.0000 | PR-US-6 |
| 20221.0000 | AR-US-3 | 20622.0000 | G-US-3 | 20932.0000 | G-US-3 | 21834.0000 | N-US-3 | 22043.4000 | M-WW-1 |
| 20225.0000 | G-WW-3 | 20624.0000 | G-US-3 | 20936.0000 | N-US-4 | 21840.0000 | G-US-3 | 22044.8000 | M-WW-1 |

| | | | | | | | | | | | |
|---|---|---|---|---|---|---|---|---|---|---|---|
| 22046.5000 | M-WW-1 | 22347.5000 | M-US-3 | 22648.7000 | M-WW-1 | 23035.0000 | SP-US-1 | 23952.0000 | G-US-3 | | |
| 22049.6000 | M-WW-1 | 22348.5000 | M-US-3 | 22654.9000 | CG-CAN | 23068.5000 | G-CAN-3 | 23964.0000 | G-US-3 | | |
| 22052.7000 | CG-US-8 | 22360.0000 | PR-US-6 | 22654.9000 | M-WW-1 | 23098.5000 | PR-US-6 | 23975.0000 | G-US-3 | | |
| 22052.7000 | M-WW-1 | 22366.5000 | M-US-3 | 22658.0000 | M-WW-1 | 23100.0000 | G-US-3 | 23982.5000 | G-US-3 | | |
| 22058.9000 | M-WW-1 | 22368.0000 | CG-CAN | 22658.0000 | M-WW-2 | 23100.0000 | PR-US-6 | 23985.0000 | N-F | | |
| 22062.0000 | M-WW-1 | 22372.4000 | M-1 | 22661.1000 | M-WW-1 | 23101.5000 | PR-US-6 | 23995.0000 | G-US-3 | | |
| 22065.1000 | M-WW-1 | 22378.0000 | M-1 | 22554.2000 | M-WW-1 | 23104.0000 | G-US-3 | 24000.0000 | G-US-3 | | |
| 22068.2000 | M-WW-1 | 22384.0000 | M-WW-2 | 22667.3000 | M-WW-1 | 23106.0000 | G-US-3 | 24005.0000 | G-US-3 | | |
| 22071.3000 | M-WW-1 | 22387.0000 | CG-CAN | 22670.4000 | M-WW-1 | 23132.0000 | AF-US-10 | 24012.5000 | AR-US-3 | | |
| 22074.4000 | M-WW-1 | 22407.0000 | CG-US-3 | 22673.5000 | M-WW-1 | 23157.0000 | G-US-3 | 24019.0000 | AF-US-8 | | |
| 22077.5000 | M-WW-1 | 22413.0000 | M-WW-2 | 22676.6000 | M-WW-1 | 23177.0000 | N-US-1 | 24040.0000 | G-US-1 | | |
| 22080.6000 | M-WW-1 | 22419.0000 | M-WW-2 | 22679.7000 | M-WW-1 | 23181.0000 | G-US-3 | 24050.0000 | AR-US-3 | | |
| 22083.7000 | M-WW-1 | 22425.0000 | M-WW-2 | 22682.8000 | M-WW-1 | 23191.0000 | BC-WW-1 | 24060.0000 | N-F | | |
| 22085.0000 | N-AFS | 22429.4000 | G-US-9 | 22683.0000 | SP-US-1 | 23206.0000 | AF-US-2 | 24072.0000 | G-WW-3 | | |
| 22086.8000 | M-WW-1 | 22431.0000 | M-US-3 | 22685.9000 | M-WW-1 | 23210.0000 | G-US-3 | 24088.5000 | CG-US-6 | | |
| 22089.9000 | M-WW-1 | 22446.0000 | M-WW-2 | 22689.0000 | M-WW-1 | 23211.0000 | G-US-10 | 24110.0000 | G-WW-3 | | |
| 22093.0000 | M-WW-1 | 22458.0000 | M-US-3 | 22692.1000 | M-WW-1 | 23220.0000 | AF/N-G | 24110.0000 | G-CAN-1 | | |
| 22096.1000 | M-WW-1 | 22467.0000 | M-US-3 | 22695.0000 | G-US-4 | 23224.0000 | N-US-1 | 24170.0000 | N-US-4 | | |
| 22099.2000 | G-US-9 | 22472.5000 | CG-US-2 | 22695.2000 | G-US-9 | 23227.0000 | AF-US-3 | 24197.5000 | AR-US-3 | | |
| 22099.2000 | M-22-1 | 22473.0000 | M-WW-2 | 22695.2000 | M-WW-1 | 23227.0000 | N-US-1 | 24240.0000 | SP-US-1 | | |
| 22099.4000 | G-US-9 | 22476.0000 | CG-US-1 | 22698.3000 | M-WW-1 | 23236.0000 | AF/N-G | 24266.5000 | N-US-4 | | |
| 22102.3000 | M-WW-1 | 22476.0000 | M-WW-2 | 22700.0000 | G-US-9 | 23240.0000 | G-US-10 | 24274.0000 | AF-US-10 | | |
| 22105.4000 | G-US-9 | 22479.0000 | M-US-3 | 22701.4000 | G-US-9 | 23243.0000 | AF-US-10 | 24300.0000 | G-US-3 | | |
| 22105.4000 | M-WW-1 | 22485.0000 | M-US-3 | 22701.4000 | M-WW-1 | 23250.0000 | AF-CAN | 24330.0000 | G-US-3 | | |
| 22106.4000 | G-US-9 | 22487.0000 | M-US-3 | 22701.5000 | G-US-4 | 23254.0000 | AF-F | 24350.0000 | N-US-4 | | |
| 22108.5000 | M-WW-1 | 22487.5000 | CG-US-3 | 22704.5000 | M-WW-1 | 23272.0000 | G-US-3 | 24460.0000 | N-F | | |
| 22109.4000 | G-US-9 | 22502.0000 | M-WW-2 | 22707.6000 | M-WW-1 | 23281.0000 | SP-US-1 | 24483.0000 | AF-US-10 | | |
| 22111.6000 | M-WW-1 | 22513.5000 | M-ARG | 22710.7000 | M-WW-1 | 23287.0000 | AF-AFS | 24512.0000 | SP-US-1 | | |
| 22114.7000 | M-WW-1 | 22515.0000 | M-US-3 | 22711.0000 | G-US-3 | 23287.0000 | N-US-1 | 24530.0000 | SP-US-1 | | |
| 22117.8000 | M-WW-1 | 22518.0000 | M-US-3 | 22713.8000 | M-WW-1 | 23315.0000 | N-US-1 | 24560.0000 | AR-US-3 | | |
| 22120.9000 | M-WW-1 | 22521.0000 | M-US-3 | 22716.9000 | M-WW-1 | 23325.0000 | SP-US-1 | 24573.0000 | G-US-3 | | |
| 22124.0000 | M-WW-1 | 22527.0000 | CG-US-1 | 22720.0000 | PR-US-6 | 23331.5000 | G-US-10 | 24573.0000 | G-SU3 | | |
| 22124.0000 | PR-US-6 | 22527.0000 | CG-US-4 | 22722.0000 | G-US-3 | 23337.0000 | AF-US-1 | 24573.5000 | AF-US-8 | | |
| 22126.0000 | AR-US-1 | 22527.0000 | CG-US-3 | 22722.0000 | M-US-5 | 23342.0000 | G-US-3 | 24612.0000 | G-US-3 | | |
| 22127.0000 | M-WW-1 | 22533.0000 | M-WW-2 | 22723.0000 | AF-US-10 | 23345.0000 | G-US-10 | 24761.5000 | AR-US-3 | | |
| 22130.2000 | M-WW-1 | 22536.0000 | G-WW-3 | 22729.0000 | G-US-3 | 23350.0000 | PR-US-6 | 24780.0000 | SP-US-1 | | |
| 22133.3000 | M-WW-1 | 22539.0000 | M-WW-W | 22732.0000 | G-US-3 | 23355.0000 | G-US-3 | 24793.0000 | G-US-1 | | |
| 22136.4000 | M-WW-1 | 22544.0000 | AF-F | 22738.0000 | SP-US-1 | 23355.0000 | N-US-4 | 24800.0000 | G-US-3 | | |
| 22162.1000 | CG-US-5 | 22545.0000 | N-F | 22740.0000 | G-US-3 | 23360.0000 | AF-US-6 | 24806.0000 | G-US-3 | | |
| 21262.7000 | CG-US-5 | 22545.0000 | CG-US-1 | 22755.0000 | SP-US-1 | 23363.0000 | G-US-3 | 24860.0000 | AR-US-3 | | |
| 22163.0000 | CC-CAN | 22545.0000 | CG-US-4 | 22793.0000 | G-US-3 | 23370.0000 | G-US-3 | 24914.0000 | SP-US-1 | | |
| 22163.3000 | CG-US-5 | 22545.0000 | CG-US-3 | 22796.0000 | G-US-3 | 23376.0000 | G-US-3 | 24930.0000 | PR-US-6 | | |
| 22178.0000 | PR-US-6 | 22557.0000 | M-US-3 | 22800.0000 | G-US-3 | 23385.0000 | G-US-3 | 24933.0000 | AF-US-10 | | |
| 22184.0000 | CG-US-3 | 22561.0000 | CG-US-3 | 22809.0000 | G-US-3 | 23402.0000 | G-US-12 | 25005.0000 | PR-US-6 | | |
| 22192.0000 | CG-US-8 | 22563.0000 | CG-US-4 | 22810.0000 | N-SAM | 23403.0000 | CG-US-4 | 25035.0000 | G-US-3 | | |
| 22194.0000 | CG-US-4 | 22563.0000 | CG-US-3 | 22818.0000 | G-US-3 | 23403.0000 | PR-US-6 | 25041.0000 | G-US-3 | | |
| 22198.0000 | CG-US-1 | 22567.0000 | CG-US-1 | 22818.0000 | N-ARG | 23413.0000 | SP-US-1 | 25085.8000 | CG-US-1 | | |
| 22198.0000 | CG-US-3 | 22567.0000 | CG-US-4 | 22820.0000 | G-US-3 | 23419.0000 | AF-US-1 | 25100.0000 | PR-US-6 | | |
| 22198.0000 | CG-US-4 | 22567.0000 | M-WW-2 | 22822.4000 | CG-US-6 | 23425.0000 | G-US-3 | 25130.0000 | AF-US-10 | | |
| 22198.5000 | M-US-3 | 22567.5000 | M-US-3 | 22823.0000 | CG-US-6 | 23442.5000 | G-US-3 | 25130.0000 | SP-US-1 | | |
| 22201.5000 | G-US-9 | 22570.5000 | G-US-9 | 22830.0000 | G-US-3 | 23460.0000 | G-US-3 | 25130.5000 | M-ARG | | |
| 22202.0000 | CG-US-1 | 22571.0000 | CG-US-1 | 22834.5000 | G-CAN-3 | 23474.0000 | G-US-10 | 25147.5000 | G-CAN-3 | | |
| 22202.0000 | CG-US-4 | 22571.0000 | CG-US-4 | 22848.0000 | G-US-3 | 23478.0000 | G-US-3 | 25149.0000 | G-CAN-3 | | |
| 22202.5000 | M-US-3 | 22571.5000 | M-US-3 | 22880.0000 | G-US-3 | 23479.0000 | SP-US-1 | 25150.5000 | G-CAN-3 | | |
| 22203.5000 | CG-US-2 | 22574.5000 | CG-US-4 | 22886.0000 | G-US-3 | 23485.0000 | SP-US-1 | 25161.0000 | SP-US-1 | | |
| 22203.5000 | CG-US-8 | 22574.5000 | CG-US-3 | 22903.0000 | PR-US-6 | 23490.0000 | N-F | 25198.0000 | SP-US-1 | | |
| 22205.5000 | CG-US-1 | 22576.5000 | G-CAN-4 | 22913.0000 | AF-US-10 | 23560.0000 | G-US-3 | 25227.0000 | AF-US-10 | | |
| 22205.5000 | CG-US-3 | 22590.0000 | CG-CAN | 22925.0000 | BC-WW-1 | 23570.0000 | G-US-3 | 25230.0000 | PR-US-6 | | |
| 22205.5000 | CG-US-4 | 22590.0000 | G-CAN-4 | 22926.0000 | G-US-1 | 23584.0000 | G-US-3 | 25230.0000 | CC-US-1 | | |
| 22207.5000 | G-CAN-4 | 22596.0000 | M-WW-1 | 22930.0000 | BC-WW-1 | 23635.0000 | G-US-3 | 25245.0000 | SP-US-1 | | |
| 22221.0000 | G-CAN-4 | 22599.1000 | M-WW-1 | 22940.0000 | AF-US-10 | 23642.0000 | G-US-3 | 25343.0000 | PR-US-6 | | |
| 22224.5000 | CG-US-8 | 22602.2000 | M-WW-1 | 22947.0000 | AF-US-8 | 23643.0000 | AF-US-10 | 25350.0000 | BC-WW-1 | | |
| 22226.0000 | M-US-3 | 22605.3000 | M-W-1 | 22950.0000 | AF-US-8 | 23651.4000 | G-US-9 | 25380.0000 | CG-US-1 | | |
| 22226.0000 | PR-US-6 | 22608.4000 | M-WW-1 | 22950.0000 | G-US-3 | 23661.0000 | SP-US-1 | 25380.0000 | CG-US-4 | | |
| 22226.5000 | CG-US-4 | 22608.4000 | M-WW-2 | 22955.0000 | G-US-3 | 23675.0000 | G-US-12 | 25380.0000 | CG-US-3 | | |
| 22226.5000 | CU-US-3 | 22611.5000 | CG-US-8 | 22960.0000 | N-BEL | 23680.0000 | BC-WW-1 | 25416.0000 | CG-US-4 | | |
| 22232.0000 | CG-US-1 | 22611.5000 | M-WW-1 | 22960.0000 | G-US-3 | 23687.0000 | AF-US-10 | 25433.0000 | AF-US-10 | | |
| 22234.0000 | CG-US-1 | 22514.6000 | M-WW-1 | 22964.0000 | G-US-7 | 23690.0000 | G-US-3 | 25461.0000 | M-WW-2 | | |
| 22245.0000 | G-WW-3 | 22617.7000 | M-WW-1 | 22966.5000 | G-U8S-3 | 23826.0000 | BC-WW-1 | 25490.0000 | G-US-1 | | |
| 22251.5000 | G-US-9 | 22623.9000 | M-WW-1 | 22970.0000 | BC-SS-1 | 23840.0000 | SP-US-1 | 25500.0000 | G-US-3 | | |
| 22252.0000 | G-US-9 | 22627.0000 | M-WW-1 | 22975.0000 | G-US-1 | 23856.0000 | BC-WW-1 | 25516.0000 | N-SAM | | |
| 22253.0000 | CG-US-1 | 22627.0000 | M-1 | 22990.0000 | SP-US-1 | 23860.0000 | G-US-3 | 25590.0000 | G-WW-3 | | |
| 22270.0000 | CG-US-1 | 22630.1000 | M-WW-1 | 22996.0000 | G-US-3 | 23862.0000 | AF-US-8 | 25597.0000 | SP-US-1 | | |
| 22284.0000 | CG-US-1 | 22633.2000 | M-WW-1 | 23000.0000 | G-US-3 | 23862.5000 | G-US-3 | 25599.0000 | G-US-3 | | |
| 22292.0000 | CG-US-3 | 22636.3000 | M-WW-1 | 23001.0000 | G-US-3 | 23863.5000 | AF-US-8 | 25600.0000 | PR-US-6 | | |
| 22324.5000 | M-US-3 | 22639.4000 | M-WW-1 | 23014.0000 | G-US-3 | 23865.0000 | AF-US-8 | 25800.0000 | PR-US-6 | | |
| 22330.5000 | N-ISR | 22642.5000 | M-WW-1 | 23030.0000 | G-US-3 | 23865.0000 | G-US-3 | 26131.0000 | G-US-3 | | |
| 22336.5000 | AF-CAN | 22645.6000 | M-WW-1 | 23035.0000 | AF-US-10 | 23875.0000 | G-US-12 | 26356.0000 | SP-US-1 | | |
| 22345.0000 | G-US-12 | 22648.7000 | CG-US-8 | 23035.0000 | G-US-8 | 23940.0000 | SP-US-1 | 26389.0000 | SP-US-1 | | |

| Frequency | Code |
|---|---|
| 26450.0000 | G-US-3 |
| 26490.0000 | BC-WW-1 |
| 26515.0000 | AF-US-3 |
| 26515.0000 | SP-US-1 |
| 26620.0000 | AF-US-7 |
| 26640.0000 | BC-WW-1 |
| 26660.0000 | G-US-3 |
| 26680.0000 | BC-WW-1 |
| 26684.0000 | SP-US-1 |
| 26703.0000 | G-US-1 |
| 26710.0000 | AF-US-3 |
| 26760.0000 | G-US-3 |
| 26800.0000 | G-US-3 |
| 26824.4000 | CG-US-6 |
| 26825.0000 | CG-US-6 |
| 26877.5000 | N-US-4 |
| 26905.0000 | G-US-1 |
| 26910.0000 | AF-US-8 |
| 26965.0000 | PR-US-7 |
| 26975.0000 | PR-US-7 |
| 26985.0000 | PR-US-7 |
| 27005.0000 | PR-US-7 |
| 27015.0000 | PR-US-7 |
| 27025.0000 | PR-US-7 |
| 27035.0000 | PR-US-7 |
| 27055.0000 | PR-US-7 |
| 27065.0000 | PR-US-7 |
| 27065.0000 | SP-US-1 |
| 27075.0000 | PR-US-7 |
| 27085.0000 | PR-US-7 |
| 27105.0000 | PR-US-7 |
| 27115.0000 | PR-US-7 |
| 27120.0000 | PR-US-3 |
| 27125.0000 | PR-US-7 |
| 27135.0000 | PR-US-7 |
| 27155.0000 | PR-US-7 |
| 27165.0000 | PR-US-7 |
| 27175.0000 | PR-US-7 |
| 27185.0000 | PR-US-7 |
| 27205.0000 | PR-US-7 |
| 27215.0000 | PR-US-7 |
| 27225.0000 | PR-US-7 |
| 27235.0000 | PR-US-7 |
| 27245.0000 | PR-US-7 |
| 27265.0000 | PR-US-7 |
| 27275.0000 | PR-US-7 |
| 27285.0000 | PR-US-7 |
| 27295.0000 | PR-US-7 |
| 27305.0000 | PR-US-7 |
| 27315.0000 | PR-US-7 |
| 27325.0000 | PR-US-7 |
| 27335.0000 | PR-US-7 |
| 27345.0000 | PR-US-7 |
| 27355.0000 | PR-US-7 |
| 27365.0000 | PR-US-7 |
| 27375.0000 | PR-US-7 |
| 27385.0000 | PR-US-7 |
| 27395.0000 | PR-US-7 |
| 27400.0000 | G-US-3 |
| 27405.0000 | PR-US-7 |
| 27410.0000 | PR-US-6 |
| 27440.0000 | PR-US-6 |
| 27440.0000 | CC-US-1 |
| 27440.0000 | PR-US-6 |
| 27540.0000 | CG-US-8 |
| 27542.0000 | CG-US-4 |
| 27575.0000 | G-US-7 |
| 27575.0000 | G-US-10 |
| 27585.0000 | G-US-7 |
| 27585.0000 | G-US-10 |
| 27590.0000 | CG-US-4 |
| 27625.0000 | G-US-10 |
| 27655.0000 | G-CAN-2 |
| 27665.0000 | G-US-10 |
| 27720.0000 | SP-US-1 |
| 27736.0000 | AF-US-8 |
| 27738.5000 | PR-US-6 |
| 27740.0000 | G-US-12 |
| 27740.0000 | PR-US-6 |
| 27741.5000 | PR-US-6 |
| 27785.0000 | CG-US-4 |
| 27829.0000 | AF-US-8 |
| 27845.0000 | G-WW-3 |
| 27870.0000 | AF-US-1 |
| 27877.0000 | AF-US-8 |
| 27900.0000 | N-US-4 |
| 27950.0000 | N-US-4 |
| 27963.5000 | N-US-4 |
| 27975.5000 | N-US-R |
| 27978.5000 | AF-US-8 |
| 27985.0000 | AF-US-8 |
| 27991.0000 | AF-US-8 |
| 27994.0000 | AF-US-8 |
| 27998.0000 | PS-WW-1 |
| 29520.0000 | PR-US-7 |
| 29540.0000 | PR-US-7 |
| 29560.0000 | PR-US-7 |
| 29600.0000 | PR-US-7 |
| 29620.0000 | PR-US-7 |
| 29640.0000 | PR-US-7 |
| 29660.0000 | PR-US-7 |
| 29680.0000 | PR-US-7 |
| 29700.0000 | PR-US-6 |
| 29701.5000 | PS-WW-1 |
| 29720.0000 | PR-US-6 |
| 29720.0000 | CC-US-1 |
| 29892.0000 | CG-US-4 |
| 29909.0000 | CG-US-4 |
| 29920.0000 | CC-CAN |
| 29928.5000 | PR-US-6 |
| 29930.0000 | PR-US-6 |
| 29930.0000 | CC-US-1 |
| 29930.0000 | PR-US-6 |
| 29931.5000 | PR-US-6 |

# THE LISTENER'S LOG

The LISTENER'S LOG accommodates entries for any type of transmission heard at any frequency, and includes column headings of all of the most useful information collected by monitoring stations.

FREQUENCY:
   The carrier frequency of the station should be indicated in kilohertz or megahertz.

AGENCY:
   VOA, USN, ham, some way of identifying the nature of the station service.

LOCATION:
   Atlanta, Cuba, Atlantic Ocean, deep space...whatever!

CALLSIGN:
   Identifications like KKN50, WA4PYQ, Houston Control, Unit 2, Alfa Tango.

EMISSION:
   Use simple indications like AM, SSB, CW, RTTY, FAX. For technical accuracy, use official designators.

SIGNAL:
   A received signal has many qualities, including strength and intelligibility. Two common methods are used for the assessment of signals:

   1) SINPO CODE-The most comprehensive method of rating the quality of a signal is shown below:

| Rating | Signal Strength | Interference from stations | Noise (static) | Propagation Fading | Overall Merit |
|--------|-----------------|----------------------------|----------------|--------------------|---------------|
| 5 | excellent | none | none | none | excellent |
| 4 | good | slight | slight | slow | good |
| 3 | fair | moderate | moderate | moderate | fair |
| 2 | poor | severe | severe | fast | poor |
| 1 | barely audible | extreme | extreme | very fast | unusable |

   2) RST SYSTEM - Readability, signal strength and tone quality are used to rate CW signals; voice signals are rated using only the R&S scales.

| Readability | Signal Strength | Tone (Morse code) |
|-------------|-----------------|-------------------|
| 1. unreadable | 1. barely perceptible | 1. extremely raspy sound |
| 5. perfectly readable | 9. extremely strong | 9. pure, clean pitch |

TIME/DAY/DATE:
   You should record your reception time in 24 hour Universal Coordinated Time (UTC). A 24 hour clock is helpful. The day of the week is important because most broadcast programming and communications network schedules follow the calendar. Log the date as well for reference and remember, it will be tomorrow on your log before it is midnight in your time zone because of the difference between local and Universal time!

ADDITIONAL INFORMATION:
   Whether the signal heard was an aircraft, ship, on a rescue mission or whatever, record helpful information in this last column.

# LISTENER'S LOGSHEET ©

| FREQUENCY kHz / MHz | AGENCY OR SERVICE | LOCATION OR BEARING | CALL SIGN OR IDENTIFICATION | EMISSION MODE | SIGNAL QUALITY | TIME/DAY/DATE | ADDITIONAL INFORMATION |
|---|---|---|---|---|---|---|---|
| | | | | | | | |
| | | | | | | | |
| | | | | | | | |
| | | | | | | | |
| | | | | | | | |
| | | | | | | | |
| | | | | | | | |
| | | | | | | | |
| | | | | | | | |
| | | | | | | | |
| | | | | | | | |
| | | | | | | | |
| | | | | | | | |
| | | | | | | | |
| | | | | | | | |
| | | | | | | | |
| | | | | | | | |
| | | | | | | | |
| | | | | | | | |
| | | | | | | | |
| | | | | | | | |
| | | | | | | | |
| | | | | | | | |
| | | | | | | | |
| | | | | | | | |
| | | | | | | | |
| | | | | | | | |
| | | | | | | | |
| | | | | | | | |
| | | | | | | | |